ALSO BY BARRY LOPEZ

Nonfiction

About This Life: Journeys on the Threshold of Memory (1998)

Apologia (1998)

The Rediscovery of North America (1990)

Crossing Open Ground (1988)

*Arctic Dreams: Imagination and Desire
in a Northern Landscape* (1986)

Of Wolves and Men (1978)

Fiction

Outside (2014)
with engravings by Barry Moser

Resistance (2004)
with monoprints by Alan Magee

Light Action in the Caribbean (2000)

Lessons from the Wolverine (1997)
with illustrations by Tom Pohrt

Field Notes: The Grace Note of the Canyon Wren (1994)

Crow and Weasel (1990)
with illustrations by Tom Pohrt

Winter Count (1981)

River Notes: The Dance of Herons (1979)

*Giving Birth to Thunder, Sleeping with His Daughter:
Coyote Builds North America* (1978)

Desert Notes: Reflections in the Eye of a Raven (1976)

Anthology

Vintage Lopez (2004)

HORIZON

Barry Lopez

HORIZON

RANDOM HOUSE CANADA

PUBLISHED BY RANDOM HOUSE CANADA

www.penguinrandomhouse.ca

Random House Canada and colophon are registered trademarks.

Grateful acknowledgment is made to Far Corner Books for permission to reprint an excerpt from "Kindness" from *Words Under the Words*: Selected Poems by Naomi Shihab Nye, copyright © 1995. Reprinted by permission of Far Corner Books.

Library and Archives Canada Cataloguing in Publication

Lopez, Barry Holstun, 1945–, author
Horizon / Barry Lopez.

Issued in print and electronic formats.
ISBN 978-0-307-35599-7
eBook ISBN 978-0-7352-7746-5

1. Lopez, Barry Holstun, 1945– —Travel. 2. Travel—Social aspects.
3. Tourism—Environmental aspects. 4. Natural history. I. Title.

PS3562.O69Z465 2019   813'.54   C2018-904432-2
                                               C2018-904433-0

Maps and globe illustrations copyright © 2019 by David Lindroth Inc.
Jacket photograph by artpartner-images.com/Alamy
Jacket design by Carol Devine Carson
Endpapers: *Remember,* by Nicholas Roerich

Printed and bound in the United States of America

10 9 8 7 6 5 4 3 2 1

Penguin
Random House
RANDOM HOUSE CANADA

For Debra

and for
Peter Matson and Robin Desser,
with profound gratitude for the years of support

To travel, above all, is to change one's skin.

—Antoine de Saint-Exupéry,
in *Southern Mail*

Site Maps

Johan Peninsula Area   134

Alexandra Fjord Lowland   140

Skraeling Island   154

Great Rift Valley   315

Ross Ice Shelf Region   440

Ross Island   454

The Dry Valleys   455

Brunswick Peninsula and the Strait of Magellan   495

Contents

Author's Note   *xiii*

Prologue   3

Introduction: Looking for a Ship   7

1. Mamaroneck   9
2. To Go/To See   17
3. Remember   29
4. Talismans   34

CAPE FOULWEATHER   49

Coast of Oregon
Eastern Shore of the North Pacific Ocean
Western North America

SKRAELING ISLAND   131

Mouth of Alexandra Fjord
East Coast of Ellesmere Island
Nunavut
Canada

PUERTO AYORA   205

Isla Santa Cruz
Archipiélago de Colón
Eastern Equatorial Pacific

JACKAL CAMP   267

Turkwel River Basin
Western Lake Turkana Uplands
Eastern Equatorial Africa

PORT ARTHUR TO BOTANY BAY   345

State of Tasmania
Northern Shore of the Southern Ocean
Southeastern Australia

———

State of New South Wales
Western Shore of the South Pacific

GRAVES NUNATAKS TO PORT FAMINE ROAD   425

Queen Maud Mountains
Central Transantarctic Mountains
Northern Edge of the Polar Plateau
Antarctica

———

Brunswick Peninsula
Shore of the Strait of Magellan
Southern Chile

Notes   513
Selected Bibliography   519
Scientific Binomials   531
Overview Maps   539
Acknowledgments   545
Index   551

Author's Note

Horizon is an autobiographical reflection on many years of travel and research, in Antarctica and in more than seventy countries. Some of these travels I financed myself, others I sought grants for or received fellowships to fund. I made several trips on assignment for magazines, and with others I was simply invited to come along. The details, and my expressions of gratitude for those who assisted me over the years, are included in the Acknowledgments.

Most of the journeys described here I made in my forties and fifties. I traveled to the Galápagos Islands, however, and to Australia and Antarctica, on several occasions and at different points in my life. The least complicated way to chronicle these experiences, it seemed, was simply to tell the story, not to try to explain any juxtapositions in time. It might help to know, however, that when I traveled to Cape Foulweather in order to encounter the winter storm I was forty-nine; that I was in my early forties and had just published a book about the North American far north, *Arctic Dreams,* when I flew into the archeological camp on Skraeling Island; and that I was fifty-four when I made the trip to Graves Nunataks in the Transantarctic Mountains.

As *Horizon* is meant to be an autobiographical work, I should emphasize that there was a long learning curve inherent in all this sojourning. I've not tried to be explicit about what was learned (or unlearned) or when, in part because it hasn't always been clear to me what changes might have occurred. The young man visiting the archeological site on Skraeling Island is the same fellow who at the end of the book encounters a stranger on the road to Port Famine, but also not.

HORIZON

Prologue

The boy and I are leaning over a steel railing, staring into the sea. The sun is bright, but shade from a roof above us makes it possible to see clearly into the depths, to observe, quivering there, what's left of the superstructure of a battleship sunk seventy-two years before.

My grandson is nine. I am in my sixty-eighth year.

The memorial terrace on which we are standing, alongside my wife, has been erected above the remains of the USS *Arizona,* a 608-foot Pennsylvania-class battleship overwhelmed at its moorings on the morning of December 7, 1941, by Japanese dive bombers. It sank in minutes. The flooded hulk, a necropolis ever since, holds the remains of many of the 1,177 sailors and marines killed or drowned on the ship that morning. I'm explaining to the boy that sometimes we do this to each other, harm each other on this scale. He knows about September 11, 2001, but he has not yet heard, I think, of Dresden or the Western Front, perhaps not even of Antietam or Hiroshima. I won't tell him today about those other hellfire days. He's too young. It would be inconsiderate—cruel, actually—pointedly to fill him in.

Later that morning the three of us snorkel together on a coral reef. We watch schools of tropical fish bolt, furl, and unfurl before us, colored banners in a breeze. Then we have lunch by a pool at the hotel where we are guests. The boy swims tirelessly in the pool's glittering aqua-tinted water until his grandmother takes him down to the beach. He runs to jump into the Pacific.

He can't get enough of swimming.

I watch him for a few minutes, flinging himself into the face of wave after wave. His grandmother, knee-deep in the surf, scrutinizes him without letup. Eventually I sit down in a poolside chair with a glass of iced lemonade and begin to read a book I've started, a biography of the American writer John Steinbeck. I glance up once in a while to gaze at sunlight shuddering on the surface of the ocean, or to follow flocks of sparrows as they flee the tables of the hotel's open-air restaurant, where they've been gleaning crumbs. For prolonged uninterrupted minutes I also watch, with a mixture of curiosity and affection, the hotel's other guests, sunning on lounges around the pool or ambling past, completely at ease. The clement air and the benign nature of the light dispose me toward an accommodation with everything here different from myself. When I breathe, I'm aware of a dense, perfume-like scent—tropical flowers blooming in a nearby hedge. Is it bougainvillea?

The exuberance of my grandson has also enhanced this sense of tranquility I feel.

Most of the guests here are Asian. I recognize in particular the distinctive cast of Japanese and Chinese faces. Strolling through the poolside restaurant in expensive clothes, discreetly signaling a pool attendant for a towel, snapping copies of *The Honolulu Star-Advertiser* to straighten the pages, they all seem to have the bearing of people familiar with luxury, as I imagine that state.

I return to the biography. In the paragraph I'm reading, the writer is describing a meeting Steinbeck once had at his home in Pacific Grove, California, with the historian of mythology Joseph Campbell. The night before this, Steinbeck, the composer John Cage, Campbell, Steinbeck's first wife, Carol, and a few others had all enjoyed dinner together in the Steinbeck home. Campbell has now come out onto the patio to inform his host that he has fallen in love with Carol. He accuses Steinbeck of treating her shabbily, and says that if he won't change his ways then he, Campbell, is prepared to ask Carol to marry him and to return with him to New York.

I look up abruptly from the book, recalling that I'd been in summer camp in 1956 with both of Steinbeck's sons, Thom and John. It had been a memorable encounter for me. I was eleven and I met their father at the same time. I marveled at the burly reification of this person who'd written *The Red Pony*. (I was introduced to his third wife, Elaine, then, too. She was cool. Dismissive.)

I pick up again where I've left off, keen to follow this unanticipated triumvirate—Steinbeck, John Cage, Joseph Campbell.

Pages later, I am feeling the westering sun burning hot on my right cheek. Another tight flock of sparrows hurtles by my head and suddenly I wonder whether I'd done absolutely the wrong thing that morning at Pearl Harbor, before we'd all gone to see the *Arizona*. I'd walked my grandson through the interior of a World War II American submarine, explaining the architecture, the periscope in the conning tower, the forward torpedo tubes. He had touched the sleek torpedoes gingerly, a lingering caress, his small hands cupping the warheads.

Just then a handsome Japanese woman striding along the pool's edge makes a graceful, arcing dive into the water. An impulsive act. A scrim of water rises around her like the flair of a flamenco dancer's skirt. The pool water shatters into translucent gems.

In the beauty of this moment, I suddenly feel the question: What will happen to us?

I stand up, a finger marking my place in the book, and search the breaking surf beyond a hedge of sea grape for my grandson. He waves hysterically at me, smiling from the slope of a wave. Here, Grandpa!

What is going to happen to all of us now, in a time of militant factions, of daily violence?

I want to thank the woman for her exquisite dive. The abandon and grace of her movement.

I want to wish each stranger I see in the chairs and lounges around me, every one of them, an untroubled life. I want everyone here to survive what is coming.

Looking for a Ship

1

Mamaroneck

A history, one purporting to depict the life trajectory of the grandfather reading by the pool, could easily begin sixty-five years before that moment in Hawai'i, in an embayment of Long Island Sound called Mamaroneck Harbor. Here is a stretch of sheltered water, a surface barely roughened that day by a wind blowing westward from the direction of Crane Island. A boy who cannot yet swim wades steadily farther out into the salt water, under the shepherding gaze of his mother. She's hardly fifty feet away, a dark-haired woman in her middle thirties, her legs tucked beneath her, her belly round with a second child. She's sitting on a wool blanket, embroidering a needlepoint image of field flowers erect in a vase. It's 1948. She's conversing with a friend underneath a large white oak tree on Orienta Point, on the Westchester County coast of New York.[1]

The boy halts when he reaches water up to his chin. She watches him steadily now. He wants to go farther, to swim out past Turkey Rock, out farther even, out beyond the Scotch Caps, two islets on the distant rim of the sound. Past that lies a horizon of water. A blank page.

He turns for shore, scuttling sideways like a crab in ripples that break over his small shoulders.

A few months later, with the approach of a New England winter and following the birth of his only sibling, the boy moves with his family to a valley in Southern California, an irrigated expanse of farmland. Groves of oranges and walnuts, fields of alfalfa. Peach orchards. The irrigated San Fernando Valley. This Mediterranean plain is bounded to the south by the Santa Monica Mountains, to the north by the snowcapped San Gabriels. A different life for him now. A different geography. An unfamiliar climate. Different races of people.

One day, a couple of years after the family arrives, the father leaves. He returns to his first wife, living in Florida with their son, and the boy and his mother and younger brother begin together another sort of existence. His mother teaches home economics at a junior high school in Northridge and, at night, dressmaking at Pierce Junior College, near Calabasas. Other evenings she works at home, creating couture clothing for her clients. The father writes from Florida. He promises to send money but never does. The three of them, anyway, seem to have all they need. The boy is curious but wary. A suburban crow. He makes friends with other boys in his neighborhood and with his classmates at Our Lady of Grace, a Catholic grade school in Encino. He gets to know a few of his mother's students, the sons of braceros working in the vegetable fields north and west of their house in Reseda.

He learns to ride a bicycle. He rides and rides, as far north in the valley as Granada Hills and west all the way to Chatsworth.

Their mother takes the boys out into the western Mojave Desert, to the eastern Mojave and the Grand Canyon, and south to the San Diego Zoo and across the border into Mexico.

One afternoon the boy stands on the shore of Topanga Beach, fronting the great Pacific just east of Malibu. He watches comber after comber crash the strand, stepping clear of the waves' retreating sweeps each time, as his mother has asked. He understands that this foaming storm surf has arrived on the beach from someplace else. Here, temperate air embraces him; an onshore breeze softens the burn of the sun's rays on his white skin. Its light splinters on bits of quartz in the sand at his feet.

This, too, is new to him, a feeling of being cradled in harmless breezes and caressed by light. Years later, walking alone in faraway places, he will remember and long for this sensation.

A friend of his mother, a man the boy hopes will one day become his father, is accompanying the family that day at Topanga Beach. He tells the boy that far off across the water, farther away even than the storm that makes these waves, is the extremely ancient country of China. The boy has no image of China. The tall, long-fingered, long-legged, soft-spoken man in khaki trousers moves through the boy's mind with the hesitating grace of a flamingo. The boy imagines that the man knows many things. He works at the Santa Barbara Botanic Garden and some days takes the boy with him to work. His name is Dara. He points out differences among the plants; he pots with the boy in the greenhouses. He explains how a large flowering plant like a jacaranda grows from a small seed.

The boy's most favorite trees now are eucalypts, the tall river red gums and blue gums that flank Calvert Street in Reseda where he lives. He likes the royal towering of them; the shedding boles, slick beneath his hands; the fragrance of the hard gumnuts. He carries a few of these buttons in his pockets wherever he goes. He likes the defiant reach of these trees, how they crowd and rake the blue sky, and how the wind chitters in their leaf clusters. He feels safe hiding in their shadows. Dara tells him that around Los Angeles they're called "skyline trees." He likes that. Originally from Australia, he says, but they grow all over the world, wherever the right conditions can be found. It's the same for the frangipani trees and bougainvillea vines growing at the botanic garden. Those two, along with the eucalypts, says Dara, are now found everywhere "in the colonial subtropics."

The boy can't picture Australia, but he is transfixed by the idea that some trees are carried off from their first country and then grow happily in other places.

When he lies in bed at night, imagining the future he wants, a strategy he uses to probe the vague precincts of his dreams, the boy envisions the botanic garden and thinks about Dara, how gently Dara's hands handle plants. By now, though, he has also learned about some things less comforting. More threatening. He circumspectly regards the lives of poisonous black widow spiders living in the garage alongside his house, red hourglasses gleaming on the females' tummies. When he talks to adults about the rattlesnake that startled him and his friend Thair while they were walking in the Santa Monica Mountains one morning, hunting for alligator lizards, he enjoys the way adults attend closely to his story.

The snake had snapped at them when they teased it. He doesn't tell his listeners that he and Thair beat it with a stick until it was dead.

One weekend at Zuma Beach the boy is stung by a wounded Portuguese man-of-war, a large jellyfish, foundering in the surf. An ambulance comes to take him, vomiting and shivering, to the hospital.

He trusts the shelter of the towering gum trees and wonders about the power of Portuguese man-of-wars. The two things are now entwined in his mind.

He is ashamed of having killed the snake and of his silence about it.

MOST EVERY SATURDAY the boy goes with his mother and brother to the Farmers Market in Los Angeles, at Third and Fairfax, driving over from the valley in his mother's dark green Ford coupe. He loves the shine and heft of the fruit. He has to reach higher than his head in order to feel within the tilted boxes for greengage plums, for kumquats and nectarines. He likes to heft the Belgian endives, to feel the brush of wetted carrot tops across his forehead, to grip a cassava melon in both his hands. They're like his first pets.

A friend of his mother owns an avocado ranch near Fallbrook. Her husband, a DC-6 pilot who flies every week to Honolulu and on to Tokyo for American Airlines, is not much interested in answering the boy, who wants to know how this actually occurs, Los Angeles to Honolulu, then to Tokyo. The boy has considered that he will one day have a ranch something like the one this couple operates. He'll raise avocados and perhaps Asian pears, which break as sharply against his teeth as McIntosh apples. This life appeals to him. He'll truck his produce and buckets of cut flowers—snapdragons, carnations, irises—to the market. He'll keep bees to pollinate his flowers and fruit trees, possibly offer their honey for sale, along with fresh eggs, asparagus, and pomegranates at a stand by the side of the road, like the fruit and vegetable stands his mother shops at on the drive home from school every day.

Most nights the boy consoles himself as he falls asleep with the certainty of the destination he has chosen. He will operate a tractor, dragging a harrow to break up the clods of dirt left behind after he discs the field where he will grow annuals. He'll determine exactly how to set out the sprinklers to irrigate the varieties of roses in his gardens. He'll light smudge pots on cold winter nights to keep the orchards from freezing.

The more he imagines a truck farm, the less anxious he feels about the strange man who has come into his life, a man who is not like Dara.[2]

ONE WINTER AFTERNOON the boy follows his mother into the post office at Canoga Park. While she waits in line he studies a fourteen-by-seven-foot mural on the east wall, *Palomino Ponies*. He's mesmerized. Years later he will misremember the image when he discovers more work by the same painter, Maynard Dixon. He will think of it, wrongly, as a tableau of American Indian faces in profile, high cheekbones, the burnt sienna and ocher tones of their skin. But there are no Indians in this mural of a California vaquero of the 1840s, racing across a golden grassland behind seven palomino horses. The boy will have conflated the image in the post office with the memory of a better-known painting by Dixon, *Earth Knower,* and he will have further confused the misremembered image with a childhood recollection of having once encountered Indians on a train platform in Needles, late one ninety-degree summer night in the eastern Mojave. He was eight. He and his brother had boarded an overnight train in Los Angeles with a friend of their mother, bound for the Grand Canyon. The boy had stepped out onto the platform in this small California town on the west bank of the Colorado River after midnight, later than he'd ever been up. He saw a dozen Mohaves milling around, or maybe they were Havasupais from the canyon, waiting for family members to board or disembark. Despite the heat, they'd all pulled shawls forward over their heads or they were peering out from the cowls of trade blankets. He couldn't decipher the nearly inaudible sounds of the words they spoke.

He never forgets the austerity of this scene. The foreignness of these figures.

In the post office that day, after he takes in the poise of the rider, the fleet jeté of his mount, the muscular exuberance of the palominos, he remarks to his mother that one day he intends to become a painter.

In the moment, perhaps all he really wants is to become a dashing vaquero.

AND THEN SUDDENLY his mother is married again, to a businessman from New York. California is over. The boy moves with his new family to Manhattan. Louder, taller, faster country than his home ground. A different color to its winter sky. Colder weather, fall leaves turning pale yellow on London planetrees, which he initially confuses with California sycamores. When his stepfather points out "Indians," dining across from them in a restaurant, he means people from another continent, not this one.

That first summer in New York he's sent with his brother to Camp St. Regis on Long Island's South Fork, near East Hampton. There he meets John, a boy he believes is from California. They share a cabin with four other eleven-year-olds. John's father, the boy learns, has written some books about California, set in the Central Valley, a place much like the San Fernando Valley in the boy's mind. He's actually read one of these books, a collection of pieces called *The Long Valley*. On Parents' Day, the California author arrives by cabin cruiser to visit his children. He anchors the boat just off the beach where he won't have to encounter the other parents. He rows a pale green dinghy ashore to fetch his sons. They spend the afternoon together with their parents on the cabin cruiser. After his own parents leave, the boy sits on the beach and watches the boat.

He waits.

The boy who waded in Mamaroneck Harbor and then moved to Southern California, and once thought he wanted to grow avocados or become a painter, now lives in a brownstone in the Murray Hill neighborhood of Manhattan. In the fall he will enter the seventh grade at a private Jesuit school on East 83rd Street and begin serving Mass as an altar boy at the Church of Our Saviour on East 38th Street, around the corner from his home.

It will take him a while to fit himself to the place.

On that July afternoon at St. Regis he waits, staring at the white vessel. It seems mute to him, with its curtained windows and no one visible on the flying bridge or at the stern. Young John has informed him that his parents have motored over from their home in nearby Sag Harbor, an old whaling town. The boy remembers the name, Sag Harbor. An image of it anchors his growing awareness of the immensity and quietude of whales, and of the enormity and violence of their slaughter.

It bothers the boy, years later, that he cannot pry loose a single memorable detail from the opaqueness of the Steinbeck boat, even after having scrutinized it for an hour. Only the pale green dinghy, hanging crookedly from davits at the stern, stands out. The boat sits almost broadside to him that afternoon on a slowly rising tide. Nothing stirs. He wants to go on reminiscing with John about days in California, but just then, he wants to swim out to the boat and tell the older John that he has read "The Red Pony," that he thinks it very good. He wants to be a part of the family having a conversation on that boat.

Suddenly the writer, with his large, balding head, is in the stern of the cabin cruiser, lowering the dinghy to bring the boys back to shore. In the diffused light that penetrates a late afternoon fog, the dinghy and its passengers appear wraith-like as they approach. The boy has yet to hear of the River Styx or of Charon, but in the years afterward, it is these images that will rise up in his memory when he recalls the moment.

That evening in their bunks the boy asks John how he thinks his father has been doing here, in New York City, having moved all the way across the country from California to East 72nd Street. He listens closely, hoping to hear what his bunkmate might have gleaned, having himself already made this adjustment. He hopes to make this same change successfully himself, but senses large undefined obstacles. He feels a potential for disappointment in his expectations.

He is unaware that his bunkmate John did not grow up in California.

In the years following, in the silence before sleep comes, the boy sometimes recalls the anonymous cabin cruiser and the afternoon mist obscuring the horizon beyond. He thinks about the California beaches, Zuma and Point Dume on Santa Monica Bay, west of Los Angeles, and about the man his mother decided not to marry, who told him about China, and about jacarandas and eucalypts. He believes there is something he must see one day in China. Or in Japan. Or somewhere far off. This repeated sensation elicits in him a now familiar yearning.

Once it came from looking at avocados motionless in his hands, or from hearing the eucalypts on Calvert Street clattering in the wind. Now it comes more often from a desire simply to go away. To find what the skyline has cordoned off.

THE BOY IN Mamaroneck Harbor is myself, and I am the grandfather speaking with his grandson in Hawai'i about catastrophe. I have been thinking for a while about the time between those two moments, wondering what transpired in the years in between, during which I saw senseless death and became a witness to the breaking of every commandment I'd learned as a boy, and during which I beheld things so beautiful I couldn't breathe.

A few scenes like the ones I've recounted above, broken off from an early life—Mamaroneck Harbor, Zuma Beach, a railroad platform in Needles—are but one way to embark on a larger story about someone who afterward would go off repeatedly to look at the rest of the world. Only a sketch, then, this, but one I feel makes reasonably good sense. No life, of course, unfolds quite this neatly and comprehensibly around any such rosary of memories. A long life might be understood, however, as a kind of cataract of imperfectly recollected intentions. Some of one's early intentions fade. Others endure through the inevitable detours of amnesia, betrayal, and loss of belief. Some persist over the years, slightly revised. Unanticipated trauma and other wounds certainly might force the car off the road at any moment, maybe forever, one's final destination lost. But, too, the unfathomable sublimity of a random moment, like the touch of a beloved's hand on one's burning face, might revive the determination to carry on, and, at least for a time, rid one of life's weight of self-doubt and regret. Or a moment of staggering beauty might reignite the intention one once had to lead a life of great meaning, to live up to one's own expectations.

My driven life has been one of occasional ecstasy and occasional sorrow, little different, in that, from the lives of many others except perhaps for the compelling desire I've had to travel to far-off places, and for what acting on that yearning with such determination has meant for me and for those close to me.

I became, almost unintentionally, an international traveler, though not a true wanderer.[3]

MANY YEARS AFTER my adolescence in New York, embarking on this autobiography, I wrote to the manager of the Orienta Apartments in Mamaroneck. I wanted to know something more about where I came from and trusted that the building was still standing and that such a person might exist. I described how, as a three-year-old, I had walked a certain path from the elevator to our apartment on the second floor. Could he or she determine the apartment number, having only this information? The manager wrote back right away, including with his letter a few drawings of the building's grounds and some photographs. On one of the drawings he'd marked out the small garden plot where, I'd told him, my mother had grown roses, tulips, and irises.

The apartment number, he said, was 2C.

2
———

To Go/To See

After the move east to New York in 1956, and after I'd graduated from a prep school on the Upper East Side, I went away to college in the Midwest. The boy who once thought he wanted to operate a truck farm had decided to major instead in aeronautical engineering, a career about which he knew next to nothing. The part of Southern California I'd grown up in, however, had been at that time, right after the end of the Second World War, a burgeoning center for aircraft design, and for the testing and assembly of airplanes. That way of being in the world was in the air I breathed as a child, and the nature of that work had been vividly represented for me by my mother's first husband, Sidney van Sheck, whom I met for the first time in California, years after they divorced.[4]

Sidney was a Czech immigrant. He married my mother in Alabama in 1934, I believe, and shortly after they divorced he moved to Southern California. In the 1950s he was living only a few miles from us, in the Santa Monica Mountains above Malibu with his second wife, Grace. During the financially strained years when my mother was supporting us with several jobs, Sidney was employed as an aeronautical engineer at Hughes Aircraft in Culver City. He was designing satellites then, some of the earliest ones, after several years of work on planes like Howard Hughes's Spruce Goose. He and Grace befriended the three of us in myriad ways. I'm sure he gave Mother money, and he would often invite me to sit quietly on a stool in his home workshop and watch while he fashioned metal and wood parts for the model aircraft he always seemed to be building.

I was attracted to Sidney's intensity, to his confidence with hand tools, and I was fascinated by the improbability of "aeroplanes," around which he was so at ease. Also, he drove a very quick-on-its-feet British sports car, an Austin-Healey convertible. A two-seater. When he drove me very fast along the Pacific Coast Highway with the top down, I heard the smooth clicking of the gearbox as he shifted flawlessly in the turns and accelerated out of them, the hood of the car rising slightly each time. A lunging animal. With the wind full in my hair, glancing at his nimble feet double-clutching as he shifted, I felt he was inviting me into the visceral experiences of his world.

Sidney, like Dara, was the image of the father I wanted. He finished a degree in painting in the 1920s at the École des Beaux-Arts in Paris. Later he completed a master's degree in aeronautical engineering at the Massachusetts Institute of Technology. Before divorcing my mother and moving to California with Grace, he was employed by Bechtel-McCone-Parsons, a construction company with a military aircraft division in Birmingham, Alabama. His work there included refining the wing designs of two bombers, the B-24 Liberator and the B-29 Superfortress, and designing the armament system for a fighter plane, the P-38 Lightning. (It was a P-38 that the legendary pilot and writer

Antoine de Saint-Exupéry was flying when he disappeared over the Mediterranean in 1944, a fact I would learn much later, after I became enamored with the persona of Saint-Exupéry as it emerged for me, in particular, in *Southern Mail* and *Night Flight*.)

During the First World War, Sidney piloted a SPAD S.VII, an early fighter plane, for the French. He was shot down in 1919 over the French Alps (by the Red Baron, Baron von Richthofen, according to my mother). The crash left him with fused cervical vertebrae and other injuries, but he went on flying single-seat aircraft until he crashed again, this time in North Carolina in the 1920s, an accident he also managed to walk away from. After gaining American citizenship, finishing his engineering degree, and securing employment with Bechtel-McCone, Sidney engaged his second passion more fully by teaching art at Auburn University. My mother, a junior at nearby Montevallo College, met him there in 1933 and soon began studying painting with him. They married shortly after she graduated. The following year, Sidney was commissioned by the Works Progress Administration in Washington to design what would become the largest WPA mural in the American South, a championing of the dignity of manual labor and a warning about the ruthlessness of corporate exploitation. It covered the legs and extensive lintel of a proscenium arch in the auditorium at Woodlawn High School in Birmingham.[5]

IT WAS WITH the unconscious idea of emulating the cosmopolitan artist and engineer Sidney van Sheck, probably, that I entered college believing I had a calling to become an aeronautical engineer. My enthusiasm for airborne flight, for the adventure of it—Mary S. Lovell's *Straight on Till Morning*, about Beryl Markham; Markham's own *West with the Night;* Amelia Earhart's life, as reported in newspaper accounts; and magazine stories about Alaska's bush pilots—had no doubt also canted me in that direction. In the middle of my freshman year, however, awakening from this misapprehension about a career in engineering, I embarked instead on a liberal arts curriculum—literature, philosophy, anthropology, history, theater—and quickly grew more comfortable there.

As I look back, I can see how determined I was as a college freshman to immerse myself in any one of several artistic pursuits. I felt the desire acutely as a fledgling actor, responding to complicated blocking directions from a director. (In theater classes, I marveled at how the emotional underpinning of a play could be made *visually* apparent by establishing certain patterns in an actor's movements.) I felt this same affinity with patterns in my initial efforts to write fiction, and with photography, as I began making images of landscapes in the countryside around me in northern Indiana and southern Michigan. I pored over the black-and-white images of Minor White and Harry Callahan, of Edward Weston and Wynn Bullock, struck by their completeness, by the cleanness of the compositions. I hoped to emulate them, and also to embrace the compassion of photographers like Walker Evans and Dorothea Lange, though I could not imagine intruding on people in order to reveal human suffering the way they did.

I can appreciate now, fifty years on, of course, that several other things were also in play when I decided not to pursue a career in aeronautical engineering. When I was nine, a friend of the family gave me eight tumbler pigeons. I spent hours out of mind with those birds, mesmerized by their arrowing and wheeling across the sky, by the way they deliberately lost aerodynamic lift and fell through the air, tumbling end over end for hundreds of feet, falling as though felled by birdshot, only to swoop out of it acrobatically just before they hit the ground. I remember the way tiny pressure gradients in the atmosphere caused flocks of them to teeter in flight, making transparent air masses visible. Watching them, I felt incredulous, gleeful.

In those same years I also put together dozens of model planes. I suspended them from the ceiling of my bedroom with sewing thread, fighter planes and bombers like the P-38 and the B-24, but, too, "flying boats" like the PBY Catalina, the Martin PBM Mariner, and the Martin M-130. These large planes could reach destinations in the world where no runways had yet been built, by putting down instead on bays and lagoons. When I woke in the night and looked up, these planes were arrayed above me like constellations. They were as alluring as any arrangement of stars I knew.

Some nights I would imagine myself in the cockpit of one of these airplanes, an aircraft without armament, without bombs. I'd roll out over the moonlit San Fernando Valley and head inland over the mountains, bearing off over Mount Whitney and heading south into Mexico. I'd push on through the night like Beryl Markham, just a few thousand feet above the skin of the earth. Coming home over the western Caribbean and then the Gulf of Mexico, I'd see the first rays of the sun in the east, an hour before they crossed the Sierra Nevada and lit up the crowns of eucalyptus trees in the San Fernando Valley.

THE AESTHETICS OF airborne flight I reveled in collided, in my freshman year, with instruction in the tensile strength of aluminum, with wind tunnel mathematics, and with the empiricisms of aeronautical engineering. I could not locate, either in chemistry or in physics, anything like the tachycardia, the pounding heart, that rose up in me whenever I turned my pigeons out against the tropical blues and cloud banks of a California sky, or saw thirty or forty of them bursting from the crown of a gum tree, triggered by some signal not apparent to me. I could not find anywhere in my coursework in calculus the headlong spirit of Saint-Exupéry, scudding over dune crests in Western Sahara beneath a bejeweled sky. The meaning of Icarus's defiant and incautious bravado was never addressed in my physics seminar.

As a seventeen-year-old, I longed for direct experience with the world. Most of my impulses, however, were purely metaphorical, without shape or purpose. Like so many immature boys, desperate to achieve some kind of standing, I floundered—inarticulate, self-conscious, and defensive.

I LEFT THE UNIVERSITY itself regularly during those years to explore landscapes in the upper and lower Midwest, driving a 1951 Buick Roadmaster, my first car, which I kept illegally off campus. I drove hundreds of miles to see whatever might be there in northern Michigan or in trans-Mississippi Iowa. Traveling, I came to understand, assuaged something in me. After graduating from prep school in 1962, I'd spent two months being driven through western Europe in a compact Fiat bus with fifteen of my male classmates and a couple of tutors. We drove eastward from Portugal across Spain and France, over the maritime Alps into Italy, and south as far as Rome, then came back north through Switzerland, Liechtenstein, Austria, and West Germany, arriving again in France, in Lorraine, and going on to Paris. We crossed the Channel to Dover from Calais and took a train to London. On our last day in Ireland, I rowed a stretch of the River Shannon in a punt, alone, not wanting this luminous journey—from the art galleries of the Prado in Madrid to the bleakness of the Brenner Pass; from the fields of crosses and Stars of David in cemeteries across Artois and Picardy to the austere Cliffs of Moher in County Clare—ever to be finished.

The stimulation of that journey—the geographies, the art, the food, the conversations with tradespeople—was intoxicating. I wanted this stimulation somehow to frame my way in the world.

BY THE TIME I was in my early twenties, I'd spent one summer wrangling horses in Wyoming and another in summer stock theater in Helena, Montana. I'd driven across all but one or two of the lower forty-eight. I'd returned to Europe, to England and to my stepfather's ancestral land in Asturias, in Spain, and I'd published my first stories. I was still deeply uncertain, however, about what to do. Before I married, I visited a Trappist monastery in Kentucky, thinking this might be my life's destination. It wasn't. (The monk Thomas Merton was living there at the time. His autobiography and other of his books had been inspirational for me in prep school and college.) In 1968, married now and in possession of a master's degree, I moved to Oregon to begin work on a second graduate degree, a master of fine arts in creative writing, thinking I would do best to follow a career in teaching. I was quickly disillusioned by the program but matriculated at the University of Oregon for several more semesters, studying folklore, journalism, and Native American culture. By that time, however, life in a university had come to represent, for me, mostly domestic comforts and unconscious detachment from the workaday world. Life in classrooms began to seem intolerably hermetic, an unsafe place in which to remain, I thought, in spite of the intense stimulation that far-ranging, learned conversations there always seemed to provide.

I began to travel more after that, to travel specifically, almost incessantly, throughout the American West. I left behind any lingering aspirations I still had to work in the theater, and after some modest success as a landscape photographer, I put my cameras down as well. I wanted to see and write about landscapes I thought I could have an informing conversation with, and about the compelling otherness of wild animals.

These trips away from home in the early seventies—home by then was the west slope of the Cascade Mountains in western Oregon, a two-story house on a white-water river, where I still live—would eventually encompass traveling with Aboriginal people in the Northern Territory in Australia and working with a group of Kamba men in Kenya searching for hominid fossils. I would travel up the Orinoco River in Venezuela, through the Queen Maud Mountains in Antarctica, and down the Yangtze from Chongqing to Wuhan. I'd explore the cliff walls at Bamyan in Afghanistan, where two massive Buddhas, a husband and wife, once stood for 1,500 years as *genii loci* before being destroyed by cultural extremists. I'd travel in northern Japan, the Middle East, and the South Pacific.

Initially I thought of myself on these journeys as a reporter, traveling outward from a more privileged world. I believed—as well as I could grasp the idea back then—that I had an ethical obligation as a writer, in addition to an aesthetic one. It was to experience the world intensely and then to put into words as well as I could what I'd seen. I was aware others could see better than I, and also that other people were not able to travel in the way I had begun to, going away habitually. And whatever a reader might make of what I tried to describe, I already understood that their conclusions might not match my own. I saw myself, then, as a sort of courier, a kind of runner come home from another land after some exchange with it and its denizens, carrying, by way of a story, some incomplete bit of news about how different, how marvelous and incomprehensible, really, life was, out beyond the pale of the village in which I had grown up.

Looking back, I see that this ideal—to imagine myself in service to the reader—had me balanced on the edge of self-delusion. But it was at the time my way of working. It didn't occur to me that taking life so seriously might cause a loss of perspective. How else, I would ask, *could* you take it?

IT MIGHT HAVE BEEN the artist Saul Steinberg who once described himself as a writer who draws. For a while, after I put my cameras down in 1981, I thought of myself, pretentiously, certainly, as an artist who wrote. I was someone alert to visual imagery and drawn to movement through, and to arrangements made within, different volumes of space. I attended to this kind of thing in my written work much as I had done in my early photographs. I juxtaposed, emphasized, and hoped delicately to balance in these written compositions whatever the components happened to be.

Somewhere along this path, writing essays and stories, and many years into my work, I began to sense the ways in which I had changed as a writer over time. I wondered then whether it would be instructive to return to some of the places I'd visited earlier, to see how much could be learned from what would now, obviously, be different circumstances. I believed I'd reported carefully and accurately on what I'd first encountered; but I wanted to experience these places all over again, to go back, for example, to the High Arctic, to return to Galápagos, to make another trip to Antarctica. (In fiction, too, I'd developed scenarios set in specific landscapes—the agricultural California of my boyhood, the streets of Manhattan, the temperate rain forest that became my home in 1970, the Jimbocho district of Tokyo, but the imperative here, to revisit, was not so strong.)

I'd missed a lot, I knew, on my first passes through these places. On a second pass, whatever I might take in, I had faith the overall experience would affect me differently. I'd overnight in different spots; the weather would not be the same; and there would be the influence of books I'd read in the interim. And the illuminations and failures of my own life that had come along since would certainly reshape earlier perceptions.

One can never, even by paying the strictest attention at multiple levels, entirely comprehend a single place, no matter how many times one might travel there. This is not only because the place itself is constantly changing but because the deep nature of every place is not transparency. It's obscurity. I've never been drawn to the idea of writing definitively about anything, especially the Heraclitean nature of cultural geographies. In revisiting these places, then, I was more interested in how, in reviewing my previous experience of that location, I might find *another* truth, one different from the one I first wrote about. I was also interested in how my memory of a place might trigger new emotions, and in how the truth of such emotions might differently inform the facts I had once so carefully gathered. The anthropologist Carl Schuster, speaking about comparing cultural epistemologies, people's ways of knowing, once wrote, "Nobody has the vaguest notion of what this world is really like; the only thing that can be safely predicted is that it is very different from what anybody supposes." Schuster was raising an objection to the sometimes condescending positions scientists and academics take in speaking about reality and human fate. He was advocating for the sort of emotional and spiritual relationships all cultures experience in their encounters with their places, and which many of these cultures still enshrine alongside their more empirical, or analytical, responses to those same places, finding those perceptions equally valid in furthering an understanding of what is, finally, beyond understanding.

As the years went on, I felt I wanted to look again at nearly everything I had seen.

MY INTENTION IN rereading my field notebooks and drafting *Horizon* was to walk the distance between that moment in 1948, standing in the harbor shallows as a boy amid the sailboats of wealthy residents living at the Orienta, and a winter day in 1994 when I visited, for perhaps the tenth time, Cape Foulweather, a headland on the Pacific coast of Oregon, the locale of Captain James Cook's first landfall in North America. What does the man who made camp that day on the flank of Cape Foulweather, to wait for a late winter storm, hope to find, recalling some scenes from his boyhood while at the same time trying to imagine Cook's *Resolution* there in front of him, approaching on a shoreward tack, the ship having emerged first as a pindot on the horizon and then grown in a few hours into a full-blown three-masted square-rigger, half its sails reefed, rust stains bleeding from its scuppers and staining the black sides of its hull?

On that long-ago morning in March of 1778, the forest-shrouded mountains of Oregon's Coast Range had loomed dark beneath low-lying clouds. The wind was whipping layers of rain through the rioting air and Cook's ship, close-hauled a few miles offshore, was plunging and yawing through cross seas. Over the course of several days, the storm would bully the *Resolution* off miles to the southwest before its crew could bring it about and begin beating northward again. By then the *Resolution* had been shoved so far offshore the lookouts would miss seeing the mouth of the Columbia River, two days later. That wouldn't happen—for Europeans—for another fourteen years.

How far had I traveled between a boyhood longing to go and this reflective time on the flanks of the cape, having gone? And having seen so many parts of the world, what had I learned about human menace, human triumph, and human failure? Or about my own failings and fallibility? I tumbled these questions through my fingers regularly on Cape Foulweather, like familiar coinage.

There is no originality in this, of course. We, all of us, look back over our lives, trying to make sense of what happened, to see what enduring threads might be there. My further desire in planning this book was to create a narrative that would engage a reader intent on discovering a trajectory in her or his own life, a coherent and meaningful story, at a time in our cultural and biological history when it has become an attractive option to lose faith in the meaning of our lives. At a time when many see little more on the horizon but the suggestion of a dark future.

I SPENT MANY DAYS over a period of about ten years camped on the heights of Cape Foulweather—the name Cook gave it that day, the seventh of March, 1778—absorbing the moods of the Pacific as they changed. From the shoulder of the cape, the ocean's broad back is a vastness not to be had in a single view, any more than a sidelong view of the beloved's cheek can carry anything like the full impact of the lover's straight-on gaze. Could I, I asked myself once, within the protean and stage-like expanse of that sea, imagine, in the same moment of looking, *another* vastness—a sere sand plain in the Namib Desert in Africa, say, trembling there over the water's opaque surface and carrying a barely discernible herd of six oryx traveling? Or could I conjure, in that same volume of oceanic space, a boyhood memory—an afternoon in the Mojave Desert, searching for coyotes, disoriented in a vastness of creosote bush—without losing either image, the real one before me or the remembered one? Or, watching a fresh wind raise the hackles of the sea, could I hold simultaneously in my memory a night of breezes easing occasionally through a hotel-room window in Mindanao, soft as a horse's sigh, and then the screeching, predatory wind that for hours thrashed the tent wall by my head one subzero night in Antarctica?

What had *changed* for that boy in Mamaroneck Harbor, whose mother, seated in the shade of an oak, looked up regularly from her needlepoint to find him once more in sunlight coruscating on the water?

WHEN I BEGAN to visit Cape Foulweather in the early nineties, it was for no particular reason beyond my admiration for James Cook. The cape wasn't far from my home and I loved watching the birds, the fishing boats, the changing weather. The view of the ocean alone, seen from the rise of the cape, high above a rampart of sea cliffs, was often dramatic. Some days the sea was so side-lit, so serene, that for dozens of square miles the water seemed to be a pane of ribboned glass, the light reflecting from its surface so molten the pupil of my eye could not close down tightly enough to produce any texture. On certain summer nights, the air was transparent enough for me to make out detail in the opposite direction, twenty miles to the east, an inland mountain range bathed in lunar light. At the same time, I was able to see off to the north, opposite the arc of the moon's course, an immeasurable, glittering field of undimmed stars.

I periodically spent a sequence of mostly idle days on the cape, camped each time in the same recovering clearcut. Apprenticeship was what this had turned into. Occasionally, sitting amid young trees in the clearcut, I'd pick up some small thing, a Sitka spruce cone or the translucent wing of a dragonfly, and attempt to sketch it. I failed repeatedly to create with my pencil anything worth a second look; but in that hour of drawing I would gain insight, not only into the shape of the object but into its overall form, its third dimension. I'd grasp the temporality of it, or on occasion, the fractal scaling of its parts, or in some other way be drawn into intimacy with it.

These innocuous palm-sized bits of life were as provoking of thought and emotion for me as the sudden appearance of a mountain lion might have been. I reached for small objects to feel their contours, to get the heft or the texture of them. Or I'd rotate and hold them in such a way as to get sunlight to refract through their crystals, as you might with a feather, or so the sunlight would illuminate deep shadows in a bit of bone.

Embedded in the system of belief that over the years came to replace (or perhaps augment) religion for me is a conviction that the numinous dimension of certain inanimate objects is substantial, as real as their texture or color. This is not, I think, an illusion. One might not be able to "squeeze meaning" from a stone, but a stone, presented with an opportunity, with a certain kind of welcoming stillness, might reveal, easily and naturally, some part of its meaning.

I spent hours on the cape emptying my mind of analysis, suspending its incessant quest for essence, and regularly encountered in doing so William Blake's enduring metaphor, that the entire world is rendered for us in a single grain of sand.

As I drove again and again up the little-used cape roads to reach the old log landing where I made my camp, I came to feel, incidentally but pertinently, an admiration for the aging vehicle I was always driving. On some steep sections of the narrow roads, I crept along in first gear in four-wheel drive, so as not to gouge the roadbed and so invite erosion. I was able to push through wet snow and deep mud in winter, at spots where heavy equipment, long gone now, had cratered the ground. When large trees fell across the road, I had to cut them into sections and pull them aside with tow chains to get past. And every time I did these things, a question arose about the propriety of doing what I was doing. Shouldn't I have just allowed this healing land to heal? Was my infatuation with my speculations, my own agenda, more important?

Was there no end to the going and the seeing?

3

Remember

One rainy autumn day in 2009, I went to visit the Nicholas Roerich Museum, a five-story brownstone at 319 West 107th Street in New York City. Roerich was a cultured Russian painter (1874–1947), a set designer for the Moscow Art Theatre, a philosopher with deep, wide-ranging interests in archeology, religion, and language, and also a gifted colorist, an artist who used color in a skillful or distinctive way. He fled Russia for America at the age of forty-six, and after a few years in New York City, left in 1923 to sojourn and paint in the Himalayas, India, and Mongolia. He returned to New York in 1929 with 500-some paintings. He and his wife of many years, Helena, eventually moved to the Kullu Valley, at the foot of the Himalayas in Himachal Pradesh, in northern India, where they pursued their mutual and individual interests in art, religion, music, and science. He died there at the age of seventy-nine.

Many of Roerich's Himalaya paintings are hanging in the Roerich museum, and I was headed there to see them mostly because I knew so little about him and felt compelled to learn (as had once been the case for me with an American painter that Roerich's life and work reminds me of, Rockwell Kent). I had seen paintings of his over the years in books and magazines and had sensed there was something in his work that would speak to me if I could see the paintings full size in a museum setting. And there was. It emerged from a vivid, 34×46-inch tempera-on-canvas painting entitled *Remember*.

The painting stopped me abruptly. Not because it was more impressive than the other paintings hanging nearby, but because it riveted me like a vision. On the far left is a lone male in dark clothes wearing a yellow vest and astride a white horse. He has risen up in his stirrups to look back while the horse waits. A traveler. On the far right of the painting is a large dwelling—the rider's home, one might assume. Prayer flags flutter on thin poles above these quarters, and two women, one bearing a water jar on her head, stand before the house looking toward the rider, perhaps a wife and daughter. All else is space—the bare ground between the rider and the women and the spectacular, rising blue walls of the Himalaya, a backdrop of vertical land, its jagged peaks white with snow. The painting is about space as much as it is about departure, and few pieces I've ever seen say so poignantly how one's memory is activated by leave-taking. The rider has turned to gaze back at the women and the dwelling. The waiting horse faces in the direction of the rider's destination. The middle ground of the painting is rendered imprecisely, almost abstractly. The serigraphed foothills suggest the great depth of this particular landscape, which finally towers in the distance.

Whether Roerich intended for us to reflect on how memory fixes upon those elements of home the departing traveler will most keenly, or most emotionally, recall, or whether his title is an admonition to the rider not to forget those he is leaving behind, I don't know. It's enough for me to sense, when I look at this image, that I am being brought immediately into the predicament of departure—the desire so strong to head out, yet at the same time feeling a breach opening, the breaking of a bond that can be repaired only by returning.

What experience might be discovered on the far side of that breach to somehow justify the leaving?

When in 1979 I encountered a traditional group of people for the first time on their home ground, at a small Nunamiut Eskimo village called Anaktuvuk Pass, in Alaska's Brooks Range, I had among my first thoughts an obvious question: Why did I know so little about these people? I didn't mean knowledge about their material culture or their hunting techniques or the way they were able to survive in the harsh landscape they'd chosen to live in, but about the way they understood the world. What did they find mysterious but still worthy of their full attention? And whatever that was, did they leave it be or did they pursue it analytically? Were the difficulties and paradoxes of leading a just life the same for them as they were for me? Why was it never mentioned in the good schools I attended that such people saw as deeply into the physical world as the Greek philosophers we were asked to read?

Did they possess attitudes and approaches necessary for survival, which my own culture might have unknowingly thrown away with the onset of modernity—or not even considered to begin with? Why weren't their insights into life's predicaments a larger part of the growing international discussion about human fate? Why were their metaphors considered less empirical, less sophisticated, by most people in the cultural West?

My anxiety about this gradually created a sense of urgency. Wherever I have traveled since those first days at Anaktuvuk Pass, I've wondered, What is going to happen to us? What is our fate if we do not learn to speak with each other over our cultural divides, with an indifferent natural world bearing down on us?

In writing this book, and in recalling the Roerich painting, I had in mind recounting my experience in five separate places, and believed that this journey through recollection would start at Cape Foulweather. As I began work, however, I felt the insistence of three other places, in each one of which I'd felt an identical and peculiar sense of urgency about humanity's fate.

It was the same sense of urgency that I imagined the rider in the Roerich painting to be feeling.

IN THE SPRING of 1987 I was traveling down the Yangtze River from Chongqing to Wuhan with a delegation of American writers. The ferry stopped late one night at Yueyang, and most of the passengers—several hundred people—disembarked to purchase food and other items at a public market that had remained open to receive the travelers. The route to the market from the riverbank proceeded up a large, brightly lit cement staircase with dozens of steps. I'd noticed another staircase on disembarking, however, hardly lit at all, but which appeared to lead to the same place. I took that route. The newer staircase must have been constructed to replace the crumbling one I began ascending, and down which a stream of foul water was meandering. I'd gone only a short way before I realized I was climbing through a stream of sewage.

About halfway up the hill I paused before a framed opening larger than a doorway in the wall of a building. On the far side, in a room lit with large candles, a group of six or seven naked men were readying themselves for bed. One man was standing upright in a washtub while another poured rinse water over his head from a metal pitcher. Others were smoking cigarettes and mending clothes. It was a humid night and the bodies of all these men—sinewy, lean, hard—glistened in the candlelight. Bunk beds were arranged on the walls in tiers of three, and several men had already retired. Stevedores, I thought. I could hear water splashing in the tub, the trickle of sewage by my shoes, spilling softly down the steps, and the murmur of conversation coming from the room. It was a scene of human laborers at day's end, but one that originated in another century. The candlelight in the room did not spill far, and the men, I knew, could not see me standing in darkness on the staircase.

At the top of the stairs I entered the night market. Passengers were haggling over root vegetables—turnips, onions, potatoes—and merchants were shouldering their way through with plastic buckets of butchered meat. Others were carrying strings of ulcerated fish from the Yangtze, water in which I had seen all manner of waste floating (and to my astonishment two endangered Yangtze river dolphins). Live monkeys and other small mammals, hedgehogs among them, stared out from the confines of screened metal cages. In one booth, wicker trays of dead crickets and heaps of caterpillars were on display, beneath a kind of clothesline from which dozens of sparrow-like birds hung by their feet. This was more than the atavistic scenes of medieval meat markets that Pieter Aertsen painted in the sixteenth century. It was the future, the years to come, when we would begin killing and consuming every last living thing.

IN AUGUST 2012, I was serving as a guide and lecturer aboard a Canadian ecotourism vessel in the High Arctic. My habit each day was to get up at five and to take a cup of coffee up to an open deck above the bridge where I could watch birds. I regularly met a couple there who had the same habit, but who were much better birders than I was. The morning I'm thinking about, the ship had turned out of the Parry Channel a few hours before and was heading south into Peel

Sound. We were bound for Bellot Strait, a narrow waterway that marks the northernmost shore of the North American mainland. We'd told the passengers that we had a good chance of seeing polar bears there. For some reason it had not yet registered with me how really unusual the scene before me was—entering Peel Sound without an icebreaker escort. In the historical literature of the Arctic, explorers have repeatedly emphasized that Peel Sound is simply not navigable for an unescorted ship, even in summer. It's always heavily jammed with multiyear ice.

I joined my companions. Neither of them spoke a word of greeting. Nor were they scanning with their binoculars. They were staring blankly into the sound. Three cups of coffee steamed on the small shelf in front of us. I knew this older man and woman had read as much Arctic history as I had, and now I realized what had made them silent. There was not a single ice floe in the waters ahead. Not a scrap of ice. We saw numerous ringed seals and bearded seals swimming there, but the polar bears we'd been certain we'd find hunting those seals were nowhere to be seen. Their hunting platforms were gone.

I thought of the passengers below, who had been asking from the start of our journey in west Greenland whether we were going to see any evidence of global climate change, about which the Greenlandic Eskimos had expressed such consternation.

IN THE SUMMER of 2007, I was traveling in Afghanistan. I'd gone to Kabul to visit a woman I'd met at a conference in Ubud, Bali, the year before. She was the head of the Red Crescent there and had invited me to stay with her family if I ever came through. One day she took me to her offices at the edge of the city. I visited with some of the people being cared for there, many of them victims of the war. At some point she introduced me to a man about my own age and then returned to her office. He and I continued to walk around the grounds of the compound, talking about the plight of the people there and about his work. We had no particular destination, we were just walking. I assumed we would end up back at his office, where my friend might be waiting for us.

At one point he opened the door to a large building and we entered.

Perhaps a corridor here provided a shortcut back to his office. It was quiet in the building's lofty hallways, which were lit by sunlight from the clerestory windows. As we entered, I saw a woman standing by herself in a broad corridor off to the left. She was wrapped in a bed-sheet and leaning against the wall. When she saw us, she began running toward us, the sheet floating behind her like a luffing sail. She was naked, a woman in her fifties, her face one of incomprehension, of both disbelief and wonder. Her mouth worked soundlessly, like a fish out of water. Suddenly she halted. She and I stared at each other without moving. Then she turned and ran back down the corridor.

The man and I walked on. He said those who lived here had been driven mad by the war, mostly women who'd lost children and husbands. Occasionally they manage to get out of their rooms, he said. He seemed to be ashamed and embarrassed, grieved by what we had seen. He hadn't wanted me to see it.

But I did, and I remember her face to this day.

4

Talismans

Over the years I've carried home a handful of mementos that signify for me, each one taken from a moment or an event that might have seemed innocuous at the time to someone else looking on. A dozen or so of these sit atop a tall Japanese tansu in my home. I've arranged them there to make intuitive sense together, the way you might arrange scenes in a short story. In this matrix they suggest for me some deeper truth about life, one that always lies just beyond my reach.

Over time the mementos on the tansu have come to include a set of four shells of *Cardita megastropha,* a clam-shaped mollusk with no popular name I am aware of in English (in Spanish it's referred to as *concha corazón*). The shell is commonly found in cool inshore waters in the eastern South Pacific. The surface of each shell is ribbed, a radiating pattern that suggests the structure of a folding fan. Each is different in size (meaning it differs in age from the others), and each carries a different version of a predominant graphic design, one composed of medium-brown chevrons. The saturation of the hues and the spacing of the chevrons in the design vary from shell to shell, a phenomenon systematists call phenotypic variation. *Cardita megastropha* makes its unheralded way in the world by constantly evolving in response to physical and chemical changes in its intertidal saltwater environment. The distinctive nature of each shell is a reminder of the astonishing and unpredictable range of individual expression within a species—the many "phenotypic expressions of a genotype," as an evolutionary biologist might put it. Within any given set of animals that at first glance might all appear to be identical—a herd of grazing impala, a school of mackerel, a flock of doves—are numerous individuals, each with a different history, a different potential. To assume otherwise would be to foreclose on evolution, and to limit one's appreciation of the moment in which aggregations like these are seen.

One morning in April 1987, at an archeological site in Xi'an, in Shaanxi Province, China, I stared through my binoculars into a series of parallel archeological trenches. Arrayed in these excavations, in strictly ordered military ranks, were hundreds of terra-cotta foot soldiers, preceded and followed by dozens of terra-cotta cavalry horses and chariot horses. All these figures, discovered by well diggers in 1974, were slightly larger than life size. Studying each human face, one by one, I saw that no two were exactly alike. The same was true for the horses. The presence of such slight variations suggested to me something about the place of tolerance within the otherwise rigid social organization of contemporary Emperor Qin Shi Huang's palace guard, in Qin Dynasty China (221–206 BCE). Also, perhaps, that the Chinese of this early period already recognized that diversity is an ineluctable component of every successful attempt to establish order.

This lesson is reiterated for me in the four *Cardita* shells resting on the tansu.

ALONGSIDE THE SHELLS on the tansu's polished paulownia wood top lies a thin sheet of greenschist, a volcanic rock about the size and shape of a thin slice of bread cut from the middle of a baguette. Over a long period of exposure, the rock has weathered to a red-orange hue, its surface brindled with streaks of black from iron deposits. I picked it up one day in a dry watercourse in the Jack Hills of Western Australia, an isolated section of semiarid country with no permanent roads. I had carried a hand-drawn map with me that day as I searched for a site where geologists had recently located the oldest intact geological objects on the planet, minute zircon crystals embedded in a chert-pebble conglomerate, a coarse-grained type of sedimentary rock. Some of the crystals, formed shortly after the planet solidified into a sphere, are 4.27 billion years old.

The reason I'd traveled to that part of the Jack Hills was to *see* these zircon crystals undisturbed in their native place. What did the landscape around them have to say? I wanted to know what its colors were and what forbs, species of grasses, and trees were nearby. How did the soil here respond beneath the press of my foot? Which birds were flying through? In which trees might they alight, and what tones comprised their calls? Any of these things might clarify the nature of the zircon crystals in a way different from the articles I had read in *Nature* and *Special Publication/Geological Society of Australia,* which had first tripped my desire to search them out. To have walked away with a piece of the conglomerate containing the zircon crystals would have been unethical, a betrayal of a place deliberately left vague in the scientific journals; and a betrayal, too, of the geologist who'd drawn the map that showed me how to get there. Instead, I took this piece of greenschist, fragments of which lay all around, a common piece of the strata of rock that underlies the zircon-bearing conglomerate.

A date of 4.27 billion years for the crystals places them in the early

Archeozoic period of the Precambrian era, more than four billion years before the emergence of the first dinosaurs.

ALONGSIDE THE PIECE of greenschist are two eucalypt buttons. I picked them up from the ground at the top of what some refer to as "the suicide cliff" at Point Puer, in southeastern Tasmania. Early in the nineteenth century a few buildings were erected here to provide housing for adolescent male prisoners incarcerated at a British transport prison called Port Arthur. Building the dormitories was part of a plan the resident commandant developed to protect the boys from sexual predation by adult males living around them in the prison compound.

Like other transport prisons in Australia during the nineteenth century, Port Arthur boarded the psychopathic and the deranged indiscriminately with the innocent and the unlucky, and jailers made little effort to keep the former from preying on the latter. Some of the Port Arthur boys, it is said, desperate to escape the daily rounds of sexual abuse and physical punishment, came to these cliffs at night, held hands, and jumped, falling more than a hundred feet into the frigid waters of Carnarvon Bay.

Standing there that day at the crest of the cliffs, rotating two eucalypt buttons in my hand like a pair of dice, I imagined that the boys were driven to this fatal act by the certain knowledge that they were trapped in circumstances beyond their control, and that they would be snared like this, caught in some pedophile's net, for years to come. The emotions on which they chose to act were emotions I once knew and which I can readily recall. But I was not entangled as a boy as hopelessly, as fatally, as they were.

NEXT TO the eucalypt buttons sits a small water-polished stone, a dark piece of basalt. I found it on a pocket beach among thousands of nearly identical rocks at Cape Horn, on a chilly, fog-drenched, austral summer morning in January 2002. I took three of these stones, each about the size and shape of a Brazil nut. One I sent to my younger brother, living on the coast of Maine, who had an affinity for the sea; the other went to my half brother in Northern California, a retired naval officer who had become a traditional healer.

The full nature of the stone that I kept for myself is not immediately apparent. Beneath the black patina on its surface is a dark gray, fine-grained volcanic rock called andesite, named for a chain of mountains, the Andes, which, as the Darwin Cordillera, plunges into the Southern Ocean at the tip of South America. The black coating, an encrustation of iron-manganese oxides, was created by a community of diatoms and other microorganisms, which millions of years before had lived on the surface of the stone.

Perhaps, over time, my brothers lost track of their companion stones, but the one I keep recalls these two men for me, and the thousands of sailors I'd read about who lost their lives trying to double that cape in sailing ships.

ON THE OTHER SIDE of the cinnamon-colored greenschist is a spent 7.62mm NATO cartridge casing I retrieved on the grounds of a cemetery at a decommissioned Norwegian whaling station called Grytviken, on the island of South Georgia. South Georgia is one of several large, sub-Antarctic islands in the Southern Ocean that James Cook claimed for England during his second around-the-world voyage, in 1772–75. South Georgia, along with the South Sandwich Islands, both former Falkland Islands Dependencies, are today British Overseas Territories. Great Britain's claim to the Falklands is founded on an opinion most historians share, that the English navigator John Davis was the first European to see these islands, in 1592. At various times Spain, France, Chile, and Argentina have also claimed them. When Argentina, off whose coast the Falklands lie, decided to press its own claim by occupying them in 1982, Britain responded with military force, swiftly ending the so-called Malvinas, or Falkland Islands, War. And the occupying British troops stationed on South Georgia went home.

I picked the empty brass casing up a few paces from the grave of Sir Ernest Shackleton, on the shore of a harbor that once served this now-long-abandoned whale processing site and Southern Ocean entrepôt. Dozens of 7.62mm casings gleamed around me that day in the pale sunlight. They were scattered like grain across a graveyard of explorers and whaling men, and they shone like spilled bits of mica along footpaths that wound through the partially collapsed, bullet-riddled structures of the whaling station. The casings told a provocative story, for me, about *pro patria mori* emotions, about persistent colonial claims to such remote bits of bleak, virtually unoccupied land in the modern era, and about humanity's enthusiasm for the violent enforcement of strongly held political beliefs.

The once enormous populations of large whales in the nearby waters—blues, southern rights, seis—have yet to recover from a spree of killing that lasted well into the twentieth century, the expression of another related persistent human desire—to take possession. To put whatever was discovered in a new place to "better use."

THESE MEMENTOS OF travel sit apart from one another on the tansu. The generous space I've left around each is meant to leave each room for its aura. As I pass them by, year after year, going back and forth to a room where I work, each object remains piquant for me, eloquent in its silence. The staggering diversity of life, the stony flesh of the ancient planet, the lethal violence of human behavior, the growing inutility of war in the modern era.

I glance at them because I know I am prone to forget.

I'VE CLEARED SMALL SPACES around our house for other talismans. I engage with these, too, as if they were votive candles I'd lit. Here are bits of volcanic scoria and water-tumbled seashell from Point Venus in French Polynesia, the spot on Tahiti's northern shore where Cook tried, successfully, to observe the transit of Venus across the face of the sun, in 1769. Beside these, a fist-size piece of raven-black dolerite, its sheer surfaces intersecting as neatly as the sides of a pyramid. A ventifact, a thing made by the wind, brought home from the Wright Valley in Antarctica's southern Victoria Land.

Two other objects hold a particular place in this olla podrida. I keep one by my bedside, wherever that happens to be, and the other on my writing desk. Next to my bed is a sand-cast silver harpoon tip, a stylized replica of a toggling implement that Eskimo hunters have used for centuries to secure and retrieve seals. A gift from my wife. To provide food for one's family, whether it is seal meat or a sack of grain or the flesh of an avocado, is to encounter again an unsettling question about the way in which death provides life. To act here is to face one's own complicity, to choose to take life in order that one's own kin might continue to live. When I lie down to sleep far from home, I place this small work of art close by on a folded scarf. It was crafted by a man named Jimmy Naguogugalik, an Inuit artist and hunter from Baker Lake, in Nunavut, Canada. It reminds me of the centrality of the symbolic in human life, and of both the *consequence* of providing and of the *obligation* to provide.

The object on my writing desk is a stark reminder of a connection I feel, though a tenuous one, with a murderous period in Western history. A *real de a ocho,* an eight-real silver coin, crudely minted in Mexico City sometime between 1630 and 1641, during the reign of Philip IV of Spain. It comes from a large cargo of bullion and coinage carried by the Spanish galleon *Nuestra Señora de la Pura y Limpia Concepción* when it set sail from Veracruz, Mexico, on July 23, 1641. Later that summer, probably after making a port call at Havana, the galleon encountered a hurricane and was dismasted, somewhere south of the Turks and Caicos Islands in the western Atlantic. The crew, it is thought, were trying to reach the harbor at San Juan, Puerto Rico, when their ship, heavily laden with gold and silver ingots and with bags of silver coins, ran aground on Abrojos Reef, about eighty miles northeast of Cabo Isabela, Hispaniola (the Dominican Republic today). The shipwreck was located during a search in 1687 and a portion of its cargo salvaged. The position of the *Nuestra Señora,* however, was not accurately fixed at the time and its whereabouts remained unknown for another three hundred years, until November 28, 1978. The coin on my desk comes from this second salvage.

For me, and for some members of my stepfamily, the story behind this coin has a discomfiting personal dimension.

In 1521, Hernán Cortés ordered four brigantines built on Lake Xochimilco for the invasion of Tenochtitlán (Mexico City). In 1524, his shipbuilder, Marín (or Martín) López, received a land grant in Pinar del Río, a region of western Cuba, from the Holy Roman Emperor Charles V as a reward for his having built the brigantines. Members of the López family took possession of the land in Cuba at the time but continued, for the most part, to occupy and maintain their former lands in northern Spain, in a region of the Iberian peninsula called Asturias. (Asturias is still referred to today by politically conservative Spaniards as "the principality of the kings," partly because it was the homeland of Rodrigo Díaz de Vivar, the man more widely known as El Cid. Asturias is also the only region of Spain not subjugated historically by "foreigners"—that is to say, by Rome or by the Moors. It's regarded today as a citadel of Spanish *pur sang.*)

Pinar del Río eventually became the region of Cuba most preferred by tobacco growers. In the 1850s, after Spain relaxed its onerous export tariffs on tobacco, the López family emerged as one of the three or four most important cigar-manufacturing families in the country. Later my stepfather's branch of the family, with money from their tobacco interests, purchased a walled estate overlooking the Asturian coastal village of Cudillero. The compound was a "casa del Indio," an estate built on wealth from the New World.

According to my stepfather, male members of his branch of the López family are best viewed historically as hidalgos, as "near royalty." In 1900 my stepfather's father, Don Eugénio López Tréllez y Albierne de Asturias y Vivar, was appointed the Spanish first secretary to the Court of St. James's by Alfonso XIII. In 1908, two years after my stepfather was born in Southampton, Hampshire, Don Eugénio resigned his appointment and returned to America. He had departed the States for Asturias at the start of the Spanish-Cuban War (Spanish-American, to most Americans), in 1898. Once back in New York City, he again took up representing the family's tobacco interests in the United States.

I sought out the silver *ocho-real* from the *Nuestra Señora* in 1997 on a visit to Christiansted, in the U.S. Virgin Islands, not so much because of its link to the early activities of my stepfamily in the New World but because I wanted this manifest symbol of unrelenting pathological exploitation around me while I wrote, a thing smaller than, say, a bale of rubber from Prince Leopold's Congo. It recalls for me the extent of international indifference to catastrophic human suffering, then and now, a worldwide indifference to the fate of human beings that has persisted through numerous slaughters, including in my own lifetime those in Siberia, in Cambodia, in Iran under the Shah, in Liberia under Charles Taylor, and in Chile under Pinochet.[6]

The temptation for someone like me with this silver coin, someone with an active objection to the mistreatment of indigenous people, is to consider myself apart from the mistreatment, not implicated in these subjugations and exploitations, beginning, say, with the Black Legend of the Spanish conquistadores, and with the English financial investment in, and development of, the Atlantic slave trade. I'd be on secure moral ground absolving myself of direct responsibility in all this, but for me—and for many others, I must think—taking this position would leave unaddressed the ethical responsibility to object. It would be to hear resonating within oneself the shouting of the Mothers of the Disappeared in the Plaza de Mayo in Buenos Aires in the 1970s, and to turn instead to other things.

Sometimes I've been able to rise to these ethical challenges and craft what I hope will be an eloquent objection. Other times, I am ashamed to admit, I step into the next room. I shut the door. Who can change this? I say to myself. The horrors—ethnic cleansing, industrial rapine, political corruption, racist lynching, extrajudicial execution— once identified and then denounced, always return, wearing different clothes but with the same obsessive face of indifference. We denounce those who order it, we condemn the people who carry out the policies, calling them inhumane. But the behavior is fully human.

We are the darkness, as we are, too, the light.

OF THE TALISMANIC REMINDERS set about in nearly every room of our house over nearly five decades, like pages from a psalter, I want to describe one final piece, an object that still unsettles me because it reminds me when I write to trust the reader to apprehend the injustice in what I try to describe. I do not need always to parse it.

Historians mostly agree that Christopher Columbus's first landfall in the Americas was at a Bahamian isle called Samana (or Atwood)

Cay, located northeast of Acklins Island, about forty miles north of the Plana, or French, Cays. For most of the twentieth century, however, it was generally believed that his first landfall was at San Salvador, eighty miles north-northwest of Samana Cay. (In 1926, the British name for this isle, Watling Island, was changed back to San Salvador, the name Columbus originally gave it on October 12, 1492. Local Lucayan people, according to Columbus, referred to the place as "guanahani.")

In the spring of 1989—I was forty-four at the time—I traveled to San Salvador with a friend, Tony Beasley. I wanted to dive the island's reefs and also to see a monument honoring Columbus, erected on the bottom of Fernandez Bay. One very hot afternoon on a walk along the island's shore, Tony and I found ourselves at Fernandez Bay but unprepared. We had no snorkeling equipment with us. On an impulse I stripped off my clothes and bolted, naked, for the water. (We were alone at siesta time on an otherwise deserted beach, out of public view.) I swam furiously toward the place I anticipated the monument would be, swam until I was so winded I felt in danger of drowning. Anger had suddenly flooded my senses on the beach. Unresolved anger over the behavior of my stepfather's ancestors, and of the other hidalgos—Pizarro, Gonzalo de Sandoval, Diego Velázquez, Andrés de Tapia—the second-son conquistadores; anger about the loss of so many unchronicled cultures, the consequence of colonial genocide and exploitation; frustration with imperial incursions of all sorts, in nearly every freshly discovered place in the world, over the centuries; fury over licentious behavior forcing its way into the hinterlands of every political empire, perpetrated by people imbued with a sense of divine right as they redesigned societies, burned out spiritual practices, and restructured economies to serve their own ends. At that particular time this was, for me, Shell Oil operating in Nigeria, Rio Tinto mining in Western Australia, the Chinese boot heel crushing Buddhist culture on the Tibetan Plateau. I was furious about the impoverishment and hopelessness of people I'd seen eking out an existence in places like São Paulo, about those separated from their homes and living in refugee camps all over the world and dying in war zones in Angola, Sri Lanka, and Indonesia. The Japanese use a word, *hibakusha,* to describe those who physically survived the bombings at Hiroshima and Nagasaki but who subsequently lost their minds.

These individuals are "explosion-affected people"—uncomprehending, disoriented, catatonic with grief. They're everywhere now, from the Lakota Indian reservation at Pine Ridge, South Dakota, to IDP (Internally Displaced People) camps in Eritrea and South Sudan. They are people capable only of existence, not recovery. For them, the damage has gone too deep.

In that moment that afternoon on San Salvador, all such genocidal horror—in Tenochtitlán, in the American West, in Sarajevo—seemed rooted for me in the same insane and seemingly ineradicable desire: to eliminate strangers and take possession of whatever they had.

I burned up my anger in the long, exhausting swim. Treading water, I could see below me the pale monument to Columbus rising distorted through a lens of clear tropical water. Voiceless. Adamant.

I swam back to shore, standing up when my toes finally touched bottom. Tony was watching from the beach, a hesitating, quizzical look on his face. Some moments passed while I stood in the shallows, catching my breath. As I waded toward the beach I began to speak aloud in disconnected sentences, enunciating the familiar principles of justice, proclaiming sorrow and regret, asking the pardon of every animate thing before me—the trees, the clouds, the broken shells washed up on the beach. Stepping clear of the water, I knelt on the beach and bent forward to rest on my palms, stupefied by the heat, squinting into glare off the sand, startled by my own outburst. Just in front of me, inches away, was a piece of chalk-white sandstone, exactly the shape and precisely the size of a human tongue.

I picked it up.

Tony and I walked together back to Cockburn Town, to our air-conditioned hotel room. He didn't say anything about what I had declaimed, words I was too self-conscious to try to recall. I lay on my bed wondering if the fury I'd felt had actually been ignited not by history but by the reawakening of my own feelings of impotence.

TAKEN TOGETHER, the objects I've described represent a kind of strategy I've used to remain connected to the disorderly world, with its numerous paradoxes and inconsistencies. Further, they point me toward an overriding and fundamental issue—the importance of preserving the human capacity to love. These objects are, too, reminders of my own unconscious presumptions and impositions, according to which I occasionally organize the world I encounter in such a way as to feel safe in it. I read daily about the many threats to human life—chemical, political, biological, and economic. Much of this trouble, I believe, has been caused by the determination of some to define a human cultural world apart from the nonhuman world, or by people's attempts to overrun, streamline, or dismiss that world as simply a warehouse for materials, or mere scenery.

It is here, with these attempts to separate the fate of the human world from that of the nonhuman world that we come face-to-face with a biological reality that halts us in our tracks: nature will be fine without us. Our question is no longer how to exploit the natural world for human comfort and gain, but how we can cooperate with one another to ensure we will someday have a fitting, not a dominating, place in it.

What cataclysm, I often wonder, or better, what act of imagination will it finally require, for us to be able to speak meaningfully with one another about our cultural fate and about our shared biological fate?

As time grows short, the necessity to listen attentively to foundational stories other than our own becomes imperative. As I've encountered other human cultures over time, especially those radically different from my own, each one has seemed to me both deep and difficult to comprehend, not "exotic" or "primitive." Many cultures are still distinguished today by wisdoms not associated with modern technologies but grounded, instead, in an acute awareness of human foibles, of the traps people tend to set for themselves as they enter the ancient labyrinth of hubris or blindly pursue the appeasement of their appetites.

It is nearly impossible for wise people in any culture to plumb the depths of their own metaphysical assumptions, out of which they have fashioned a world view. It is also difficult to listen closely while some other people's guiding stories unfold, or to separate successfully the literal from the figurative in those stories, the fact from the metaphor. And yet if we persist in believing that we alone, living in whatever culture we're from, are right, and that we therefore have no need to listen to anyone else's stories, stories that we often can't quite understand and so are unwilling to discuss, we endanger ourselves. If we remain fearful of human diversity, our potential to evolve into the very thing we most fear—to become our own fatal nemesis—only increases.

The desire to know ourselves better, to understand especially the source and the nature of our dread, looms before us now like a specter in a half-lit world, a weird dawn breaking over a scene of carnage: unbreathable air, human diasporas, the Sixth Extinction, ungovernable political mobs.

IN *THE WISDOM OF THE DESERT,* the Trappist monk Thomas Merton, considering the moral obtuseness of the conquistadores, writes, "In subjugating primitive worlds they only imposed on them, with the force of cannons, their own confusion and their own alienation." If this colonizing impulse in our heritage is still with us, a need to dominate, must we continue to support it? Must we go on deferring to tyrants, oligarchs, and sociopathic narcissists? The French poet, diplomat, and Nobel laureate Alexis Léger, in his epic poem *Anabase,* asks where the troubled world is to find its *real* protectors, warriors so dedicated to protecting the welfare of their communities that they can be depended upon "to watch the rivers for the approach of enemies, even on their wedding nights."

Where, today, can the voices of such guardians be heard over the raucous din in support of economic growth?

In her poem "Kindness," the Palestinian American poet Naomi Shihab Nye writes that to learn the kindness required to ameliorate the cruelty and injustice the real world presents us with,

> *you must travel where the Indian in a white poncho*
> *lies dead by the side of the road.*
> *You must see how this could be you,*
> *how he too was someone*
> *who journeyed through the night with plans . . .*

In which national parliaments and legislatures today can we find deliberations characterized by such a measure of humility? In which congresses might questions of ethical irresponsibility be successfully raised for discussion? In which Western nations does a determination to address the mental, spiritual, and physical health of children override indifference toward their fate? Or are these questions now thought to be anachronistic, questions no longer relevant to our situation?

IT IS NOT POSSIBLE, of course, to live up to one's own standard of good behavior every day. Distraction and indifference always offer us a way out of dilemmas otherwise too exhausting or harrowing to face. Still many, in every corner of the world in my experience, press on through such discouragement and defeat, bind up their wounds, and tend to the needs of others, like the Aparajitas of Bangladesh, the "women who never accept defeat." Most anyone today can imagine the biblical horsemen of the Apocalypse deployed on the horizon, pick out each one and characterize him. Anyone, too, facing this frightening horizon, might opt to turn away, decide instead to become lost in beauty, or choose to remain walled off from the world in electronic distraction, or select catatonic isolation within the fortress of the self. But one can choose, as well, to step into the treacherous void between oneself and the confounding world, and there to be staggered by the breadth, the intricacy, the possibilities of that world, accepting its requirement for death but working still to lessen the degree of cruelty and to increase the reach of justice in every quarter.

For many years the need for this kind of heroic effort—essentially to learn to cooperate with strangers—has been calling to modern people. I've wondered, watching economically powerful nations scrambling in the world's remote corners for the last large deposits of copper, iron, bauxite, and other ores, or reading about the failure of once-dependable ocean fisheries, or about cynical corporate maneuvering to secure the last large reservoirs of potable water, whether an unprecedented openness to other ways of understanding this disaster is not, today, humanity's only life raft. Whether cooperation with strangers is not now our Grail.

———

I LOOK BACK at an unsuspecting boy, a child beside himself with his desire to know the world, to swim out farther than he can see. The boy, I know, will live his life like this, always searching, even though he doesn't really know what to look for. It will be many years before he understands that this continuous search for meaning is most everyone's calling. Facing chaos, we're sometimes prone to insist that we're only ardently searching for *coherence,* for a way to fit all the pieces of our life experience together into a meaningful whole, to find a direction in which to continue. Gaining that, we say, we can expect to find relief from some of our pursuing anxieties.

It has long seemed to me that what most of us are looking for is the opportunity to express, without embarrassment or judgment or retaliation, our capacity to love. That means, too, embracing the opportunity to *be* loved, to ferret out and nurture the reciprocated relationships that unite people, that bring people and their chosen places, both the raw and the built Earth, together into one agreement, without coercion or sentimentality. If someone was to suggest that the evidence for how things can and do go wrong is only evidence of the repeated failure to love, even the boy, I believe, would agree. He would lean into the belief, as he grew older, that the failure to love or to be loved explains most of the mental pain people endure. The failure to love explains the burden of human loneliness, which each person prays or hopes or works hard to be rid of.

The boy who wanted to go and see, and then to return home with a story, came to see that he'd never be able to carry a story forward very far by himself. He believed, though, that others might, those who were able to see, with a different clarity of mind than his own, the things that are now at stake for everyone.

Cape Foulweather

Coast of Oregon

Eastern Shore of the North Pacific Ocean

Western North America

44°47'00" N 124°02'38" W

A light winter rain descends in weak pulses over the ocean, is buffeted across a flattened tide-built beach by a fresh wind, and rolls up into the mountains. A female rain. A swirling mist. Farther north a heavier rain, what the Navajo call a male rain, is falling hard and trundling this way, southward out of the Gulf of Alaska.

I start off searching along the high tide line.

I'd learned about the storm last night, as it was starting to build below the arc of the Aleutian Islands, bringing wind-slanted sleet and fifty-foot seas to the gulf. A few hours ahead of it, trawlers were hauling their nets and battening hatches. The next morning, as it was bearing off farther to the south, I'd put a few things in my truck and driven to the coast, 150 miles through the mountains west of my home. I wanted to be in it, to feel it thrashing Cape Foulweather, to know the punch of it against my back, to inhale the ionized air, infused with the smells of fish and trees.

I'm straddling a serpentine wrack line of kelp fronds just now, which binds together broken bits of razor clamshell, scraps of salt-encrusted, sunburnt plastic, gull feathers, empty water bottles, kelp bladders, and the vacant carapaces of shore crabs. I hope one day to discover a glass float from a Japanese fishing net in these wrack lines, but it won't be today.

Another hundred yards farther along, ducking down from the humid air accumulating like dew on my face, I retrieve a ball cap. In script across the plastron are the words *Calico Enterprises.* My mother taught me to recognize calico cloth when I was young, a plain, serviceable cotton the British once exported from India. Before that, I'd assumed the word referred only to the patterned color of a horse or a cat.

I recall her instructing me about the many fabrics she used: worsted wool, chambray, complicated Jacquard weaves like damask and brocade. She spoke of what she called the "hands" of these cloths, the fine, silky feeling of batiste, the stiffness of organdy, the coolness of linen. I saw these textures everywhere in nature later—a late-recalled gift from my mother's concern about my education.

I carry the hat absent-mindedly another few hundred yards, then place it back on the wrack line for someone else to find before a spring tide takes it away. I pick up a half dozen other things to scrutinize— a doll's eyeless head, a cormorant's long primary feather—but hold on to nothing. The informing history behind flotsam and jetsam notwithstanding, an hour more of this flânerie and I turn around, putting my back to a damp north breeze. I cover the same ground once more, then walk up the shallow slope of the beach, fists in my pockets, to a paved parking area where my dark gray truck sits by itself.

I head south on the Oregon Coast Highway. A few miles on, I turn left and drive east into the mountains on a gravel road that parallels a creek. A mile in, I cross the creek on a stout timber bridge and enter a maze of narrow logging roads. The roads I follow climb hillsides through deep-shadowed copses of spruce and cross over tracts of upended land shorn of its trees.

The last spur I turn onto climbs up a steep grade to a landing, part of an old logging operation below the heights of the cape, about half a mile inland and six hundred feet or so above the Pacific. The slope below me is planted to young Douglas-fir trees, five or six feet tall now. I guy my tent in the lee of the truck and begin making dinner on the tailgate. The wind is fitful, sodden, but real rain hasn't started to form yet. Harsher weather is still hours off, its leading edge probably somewhere near the Olympic Peninsula in Washington now.

On the way in here something strange caught my eye, a white thing in the logging debris of the scabland. I shut the truck's engine off and walked toward it, leaving the door ajar, the vehicle blocking the road. I threaded my way through the clearcut slash and past debarked stumps glaring skyward from the cratered earth.

I stood before a white brassiere, as incongruous an object as I could imagine here. Its straps were stretched around the wide face of a stump and secured there with pushpins. Each cup had been painted with concentric orange rings. Each was punctuated with a half dozen bullet holes. I pulled the bra free, shoved the wad of it deep into the weedy undergrowth, and started back toward the truck. No. I returned, retrieved it, and stuffed it under the driver's seat, thinking to put it in a trash receptacle in the town of Newport.

Is it good, I wondered, to be drawn off by such things, mute evidence of the malign cast of some human mind? Is it useless, perhaps even wrong, to hide the evidence? Should one just give misogyny its quarter? Is it hopelessly naïve to think that by preventing others from being confronted by such things, there will be fewer imitators? I also wondered whether you might ever find such signs of degeneracy in the rural clearcuts of Kalimantan or Sarawak. I suspect not.

The incident rankled me. Unduly.

I'VE CAMPED IN this place before. From here, my view of the white-capped ocean and, away to the northeast and southeast, the dark hills and old mountains of the Coast Range, is unobstructed.

I fix dinner, watching the sea. It's just beginning to heave. "The farspooming Ocean," Keats once wrote.

In the years ahead, I will listen to elders in the mountains of Tajikistan talk about 80 percent unemployment in their villages, seventeen years after the collapse of the Soviet Union. I will visit the province of Aceh in northern Sumatra, in the wake of the Boxing Day tsunami of 2004, which in just a few minutes killed more than 175,000 Indonesians. In an effort to understand lethal human tragedy, the murderous way we're capable of behaving toward each other, I will one day follow a guide through abandoned cellblocks at the New Mexico State Penitentiary outside Santa Fe. On February 2 and 3, 1980, nearly forty people were executed here by rioting prisoners, many of them using blowtorches and hammers. (Some, burned beyond recognition, died in anonymity.) In the spring of 2014, I'll walk the western waterfront of Singapore looking for a man who told me he would take me north from there, for a fee, to see the Strait of Malacca, for more than five hundred years an ambush site used repeatedly by thieves. A hero of mine, the British navigator and explorer John Davis, was killed here in 1605 by Japanese pirates. My desire was only to see those waters, to put myself in the place where he died. It's a pursuit of the kind of sensation that feels like insight. Often, though, it is no more than seeing the way things are in the world.

With the approach of dusk, dinner now finished, I follow long dark lines of returning Brandt's and pelagic cormorants with my binoculars, rookery bound, skimming the ocean's surface. I lean into the hood of my truck, propping my elbows there to steady my view in order to distinguish between the look-alike species.

AT DUSK the coming rain is still little more than mist beading up on my cheekbones, a chill on the back of my hands. In the agitated air, droplets of water tremble on my fingernails when I hold my hands still.

Somewhere, once, someone must have composed a list of the gradations of color I now see before me in the sky: the grays of pigeon feathers, of slate and pearls; in one sector the puce of a fresh bruise, in another the whites of eggshells. In the taxonomic language of meteorology, the sky is heaped with nimbostratus and cumulonimbus. Tiers of clouds are decked shoulder to shoulder in every direction.

This particular February storm doesn't have enough violence in it to have been given an official designation; nevertheless, in its details it differs from every other late-winter storm that has ever come down this way off the North Pacific, lumbering southeast over this very old water, this modern-day child of one of Earth's early oceans, the Panthalassic of the Permian. This Pacific, this once-upon-a-time Orientalis Oceanus, rolling and unrolling under the carry of clouds. Hessel Gerritsz's Mar Negro. Paolo Forlani's Golfo di Tonza.

This great disturbance of air has brought forth red signal flags in the harbors of Oregon coastal towns. Strangely, I know I'll experience a measure of grief after it passes through, that sense of loss one some-times feels when a brief, intense relationship with someone encountered on a plane or at a café ends. I'll *feel* its absence, because the nature of the storm is to be emphatic, though it is indifferent to all life. Managing the force of it is beyond the reach of any machinery. It can be sketched in isobars, shifting through time, contoured around the points of a compass, but it cannot be contained or pinned down, even by the most precise numbers.

It's entirely free. Its own idea.

WHAT'S LOCALLY CALLED Cape Foulweather is actually a distinct coastal ridge, a gently rising prominence about two miles long, bow-ing outward into the Pacific.

On his third around-the-world voyage, James Cook made his first landfall on the west coast of North America at this place. He was thirty or so miles out to sea when his lookouts, standing on the ship's tops, spotted the twin crests of what today are called Cape Perpetua, to the south, and Cape Foulweather. Very early on the following day, a late-winter storm, bearing down out of the north, intensified. Even as it raked HMS *Resolution* and Cook's consort, HMS *Discovery*, he tried to crab in closer to this unknown (to Europeans) lee shore. Twice on the ninth he got within a few nautical miles, dangerously near the sea stacks and reefs, before standing out to sea again. On March 13, after four straight days of being "unprofitably tossed about," as he wrote in his journal, he departed, setting "more sail than the ships could safely bear" to get clear of the area.

On his tactical retreat to the southwest, Cook named Cape Perpetua for the saint on whose feast day he had first seen it. He named Cape Foulweather for the rough weather that had accompanied his visit.

THE PHENOMENON OF James Cook—a determined explorer in a tran-sitional era toward the end of the Enlightenment, master of a complex "platform" for exploration, as we might term it today, a square-rigged barque—has occupied my imagination for a long while. Cook embodies both questing as an idea, a mental pursuit, and the indispensable skills of a professional mariner. His eighteenth-century search for a commercially viable Northwest Passage, of which his landfall at Cape Foulweather was an early part, is what originally brought me to this part of the Oregon coast. My thought was to become acquainted with the physical geography here, with the plants, animals, and creeks that today represent only a shadow of the place Cook saw more than two hundred years ago. The wolves and grizzly bears of his time are no longer here. Nor are the original Alsea people, their culture and traditions having been diluted and then almost entirely stripped from these hills. A cellphone tower stands today atop the 1,096-foot peak of Cape Foulweather. Invasive plants of many sorts—Scotch broom, Russian thistle, Himalayan blackberry—have moved in, along with exotic grasses. The forest soil is saturated with the residues of herbicides and other poisons once used in the wake of industrial logging to ensure the health of the artificial woodlands that replaced the original forests. At the time Cook was being tossed about offshore, a greater number of native trees flourished here, trace elements in the bloodstream of this temperate-zone rain forest—red alder, black cottonwood, golden chinquapin, Pacific yew, Pacific silver fir, lodgepole pine, bigleaf maple, Pacific madrone, western red cedar. Today, a crazy quilt of plat claims blankets the replanted acres, claims of possession many of the owners hope one day to profit from, by selling off, yet again, what once belonged to no one.

The cape is a strangely ghosted landscape now. I no longer complain, though. This is where we find ourselves today. Here, then, is a place from which to explore further, though perhaps with different ideas than those that drove Cook.

FOR DECADES I've been drawn to biographies of Cook and to revisionist thinking about his accomplishments. He made three epic voyages of reconnaissance in the later part of the eighteenth century, each one circling the planet. Early in his career he made remarkably accurate coastal surveys of Newfoundland and a defining circumnavigation of Antarctica, which produced what was then considered Earth's final continent. (Some geographers today consider New Zealand to be the high ground of a submerged eighth continental mass, Zealandia.) He deployed new meridians of longitude over the Earth's already established parallels of latitude, so that a place once found could be found again more easily. He also, to my mind, prefigured on his final voyage a type of modern-day colonial derangement not unlike that of Joseph Conrad's Kurtz, who lost his bearings in the jungles of the colonial Congo Basin. Cook does not take well to being solved, though. In my reading of his journals, he was conflicted about, and also perplexed by, the consequences of Europe's imperial reach, and by the search for material wealth that began with Prince Henry's navigators setting their sails for Vasco da Gama's South African cape, and with other Europeans embarking on the Silk Road, eastward from Anatolia in modern-day Asian Turkey.

The case for Cook as a paragon of the Enlightenment, as a representative of Progress, of precision in mapmaking, of ennobling virtue, and of the pursuit, in general, of practical improvements, has been made often in the past, most ably by his New Zealand biographer, J. C. Beaglehole. In recent decades, however, some biographers have made a case against Cook. He's been described as an "unreasoning, irrational, and violent" man, the prototype of the Old World's imperial questers—Columbus, Bougainville, Cortés—a person intent on conquest, and restricted in his thinking by a narrow frame of reference. Among the better known of these revisionist histories is *The Apotheosis of Captain Cook: European Mythmaking in the Pacific,* by Gananath Obeyesekere, a writer more sympathetic than most of Cook's biographers to the fate of those cultures across the Pacific that felt the brunt of Cook's determination "to know," to mathematically corral the planet's last unknown reaches. Some recent works, examining the aftereffects of Cook's various unbidden visits, have been brutally straightforward in their description and assessment of the costs to non-Western peoples. Among the most chilling is *Before the Horror: The Population of Hawai'i on the Eve of Western Contact,* by David Stannard. Stannard's primary focus is on the impact of those diseases Cook's party brought to the Hawaiian Islands, among them smallpox, venereal disease, tuberculosis, and an influenza virus.

Collateral damage, a military locution, is a term often used today to describe the unintentional harm done to innocent people as a result of sixteenth-, seventeenth-, and eighteenth-century "exploration," the aggressive economic strategies of exploitation that followed, and the international struggle afterward for political leverage and control in Europe's colonies. People in power today generally don't like to reconsider such damage; and by and large, ordinary people fear the consequences of confronting modern tyrants who still stand behind such schemes, in quasi democracies as well as in dictatorships and police states.

Inarguably, Cook was the great nautical chart maker of his time and a relentless prober of the Pacific Ocean, the last great unexplored geographic space on the surface of Earth in the eighteenth century. In my mind, though, he was also a person quietly but profoundly conflicted about the consequences of his work. Like many other readers of books about him, I find myself stepping back from hagiographies like Beaglehole's *The Life of Captain James Cook;* and while glad for the necessary correctives that books like Obeyesekere's provide, I remain reluctant to demonize the man. It is ever helpful, it seems to me, to view someone like him as an unwitting collaborator with historians arguing for their own interpretation of a particular sequence of historical events.

If on some occasions Cook was gruff, insensitive, petty, obstreperous, or tyrannical, an exceedingly strict officer with a hasty temper, he was also at other times selfless, moral, and gracious. The thing worth considering in our time is what he bequeathed us: the fruit of his keen desire to know Earth's oceans and shores. Besides the east coast of Australia, which had never been visited by a European, and continental Antarctica, he gave us the Hawaiian Islands (which might have first been seen by sailors in Spanish galleons), New Caledonia, and the Cook Islands; also the empirical education of Sir Joseph Banks, who went on to become the iconic longtime president of the Royal Society; and his official no to the existence of a western entrance to the Northwest Passage. He gave us, I believe, the first three-dimensional sense of Earthly order, something no one in the world before him had ever provided. In a time long before the modern era's forced rearrangements of political geography, his was a stupendous accomplishment.

After Cook, we were able to picture the entire planet, the whole of it at once, a sense of open space that, in the centuries of Western exploration before him, had eluded us. After Cook, the old cartographer's admission of ignorance, *Here Be Dragons,* disappeared from the perimeter of world maps. Like his contemporary Carolus Linnaeus, the Swedish botanist who revolutionized field biology by assigning a scientific binomial—a genus and species—to every known organism, and then placed each genus within a matrix of families, orders, and classes, Cook gave us a system that effectively organized what had once been geographic speculation.

After Linnaeus established his categories of scientific description, human beings were, specifically, *Homo sapiens.* Our very distant mammalian arctic relative the narwhal was no longer a type of unicorn but *Monodon monoceros.* The delicate deer's-head orchid of the Pacific Northwest, *Calypso bulbosa,* was no longer to be confused with its relative the phantom orchid, *Cephalanthera austinae.* And the African hunting dog, *Lycaon pictus,* was a remote, not close, relative of the wolf, *Canis lupus.*

Once terrestrial geographic order was established, plans to further explore, explain, and speculate about all the geographic lacunae that were still left could be made with greater confidence. After Cook, we had a better sense of where the last blank spots on the map might be.

Cook provided an empirical reference for the perennial figurative question, Where are we going?

MISSING FROM MUCH of the scholarly and popular exegesis about Cook, strangely, is any reference to the many places he visited that goes much beyond what Cook and his companions themselves reported. The assumption, I assume, has been that the physical place, the actual place, is of no more consequence than the scenery behind a group of actors, something to keep the narrative going while imported ideas unfold against a backdrop. But physical places, it is my belief, do shape the attitudes of visitors arriving from distant homelands with an outlander's mindset. The nature of the visited place affects the very tone of a journal entry. It influences the selection of the facts one chooses to jot down about that place. In short, the historian who visits a place writes a different history than the historian who stays home, satisfied to read about a place someone else once visited. I'm no historian, nor any biographer of Cook, but over time I unconsciously came into the habit of trying to see the places where Cook had disembarked. I felt to do so might prevent some kind of presumption in me about what had actually happened there or about how it might have happened.

If I could actually see what he saw and linger there, whatever the weather or the season might be, I knew I would know more. Each place on Earth goes deep. Some vestige of the old, now seemingly eclipsed place is always there to be had. The immensity of the mutable sea before me at Cape Foulweather, the faint barking of sea lions in the air, the nearly impenetrable (surviving) groves of stout Sitka spruce behind me, the moss-bound creeks, the flocks of mew gulls circling schools of anchovies just offshore, the pummeling winds and crashing surf of late-winter storms—it's all still there.

ONE BALMY, SUMMER AFTERNOON at Botany Bay, on the southeastern periphery of greater Sydney, Cook's first landfall on the east coast of Australia, I pondered the sense of compassion I'd developed for Cook. I was prompted to do this by the bright riot of afternoon sunbeams ricocheting from the calm surface of the bay, by the distant clatter of dry eucalypt leaves roiled by the wind, and by the towering, fair-weather cumulus clouds above, with their convoluted cauliflower heads. Together, these framed for me a prelapsarian scene. I sensed in all this an absence of violence, and of malicious intent.

As I strolled through a public park on the southeast shore of the bay he entered in 1770, I probed my sympathy for Cook. He did, of course, lay the groundwork for the colossal abuses of colonial exploration, but this was indeliberate and it was preceded by centuries of French, Spanish, English, Dutch, and Portuguese barbarism. Cook was no King Leopold, with ten million dead in the Congo, a Lord Kitchener belligerent and imperious at Omdurman, near Khartoum. Yet Cook was murdered for his own unholy transgressions (as they were perceived on February 14, 1779, by native Hawaiians at Kealakekua Bay, on the island of Hawai'i). There is no grave to visit because the Hawaiians took his corpse away immediately and cut it up into pieces. What few parts of Cook's body his men were able to retrieve were buried at sea shortly after his death. Later, perhaps because of Cook's "martyrdom," his achievements were praised by colonizers and missionaries eager to advance enterprises he might very well have wanted no part of.

I spent that day at Botany Bay mostly wandering around Cape Solander in the shadow of Sydney's major desalinization plant, another sightseer, an amateur untutored, among other things, in the psychology that informed Cook's personal quests and private needs, but wondering as well what he might have meant by his life. In the end, there being no obvious answer in front of me at Botany Bay, I tried to imagine the new world that arrived for him on his voyages nearly every day, the spectacular primacy of his time at sea, which he so enjoyed. In my late forties at the time, I was no longer able to summon the indignation needed to vilify such a man. He had led a dedicated life and, like so many, caused others pain he did not mean to inflict. As the years closed in, however, I believe he found anger growing in himself. And greater self-doubt.

I enjoyed the salutary weather that afternoon at Botany Bay, the pink galahs flying in flocks across the face of white clouds and then into a patch of blue sky, where suddenly they winked dark. I experienced from this ambience a generosity of spirit in myself I cannot always find. An uncomplicated love of the world.

On another "Cook occasion," in Tahiti, I hired a car in Pape'ete and drove east a few miles to Point Venus, where I walked the public beach at Matavai Bay past topless sunbathers. A small contingent of eighteenth-century scientists, in service to His Majesty George III and sailing aboard HMS *Endeavour* on Cook's First Voyage, made their observations of Venus transiting the face of the sun here, on June 3, 1769. (This was Britain's part in a worldwide European effort to determine the distance of Earth from the sun and from the other planets.)

Long before my visit to Point Venus and that afternoon at Botany Bay, I was able to travel some hundreds of miles of Cook's sea route along the northwestern coast of Alaska aboard a NOAA research vessel, finally sailing through Bering Strait from the Chukchi Sea into the Bering Sea. Twenty years after that, I rode out a storm in the Southern

Ocean's Drake Passage aboard another ship, pitching and rolling on all three axes in forty-foot seas and screaming winds—the kind of seas Cook saw in the very same place on his Second Voyage. On yet another occasion, I stood ashore, trying to compose a silent eulogy to Cook below the cliffs at Kealakekua Bay, trying to choose phrases neither insipidly deferential nor disdainfully sophisticated to express my gratitude for the example of his determination.

I wanted in all this, however obsessive it might have been, to feel contemporaneous with him. To feel empathetic.

AS I UNDERSTAND THEM, Cook's journals reveal a man in a private perpetual state of wonderment, despite his vigorous efforts to remain objective in approaching other civilizations and in examining geographies other than Britain's. I read with admiration his assessments of the indifferent and gnomic character of the seas on which he sailed. His voyages were not like Parsifal's, a quest for the Grail, though some historians want them to be; nor were his merely the dispassionate, disciplined ambulations of a curious, rational, and self-made Englishman. Like Meriwether Lewis, I believe, he was also exploring things so fragile and tenuous he would never write explicitly about them, because he found the task intimidating, and describing the experience itself too tenuous, too indefinite an exercise. Here were suggestions of his thoughts about the deeper consequences of his excursions, the ramifications of his trespass.

Toward the end of his Third Voyage—his consort, Charles Clerke, would finish it after Cook was killed—Cook started to show signs of unraveling. He made uncharacteristically rash decisions. He was strangely inattentive to the need for careful navigation in unknown near-shore waters off the coast of southern Alaska, in June 1778. He seemed, too, oddly attracted to the idea of giving up entirely on the Third Voyage. Maturing in him, as I read his journal entries from the final months, was a realization of what he might actually have wrought by toppling the last icons of classical geography, and by having imposed as an explorer on other cultures—cultures profoundly, even eerily, different from his own. By this time, in my mind, he could not free himself from the driving force of his own fame, nor from the responsibility that he believed came with it. One had to consider, too, not incidentally, that although he had made major revisions in the geography of the world that had come down to him from Strabo, Ptolemy, and Eratosthenes, he spent his days in the company of sailors for whom such ideas were inconsequential. Lionized at home, feared at sea, a stranger to his wife and children, he had become over the course of the three voyages a person hardly known to anyone.

JOSEPH BANKS, the supernumerary of note aboard Cook's *Endeavour* on the First Voyage, was an aristocrat. He assumed the privileges of his social class during the voyage and expected deference be shown to the elevated position he held in London society. He has long served historians as an emblem of the person who feels superior to all he encounters, immensely and genuinely curious though Banks was. When I read Banks's journal of the voyage alongside Cook's journal, I found their temperaments strikingly different. Banks's formal education, together with his lack of interest in navigation or the character of the sea, his peerage and his gregariousness, immediately set the two men apart. And of course, by journey's end, each had come to a different conclusion about what was to be valued about the experience.

Comparing the journals makes clear the nature of each man's ethnocentrism and makes apparent, too, how differently they evaluated issues of race and social rank. The journals share an unconscious deference toward Western logic and the metaphysical assumptions of Western philosophy, but a prominent contrast between the two is the way each one regarded the physical world through which they moved. Banks was primarily interested in the cultures and terrestrial geography of the islands they visited. He could not make much sense of the seas between his landfalls. Cook was equally curious about the elements of each island—their botany, anthropology, and topography—but his attention to the ocean was just as intense. For him there was no "emptiness" between one island and the next. He thought it actually possible to *define* this apparent void. Even though he could navigate through it with a certain precision, he appreciated the fact that it was—and perhaps would ever remain—unmarked. Its only boundary was the horizon, the sill of the sky, separating what the eye could see from what the mind might imagine.

Cook, I think, did not entirely trust the assumptions behind the Enlightenment principles that urged him to measure, to record, and to define the world. He did not completely concede the authority that lay behind gradations of social rank, perhaps even naval rank. He spent his life charting raw space, putting down grids and elevations, but he also understood what could not be charted, the importance of the line that separated the known from the unknown. He understood what occurred in the silence between two musical notes. He also knew, I believe, the indispensability of this.

WHENEVER I TRAVELED to Cape Foulweather, I took a pair of binoculars with me. Sometimes I brought along a catadioptric telescope. Since early childhood I've been enchanted by the possibility of seeing farther, enthusiastic about the way a few convex and concave glass surfaces can resolve a distant scene into a crisp, isolated image.

Sometimes when I made the journey from home to the cape, I was forced to set up camp in rainy or snowy weather, and learned what that taught. It was easier to hike in the surrounding hills during stretches of dry, clement weather; and of course it was also possible to see farther then, if the atmosphere was clear. On starry nights I inspected the surface of the moon with the telescope, crater by crater. I picked out distant ships, trailed by their dark wakes in the moon's beryl light. During the day I followed migrating gray whales and flocks of migrating white-winged and surf scoters offshore. With my binoculars I could inspect minutely the slope in a recovering clearcut north of me and resolve the nearly uniform emerald green there into greens of lesser and greater saturation, separating the leaves of salmonberry, for example, from those of red flowering currant or great hedge nettle.

The binoculars and the telescope tunneled the space around and above me, making the indefinite more distinct, the general more specific.

They made the space I was in larger, and my intimacy with that space greater.

THE CLEARCUT I usually camped in was about ten years into its recovery. The felled trees had been hauled away, the limbs, stumps, and rotting logs bulldozed together and burned. The exposed soil had mostly been replanted to a single species of tree, the commercially valuable Douglas-fir, and something of a forest was in the making again, as volunteer Sitka spruce, red alder, and shore pine (also called lodgepole pine) began to emerge in the understory. Red elderberry, evergreen huckleberry, and salal stood out in the ground cover, and where the ground was wetter, sword fern and sedges were growing. In summer, smooth yellow violets, wild strawberry, fireweed, pearly everlasting, yarrow, and native dandelion brightened the scene, as did the small fruits of Oregon grape and domestic holly.

It is reassuring to know the names of these native plants, and to be able to distinguish them from so-called invasive species like woolly mullein and European beachgrass, all this in order to appreciate the character of the place; but this is not necessary. The urge to distinguish between native and non-native plants has become, for some, a xenophobic pastime. Camped here, I've often thought my long-standing aversion to clearcuts is no longer really warranted. Aesthetically, fresh clearcuts are as off-putting as patches of mange on a dog. And the evidence of an indifferent and rapacious harvest in such places is sometimes offensive, the battlefield aftermath from the felling of enormous trees, the ashes and masticated earth that industrial logging leaves behind. One might wonder, picturing Cook cruising offshore here two centuries and some decades ago and exchanging shouted communications with his lookouts, whether he speculated at all about an avarice like that of the conquistadores coming to life here one day.

However it might be viewed, the throttled Earth—the scalped, the mined, the industrially farmed, the drilled, polluted, and suctioned land, endlessly manipulated for further development and profit—is now our home. We know the wounds. We have come to accept them. And we ask, many of us, What will the next step be?

One Sunday afternoon in Lebanon, I rested my forehead against the trunk of a cedar in a protected, remnant grove of them in the Barouk

Forest, in the Al-Shouf Cedar Nature Reserve. These were the fabled cedars of Lebanon (*Cedrus libani*), in the mountains southeast of Beirut. I did not feel the grief I thought I might over the nearly complete disappearance in Lebanon of this species of tree. I felt only respect for these few that were left. The bark of the tree I stood beside had been polished to a sheen by tens of thousands of handstrokes, each person's caress a mixture of their own regret, affection, and forbearance.

This is what we have now, not the cedars of Lebanon some of us once imagined, reading the ancient texts.

To enumerate the species of native plants that are left today on the deforested flanks of Cape Foulweather would be to provide a vocabulary without a syntax. To write the history of plants here, a chronicler would have had to reside among them for many decades. It's work few have any time for now.

And the value of possessing such a coordinated litany has become problematic.

I DROVE OUT to the margin of the Pacific Ocean that day in February when I heard the storm was coming, anchored my tent in the lee of the truck in case the storm came hard, and went to sleep after I had supper. At first light, I stood with my steaming coffee in a stir of mist marking the advance of the storm, and thought again of Cook and of all I had seen from this platform high above the ocean.

Seven miles to the south, the navigation light at Yaquina (yah-QUIN-ah) Head flashed its warning. Three miles to the north, fishing boats were wallowing in freshening seas off Depoe Bay, the water there soon to become more treacherous. From a certain perspective, storm vigils like mine—waiting for the mist to harden into rain, for the wave troughs to deepen under the surging wind, trying to read the surface of the sea where, for hours on end yesterday, nothing much seemed to be going on—all of this might strike some as a setup for boredom. I've found myself blanking out here, even on bright spring days when warm breezes enlivened the land and thousands of nesting common murres, double-crested and pelagic cormorants, and Cassin's and rhinoceros auklets were diving and swooping the water. On those days when I couldn't focus, I'd retreat to the front seat of my truck to read or doze. I was soon back, though, weirdly attentive to the apparently uneventful landscape.

It's been my experience that these hours of perusing the water, here or while at sea—taking in the occasional bird or surfacing whale, watching light shift on the surface—induce an awareness of another sort of time, a time that fills an expansive and undifferentiated volume of space, one not easily available elsewhere. On those days, such a seemingly mindless vigil offers relief from the monotony of everyday experience.

During certain periods of uninterrupted vigilance at the edge of the sea, I've also had the sense that there is some other way to understand the ethical erosion that engenders our disaffections with modern life—the tendency of ruling bodies, for example, to be lenient with entrenched corruption; the embrace of extrajudicial murder as a legitimate tool of state; the entitlement attitudes of those in power; the compulsion of religious fanatics to urge other humans to embrace the fanatics' heaven. The pervasiveness of these ethical breaches encourages despair and engenders a kind of social entropy; and their widespread occurrence suggests that these problems are intractable.

I can't say what this other way of looking at these situations is, how a huge domed space like the daylit ocean, a space almost entirely free of objects and offering a different sense of time passing, might provide a perspective to make banal human failure seem less enduring, less threatening; but taking in this view, I always sense that more room for us to maneuver exists. That what halts us is simply a failure of imagination.

The history of art in the West, I believe, can be viewed as the history of various experiments with volumes of space and increments of time, with frequencies of light and of sound. Art's underlying strength is that it does not intend to be literal. It presents a metaphor and leaves the viewer or listener to interpret. It is giving in to art, not trying to divine its meaning, that brings the viewer or listener the deepest measures of satisfaction. The authority of art, its special power to illuminate, was partially eclipsed in Western culture by the Scientific Revolution. After that, art's place in everyday life became increasingly more decorative, its influence undermined by science's certainty, its insistence on authority given little more than polite notice. The history of the separation of art from the natural world is older than the history of the separation of art from the world of reason, but this breach, too, has had a staggering effect on how humans grapple with their fate. Art does not aspire to entertain. It aspires to converse. It, too, like Clausius's statement of the second law of thermodynamics, the one about entropy, is about fated life.

When the coastal storm I've been waiting for finally comes, it will bring its musics, the active colors of its pummeled skies, and wind to choreograph the movements of the clouds. It will crack land and sea with its pellets of rain. It will dim the sun. If the response is awe, not analysis, that, really, is all that is needed.

IT WAS AT Cape Foulweather one day, sitting on a chair on the roof of my truck with a pair of binoculars, that I realized I have never in my life gotten quite enough of the Pacific. With its forever-changing surface, its rafts of sea ducks and reflections of sky, its ferocious surf, its handling of intrusions by the dry land, it seems knowable, even definable. Below its surface are its invisible parts—its volcanoes, canyons, and abyssal plains, mapped but still largely obscure. It's from there, from this immense crater, that the primordial Earth, according to many, flung off the material that became its single moon.

It is my hope one day to be able to descend into this wilderness, to aim a spotlight into the extensive darkness.

On January 23, 1960, at a little after one in the afternoon local time, two men seated on small stainless-steel boxes inside a forged steel-alloy cabin, a sphere slung like an udder beneath a ballast chamber filled with gasoline, settled gently onto a patch of ivory-colored silt on the floor of the Mariana Trench, a spot in the western Pacific later to be named the Vitiaz Deep. They were at a depth of 35,800 feet. The pressure on the walls of the observation sphere was 7.97 tons per square inch. In that moment, less than seven years after a Sherpa mountaineer named Tenzing Norgay and the New Zealand mountaineer Edmund Hillary reached the summit of Mount Everest (29,035 feet), Don Walsh, an American naval lieutenant and submariner, and a Swiss oceanographic engineer named Jacques Piccard reached Earth's other vertical extreme.

Today, Everest—the Nepali word for it, *sagarmatha,* means "forehead of the sky"—has become a crowded and occasionally deadly tourist attraction. The stretch of ponded sediment where Walsh and Piccard spent twenty minutes together remains a Styx-like corner of humanity's inquiry, a ghostly scene in the ocean's hadal depths.

Don Walsh lives now on the edge of the Pacific near Myrtle Point, Oregon. He is an unassuming man, a retired naval captain with a broad background in oceanography and exploration, an engaging and authoritative individual given to self-deprecating humor. When I visited him at his home, he recounted for me the details of the dives he had made in the *Trieste,* the bathyscaphe that took him and Piccard to what Piccard called "the basement of the world." It was apparent from the way he tried to render these moments for me that the experience had affected him deeply. Decades later, the achievement remained nearly beyond description for him. He had descended through an Earthly region, one without sunlight and without weather, to land on a soft plain of talc-like powder called marine snow, an accumulation of bony detritus, the remains of billions of creatures that had died over eons in the nearly seven miles of water above. More than anyone before them, he and Piccard had been "in" the Pacific. They had reached its other, lower surface.

On its descent, it's possible the *Trieste* passed through one of the Pacific's still poorly understood sonic tunnels (the *Trieste* wasn't outfitted with the instruments needed to detect one), corridors through which the sound of an erupting submarine volcano, for example, travels very fast for thousands of miles, undiminished in intensity and relatively undiffused. The nature of these transmission tunnels was of great interest at the time to the U.S. Navy, and their characteristic makeup—a specific combination of salinity, temperature, and pressure—was one of many things about this pelagic frontier turning over in Walsh's mind on the way down to the floor of the Vitiaz Deep. No one really knew then how foreign (or familiar) this realm of the planet might prove to be.

If Walsh could have seen into the darkness beyond the bright cones of his searchlights on that descent in 1960, he would have seen dark scarves of turbidity snaking off a continental shelf near Guam and streaming across the plain below, evidence of the complexity of deep subsurface oceanic currents, currents that belie the notion of order that two-dimensional maps of the ocean's *surface* currents often convey, with their neat boundaries and points of convergence—just here, say, is where the Humboldt Current meets the South Equatorial Current, and just there is where the Gulf Stream and the Labrador Current meet. As is the case with many maps covering large areas, the greater the scale, the less reliable the information, no matter how elegantly or beautifully presented.

Walsh described his historic dive in striking detail—the khaki-colored naval uniform he wore that day, the brown dress oxfords he had on, the small American flag he brought with him. It was chilly inside the cramped, unheated sphere, and there were long silences during the four-hour and thirty-eight-minute descent and the three-hour and twenty-seven-minute ascent, partly because the taciturn Piccard hardly spoke. The sound Walsh recalls most frequently is the creak and groan of the steel sphere as it adjusted to changes in pressure and temperature. At the bottom, in 38° F water, they saw a single sole-like fish (*Chascanopsetta lugubris*) about a foot long, undulating across what had long been thought a lifeless plain. It swam slowly out of Walsh's view of it through the bathyscaphe's single miniature window, four inches across and five inches thick.

Piccard once described the floor of the Pacific as "a vast emptiness beyond all comprehension." In the years since he and Walsh made their dive, scientists have embraced the theory of plate tectonics; discovered deep-ocean thermal vents and their ecosystems of sulfur-based life (life chemically dependent on sulfur, not solar radiation, for its metabolism); discovered hundreds of new species of deep-ocean dwellers; and mapped much of the planet's deep-water currents. The emptiness Piccard alluded to has largely been filled in, though his essential impression of the existence of a vastness too extreme to comprehend is neither uninformed nor outdated. This concept of seeing an unbounded emptiness is a recurrent observation today in cosmology, as it was, too, in Piccard's time, in French existentialism. Cook faced the same "eternity" on the *surface* of the Pacific once he sailed north of the Marquesas, leaving behind the then lightly reconnoitered precincts of the South Pacific. But it was not solely the *size* of the Pacific that gave Cook pause; it was his intuition that this ocean could not be rendered with the tools with which he was familiar. No mathematics he knew could make it comprehensible. Even if he had had a globe before him, even if that globe was only the size of a marble, he knew it was impossible to take in the entirety of the Pacific with one look. You had to rotate the sphere to see it all.

Walsh's one regret about his stupendous feat of exploration was that amid his monitoring of all the gauges in the *Trieste* and the slight distraction of Piccard's unfriendly preoccupation with tables of data and his enormous Swiss flag, there had not been sufficient room for astonishment, for an expanded sense of appreciation in the moment. Walsh wrote an article for *Life* (Piccard wrote one for *National Geographic*) and filed the required reports with the Navy, but he could not find a way to satisfactorily express the breadth of his wonderment. He works hard to explain to a stranger the realization of his vision of the absolute depths of the sea. He told me that as a submarine captain, he had not actively considered the character of the ocean below 400 feet, not until that singular universe within Earth's deepest declivities suddenly filled his awareness of where he was and who he was, an explorer craning his neck to peer out a tiny window at 11°18'30" N and 142°15'30" E, nearly six Grand Canyons deep on the utter bottom of the western Pacific, a scene no person before him had been able to experience.

It was important to Walsh that a human being, not a rover or a probe, first saw the bottom of the Pacific. "You can't surprise a machine," he said to me. And it is this capacity to *appreciate* the unknown, to be surprised by it, he believes, that will always set the human explorer apart from the machine. The moment of surprise informs you emphatically that the way you once imagined the world is not the way it is. "To explore," he says, "is to travel without a hypothesis."

As he walked me to my truck, in an effort to convey how completely I appreciated his evocation of the Pacific, I asked Walsh if he happened to know anything about ocean striders. He stopped and exclaimed, "*Halobates!*"

I was glad to have someone to share my enthusiasm with, someone familiar with these obscure small wild animals.

The life of ocean striders, species in the genus *Halobates,* is one of interminable exploration. Aquatic insects, they skitter, stride, and skate over the active surface of the open ocean, searching for food and mates, which they locate by some as-yet-unknown method in order to perpetuate themselves. They survive pelting rainstorms, terrific winds, and chaotic seas. If death does not arrive in the shape of a seabird, a fish, or a turtle, they sink alone into the abyss when their time comes, as gently as a fingernail clipping. A life lived alone, and for some entirely out of sight of land.

ON CLEAR SUMMER NIGHTS at the cape, I'd sometimes set up the catadioptric telescope to explore the night worlds above. I've no real skill with this, navigating the night sky; it was little more than a way to sense the great umbrella of space I was under. Most often I went to well-known stars like Polaris or to the constellations of my childhood like the Big Dipper in Ursa Major, or to curiosities like the Horse-head Nebula in Orion. Other times I felt no challenge to locate, say, a particular star in a complicated constellation like Perseus, but drifted instead toward constellations completely unfamiliar to me—Auriga, Boötes, Lyra. How might the suns in these constellations be connected in a way to suggest, by lines and dots, respectively, a charioteer, a herdsman, and a musical instrument? And whose were these ideas, anyway, that I was to begin with?

One night I was trying to resolve the stars between Deneb and Albireo into the shape of a swan, the form in which Zeus seduced Leda. I'd first come to know these stars as the Northern Cross, a counterpart to the Southern Cross visible in the opposite hemisphere. Deneb, the brightest star in this constellation, marks the top of this cross. To Inuit people in northeastern Canada, Deneb is an unaffiliated brightness, a lone primary guiding star, which they call *nalerqat.* They perceive no Latin cross there. No swan, either. Their "constellations" are, for example, *tukturjuit* (part of our Ursa Major), representing a few caribou. And *udleqdjun,* hunters pursuing a polar bear, the bear represented by *nanuqdjung* (our Betelgeuse), the pursuers by Orion's Belt, and the hunters' sledge by *kamutiqdjung* (Orion's sword). Inuit constellations, for the most part, are tableaus instead of single schematics.

In crossing star fields with the telescope—Babylonian astronomers referred to the twinkling expanse of the Milky Way as "the celestial flock"—you find it difficult to remain conscious of the fact that what you're seeing is not a flat surface, a two-dimensional chart. It's a *volume* of space, something with a third dimension. This of course complicates everything for someone trying to read the skies. The stick figures we call constellations exist only for someone looking outward from Earth. If you were to ask which is the top or bottom of anything out there, or what constitutes left and right within the starry dome, the problem of delineating constellations becomes only more confusing.

With regard to Deneb, then, what is the correct, the trusted, point of view?

For some cultures, like ours, constellations memorialize the appetites of Zeus. For others, it is the importance of hunting caribou. For all cultures, the individual stars, it seems, and the designs in which they are arranged, stand in for important elements in a fundamental guiding narrative. Like Cook's stark and reassuring designations of latitude and longitude, the celestial narratives make the ordinary vicissitudes of life manageable. Without these references, the path through life can seem confusing. The constellations soothe and affirm.

IN DAYLIGHT I sometimes used the telescope (with a filter) to pore over the surface of the sun, looking for solar flares and trying to imagine the scale of these tongues of flame spewing into the sun's corona against the blue background of Earth's atmosphere, which scatters the sun's stupendous outpouring of light. Occasionally I studied the topography of the moon with a map of its surface to hand, trying to get a third dimension out of the Ocean of Storms, the Montes Jura, the Lake of Dreams. Once resolved in the lenses of the telescope, the face of the moon has for me a vivid and charismatic presence. If one could stare back in time and see in as much detail, say, the weathered face of Marco Polo, or of Nefertiti in private contemplation, or of Montezuma confronting Cortés, it would, for me, be very like these evenings with the yellow moon.

I once experienced a direct primal connection with the Earth's lone satellite, during a walk over an ice sheet in West Antarctica. I was traveling with a friend, John Schutt, in Antarctica's southern Victoria Land, eight thousand and some miles south of Cape Foulweather, when he offered to show me the place where the first piece of the moon lying on Earth's surface was identified. On January 18, 1982, he and another scientist, Ian Whillans, spotted a meteorite the size of a golf ball on the surface of the Allan Hills Middle Western Icefield, about 940 miles from the South Pole. Seventeen years after he and Whillans collected this small space rock, with its partial tan-green fusion crust (the residue of its burn through Earth's atmosphere), we walked the furrowed surface of that blue-ice field together. The meteorite itself was long gone. (It's stored at the Johnson Space Center in Houston, with moon rocks from the Apollo missions.) In the barren desolation of the Middle Western Icefield, west of an outlier of the Transantarctic Mountains called the Allan Hills, and despite the fact that the ice that had once held this piece of the moon had since moved farther along on its trajectory to the Ross Sea, I felt the doppelgänger of it. Its ghost.

John, a geologist and mountaineer, didn't know at first that the meteorite he and Whillans found on that day was a piece of the moon. Before then, no one had imagined that a piece of rock might be blasted off the surface of the moon with enough force to land it on the surface of its companion planet. Scientists have since found more pieces of the moon (and Mars) sitting on the ice in Antarctica.

VERY EARLY ONE MORNING at Cape Foulweather, on what would become a clear spring day, I set up the telescope and intentionally began working the ocean's horizon from right to left, starting from a point beyond Depoe Bay, on the coast to the north of me, and coming around to a spot beyond Yaquina Head to the south, a sweep of about 160 degrees. (As was so often the case, I was only trying here to expand my frame of reference.) I anticipated it would take no more than a few hours to focus intently on each successive minute of that arc, the beckoning line where the dark edge of the ocean trembled against the sky. Passing beyond this line, ships disappear; on this side, they rise up from the water. This was the mapmaker's liminal line, the edge of the known. Heidegger called it "the place from which something begins its essential unfolding."

The western horizon began to emerge at first light, a little after five. At the sky's zenith, a deep blue hue was following the last half of night's dome westward into the water, leaving evaporating stars in its wake. The western horizon began to broaden both to the north and south of me, its rising pastels forming a kind of chrysalis that soon defined sharply the whole of the western horizon.

To thoroughly inspect this simple declarative line took me not a few hours but from breakfast until dusk.

The startling power of this elementary tool, the optical telescope, to resolve what is distant and unreachable into readable images might have been what initially induced me to order a book called *Hubble: Imaging Space and Time.* I'd brought it out to the cape with me once and had perused its photographs while relaxing in the greenhouse comfort of the front seat of my truck on a chilly, windy day. The images of nebulae and galaxies were wondrous. Mesmerizing. With images like these before us, I remember thinking, our direst problems as a species—desertification, collapsing fisheries, barbarism, poverty, species extinction—might shrink down into something conceivably manageable. These images of timeless creation, carefully contemplated, might unfreight a depressed soul. They prompted in me a sense of the impossible having given way to the possible, a feeling as intense as the despair I'd once felt before I looked at Fernando Botero's drawings of tortured prisoners at Abu Ghraib, or before I looked at Sebastião Salgado's photographs of broken families, victims of drought, famine, and war. I was uplifted by Botero's and Salgado's witness. As I turned the pages of the Hubble book, however, this feeling of transcendent awareness slowly faded. I became weirdly uneasy.

The imagery of the earliest Hubble photographs is immediately familiar to most Americans because it mimics the landscape paintings of Hudson River School artists like Albert Bierstadt. Their romantic portrayals celebrated the splendor of the American landscape; and the artists colored and framed their large canvases in ways to suggest that these supposedly intimidating North American wildernesses were actually more beautiful than threatening. Man, clearly, was an insignificant observer here, an intruder; but the underlying theme of these paintings was that mankind's *destiny*, which at the time was widely believed to be divinely directed, was to take possession of these places.

These were commercial landscapes then, not portraits of the unknown.

One of the curiosities of the Hubble images is that they don't actually exist as photographs, as that term is commonly understood. The authors of *Hubble: Imaging Space and Time* write that these photographs are "mediated views of the universe, and [in creating them] astronomers and image-processing specialists employ a degree of artistry to make the universe more understandable and attractive." The Hubble "photographs," in other words, are "impressions, based on scientific data." They reflect the desire of their creators "to balance science, aesthetics, and communication."

One wonders what the images might have looked like if the raw data—the streams of binary information collected outside as well as within the spectrum of visible light by the Hubble telescope—had been handed to visual artists to "balance," let alone frame and color, instead of being given to "image-processing specialists" inclined toward a version of pictorial beauty that mainstream viewers would be immediately comfortable with and find reassuring.

A sense of having been misled, however, didn't cause me to close the book. The images do represent someone's effort to make interstellar space beautiful, and they show a certain reverence for what is not known, admirable qualities in a world that in many quarters has grown suspicious of beauty and of reverence for anything not made by or for human beings. If we could see past what our eyes can take in, envision energy in realms beyond the infrared and ultraviolet, and therefore see radio waves and gamma rays coming toward us from out of what the British astronomer William Herschel once called "the shining fluid," then maybe these same manipulated images would enthrall us, and we would hold them in higher regard.

Romanticism gave Western cultures inspiring literature and art, much of it indelibly inscribed in our imaginations. What we seem to crave now, more than this kind of inspiration, is the authority of the dependable, inspiration that comes from what is authentic. And we also desire, I think, a species of beauty different from the conventional one.

The rationale behind the creation of the Hubble images seems to confuse the authority of science with the authority of art, co-opting the latter to elevate the standing of the former. To employ metaphor purely as instruction is didactic. And to treat art as utilitarian, a means to an end, is to dress materialistic ideology in a seductive disguise.

EARLIER, I DESCRIBED an eight-real Spanish silver coin I keep on my writing table, a reminder of how, in a quest for material wealth, darkness might so easily corrupt one's efforts. One of the hard lessons of travel, I think, is having to accept that the human impulse toward the sort of exploitation and fundamental dishonesty the coin represents, not just outright thievery but a tendency to, for example, acquit the well-connected or wealthy criminal, or to accept misrepresentation in the promotion of products, is prominent in nearly every developed country. The evidence forces us to think that American culture is not an exception, that the general cover-up and denial of responsibility that we see regularly in commerce and government in the United States is just as challenging to root out as we might imagine it to be in cultures more often cited for their waywardness and lack of integrity. The only question in a world like ours, starting to run short of vital supplies while attempting to support a burgeoning population, is whether these injustices will eventually be so severely condemned that they will cease to be a source of cynicism and distraction and become instead an inducement for social change. A world in which all public dealings are just, where no refugee camp is ever built, is not to be had, of course; but it is possible, I think, if the stakes are high enough, to forge a world in which there is less tolerance for the self-serving abuse government and business too often endorse.

HERE AT CAPE FOULWEATHER, as I watch the ocean for days at a time, marking the passage of loaded freighters, container ships, and fishing vessels in these heavily fished seas, the memory of the eight-real coin from the wreck of the *Nuestra Señora* occasionally comes into my mind. The coin suggests the cultural density of human history, and how little of that richness was recorded before much of it was wiped out, how judgments about who is the primitive and who the real barbarian forestalled further inquiry into the complexity of human cultural life. The coin reminds me that the urge to condemn the conquistadores and other miscreants in world history, the witless and avaricious mobs who followed the leads of the Genghis Khans, the Pizarros, and the Trujillos, might be countered today by refusing to define as evil any other culture, or even the wayward in our own. If we don't, we risk ending up in a wasteland of uninformed dogmatists, the same shortsighted, narrow-minded belligerents who rise up in every era of human history.

It might have been useful once to identify and denounce enemy cultures, those that were seen as ruthless and exploitive, obsessed with wealth and indifferent to social justice at the highest levels; but looking out over the measureless Pacific from this cape late one afternoon, sipping my coffee and waiting for the storm, I feel that this time has passed. People in every country today can identify with the very same threats to their lives and to the lives of their progeny. And many know their governments, elected or self-appointed, are too cowardly, too compromised, or too mean-spirited, to help them.

There are among these citizens some who believe it is not a technological miracle, not the emergence of a philosopher-king, that is called for here. It's something altogether different, some capacity to perceive a reality that is hinted at but not yet realized in modern *Homo*. And because most are having trouble imagining more than two or three generations ahead of them, they're wondering who will step forward right now to face the obvious threats, the obvious menace.

Are there not among us, they are asking, those who will act on behalf of all, people who will attempt to do the things that need to be done, and who do not need to be supervised as they address these problems?

———

AT THE TIME Cook raised Cape Foulweather, this particular stretch of the northwest coast of North America was a lifeground for Alsean and Tillamook Indians—Siletz people living just to the north of Cape Foulweather, and counted among the Tillamook (TILL-ah-muck), and Yaquina people, living just to the south, between what are now Yaquina Bay and the Alsea (AL-see) River people, and counted among the Alseans. Archeological evidence suggests that both these coastal tribes had "sharing traditions" with tribes living inland along the Columbia River, so cultural anthropologists have been comfortable including them with the so-called Lower Columbia tribes, the most prominent of which, in Euro-American histories, are the Chinook.

It's guesswork to say when exactly ancestors of the Siletz and the Yaquina began permanently to occupy this stretch of the Oregon coast. By about three thousand years ago, however, probably both cultures were thriving here, depending on estuarine shellfish, marine mammals, blacktail deer, fish, and the eggs of shorebirds for their food. It's not known, either, when Tillamook and Alsean people might first have become aware of Europeans. Sixteenth-century Spanish galleons en route from Manila to Acapulco might have been driven this far north off their customary route by storms, and they might have been seen by Indians or, having foundered, been salvaged by them. João Rodrigues Cabrilho and Bartolomé Ferrelo might have been offshore near here in 1542. It's unlikely, but some historians place Sir Francis Drake off the Oregon coast in 1579. Sebastián Vizcaíno and Martín de Aguilar could have pushed this far north in 1602 or 1603; but all of this is speculation. Such claims were originally advanced by the English and the Spanish to establish their rights to whatever of value might be found and appropriated here. Until the time of Cook, the geography of these coasts was of little interest to maritime nations. The primary—sometimes only—interest of the colonizing nations was to secure trading advantages and to take control of the sources of material wealth, mostly through private commercial enterprises.

The traditions of the Yaquina, the Siletz, and other coastal tribes were gradually completely undone by commercial exploitation and cultural subjugation. What was lost to humanity with the passing of these unique epistemologies and ontologies (understandings of being), though lamented by some, has been ridiculed by most in the "civilized" nations as an inconsequential loss to the collective wealth of human knowledge. The loss of an entire way of knowing, however, is a tragedy hard to reckon.

A people's way of comprehending the fundamental mystery we call "the real world" is most clearly and succinctly obvious in the vocabulary, syntax, and tropes of their language. The linguist K. David Harrison has written that the sibilants and clicks, the fricatives, tones, and ejectives of each human language comprise "a singularity of conceptual possibilities." Some languages are so place specific that it is not possible, he tells us, even to speak them intelligibly apart from the landscapes in which they arose. He emphasizes that languages are more than mere words and grammar, that they reveal ecologies and potentialities unrecognized in other languages. He makes it clear that each language brings with it another history, another mythology, another set of technologies, another geography. In *The Last Speakers: The Quest to Save the World's Most Endangered Languages,* he writes, "We will need the entire sum of human knowledge as it is encoded in all the world's languages to truly understand and care for the planet we live on."

The loss of any human language means that, in the most difficult straits humanity has ever found itself in, one more strategy for survival has been thrown away. In the time of my own travel, of a passing acquaintance with Kamba people speaking Kikamba in Kenya, Pitjantjatjara people in the Northern Territory in Australia speaking their tongue, Ainu people speaking theirs in Hokkaido, and Pashto- and Dari-speaking people speaking their languages in Afghanistan, I've come to realize how important has been the insight that linguists and anthropologists have been stressing to lay audiences for decades: there are significant differences among human societies when it comes to assessing the seriousness of any threat to a culture's spiritual, physical, or psychological well-being. The idea here would be to forestall, by whatever approach is most effective, the onset of those feelings of despair that paralyze a people.

Everywhere I've been able to travel in the war-torn, ecologically compromised, misgoverned districts of the world, this thin hope has been something I thought worth plumbing, whether it's actually possible for us to help one another with the economic, climatological, health, and environmental emergencies now camped in our front yards. Sadly, when it comes to describing what one's own culture might offer other cultures, once the talk of increased tourism and commercial trade benefits has run its familiar course, few can say.

Cook writes in his journal, during the six days he had these densely forested mountains and the snow-covered foreshore of this coast in sight, that the smoke of campfires was apparent. He does not wonder in his journal who these people were, but he would not have thought their insights, their pharmacology, their knowledge of riverine ecology, their pathfinding skills, any match for his own.

They and their way of knowing were little more than a curiosity for him.

I AM NOT CAMPED in anything like a "pristine landscape" at Cape Foulweather. The mountain ranges running up and down the Oregon coast have, for example, been swept by natural fire for thousands of years. When I first began visiting here, most of the clearcuts apparent on its slopes were recent ones. Walking these plots of fifty acres or more, passing the inert stumps and the burned-out craters of slash piles, I most often felt grief, not anger. The damage to this once heavily forested ground is greater than what any natural fire has ever caused. Thoroughly exposed soil is vulnerable to erosion from the heavy winter rains that characterize temperate-zone rain forests. The minerals and nutrients a fire would have left behind have been hauled away in the form of merchantable logs; and the loggers and the heavy equipment they brought with them have unintentionally carried in the seeds of exotic plants, which quickly took advantage of the disturbed ground to establish themselves. Once established, some of these "weedy species" drove out native species, and the original ecosystem unraveled.

Clearcutting, as an industrial practice to effect cost-efficient logging, has not been in use long enough for timber managers to be certain exactly what the consequences of employing it might be. The timber industry likes to say that the clearcuts here have been restored or replanted, but these are not the right verbs. It would be more accurate to say that the land is now farmed, because, most often, only one species of tree, Douglas-fir, is replanted. It is favored over all the other trees that once made up the forest here because it is the most valuable of all these trees commercially. Modern lumber production has been streamlined in such a way that this one tree now works better than any of the others in the industrial corridor where wood fiber is brought to market and sold. Biogeographers have argued for years that disrupting natural ecosystems on an industrial scale—mountaintop removal to expose coal seams, clearcutting, large-scale corporate farming, and damming rivers that were historically defined by huge salmon runs—establishes new ecosystems over such a short period of time that long-term consequences are not immediately apparent. Similarly, when British colonizers arrived in India, when Alexander entered Egypt, or when Arab science and philosophy came to the Iberian Peninsula, cultures quickly changed. And with these changes, of course, strong reactionary arguments about the value of an original purity arose, with enormous political, social, and economic consequences.

After such massive biological and cultural disturbances, how is one to evaluate or even construe the meaning of any conjectured primal state of ecological purity, let alone states of cultural purity?

I cannot easily characterize the plant communities of the second- and third-growth forests that surround me at Cape Foulweather. I find it difficult, therefore, to unilaterally condemn an industry (or a culture) for what has taken place here since Cook first glassed these slopes. But it has struck me that one challenge Darwin's theory of evolution by means of natural selection holds up for us, in addition to reckoning with the meaning of the "immigrant" plants and creatures that turn up in a second-growth forest, is the more unsettling idea that *undisturbed* ecosystems still exist in a constant state of change, with some relationships changing slowly, others more rapidly. Today's so-called natural landscape is not yesterday's natural landscape. Heraclitus, insisting that permanence was an illusion, ran afoul of philosophers who believed a putative state of impermanence threatened the very foundation of "stable" societies like Hellenistic Greece. Likewise, Darwin and Alfred Russel Wallace met substantial resistance to their idea that biological life changes over time. Darwin's contemporary the geologist Charles Lyell was ridiculed for his belief that the ground itself was ever in a state of change. How else, he argued, were the seashells found embedded in the sedimentary rock of mountain peaks to be explained?

When Copernicus insisted that Earth was not the center of the universe, and then Darwin and Wallace declared that man was not its ultimate creature, and then Jung and Freud made it clear that the rational mind was not the whole story for *Homo sapiens,* theologies had to adapt, or at least react. If the real human environment in developed countries today is third-growth monocultured "forests," tar-sand petroleum, cow-burnt grasslands, and smog-like clouds of microplastics floating in oceans where fish once thrived, then human cultures need to distinguish between sentimentality about loss and the imperative to survive. They need to establish a more relevant politics than the competitive politics of nation-states. And to found economies built not on profit but on conservation.

Or so it seems to me, fingering a silver coin in my pocket and sitting on the close-cropped slope of Cape Foulweather, waiting for a storm to blow in.

WALKING THE STREAM COURSES of the cape, I rise from damp intimate hollows through patches of primordial forest, into high, natural openings from which I can see far inland and, in the opposite direction, silent water out to the very edge of the sky. The chiaroscuro faces of the inland mountains, most of them, are scarred, inscribed with a filamentary network of roads and way-tracks to accommodate logging machinery, firefighting crews, and the maintenance vehicles that service power lines and cell towers. At sea, the wind-burnished water, lifted by waves and pushed onward by currents, bears trawlers and, farther out, bulk freighter traffic. Those square miles preserve no history of what is happening, as I watch. The whitecaps on a windy day collapse, the ships' wakes disappear, the skittering takeoff track of a seabird fades out, leaving no record of its having been there. A glaucous-winged gull gliding above me takes the invisible line it writes in the sky with it as it slides downwind, a line that ties on a slant a cloudless airy cerulean sky with the dense black-green of the ocean.

An active tabula rasa, the ocean, with its leaden and bellwether skies.

These scenes, inland and oceanward, as I follow creeks and elk trails through patches of old-growth forest, or put my binoculars on some anomaly seaward, encourage two thoughts that have organized my perceptions in nature for decades. First, diversity is not a mere characteristic of life—the wands of a salmonberry bush in front of me, say, stand erect differently from the wands of sword fern beside it; or, out there, a harbor seal, a sea mammal, pursues its prey, a canary rockfish. Diversity is a condition necessary *for* life. Diversity creates the biological tensioning that makes life in general vigorous and sustainable. It's diversity that ensures perpetuity. The loss of diversity, on the other hand, threatens all life with extinction.

The second thought is that understanding strategies for successfully adapting to change in an ecosystem—change, like diversity, being part of the foundation that perpetuates life—has long been the central responsibility of wisdom keepers in every human society. Their special skill was, and is, the ability to recall multiple techniques for human survival. This library of possible scenarios represents a crucial repository in each society. Whether the change facing a people comes on swiftly, like the protean surface of the ocean for the ocean strider, or arrives slowly, so slowly that it masquerades as unchanging permanence, like trees standing in a primordial forest, the responsibility of the wisdom keeper is to recognize the early signs of significant change, to look into the past, and to locate, again, a through line to the future.

Conceding the inevitability of change is not the same as passively accepting whatever change comes along. By putting economic growth on an equal footing with the preservation of human health, by promoting a need to possess and to consume that borders on the pathological, and by permitting industries to run roughshod over landscapes in order to create financial profit, the governments of industrialized nations have supported the changes that are primarily responsible for the befouled and poisonous environment that in many places has become our heritage. What resistance humanity is able to mount to the juggernaut that many call "the economy" is essentially an objection to the indifference toward human and nonhuman life that *drives* the juggernaut. A clearcut is not the outward sign of a healthy economy but of an indifference to life. And the denigrated "weedy species" that have arrived to replace some of the native species after the trees are harvested are not lesser beings, but a sign of life's fundamental resistance to the threat of extinction.

When I pause in a clearing to glance west at the setting sun, at the horizontal reach and vertical thrust of cathedral rays there rendered visible by clouds—with their tints of pink and salmon, of tawny orange and ocherous yellow, and some evenings, dragon's-blood reds—I try not to let it pass as only the striking image of a particular waning afternoon—the clarity just then of the air, and the less uniform color of the sea—but to see the inflammation as a conflagration, and wonder whether it's only a delusion to wish for the continuation of promising human life on a planet we are remaking with the Sixth Extinction. What would happen to our plans for survival if we were no longer stymied by a belief in the virtue of permanence or no longer distracted by the hope of returning to a world that has already come and gone?

WHEN I WAS a boy most of the model planes I assembled were the planes of warfare—fighters and bombers. American boys were encouraged back then to think of life, broadly speaking, in these terms—combat, and the unending struggle necessary to make things right. The model plane that has remained longest in my life, though, the plane that I have had the greatest affinity with, is a passenger aircraft from the 1930s, a four-engine seaplane called the Martin M-130. Today I keep a 1/72-scale wood model of an M-130, the *China Clipper,* in the room where I write. It has a twenty-inch wingspan and is dressed in Pan American Airways livery. When I was young, this plane represented for me the possibility of traveling to the faraway world. This type of large amphibious aircraft, along with its sisters at that time, the Boeing 314 Clipper and the Sikorsky S-42, needed only a stretch of relatively calm open water in order to put down at seaside locales—Hong Kong, Rio de Janeiro, Sydney, Singapore, Cape Town. In my imagination all these places—the quintessence of the exotic for a California boy in the mid-1950s—were characterized by continuous balmy weather and endless sunshine.

The Martin M-130, all but forgotten now, drew extraordinary attention when it debuted. One hundred and fifty thousand people gathered at Alameda, California, to watch the *China Clipper* take off from San Francisco Bay on its first transpacific flight, on November 22, 1935. The promise of adventure these planes incorporated—a Depression-era version of the nineteenth-century Grand Tour—was short-lived, however. Their movements were soon restricted by the onset of World War II. Some believe the *Hawaii Clipper* was shot down by Japanese fighter planes over the Pacific east of Manila, on July 29, 1938. The *Philippine Clipper,* on loan to the U.S. Navy, flew into a mountainside in California, on January 21, 1943. And the *China Clipper,* based in Miami during the war, hit a submerged object on landing at Port of Spain, Trinidad. It sank on January 8, 1945. These three were the only M-130s ever built.

Despite their short lives and violent endings, these airborne "clipper ships" were potent symbols for me of an expanding (and, of course, contracting) world. They held out the possibility of pursuing a life of adventure and, at the same time, represented the risk that attaches to any such quest. As a seventeen-year-old in Greenwich, England, I walked through and clambered over as much of the dry-docked tea clipper *Cutty Sark* as the ship's guards would permit. It was the last clipper of its kind and I was desperate to get the feel of it—its shrouds, the catted anchors, the glass dome of the binnacle compass, the rows of belaying pins. I wanted to climb the mainmast to the crosstrees, to gaze down and to look out away from up there.

In my unexperienced mind, the M-130 became a symbol for me of casting off, of the potential and promise inherent in departing the known.

COOK'S *RESOLUTION,* a converted collier, was a compact 111-foot three-masted vessel with a crew of 110, including officers. Its shallow draft permitted it to sail close to shore, and its bluff bows and the relatively light armament it carried gave it storage room enough between decks to sail long distances without being resupplied (except for water, wood, and fresh provisions). The ship was square-rigged on the main- and foremasts; the mizzen was square-rigged on top, with a fore-and-aft sail on a boom below, making the ship a type of barque. Fore and aft between its three masts, and between its bowsprit and foremast, and on booms extending from its spars, the *Resolution* carried a variety of jib sails, trysails, spankers, and possibly, above the mainsails, royals. It occasionally carried, in addition, topgallants on its fore- and main-masts, and above them, conceivably, moonrakers, to wring the last bit of advantage out of a favorable wind.

This once precise and vibrant vocabulary of seaman's argot, of now-opaque terms for sheets of canvas, like spritsails, drivers, bonnets, and spinnakers, is today completely lost to most of us. They are sounds without referents, like the distinction between a ship's tender and its snow, or the purpose of a dog thimble in a ship's running rigging, or the position of a ship's dolphin striker. In Cook's time, sailors were expected to know a jib sail from a studding sail in the dark, and how to raise or strike each one quickly, in any weather, at any hour. The effective management of it all was dazzling.

At the Oregon Historical Society Museum in Portland, a meticulously constructed wood model of HMS *Resolution* is occasionally put on display. I've gone several times to the museum and sat in a chair for a few hours each time studying it. Among other things, it's possible to learn from this model that from the eleven nine-paned window panels in the captain's cabin, in the stern of the ship, nine facing aft and two facing forward on the cabin's wings, Cook was able to take in all but about 25 degrees of the horizon. The droop of foot ropes, called horses, on the spars and of ratlines on the mast shrouds contrasts sharply with the straight taut lines of stays and guys, the standing rigging that secures the masts. And a helmsman's view forward from the double wheel is partially blocked by the ship's yawl (stored inverted on the midship hatches) and by standing rigging secured at the beakhead, below the bowsprit.

The trimness of the model gives a viewer the impression that whatever the sea might throw at it, the ship will weather the blow. It appears prepared for all eventualities. On the three voyages that culminated Cook's career, masts were, indeed, sprung; dry rot spread through the ship's framing timbers, ruining them; the hull of the *Endeavour* was stove in; and storms tore the ship's running rigging loose and ripped its sails to shreds. But the men sailed on, with eerie indifference to the threats they faced, like being washed overboard or falling to their death on the spar deck. It was widely known, certainly, that work for men crewing a sailing ship was unremittingly demanding, fatiguing, and dangerous; less well known today, because illiterate sailors left so few records, was how unconsoling and unremunerative the work was. By temperament, most sailors, according to their biographers, were rebellious, strongly fraternal, belligerent, hardy, and unsympathetic. Like other naval captains who employed civilian crews, Cook had to convince them that what they stood to accomplish as explorers, by their obedience and diligence, was worth their injuries, the punishment he doled out, and the diet of terrible food they were served. I would prefer to have heard in full any one of a commanding officer's profanity-laced exhortations to his crew to serve the expedition well, than to have access only to the brief sanitized summaries of these harangues and cajoles that turn up in officers' journals, where one finds very few comments or observations from ordinary seamen on what, if anything, the discovery and charting of an unknown world might have meant to them.

Another book I once took with me to Cape Foulweather—which I read in a motel room in a nearby town, after a different storm drove me off the cape—was a *Dictionary of Disasters at Sea During the Age of Steam, 1824–1962,* a two-volume work. I could find no comparable record for sailing ships in the eighteenth century, but sensed that these volumes would nevertheless convey the gist of what might await a crew like Cook's. The books reported, briefly and bluntly, the fate of many ships lost at sea in those years, or described incidents where there was great loss of life though the ship itself did not sink. Thousands of sailors were washed overboard in storms in an era when few of them knew how to swim. Fire spread through ships after an accident in the galley, a hull was can-openered by an uncharted rock, a ship was dismasted by a typhoon or entombed by a rogue wave or overwhelmed by leaks. Ships were idled in the doldrums and ran out of water and food. The dead were shoved overboard, the living ate bits of rope and hoped for rain to fill their buckets.

This compilation of disasters was commissioned by Lloyd's of London, an English firm famous for insuring ships. Ignored in the book's extensive listings were small ships and boats, ships in which Lloyd's had no financial stake, or in which Europeans were not foremost among the victims. Still, if one is looking for some real sense of the perils of a life at sea, this severe introduction would be a place to start. The indifference of Cook's sailors, both to the threat of death or to the possibility of fame, reflects the indifference they believed the sea showed to the value of their lives.

When a registered ship went down with all hands aboard, someone in the London offices of Lloyd's recorded their names in a journal with a white quill pen. The pen was crafted from the primary feather of a mute swan, several of which were raised specifically for this purpose at Abbotsbury Swannery on the Dorset coast.

It has been the custom at Lloyd's for as long as anyone can remember.

Whenever I sat those hours in the museum with the model of HMS *Resolution,* I hardly ever considered the peril. I thought of the implacable nature of the ship and wondered where the figurative equivalent is in our time, the vessel that would convey us successfully through the myriad threats we face from the natural world and from the human world we've not so perfectly designed.

STARING DOWN AT the surf from precipitous seaside cliffs at Cape Foulweather, or watching shearwaters amassing just seaward of the flat beaches farther to the north, I've wondered at the way some of the world's coasts have been set apart, while others have remained anonymous. A historical imagination, disinterested and attuned to international trade rather than to local geography, might recall reading of the Turquoise Coast of old Anatolia, the Gold, Slave, Ivory, and Grain Coasts of West Africa in the colonial era, the Dalmatian Coast of the eastern Adriatic, North Africa's Barbary Coast, Namibia's Skeleton Coast (named for its many whale and seal skeletons, the remains from carcasses washed up from those South Atlantic commercial fisheries), or Nicaragua's Mosquito Coast (named for the Miskito people). The names resonate, but these are still imposed labels, dreamed up, for the most part, by merchants. Once in Hobart, Tasmania, I saw a rendering of the entire Tasmanian coast, an elongated fifteen-foot drawing executed in great detail. It was a continuous coastal elevation, drawn by an indigenous woman during a circumnavigation of the island. It was her portrait of the place. She'd affixed no names to the drawing, but had profoundly evoked the unique physical character of Tasmania's seaward face.

The many different sections of Oregon's coast are not set apart by any such list of names of which I'm aware, though local histories might have recorded some that local use keeps alive. To someone like me, who grew up far from here, the coastal events I've been most surprised to learn about, and for which coasts might once have been named, include the history of U.S. merchant marine ships torpedoed offshore here by Japanese submarines early in World War II, and stories of the ships—hundreds of them—that foundered and sank at the mouth of the Columbia River, trying to cross the Columbia bar. Also, the story of a Spanish galleon, possibly the *Santo Cristo de Burgos,* outbound for Acapulco from Manila in 1693. It sank near the modern-day coastal town of Manzanita, and its cargo of beeswax and porcelain continues to turn up on the beach to this day. In June 1979, what was at that time the third-largest stranding of sperm whales in the world occurred near the town of Florence, Oregon, sixty miles south of Cape Foulweather, near the entrance to the Siuslaw River. Forty-one of them died on the beach despite the efforts of some people to save them.

The coastal stories from this region that I most often recall, though, concern a group of Japanese fishermen, sometimes referred to by historians as "the shogun's reluctant ambassadors."

During the time of the Tokugawa Shogunate (1603–1867), Japan's international borders were tightly closed to the outside world. Only a limited amount of regulated trade with foreigners was conducted, solely at Nagasaki and solely with the Dutch. Under the shoguns, Japanese people were forbidden to travel beyond their own coastal waters or to build an oceangoing vessel. Small coastal vessels were occasionally dismasted by storms, and fishermen and tradesmen might then find themselves drifting rudderless or without sails in "the black river," the Kuroshio, or Japan, Current. Carried eastward across the Pacific, most of these compromised ships eventually either sank or, carried farther east by the North Pacific Current, were washed up on the shores of

North America, usually on Vancouver Island or the coast of Washington. The crews of the majority of these hulks were most often found dead, but occasionally a ship carrying food—rice, for example—whose crew had been able to catch fish and trap rainwater along the way arrived with survivors. Even with the support of sympathetic foreign nations, however, it was extremely rare that survivors were ever permitted to return to Japan.

They became, instead, the shogun's "reluctant ambassadors" to America and to the European nations that offered them refuge.

In January 1834, one of these drifting ships, the *Hōjunmaru,* washed up on the Olympic Peninsula, in Washington. Only three of its crew of fourteen had survived their fourteen-month journey. A nine-year-old boy named Ranald MacDonald, living at the time two hundred miles away at Fort Vancouver, on the Columbia River, became fixated on the fate of these three men. The obsession did not leave him until he was twenty-four, when by a clever ruse, he broke into what Melville refers to in *Moby-Dick* as the "double-bolted land" of Japan, in 1848.

If James Cook represented the apotheosis of the Enlightenment, MacDonald might be said to represent that place in Western society reserved for the marginalized, in his case a mixed-race working-class man whose great deed in life, and whose insights as an international traveler, were ignored or dismissed during his lifetime. When Commodore Matthew Perry arrived in Edo in 1853, intent on forcing Japan to open its borders to foreign trade, he was stunned to find that the emperor's advisers were able to speak English. They had been taught to do so, four years earlier, by the seaman, raconteur, and roustabout Ranald MacDonald.

MacDonald was born at the mouth of the Columbia on February 3, 1824, to a Chinook woman named Koale'xoa, the daughter of the most prominent Chinook chief of that time, Concomly. Ranald's white father, Archibald McDonald (as he spelled their surname), was a Scots clerk at Fort George, the Hudson's Bay Company (HBC) post at Astoria. MacDonald's mother died shortly after he was born. In the years following, MacDonald's father was stationed at various HBC forts in present-day Washington and British Columbia, including at Fort

Vancouver, eighty miles up the Columbia from Fort George. Ranald spent much of his early childhood with his Chinook relatives at the mouth of the river and periodically joined his father at his postings, where he got to know itinerant fur trappers in the HBC network, many of them native Polynesians from the Pacific and French voyageurs from Canada.

Archibald, who had married again, was pleased to have his son living with him in what he regarded as a more civilized environment at Fort Vancouver, which boasted a school. McDonald made clear to Ranald his hope that he would grow up to take a managerial position with the HBC, like his father, or otherwise find himself a place in the business world, perhaps in a city like Ottawa or Montreal. When Ranald finished his elementary education at Fort Vancouver, his father sent him to an HBC boarding school at Winnipeg, Manitoba. At the age of fourteen, Ranald accepted a job as a clerk in a bank in St. Thomas, Ontario, where his work was supervised by a business friend of his father.

Throughout his early years, according to his chroniclers, Ranald felt thwarted in his ambitions because of his mixed-blood heritage. It's clear from MacDonald's autobiography that his sense of cultural allegiance was conflicted as he was growing up. He felt aligned to some degree with the Chinook society into which he was born, but Chinook social organization and cultural values were changing rapidly at the time of his birth. He also identified with his father's educated white mercantile culture and, to some extent, with his father's aspirations to material wealth, domestic comfort, and executive authority. In Winnipeg, Ranald fell in love with a white girl. He was roughly confronted and informed that he had no business as a "half-breed" trying to court her. In Ontario, later, he encountered similar judgments and suspicion. Sitting at his stool in the bank every day, facing long columns of figures, Ranald felt straitjacketed by local social conventions. He grew bored and became depressed at the prospect of slowly amassing enough money to leave the bank and move on to well-paid work he was more comfortable with. He soon quit his apprenticeship and traveled east to Long Island, where he found employment aboard whaling ships, the crews of which comprised men of extremely varied backgrounds—West Africans, Pacific

Islanders, Coastal Asians, Native Americans, Scandinavians, and sailors from the Caribbean.

As a boy, Ranald had heard stories from his Chinook relatives about the *hyōryūsha,* Japanese drifters who washed up on the shores of North America during the Seclusion Era, including the three survivors of the *Hōjunmaru* who came ashore near Ozette, Washington. His relatives had told him that all the *hyōryūsha* looked like Indian people.

Over the years—neither MacDonald's own papers nor his biographers' research make clear precisely how this occurred—MacDonald came to feel a complex sense of identity with the *Hōjunmaru* survivors and other *hyōryūsha.* His sense of identity with a distant indigenous people in the Western Pacific, walled off from European culture, combined with his feelings of resentment over episodes of social rejection and racism, along with his anxiety over having chosen yeoman work as a seaman, which he believed had deeply disappointed his father, led him to feel he had to distinguish himself in some way. He had to escape from the category society had assigned him to, that of a "half blood," a mestizo, and do it by some act that would, not incidentally, impress his father.

MacDonald came to believe that Japanese people were actually related to American Indians and that they were soon to suffer the same social disintegration the Chinook had at the hands of aggressive European and American traders. They would be as powerless as the Chinook, he thought, to prevent or control it. (The British had already forced an opium-for-tea trade on a reluctant China with the first Opium War, securing the ports of Canton [Guangzhou], Amoy [Xiamen], Fuzhou, Ningbo, and Shanghai in the Treaty of Nanking [Nanjing].) By this time in his life, MacDonald had gained enough experience in Atlantic and Pacific ports to understand the enormous political and economic power driving the development of international trade, and he comprehended the effects of its momentum.

MacDonald believed someone had to warn the Japanese about what was coming. He knew that the crews of shipwrecked American whaling ships who had come ashore seeking help in Japan in recent years had been treated rudely and forcibly expelled from the country. He also knew that pressure was growing in America "to do something about Japan."

That someone, he decided, would be him. And he believed he had to get there right away.

I FIRST LEARNED about Ranald MacDonald when I came across some of his letters in the holdings of the Eastern Washington State Historical Society, in Spokane, while I was teaching at Eastern Washington University, in 1985. Later, at the Oregon Historical Society in Portland, I read more of his personal correspondence, marking his occasionally florid and affected prose and his consistently courteous and gentle tone. I read several biographies of him along with a heavily footnoted edition of his autobiography, and eventually went to visit his grave site in an old cemetery just north of the Colville Indian Reservation in northeastern Washington, near the Canadian border. In the course of other work I was also able to visit some of the places that had been important in his life, including the old whaling town of Lahaina on Maui, in Hawai'i; Astoria, at the mouth of the Columbia; the Australian goldfields where he worked after being deported from Japan in 1849; and Hokkaido, in northern Japan, where MacDonald came ashore among the Ainu, the indigenous people of the area.

I developed an affection for MacDonald because of his earnest and dignified struggle for personal recognition and credibility, a lifelong soul-wrenching effort to find out who he was. And, too, because of his capacity for self-delusion, and the admixture in his life of venal desire—his hopes for fame and fortune. These traits, for me, only made him more deeply human.

MacDonald's best biographer, Frederik L. Schodt, in his *Native American in the Land of the Shogun: Ranald MacDonald and the Opening of Japan,* writes that he's not convinced MacDonald ever sailed as crew on an (illegal by then) Middle Passage voyage aboard a West African slaver in the 1840s, though this grim experience with nineteenth-century commerce in human beings remains part of the popular perception of MacDonald. In an impressive number of ways, however, MacDonald gained experience with the dark side of a system of international trade that began to accelerate after Cook's voyages created more dependable maps. MacDonald was born into the North American fur trade, and he came of age around the early nineteenth-century competition that pitted British and American traders against each other in the Pacific Northwest. He sailed from New England ports aboard both cargo vessels and whalers. He was part of the gold rush in Victoria, Australia, in the 1850s, on lands usurped by frontier speculators, country that was taken from Aboriginal people with the support of Australian businessmen and politicians. It's possible he also crewed on coastal traders in Southeast Asia, and he might also have worked aboard British East India Company tea clippers, which hauled opium from India to Guangzhou and then freighted Chinese tea to London.

THE HISTORICAL FIGURES of Cook and MacDonald, the one famous, the other obscure, have remained linked in my mind for many years. Each in his own way is archetypal, and both found the pivot point of their personal histories in the Pacific. Both also remain to some degree enigmatic. Together, they raise the issues of race and class privilege, and questions about the history and morality of modern commerce, which resonate intensely in modern times. On their long ocean journeys, these men, each one putting to sea from a country roiled by economic change, defined a search for the ineffable. I can imagine them in conversation, speaking out of earshot of anyone who might judge them or try to explain them. One might presume that they wouldn't be able to converse much beyond pleasantries and sea stories; but each person was a driven man, and both had been wounded by an effort to achieve something memorable. They might have come to realize this in private conversation, let's say over tea on a conducive afternoon on some Pacific island lanai, each of them speaking freely, without fear of being criticized or contradicted. An observer would note that Cook was thinner and slightly taller than the other man; that where Cook had a small head and a large nose and was otherwise oddly proportioned, MacDonald was broad-shouldered, handsome, compact, and muscular. Cook had piercing blue eyes, MacDonald deep-set gray eyes, with hazel rings around the irises.

Cook grew up the son of a tenant farmer, with little ahead of him in the way of prospects but work of that sort. He set himself apart from his childhood companions, however, by entering the merchant service. He was soon a ship's master aboard a collier, freighting coal along the North Sea coast of England. He then entered the Royal Navy as a master's mate at the age of twenty-six, essentially starting all over again to prove himself at sea outside the merchant navy.

Though Cook didn't grow up with a sense of privilege, he intended to make his own way in life, not be assigned a place because he was (only) a landless working-class man. In the British navy he distinguished himself immediately by creating a set of exceptional maps and charts for sailing Newfoundland's coastal waters. He departed England on his First Voyage of circumnavigation (1768–1771) feeling gratitude for having been chosen for the post of captain. He was overcome with similar emotions when he opened his sealed orders in the South Pacific, after the observations at Point Venus had been completed, and learned the Admiralty wanted him to settle the matter of the existence of a rumored southern continent. They wanted him, further, to follow up on the work of other British mariners by laying claim to additional islands in the South Pacific. By the time of the Second Voyage (1772–1775), Cook knows from experience what is expected of him and how to achieve it. He strains his ships and drives his men hard to get the work done.

With the Second Voyage behind him, Cook is comfortable with the thought that he might have arrived at the end of his string of extraordinary accomplishments. He'd done the one thing every man of his class hoped to do—a class he was not born to, but into whose company he had moved himself with his achievements. He had become great—in his case, as an explorer. It is now another, perhaps less admirable form of ambition that takes hold of him and leads him to accept command of a Third Voyage.

He has by 1775 dispensed with the question of Antarctica. He's given the Atlantic, the Indian, and the Pacific Oceans a southern boundary, and he's fixed coordinates for the east coast of Australia. The vastness that still needs definition is the North Pacific, between the latitudes of Mexico and the Bering Strait. And if he is able to sail far enough up the coast of North America to do that, he might possibly find a western entrance to a Northwest Passage.

What keeps Cook from immediately accepting command of the Third Voyage is his recollections of the physical demands of such voyaging, and his awareness of how tiresome he has come to find members of his crews, with their lack of self-discipline, lack of ambition, coarse appetites, and penchant for creating problems ashore. Weren't these the same fellows of middling mind he had grown up with, whose company he wanted to escape?

But he agrees to go. His exasperation with his crew grows, as does his impatience with the voyage and also, perhaps, a sense that he has thrown his life into this effort for no sufficiently good reason. He doesn't know who his wife is, really. He hardly knows his children, having been away so many years. What will even greater fame now mean? And what will be the cost? And what is he to do with his vague misgivings about the rise of empire? He can share these misgivings with no one. The Navy, he knows because of how they edited his journals for publication, has made him the bearer of their belief in British exceptionalism. They will make him an avatar of empire, as will his foremost biographer, Beaglehole, two centuries later, fashioning of Cook a man steeped in Beaglehole's own prejudices.

Cook will be used for ends for which he had no sympathy and this, too, seems to be on his mind on the Third Voyage.

At the start of that voyage (1776–1780) Cook was financially secure and had the respect of his peers. What he did not have was control over the meaning of his own life. Before he accepted command of the Third Voyage, during which he would lose his grip, both as a man and as a commander, and be stabbed to death by angry Hawaiians, did Cook imagine a retirement that would allow him to write up the story of his life in his own way? Did it concern him that for all his apparent freedom of expression, he was still in service to a national vision, not really free to say what he thought or believed? And if he chose to, would he, the son of a Scots farmer, be ostracized socially, along with his family? Did it irritate him that he had built the floor upon which Sir Joseph Banks and others now pranced, people with more latitude to speak their minds than he? If he did find a western entrance to the Northwest Passage on the Third Voyage, would that make these concerns irrelevant? And finally, did he worry whether anyone was really interested in hearing his mature insights into any deeper meaning his voyages might have had? Or was he so thoroughly exasperated by the public's appetite for tales of South Seas cannibals and sexual adventure with Polynesian women that he could not muster the will to plumb his soul?

For MacDonald, the question of ambition and his own meaning in history was perhaps more tortured. He grew up unsure of his social status and not seeing any clear path to personal success. At different times and in different circumstances, he tried to imagine himself quite apart from the world of Indians, to imagine he had erased what he regarded as a social stigma by becoming an educated and cultured man, traveling easily in the world his father lived on the fringes of. At other times, during his months in Japan, for example, he identifies strongly with his Chinook heritage.

MacDonald regarded his father as a person of consequence, but knew that his filial connection offered him no certain social standing outside the HBC trade network, especially in eastern Canada, because he was a mestizo. It was important to him to know that his maternal grandfather, Concomly, was a member of Chinook royalty in that socially stratified, slave-holding society; but this was of no real help to him. Chinook society was unraveling, and the HBC's reorganization of traditional trade networks had taken from the Chinook their power to control regional trade. His grandfather was a transitional figure in Pacific Northwest history. Fifty years earlier, the grandson of Concomly would have commanded respect among the Chinook and other Lower Columbia tribes, and he would have inherited material wealth. In the 1830s and '40s, his lineage gave him no advantage.

MacDonald's father treated him like a favored nephew. His stepmother, Jane Klyne, with whom Archibald had thirteen children, treated him like a son, but she herself, it is thought, was the daughter of a Swiss voyageur and a Cree mother. She could provide him with no entrée except among the Cree and members of the Métis nation, a mixed-blood people of south central Canada. MacDonald, in short, saw no sure or promising opening for himself in North American society. He would, as it turned out, never marry or have children. He would die an unheralded isolate on an Indian reservation, where he was regarded as quaint and eccentric; and he would be buried in no place of honor. He would have no plot commensurate with the way he saw the meaning of his own life.

In the end, there is about MacDonald a certain false note. The educated locutions of his correspondence, his cultivated mannerisms, and the way he presented himself in public all seem affected, suggesting he never found out who he actually was. In his later years he was viewed by visitors to the Colville Indian Reservation as little more than a well-traveled raconteur, not as someone deserving of serious attention. The condescension he sensed was driven home for him in a parochial article in the July 18, 1891, issue of *Harper's Weekly*, written by Elizabeth Custer, wife of the foppish American brevet general, George Armstrong Custer. She belittles the courtesy MacDonald showed her when she visited him and treated his life experiences as an amusement. In a vulnerable letter written in 1892 to Eva Emery Dye, a California woman who turned out to be his first biographer, and who would later characterize him as "the strangest, most romantic and picturesque character of Northwest annals," MacDonald implores her not to be "so hard on me as Mrs. General Custer."

If they were to sit down together on that Pacific lanai I imagine them meeting on together—Cook the more reserved, MacDonald the more loquacious; Cook the nattier dresser, MacDonald the one more at ease with the waitstaff; both of them sons of Scots fathers—I believe Cook would have been amused by, but appreciative of, MacDonald's harmless bravado, and that MacDonald might have understood Cook's dilemma as a famous person. I infer from their biographies that both died without having anyone really to talk to: one a sailor with all the trappings of conventional success, eventually honored with life-size statues of himself in half a dozen Pacific ports; the other a sailor without a medal, no letter of gratitude or commendation from anyone to show to Mrs. Custer, and all but forgotten—except in Japan, where he remains widely known and celebrated.

———

IN DECEMBER 1845 at Sag Harbor, Long Island, MacDonald signs on to the American whaler *Plymouth*. He's twenty-one years old. For the next two years he hunts mostly sperm whales from her boats, visiting several mid-Pacific ports along the way, including Floreana Island (Isla Santa Maria) in the Galápagos, where the crew deposits and retrieves mail at Post Office Bay. In June 1848, in the Sea of Japan, the ship's captain honors a pledge he made to MacDonald in Sag Harbor and which he renewed with him at Lahaina. He allows MacDonald to be lowered over the side in one of the ship's boats with a cache of supplies. (The *Plymouth* is homeward bound, its hold full of barrels of sperm whale oil.) This is somewhere south of the island of Yagashiri, in the sea's eastern waters. From there, MacDonald makes his way north to Rishiri, an island off the western shore of Hokkaido, where he poses as a shipwrecked sailor among the Ainu.

The indigenous Ainu, fearful of being accused by the local daimyo's representatives of socializing with this *gaikokujin,* turn him over to Japanese authorities at the military post at Sōya, on the mainland. A series of land and sea journeys, interrupted by brief periods of house arrest, brings him three months later to the shogun's court in Nagasaki. Here, for the next seven months, MacDonald instructs fourteen men in the English language, believing they will need to master it in order to deal successfully with the British and American merchants and military personnel he's certain are coming.

MacDonald makes a good impression on his pupils, especially his favorite, a man close to his own age named Einosuke Moriyama. MacDonald makes a favorable impression on almost everyone he meets. He has the right demeanor, the right attitude for a visitor, in the eyes of the Japanese.

On April 27, 1849, MacDonald leaves Nagasaki for Hong Kong in the company of a contingent of legitimately shipwrecked American whalemen who are being expelled from the country. By 1851 he is prospecting in the Australian goldfields near Ballarat, in south central Victoria. There is no dependable record of MacDonald's whereabouts before he arrives in Australia or afterward, until he turns up again in the Cariboo region of east central British Columbia, during the 1858 gold rush. He finds work here as a chandler and horse wrangler, ferrying supplies into the Fraser River backcountry. By then, knowledge of his singular achievement seems to have evaporated completely, to have disappeared from the historical record, perhaps in part due to MacDonald's reluctance—or inability—to impress anyone.

AT A DISTANCE of some 170 years, it's difficult to appreciate the degree to which Japan was closed to the outside world during the forty-three weeks MacDonald was traveling there and teaching. Foreign ships that approached any Japanese port except the one at Nagasaki immediately drew cannon fire from shore batteries. Survivors of shipwrecks were quickly rounded up, transported to Nagasaki, and put aboard homeward-bound Dutch trading vessels. The civil treatment MacDonald met with in Japan stands out as anomalous, until one remembers his respectful temperament and the seriousness with which he undertook his mission. He struck the Japanese as amiable and deferential, in contrast to other shipwrecked American whalers, who were frequently confrontational, rowdy, disrespectful, and condescending. The oriental cast of MacDonald's facial features seemed to diffuse his hosts' own racial prejudice, and the ease with which MacDonald adapted to local customs and a Japanese diet surprised them.

In July 1853, when Commodore Matthew Perry strode ostentatiously into the emperor's court at Edo, his proposal and orders were translated for the court by MacDonald's former students. Perry's disingenuous courtesy, his preemptory demands, and his thinly disguised military threats contrasted sharply, for some of those present, with the way MacDonald had conducted himself, a man steeped in what it means to be of mixed race and culture, no believer in the immutability of economic, social, or racial hierarchies, and conciliatory by nature.

Perry would have thought MacDonald's approach to negotiation in these circumstances uninformed and weak. Moriyama, who was present, must have seen in Perry the reification of what MacDonald had warned them about.

For me, MacDonald represents a man who passed away oddly unfinished or sidetracked, a person in whose history one finds too many doors automatically closed simply because he had the wrong physical appearance, the wrong work history, the wrong ideas.

I often recall, in thinking about MacDonald, the many exemplary people I've met all over the world who, for reasons of race, religious conviction, lack of formal education, or nationality, are not likely ever to be invited to the table to discuss the fate of humanity.

AS FAR BACK as I can remember I've had a deep fear of being caught in hurricane weather or heavy seas out of sight of land, even though I've found the sea seen from shore, in almost any weather, mesmerizing and soothing. Perhaps its primary attraction has been its breadth, like a stage's, or the unbroken line of its meeting with the sky, or its inconstancy. Or the transparency of its colors, from the dark purple of prunes through tropical blues to the green of the verdigris that forms on oxidized copper. Once in Camden, Maine, walking its waterfront with a friend, the painter Alan Magee, I saw in a shop window a perfectly crafted scale model of a whaleboat, the type of longboat with a step mast that MacDonald would have worked from, though I didn't know of him at the time. I bought the model because it was beautiful, and because it had been built with near-microscopic attention to detail—the coiled lines, the oarlocks, the rigged harpoons. It was fashioned out of someone's love and perfect knowledge. I wanted it in the room where I worked, next to the model of the Martin M-130.

Today the boat resides in my workroom in a glass box, to keep dust from settling in its many small cavities. It's an image for me of courage, even of security. For a long while after I purchased it, I was not able to imagine the boat in the sort of seas I knew whalers had encountered or to regard it as anything but unsafe. One day this perception changed in the Drake Passage, that corridor of notoriously wild water that separates the tip of South America from the Antarctic Peninsula. On that day I learned about a kind of beauty I had not until then been able to grasp. I was aboard a large ecotourism vessel with 130 others, trying to reach the leeward shore of South Georgia, 750 nautical miles southeast of Port Stanley, our point of departure in the Falkland Islands the day before. The ship, the *Hanseatic*, was weathering a Beaufort force 11 storm—sustained winds over 55 knots, chaotic seas of forty-foot waves with some fifty-footers breaking over the upper decks. Hardly a spot on the surface of the water was not blanketed with sea-foam. Sometimes the bow of the *Hanseatic* was entirely buried in a wall of water. It geysered through the anchor chains' hawseholes and crashed against the windows of the bridge. For some reason I decided this was the time for me to address my old fear. I stepped outside on a lee deck, just below the bridge, with a trusted friend, the polar explorer Will Steger. Dressed in storm gear, we stood together in the caldron of soaking air, listening to the shrieking wind tear through the superstructure.

We quickly hunkered in the shelter of a companionway with our feet spread and with death grips on the railing. We watched in astonishment as albatrosses forty feet away navigated the chaotic wind like Olympic snowboarders, glancing over to make eye contact with us as they did. I turned at one point to see the stern of the 403-foot ship rise from the water and swing thirty feet to port. The only stillness here was the steel deck directly under our feet, which carried the shuddering of the ship, as it crested, into our thighs.

Some time into this spectacle I realized I was relaxed, that thoughts were unfurling in my head in a normal way, without panic or anxiety. What had for so long been an image of terror for me was now an image of something else, a kind of perfection. Here was Earth's fundamental wildness, here was William Blake's sense of the Divine in chaos. A well-traveled friend of mine, when I told him of my fear of encountering big seas offshore, had said to me, of just such a storm he'd been through in the Drake Passage, "I saw the face of God."

When I got home from that trip, I looked differently at the whaleboat in its glass case. Its oars are shipped, its sails are raised. No human figure is aboard. I now sensed the daring in its architecture, imagined the seamanship that would keep it from capsizing in heavy weather. I could appreciate more deeply its integrity, which is chiefly what made it attractive.

The subtlest memory of that hour I spent watching the storm from one of the *Hanseatic*'s upper decks is that by standing that close to a force that might easily have killed me if I became inattentive, I'd fed both a sense of gratitude for still having, at the age of fifty-seven, a life I could lead and a sense of forgiveness for the harm any random person might do to another. In those minutes of gazing at the boiling cistern of waves and watching the albatrosses addressing the storm with great seriousness, I could fix only on what I admired most often in other human beings, their enduring grace and poise.

As I look back on Cook's experiences in the Pacific and on MacDonald's, and look at the model of the M-130 sitting in my studio, and consider my fascination with the nautical details of Cook's *Resolution,* I can see that I have spent much of a lifetime thinking about such conveyances. When the time comes, what sort of person will be at the helm for us? And how will we know whether we can trust this navigator?

THE EVENING OF the day before the storm was supposed to arrive at the cape I lay sleepless in my tent, wondering why I returned to this place so regularly, as though one day I expected to find a letter here from God. What tugged at me was how well history, biology, geography, quietude, and space came together in this place—at least as I understood those things. I anticipated, I suppose, that an illuminating convergence of some sort might reveal itself here one day. But I'd break camp, unenlightened, and return home; and then, as often as not, leave the country for somewhere else—the Galápagos Islands, South Africa, Afghanistan, Prague, the Tanami Desert. Many months—sometimes a few years—later, I would return, having found some piece of insight elsewhere; and with a cup of black tea in the evening I'd watch the ancient ocean, watch the infinite variety of its surface, the nap of tweed or sheen of satin or wrinkle of crepe there, losing definition as the evening's darkening atmosphere settled over it.

I felt a peculiar intimacy at Cape Foulweather with events that were local—the history of the Alseans, the Tillamooks, the Chinook. The differing ecologies of disturbed and undisturbed lands. The summer and winter regimes of light, the neap and spring tides of the lunar months. In some ways I envied Cook the precision and order of his grid of latitude and longitude, the certainty of it, in all weathers and lights, his way to connect one thing directly to another, a dependable matrix upon which to lay out a dependable route. But his grid lacked the measure of time. And lacking a third geometrical dimension, it gave the navigator a false or incomplete sense of security.

We no longer seem to be sailing in a time of fixed stars, of accurate chronometers, and of reliable routes. I met a photographer near the cape one day, close to Otter Rock. He'd been making images of Oregon's beaches during the spring tides. He believed that the seascape exposed twice a month during extreme low tides will disappear in his lifetime, as the Pacific slowly rises. North and south of Cape Foulweather, to one way of thinking, the ocean *is* in fact dying. For prolonged periods of time now, the amount of oxygen available to organisms in the water here is not enough to sustain anything but marginally anaerobic life. And the pH of the water is dropping, as it is in all the world's oceans. As they become more acidic, they become more hostile to life. Some ocean ecologists believe that in fifty years pelagic food fish might mostly be gone from Earth, their loss representing a major part of the ongoing Sixth Extinction, the first worldwide extinction since the end of the Cretaceous, 65 million years ago.

But perhaps this particular situation will not materialize, and the calculus here is incorrect.

Where are the rutters, the compilations of sailing instructions and warnings, that we need to successfully navigate a threatening future? What will prove to be the metaphorical gridwork of latitudes and longitudes, the dependable charts for human navigation that will let us heal the rift between knowing something and feeling something, a chasm the Enlightenment created for us when it privileged the ability to know over the ability to feel? What will be the grid of metaphorical rhumb lines and meridians in a new portolano, one that will not permit the integrity and profundity of the local to slip away in order to serve a vision of the grand?

Lines of latitude and longitude speak most eloquently to the head. Mastering them can make a person feel confident and smart. In a similar way, committing to memory the myriad associations among species of animals and plants in a particular locale can make one feel competent when navigating a route over land. The Alsean hunter knew where he was and where he was going in a way different from Cook's. Cook's view was an overview, peering downward, figuratively, from a great height to take in specific details. Hawai'i, for example, was a detail. The Alsean's view was upward, from fine distinctions into the grandeur of the larger sphere that was Cook's dominant reality.

It seems a person would need both points of view to become fully informed, a knowledge of both the extreme complexity of the local (which Cook had neither the time nor the inclination to acquire) and the unbounded enormity of the grand overview. If one has a capacity for appreciating both, the customary arrangements of space and time that constrain imagination become veils. They are no longer rigid walls. The old arrangements of time and space that contribute to a sense of impossibility when we face the worst situations are no longer able to defeat our capacity to imagine.

With the indigene's acute awareness of the depth and intricacy of the local, the myriad relationships that, attended to, create the sustaining wholeness of his immediate world, and with a visionary's awareness of a fabric comprised of all these local universes, more options for humanity become apparent.

The idea that a person could be both indigenously rooted and internationally aware beggars belief, but the emergence of individual traditional elders at international forums on the future, and the world-wise lucidity of their testimony, implies the existence of more such people.

I think of Ranald MacDonald as an unfinished man, not as an elder, because the erratic path of his life suggests he was never able to determine what he meant by his life. He became, instead, a kind of poseur, an actor. This is why Elizabeth Custer did not take him seriously. To her, he was inauthentic. Cook, of course, on the other hand, mostly knew what he wanted his life to mean and knew that humanity would benefit from his having filled in most of the large geographical spaces still blank at the time of his birth; but he lacked MacDonald's intuition about menace, and MacDonald's sense, I think, that one day the world of nation-states, which Cook was helping to shape, might falter, and that then the world would need to be mapped out all over again.

If Cook was the person with an overview, MacDonald existed somewhere in that same reality but with an awareness that the local knowledge of his Chinook family was slowly becoming almost worthless. It was being erased. His effort to warn the Japanese about the cultural nerve gas that had felled his own people was prescient. He prefigured the utility of the bicultural mind in international affairs, and the conventional failures of his life seem inconsequential alongside this.

ON WHAT WERE SOMETIMES melancholy evenings for me, when I would watch the last movements of the ocean before the night's walls fell, leaving only the sound of the surf, I would occasionally recall the life journeys set out in works like Mozart's *Requiem in D Minor* or Brahms's *A German Requiem,* the human passage from abject grief about life's realities to exalted peace. My knowledge of classical music is thin, but some pieces have spoken to me so forcefully over the years they've become unforgettable. Music that evokes the darkness in our lives but which at the same time elevates a listener's emotions by transcending that darkness remains with me because it is exactly this ability to launch the heart in the face of despair that I find astonishing in certain singular people I've met. They have every reason to give up—poverty, the threat of prison, ethnic persecution, civil war, dictatorship—yet they do not falter. Something in the lyricism of this music calls forth feelings of hope, of faith in the enduring ability of ordinary people to overcome difficulty. (It does if one's cultural foundation happens to include a sensitivity to this particularly Western musical tradition, which it very easily might not of course for, say, the Igbo, the Yi Chinese, the Inuit, and so on.)

Early in the Brahms *Requiem,* the hardship and longing Western people experience in life are limned, evoking the sorrow they feel, knowing that death is inevitable; and then their capacity to transcend death is celebrated. This path out of darkness unfolds in seven sections. In the fifth section, "You [Who] Now Are Sorrowful," the soprano's voice, trembling on the highest notes, releases feelings of hope in the listener, prefiguring the phrases of profound peace with which the seventh section ends.

No doubt this music, as I say, speaks this way only to a small portion of humanity; but the arrangement of these tones, and the changes in tempo that deal with human suffering, sorrow, and death, make it possible for some who listen to rise above despair.

Other musics I've experienced in foreign settings apparently healed and inspired the strangers around me as profoundly, though I remain ignorant of the themes and intentions of these other composers. It is also true that some composers—Mahler comes immediately to mind—who address the darkness we all face are not as popular in our age because there is not sufficient lyrical relief in their work, while certain other composers—Mendelssohn would be one—are less appealing in the modern age because there is not enough darkness in their compositions.

Music has a remarkable capacity to revitalize one's expectations, music as different as Bach's *Mass in B Minor* and John Luther Adams's modern Pulitzer Prize–winning composition, *Become Ocean*.

If I remember, I always try to bring recordings of music like this with me on a trip.

THE IDEA THAT I would somehow encounter an illuminating insight at Cape Foulweather, a coming together there of complex feelings I had about injustice, about the apparent intractability of ethnic and religious factionalism, that I might discover a cause for hope, seemed thoroughly uneducated to me some days. Delusional. But I persisted. I subscribe, I suppose, to a popular notion, that "the [undisturbed] land heals," that it can bring the disheveled or distracted mind to a state of calm transcendence. Exposure to an unusually spectacular place in conducive circumstance, the thinking goes, can release one from the prison of one's own ego and initiate a renewed awareness of the wondrous, salutary, and informing nature of the Other, the thing outside of the self. But my days at the cape were sometimes too much like days in my studio. I was consciously pursuing insight, the sort of thing that might come suddenly to the writer as well as to the reader at the end of a short story. What I trusted in, and what did not exist in my studio, was what Iñupiaq Eskimo acquaintances in Alaska call (in translation) "earth and the great weather." The earth is always in a state of enduring change—ice breaking up on the rivers in spring, caribou grazing on the tundra, red foxes preying on northern red-backed voles—but it gives us an image, or an illusion, of constancy. Passing through and over the land is the weather. Experiencing this fundamental dynamic, the enduring earth and the changing weather, and not the static interior of my rooms, was the right setting in my mind for getting out of myself and trying to address the questions I had.

The cape in its entirety, then, occasionally served as "the larger scheme of things" for me.

ONE NIGHT I woke to soft sounds. I listened hard for clear definition, but the disturbance was difficult to characterize, even when I came fully awake. These were sensations occurring between a sound and a feeling, but unmistakably there. I rose to my knees, pivoted, and slowly unzipped the front flap of the tent. Five Roosevelt elk. They stopped grazing, looked about, as if there might be other beings than me staring at them, and slowly walked on. I dressed and stepped out into the cool air. It wasn't too dark to follow them a ways through the recovering clearcut. I navigated by feeling my way along the gravel road they were following beneath a clear sky and its stars. The preternatural silence suggested something imminent, or that I was at a crossroads between my own and another world.

I thought about these mute elk, about the fact that it was "the middle of the night," and about the absence of light to navigate by. I could barely make out the shape of my tent with the truck parked next to it, fifty feet away. The two loomed together like some sort of space vehicle, to which I was attached by a tether. The salient thing in that moment, however, was how the elk had just appeared. And then disappeared.

A friend of mine, a physician, once spoke to me about an expedition he had been on which had taken him to see mountain gorillas at the Virunga volcanoes in Congo, orangutans in Borneo, Komodo dragons in Indonesia, and giant pandas in China, all living "in their natural habitat." When I related this story to another friend of mine, he pointed out that these animals were protected by armed guards, and that people came to see them every day in organized groups. "These are wild animals, yes," he said, "but they are not free animals."

I'd never made this distinction, but saw that he was right. The elk that night were free animals. There was no expectation that they might entertain anyone, and they were not exotic enough to draw crowds. Their temporal schedules were their own. They browsed the mountainsides, drank from the creeks, slept where they would, gave birth in seclusion, and made do with the clearcuts and roads, even with hunters. They did whatever they wanted, whenever they wished.

The poet Robinson Jeffers often explored the meaning of freedom, by which he meant not "freedom to do" but "freedom from." For him, being free from unnecessary interruption and from scrutiny was essential to the moral, psychological, and artistic development of a human being, and of humanity in general. He believed freedom was more important than equality, a conviction that caused some critics to characterize Jeffers as a misanthrope. What he meant by taking this stance, though, I believe, was that equality was contingent on freedom. He might have been willing to go so far as to say that for humans wishing to establish and maintain a social contract, the sine qua non of a stable human society, they had to understand that this was not possible except in the hands of fully mature people, people, in the modern idiom, who had "gotten over themselves."

Perhaps I'm in a minority, feeling that the instability of my own country is partly the result of its support of an adolescent's ideal—that people should be free to do whatever they want—and its obsession with personal gratification, whatever the cost. Lives without restraint are eventually ruinous, to those individuals and to the social and physical world around them. The hedge fund manager who amasses material wealth with no thought for the fate of the pensioner he cheats ruins lives. He is a kind of suicide bomber.

The political rhetoric in my country of recent years has included the notion that America is reviled by the citizens of other countries because these citizens resent (or are jealous of) our freedom. The disturbing thing about this naïve thought is that relatively few people in America are able to lead lives in which they are truly free. Like citizens in a dictatorship, to survive they must learn to toe the lines they've been urged to toe, though they have grown to believe that this kind of regimented, obeisant, bounded existence actually represents freedom. They—we, I suppose—cannot risk thinking otherwise.

I don't hold myself apart from those who seem to me to be living with a misapprehension about freedom; but I find it alarming when someone says, for example, that their laptop is designed to do exactly what they want, when it was in fact designed to function efficiently only when the operator does what the *machine* wants. Or that the cubicles so many enter every morning (with no guarantee that their spot will be there the next day) really represent the shape of what the people in them might want for themselves and their families. Or that unsolicited phone calls, random police snooping, invasive "easy listening" music in public spaces, unnecessary searches at security checkpoints, and political and commercial microtargeting programs made possible by Big Data represent welcome intrusions.

Jeffers hated this element of unconscious imprisonment in American culture and was marginalized as an artist for suggesting that such a thing existed.

The elk I saw that night no doubt have someone's drone waiting for them in their future, a curiosity machine for its owner's idle entertainment, groping for titillating moments in a neighbor's private life or looking for elk to poach.

ON A CLEAR SUMMER AFTERNOON, gazing across the settled surface of the ocean from a high vantage point, and considering the horizontal border that runs between the blue rim of the bowl of the sky and the dark, opaque plain of the water, looking into all that unthreatening space, I sometimes thought of leaving, of a journey out across that tabula rasa of unstructured space; and I felt the questing energy inherent in that image rising up. Numerous human traditions include allegories in which a hero departs his or her homeland for an unknown land, a quest for wealth beyond the horizon—the wealth the Grail symbolizes, personal material wealth, or the wealth that redounds to a community when a hero slays a threatening monster in the outlands.

How does one calculate real wealth today, confronted with that unparsed space? The questing heroes who came to the Americas from Europe and Russia, beginning in the fifteenth century, had no wealth in mind but material wealth. It was not other kinds of knowing that they were after, knowledge of the ways other humans dealt with things that were incomprehensible; and for most of them, geographical discoveries, though much was made of these later, were an afterthought. It was the possession of material wealth or of a potential source of that wealth that moved most of these explorers to action. And it was the legitimacy of such a quest, in the minds of many, that justified whatever was needed to succeed at this—theft, mayhem, dishonesty, genocide, parsimonious stockpiling. When denounced by people like Bartolomé de las Casas, the conquistadores said they worked for their god, and Sir Francis Drake held up his letter of marque.

Where in all of this do the lives of formerly free animals, like the elk that passed by that night, fit? How do elk object to the drone? How does anyone resist the many forms of daily invasion? If one intends to run, what is the destination?

THE CAMP I usually made on the cape was about 105 crow-fly miles south of the entrance to the Columbia River. The first Western trading post built there was erected on Chinook land in 1811 by representatives of John Jacob Astor's newly founded Pacific Fur Company. (His company had the distinction of building the first American fort west of the Rocky Mountains, six years after Lewis and Clark arrived at the mouth of the Columbia.) Astor's men made use of a Chinook trade network already in place, which extended from the river's mouth north and south up and down the Pacific coast and far inland to the east along the Columbia River. The commerce in furs was lucrative for Astor, but political jockeying between Britain and the United States for the right to nationalize these fur-rich lands and to control the trade in them forced Astor to relinquish his foothold in Astoria to the North West Company in 1813 (later to merge with the Hudson's Bay Company). The settlement at Astoria became Britain's Fort George, and Astor concentrated afterward on monopolizing the fur trade around the Great Lakes and on the upper Missouri River.

John Jacob Astor amassed a staggering fortune marketing the pelts of lynx, wolverine, bear, ermine, beaver, fox, mink, marten, fisher, river otter, and wolf. When he died in 1848, he was the wealthiest man in America. In 1893 two of his grandsons built a bank in New York City at 21 West 26th Street, from which they managed part of the family's business. My stepfather happened to buy this building in the late 1940s. In the late 1950s, when I was just entering my teens in New York, I could not get it out of my head how much wealth the Astors represented, wealth on a scale with the Carnegies and the Rockefellers. Thinking like a thirteen-year-old, I became convinced that the Astors had buried bags of silver dollars or stacks of banknotes in water- and vermin-proof boxes in the bowels of this building, where they would be overlooked by thieves more intent on cracking the door to a huge walk-in safe on the first floor.

What my stepfather might have thought of my obsession with these riches I can't guess. To humor me, though, and a friend of mine I invited to come along, he took us to the building's subbasement one Saturday morning when his offices were closed. I had already located earlier a spot in the subbasement's east wall where bricks were loose in the mortar. Working a few of them free and using a flashlight, I'd also discovered there was a crawl space running north and south behind the wall. My friend and I now worked more of the bricks loose and set them aside, creating a narrow portal through which we pulled ourselves.

The crawl space was two feet high and a little more than that wide. We could see to the south that the tunnel dead-ended twenty feet away, beneath the sidewalk on the near side of West 26th Street. To the north it extended another fifteen feet and then opened up into a larger space. We hunched and squirmed forward over gravel and a thick layer of fine dust until we could see a low-ceilinged cavern opening up to the west and north of us, to nearly the full width and depth of the building. The place was festooned with derelict spider webs and subdivided by piers of cross-laid bricks supporting the joists of the basement floor above us.

We searched every bit of this cramped place, pawing through dust that had sifted in over the decades and, here and there, digging down through the gravel with a garden trowel I'd brought.

My desire to pursue this quest for a part of Mr. Astor's fortune, which we did not find, had been inflamed, I think, by my growing awareness of America's voluminous folklore of fortune hunting, particularly in the West, and by my stepfather's position on the conquistadores, which was to be tolerant. Like many other boys, I believed that the wherewithal to lead a successful life required, primarily and absolutely, a fistful of banknotes. The abject narrowness of this vision, the incompleteness of the thought, didn't dawn on me. Nor did it occur to me that forty years later I would still be wondering what constituted real wealth—the "wherewithal"—in a campsite on the shores of the North Pacific, some miles down the Oregon coast from Astoria.

Whatever real wealth might be, it was not what Mr. Astor had accrued, selling the pelts of dead animals.

I CAN REMEMBER today standing in places where strangers' lives ended horribly and feeling my heart fall down at what I saw. At the age of seventeen, I saw the fields of white crosses and Stars of David blanketing the hills near Verdun. As an adult, I stood within the walls—by then they'd turned to rubble—of isolation cells in which criminals had been warehoused in the transport prison at Port Arthur, Tasmania. I had squeezed through the claustrophobic corridors of a windowless dungeon in the basement of Block 11 at Auschwitz, and stood in the gas chambers at Birkenau. I'd walked the blood ground at Bear River, Idaho, where more than three hundred Shoshone men, women, and children were raped, tortured, burned, and shot dead over two days by a bored and frustrated group of California volunteers, a militia whose commanding officer theorized that two Shoshone men, rumored to have beaten up a local white miner, might possibly be in that camp.

One July morning in 1999 I waded into Antietam Creek with one of my stepdaughters. We wanted to work out the details of a reconciliation ceremony we intended to stage that night on the Antietam Battlefield, a somber opening for a program of uplifting music and, we hoped, elevating words about environmental awareness and action.

On September 17, 1862, 23,000 men were killed, wounded, or went missing here in about twelve hours. The date is the bloodiest in North American history, a day of staggering carnage during the American Civil War. My stepdaughter Stephanie and I waded up the creek from its confluence with the Potomac River to where it crosses the battlefield, a mile upstream. In water that was sometimes up to our waists, we tried to imagine it as a creek of blood, as historians of the war have said it had been that day.

That evening we laid out a path of luminaria (votive candles placed inside white paper sacks), a path the same width as the creek. It descended from a grassy slope where people would soon be sitting, and ran ahead, curving, to a point just beyond a small stage. We asked people as they arrived to take the souvenir boxes of matches we handed them and to walk this simulacrum of the creek, lighting a few of the five hundred or so luminaria as they went. As they did, a young man standing alone by the stage played "Amazing Grace" on his fiddle.

The physical harm humans are capable of inflicting on others—the ease with which people are baited into these lethal hostilities—has the look of something that will not quit. The years of my adulthood are filled to bursting with this kind of mayhem. Papa Doc Duvalier with his Tontons Macoutes; Ferdinand Marcos, during his murderous years in the Philippines; Joaquín Guzmán in Mexico with his sicarios; Nicolae Ceaușescu in Romania; Hissène Habré in Chad; marauding militias like the Hutu Interahamwe and Joseph Kony's Lord's Resistance Army; and haranguing fundamentalist clerics in the Middle East, with their suicide bombers. Slaughter is routinely the resort of dictators, of warlords and drug lords, of fanatics, sociopaths, and those with offended egos, operating from Paraguay to Congo to Chechnya. In July 1995, Bosnian Serbs under Ratko Mladić killed more than 8,000 Bosnian Muslims in and around the town of Srebrenica, in the Drina Valley in northeastern Bosnia, the largest mass murder in Europe since the end of World War II.

We speak now of the human toll from what are collectively called the "oil wars" in Nigeria and Ecuador, and of the "wars" to come over water and fish. At some point, say those who are adding up the sources of available protein, the storage basins of clean freshwater, and the numbers of hungry, thirsty people, we will need to accept the fact that thousands more will perish every day from lack of food and water than are dying now. Or we will look into the face of what they say looms and discover whether we have the wisdom, the imagination, and the intelligence to dismantle the apparatuses that are leading us to this end.

THE SEVERED HEAD of the Chinook chief Concomly, Ranald MacDonald's grandfather, left Oregon in 1835 in a satchel with one Meredith Gairdner, a British physician and Hudson's Bay employee who dug up Concomly's corpse the night before he sailed for Honolulu. Stealing the heads of Indian people and passing them on to phrenologists had become something of a sport for certain white people, and that night a few wary Chinooks almost caught Gairdner at it. A couple of hours after Gairdner took the head, Chinook tribal members discovered a fine spray of blood on the ground around Concomly's disturbed grave site. They connected this immediately with Gairdner, who had pulmonary tuberculosis. (The blood came from his exhalations, from the exertion of his digging.) But they could not catch him before he boarded his departing ship.

When he died at about the age of sixty-five, Chief Concomly, a short, one-eyed man with dark skin and brown hair, was the spokesman for the confederacy of Lower Columbia tribes. He'd tried for twenty years, starting with Astor's Americans and then with HBC employees, to develop a system of equitable and peaceful trade between the tribes and white merchants. He was confounded and finally defeated by the concept of ownership and by the necessity for profit that underlay the whites' system of trade. And he could never fathom the reason for funneling profits to distant owners, people who were not part of the local economy.

Gairdner, intrigued by the pseudoscience of phrenology and curious about the reasons why Concomly had been chosen a chief, sent the head he stole to another physician, a friend in London named John Richardson, asking him for an opinion. Richardson soon moved on to other projects, however, and the head was apparently never closely examined.

It languished on a shelf at the Medical Museum of the Royal Naval Hospital at Haslar for almost a hundred years before being shipped to the Smithsonian Institution in Washington, D.C. In that time the skull's skin and hair had disappeared, the maxillary teeth were lost, and the lower jaw was misplaced during a cleanup following a night of aerial bombardment in London during World War II. The Chinook were eventually successful in retrieving the skull from the Smithsonian and seeing to its proper burial at the mouth of the Columbia.

Gairdner died soon after arriving in Honolulu. He was twenty-eight. He's buried in a modest, neatly maintained church cemetery in the city, and his prominent headstone bears a long inscription. He's praised for his Christian faith, his "vigorous mind," and his "pursuit" of a knowledge of "nature's workings." The paean ends by reminding us that he was "the fond object of a mother's ceaseless prayer."

It does not say that he was involved in the international trade of the heads of American Indians.

GAIRDNER'S INDIFFERENCE toward customs not in keeping with his own beliefs and ideals and the suspension of his own moral and ethical codes when dealing with nonwhites were, of course, emblematic of the times. His act is worth reconsidering today, however, for more than the barbarism it reveals. Cultural superiority, and the superiority that has historically been claimed by races, nations, and genders, has poisoned human relations for millennia. Gairdner's presumption that there was nothing wrong with his actions because he understood his effort as a conscientious attempt to advance human knowledge derives from a kind of staggering blindness, or obtuseness, that today we are paying a more staggering price for. The unprecedented age of international cooperation (as opposed to international commerce) that some imagine lies ahead cannot bear the burden of exceptionalism if it is to be realized.

Reading historians of the European exploration of the Pacific, especially of the explorers' contact with indigenous Pacific peoples, reveals a generally coarse and cavalier disregard for the mores of other cultures.

In the light of this, part of what I find admirable about Cook, again, is that whatever his failings might have been, he made strenuous efforts to understand, even to honor, societies that initially seemed inferior to him. He conceived of them as foreign but did not write them off as worthless.

Cook is often held out as someone who embodied all that was right about the Enlightenment—informed thinking, curiosity about the world, a commitment to the ideals of humanism. But he also, of course, represented the dark side of the Enlightenment, a belief that there was only one right way to govern, to organize one's economy, to worship God, and to think. All other ways were primitive (i.e., unenlightened), and those practicing them were assumed to be far behind on Progress's inexorable path. Non-European (and later on, non-American) people the world over were to be pitied (the compassion of the humanist); to be helped (i.e., converted to the Christian faith and educated in Western-style schools); pressured to reconfigure themselves into nuclear rather than extended families; and exhorted to become gainfully and permanently employed.

The idea that other theologies, economies, diets, arrangements of empirical knowledge, and forms of social organization might be better suited, or at least as well suited, to a people and a place was regarded as benighted. Many people, into the tens of millions, died as a direct result of their resistance to a European way of knowing the world. European nations, and later the United States, became so committed to notions of Progress and Improvement, became focused to such a degree on Development, so insistent on the legitimacy of theft from "lesser" peoples, so tolerant of legal concepts like *terra nullius* (indigenous people live on land to which they have no legal claim, lands that therefore legally belong to no one, so they can be claimed by Europeans without compensation), so dedicated to the imperative to profit, to reorganize, to purge, to quash, that the British, in their endless bickering with other imperial nations over the spoils of colonization, were able to create and then to perpetuate the idea of a Black Legend, their characterization of the Spanish invasion of the New World. (The English vilified Spain for its barbarism and Catholic proselytizing at the same time that Britain itself was becoming the greatest slave-trading nation the world has ever known. With the diminishing importance of slave-based economies at the turn of the eighteenth century, a reformed Britain was even able to represent itself as an avatar of abolition, if not emancipation.) When Britain set up concentration camps in Kenya in the 1950s, primarily to corral Kikuyu people and end their resistance to British rule, the world took Britain at its word and supported their effort to stifle native resistance, believing the British were engaged in a laudable and necessary fight against Mau Mau terrorism. Jomo Kenyatta, a political leader of the Kenyan resistance, was hauled off to Lokitaung on Kenya's desolate Northern Frontier, where he was imprisoned for years. And dissolute white settlers, primarily in the Kenya highlands outside Nairobi, were rallied to arm themselves and to hunt down rebellious blacks.

Britain condemned Spain, sought to exonerate itself as a slave-trading nation, and as meticulously set out in Caroline Elkins's *Imperial Reckoning*, sought to conceal the degree of its opposition to Kenyan independence.

IN CONVERSATIONS WITH different indigenous peoples over several decades, I've found few topics more sensitive than grave robbery, the outlander's seizure of someone's body and/or the things buried with it. There is more to rectifying a theft of this nature than the humiliating legal due process required of traditional people, the degrading bureaucratic tedium one must endure to recover a few bones, then to be placed in the hands of the deceased's descendants. The primary effort of most indigenous people is to prevent complete disintegration of the body of a relative in a profane place. The long fight of colonized people against desecration of their graves is pursued to ensure, by bringing the bones of their ancestors back into the circle of their traditions and thus protecting their elders, that their culture will not be eclipsed.

What traditional cultures must face when they defy their colonizers requires almost inconceivable strength and moral authority. In most cases, these activists remain undaunted. They understand what is really at stake. Oblivion.

They do not hold with any belief in cultural exceptionalism.

TODAY THE PRESERVATION of traditional cultures and their wisdom keepers is one of the most tenuous of human projects. Many traditional people see the attempt to prevent cultural disintegration, to resist incorporation into one or another dominant culture, as useless. Others believe it is better to pass away for who you are than to try to become someone you are not, recalling the words of the Oglala Sioux wisdom keeper Black Elk: "Sometimes I think it might have been better if we had stayed together and made them kill us all."

I wonder whether in my own travels I, too, of course, haven't unconsciously behaved in some way like a grave robber, given offense where I haven't meant to, assumed rights or privileges not mine to assume—at an Afghan dinner table one night in a restaurant in Bamyan, in a Warlpiri village in the Northern Territory, in an Inuit village on Baffin Island. By my simple presence in these situations I've brought with me the possibility of (further) disintegration. The fact that I haven't taken anyone's life, haven't set up a business that took advantage of people's naïveté, haven't seduced them with intoxicants or attempted to enlighten anyone about the strengths of my religion, all seem somewhat beside the point. The nature of the transgression in these situations isn't always clear. Often it's no more than this: the white guest does not see himself as a guest. He sees himself as an emissary. Even if he sees himself as only a well-intentioned visitor, he's prone to believe that, in the long run, he knows what's best, whether it's how to sharpen a knife, how to run a bodega, or how to worship the Divine.

For centuries American and Europeans have arrived in foreign lands as though sent by a superior god. Even avowed atheists bent on making business deals now arrive regularly in foreign lands with this attitude. It's an outward sign of their success as a "superior" culture.

It kills people.

ON MARCH 25, 1960, a 136-foot refrigerated freighter with sixty-six people aboard, the *Western Trader*, sailed under the Aurora Bridge between Seattle's Lake Union and the open water in Puget Sound and set a course for Wreck Bay, in the Galápagos Islands. The passengers were all members of the Island Development Company, a cooperative intent on establishing an American colony in Ecuador's Galapagean archipelago. As a group, the colonists were idealistic and enthusiastic individuals, a mix of nuclear families and single men, many of them unaware that Ecuador already had its own plans for the archipelago, thank you. They'd left Seattle believing that Ecuadorian authorities would be delighted to have them there—developing a fishing cooperative, applying their (largely nonexistent) farming skills, and generally improving the lot of a small population of Ecuadorians at Wreck Bay, who were living, the colonists assumed, an impoverished life in a dilapidated village. They were prepared to show the villagers how to develop more productive lives.

Initially wary of each other, the ship's complement were galvanized in friendship during a terrific storm that battered the *Western Trader* off the Oregon and California coasts between March 30 and April 2. When the vessel put in at San Pedro, the Port of Los Angeles, to pick up visas, the Coast Guard ordered the *Western Trader* refitted to make it more seaworthy and also required that the ship undergo extensive repairs. The Coast Guard restricted as well the number of colonists who could reboard the vessel for the continuation of the voyage. The long delay in obtaining their visas, and the Coast Guard's lack of confidence in the *Western Trader,* unsettled some of the passengers. Shouldn't a project of this size be going more smoothly? they asked.

When the *Western Trader* finally cleared the port at San Pedro, the colonists aboard were in restored high spirits. Many were skilled laborers with backgrounds in a variety of trades, and all of them were willing workers. They agreed that there were some philosophical differences they would have to deal with in Galápagos, but exactly how the workload was to be shared and how the profits from farming and fishing were to be divided, and who had the authority to make certain decisions with regard to the ship or a now-derelict refrigeration unit that had previously been built at Wreck Bay, or what was to be done about housing—all this they'd already taken care of with a set of written guidelines. Other decisions, such as how much of their profit would go toward the improvement of the lives of local Ecuadorians, the group would address after they'd made an assessment.

News that legal title to the 64,000-acre coffee plantation on Isla San Cristóbal that they planned to revitalize was encumbered (which reached them in Los Angeles) was, for the time being, a setback. But the colonists were not discouraged. They simply rededicated themselves to achieving the larger goal—making the colony economically sustainable by capturing and freezing lobsters for export to the Ecuadorian mainland. They'd been forced to leave behind in San Pedro the equipment needed to run the freezer in Wreck Bay properly but were confident they could cobble something together until they could retrieve the machinery.

On the morning of August 19, 1960, the captain of the *Western Trader* found the dock at Wreck Bay too dilapidated to accommodate the ship. The ramshackle village they expected to see was actually a settlement in good shape; and the handful of needy peasants they'd imagined greeting were actually about a thousand relatively happy Ecuadorians. Welcoming and polite, the Galapageans were puzzled by the notion that their government thought they needed help. And when the refrigeration engineer came to report to the others on the condition of the freezer plant they expected to take over, he said it was beyond repair.

It took four or five days for the Arcadian dream of the Island Development Company to fall apart completely. It is not clear precisely what sabotaged the colonists' plan to establish a mid-Pacific utopia, dispensing the fruit of its labor to local residents and leading exemplary lives guided by egalitarian principles; but the endemic corruption of Ecuador's government, and a lack of full disclosure on the part of a couple of Ecuadorian businessmen negotiating with the American organizer of the expedition, seemed a good place to start—if one didn't concede that the plan was daft to begin with.

One of the colonists, a young man of nineteen at the time, later wrote that the enterprise was "a flight of mad fancy that captured the imagination at the cost of ignoring too many realities," such as the distance between the Galápagos colony and potential Ecuadorian markets for their lobsters.

I met with this young colonist one afternoon some fifty years later, at his home in Redmond, Oregon, and listened to him describe what he'd gotten himself into all those years ago. He'd taken many lessons away from the experience, he said—he himself had stayed on in the Galápagos, working at odd jobs, before moving to mainland Ecuador to work for a while—but the lesson he most wanted to discuss in detail with me was the group's belief that they would be helping out the Ecuadorian community of Puerto Baquerizo Moreno at Wreck Bay. He said his companions had so convinced themselves of the worthiness of this idea, were so imbued with the notion that they were performing honorable work in the world, that achieving one of their goals, to "vastly improve" the lives of the Galapageans, was for them simply a given. The reality—that the Galapageans were doing fine, better than the colonists, actually—came as such a shock that the dream of being benefactors in a remote part of the world fell apart for most of them in a matter of hours—and some of the colonists had invested their life's savings in the venture.

The man I interviewed, Stan Bettis, said the impact of the realization that they had been foolish, that they were the victims of a confidence game that depended on their innocence about the world beyond their home borders, was staggering. "We weren't going to improve anyone's life, and we were faced with the wreckage of our own," he said.

The idealistic vision of designing a human community in complete harmony with a place, of a group of people coming together to deal fairly and lovingly with one another, through good times and hard, is an aspiration deeply embedded in Western thinking about our own future. It accounts for the magnetism and popularity, at different points in our history, of concepts like the Christian's Isles of the Blest, Francis Bacon's New Atlantis, the sailor's Fiddler's Green, and James Hilton's *Lost Horizon,* with its story of the Tibetan elysium of Shangri-La, evocations of irenic sanctuaries far from the trials of human life, places without violence or greed, Edens where no one covets or disrespects. A belief in the existence of such places tinged the thinking and shaped the expectations of many explorers poring over nautical charts, in an era when considerable white space was still left to the imagination. When Cook put an end to the notion that a habitable continent might exist in the Southern Ocean, he nevertheless left a blank space behind on late eighteenth-century maps. If there actually *was* a continent within the ambit of his circumnavigation of the ice barrier he had found in all longitudes, he said it was a frigid and inhospitable place. It was; but fifty years later an Estonian explorer named Thaddeus von Bellinghausen discovered its shore, and the filling in of Earth's last large blank spot began.

AFTER COOK, we were able to locate the coordinates of our position on Earth dependably with sets of numbers—latitudes and longitudes. They gave us a sensation of precision and irrefutability, and they glow today on the screens of our handheld GPS units. We no longer define our positions by the contours of the land we stand on, the texture of the soil, the colors and density of the vegetation, the rete of gravity-driven water in rills, brooks, and rivers. We fully embrace the coordinates Cook gave us, but in everyday use they represent a kind of Esperanto.

After Cook, there were far fewer places left for men to imagine they could run off to. Popular misconceptions about life on islands in the South Pacific, which Cook inadvertently encouraged, would neverthe-less remain viable for many decades, well into the present, the idea of escaping the disappointments and burdens of quotidian life by aban-doning home and sailing for one of the Pacific's tropical destinations. (Gauguin went there to paint, Robert Louis Stevenson to write.) Cook's voyages widened the gap between rational thinkers, on the one hand, and mystics like the metaphysical poets, working with another sort of geography; and the promise of landscapes or situations that might pro-vide humanity with everlasting relief from its troubled dreams began to fade from the Western imagination. After Cook, humanity was forced to accept the absence of a boundary beyond which the prospects for humanity would surely improve, though that tradition lives on with the quest to put people on the moon and to explore the outer planets.

Humanity was left with an almost completely explored planet, faced with seemingly intractable moral and social problems, but, importantly, with a not yet fully tested imagination.

William Blake, prominently in Western history, wanted to rid the human imagination of a particular kind of darkness, the darkness that leads to despair, to hatred and war, by opening it wider to both the real and the numinous dimensions of the world. He wanted humanity to realize the immeasurable breadth of the human imagination, its capacity to rise above fatal despair, even as the world grew darker at the beginning of the Industrial Revolution.

Somewhere, Camus wrote: "The world is beautiful, and outside it there is no salvation."

Blake and Camus were asking us to set aside our cherished illusions and to engage instead with the problems they both saw coming.

ONE WARM AUGUST AFTERNOON I was trying to comprehend a technical article about the refraction of light, trying to get around the dense mathematics in it and arrive at the heart of the miracle. I was seated on the hood of my truck, leaning back against the windshield for support. The ocean lay off to the left in its silence. I had my binoculars beside me, to fetch birds that were occasionally passing through, far away; but for some minutes my attention had really been fixed on the dark line of a spruce forest, a familiar sight to me, a mile or so to the north. I knew this forest, having driven past it often over the years. It had always seemed to me impenetrable. The trees grew so close to one another that the interior of the wood appeared entirely closed to sunlight. And because so many dead limbs lay jackstrawed between them, the trees presented the visitor with a formidable navigational maze. A dirt track, a skid road, separated this sprawling forest from a recovering clearcut to the south. The wood was a thing apart.

I set the article about light aside, put some things in a rucksack, locked the truck, and took the road from the landing downhill. I briefly lost sight of the forest walking through the steep valley below my camp, regaining it when I reached a ridge. The same logging road led me directly to its edge, a ruler-straight perimeter, like the stockaded wall of a fort.

Fifty feet or so after I entered the wood, I turned around to mark the scene I'd just left. I could still see the clearcut beyond the picket line of trees lit up by a cloudless sky. I was viewing it now from a shadowed place, like the entrance to a cave. Wild animals, predators as well as their prey, often travel this way, slightly inland from light falling on a clearing, from an exposure that might compromise them. I went in deeper, glancing back occasionally, fearful of losing the scene, which grew less unified as I walked on. If I lost it, how would I get back? Which direction would I go?

At some point my anxiety increased noticeably. I'd reached an emotional limit. I could barely discern now the sunlit landscape behind me. The light appeared as tiny rays, like stars in a night sky or the sun viewed through a colander. All around me was shadow, a dimness with no flux to it to indicate the source of light. I sat down at the base of a tree, facing the clearcut, perhaps two or three hundred yards away. I felt obscured here, unknown to the world. Above me was only dimness, no shred of sky. The dark enveloped me like a gas. I could still see where I should go if I felt the onset of true terror, which might easily emerge from this untenanted wood, triggered by the surfacing of vivid memories of threat, of menace and violence, recalled from unsettling movies. I tried to prevent those images from taking shape in my conscious mind, from turning me back toward the faint wall of distant sunlight.

Eventually I found a calmness that let me remain braced against the same tree but this time facing in the opposite direction, into the deeper dark, a place I knew was Earthly but which I imagined to be like the darkness at the edge of the expanding universe, the unbounded Empty Nothing that the universe was hurling itself into at the speed of light.

I strained to hear any sound, but detected none. This deep in the wood, there was no hint of movement. And then, without warning, I encountered anguish. The wall of darkness in front of me seemed suddenly dense with the kind of agony and despair I perused nearly every day in international reports about the fate of nameless thousands of the defeated, in the Horn of Africa, in South Sudan, and in Syria, their lives falling like black snow in some bleak corner of the world, these lives made expendable by the indifference of those who've never heard of them, or who, if they have, have looked away. In the gloom before me I saw the vast terrain of the defenseless, murdered.

I had wanted, I suppose, to frighten myself by walking so far into the forest. Instead, I let my head drop to my chest and felt impotent compassion, the weight of the horror we force on one another in our manic quests for greater satisfaction.

I left this spot, a watchtower from which I had looked into an absence of light, a space seemingly infinite in every direction, and step by step reentered a world of differentiated objects, conscious again of the force of gravity, pulling my booted feet back to the soft duff of the forest floor. Moment by moment, I waded into the growing light. Whatever might have compelled me to run from the darkness behind me, to run from whatever ghouls it might have harbored, was no longer present. I felt weirdly cleansed of the ancient cowardice that causes us—for reasons of self-preservation, they say—to turn away from the suffering of strangers.

I came into the clearing unable to feel any indignation about the butchery apparent in the clearcut before me, unable to condemn anyone for the failure of spirit that led them to feeling indifferent where the life of a stranger was concerned. I wanted only the flow of water down my throat, and to return home, not caring how long this might take, not fearing nightfall with its obliterations and demons.

I COULD SEE CHANGES in the skies to the north at dusk that told me the storm was now on the doorstep. I made another cup of coffee and my supper and checked the tension in the guylines anchoring the rain fly over my tent. I could tell from the flocking behavior of some of the birds, or so I thought, coming in off the ocean or headed in larger groups than usual for shelter, that the storm was only hours off.

ONE BRIGHT MARCH DAY I sat at the edge of cliffs that mark the foot of Cape Foulweather. I had spread open on my lap a sketch map of Cook's 1778 approach. I tried to impose it on the ocean before me. No whitecaps this afternoon, as there had been for him, just the sea's hard metallic scales, rippling like chain mail over the shoulders of shallow rollers arriving in sets of five from the west. Cook approached that morning to within about ten miles, according to the research behind this sketch, then crossed back over his own wake and stood out to sea for the night. After several days of approaching the coast and retreating because of heavy weather, Cook departed, bearing away to the southwest with his consort, HMS *Discovery,* and altogether some two hundred officers and sailors, having had a glimpse of the last temperate rain forest on the planet to be discovered. Ashore, perhaps a few Alsean hunters studied the thing that had come so close during those four days, perhaps as close as three miles. They wouldn't have been able to make out the two barefoot sailors standing at the head of the lower mainmast, straining to read the near-shore waters for possible reefs. Like the other ratings aboard, these men would have been wearing short breeches tied below the knee—petticoat trousers—and short double-breasted woolen pea jackets over collarless shirts and a red waistcoat. Around their necks, black silk scarves; on their heads, brimless black three-cornered hats. On the right foot of one, perhaps, was the tattoo of a rabbit, on his left foot a rooster, charms to keep him from drowning. What the Alsean hunters would have appreciated most about the two men was the unbroken intensity of their scrutiny of the near-shore waters.

A "canoe" of this size would not have been incomprehensible to the Indians. What would have seemed unusual was the absence of paddles and the spider-web maze of lines about the main deck—braces, stays, clew lines, lifts, tacks, nave lines, leech lines, halyards, and sheets. And the towering masts, with a dozen canvas sails full of wind. And the single man at the double helm, seemingly alone in controlling the direction of travel. And the clear glass windowpanes of Cook's quarters, above the wood transom in the stern.

Or maybe, like Aborigines seen walking along the beach as Cook approached the eastern shore of Australia in April 1770, they glanced only briefly to seaward before returning to the pace and importance of what they were doing, paying the ship no further mind.

Twenty-seven years after Cook's landfall at the cape, Lewis and Clark would arrive at the mouth of the Columbia. Nineteen years after that, Ranald MacDonald would be born at the river's mouth, on the south bank.

THE STORM ARRIVES late in the evening. It batters the tent with pellets of rain the size of honeybees. The wind blows hard through the night and into the following afternoon, yanking at the crowns of trees, hurling rain in sheets I can hear breaking over the metal body of my truck. The following afternoon the wind tapers off to squalls crashing through, and to spiraling vacuums of mist that cause my inner ears to ache. The sun sets behind aubergine clouds and darkness overcomes the clearing, absorbing the truck and the tent huddled in its lee. No stars are visible. I pass another damp night. In the morning I pin a few things to a line, which begin to dry in the pale light. Maybe they actually will. Before the day is out, I will have hiked down into one of the valleys of the Siletz and returned from the enlivened creek flowing there. Even in this logged-over landscape, soaked and gleaming, contradicting the apparent desolation of the clearcut, where stillness now accompanies the silence, I can imagine something like the original creation, however mythic that thought might be. Or the blueprint of another creation, unknown and unplanned.

The wildness around me here, the clearing where I camp and the stands of undisturbed old-growth Sitka spruce beyond, within which the brightest light at midday is still only crepuscular, is not a point of arrival for me. It is my point of departure.

Skraeling Island

Mouth of Alexandra Fjord

East Coast of Ellesmere Island

Nunavut

Canada

78°54'02" N 75°36'39" W

'm forty-two years old and have not been back in the Canadian High Arctic for six years. I've missed it, the look of this country in summer especially; but it has not missed me, of course, an occasional visitor, someone inspired by its line and color, by its immensity. If the harassments of everyday life, the emotional tangle and constricted spaces we all must navigate daily take a toll, here is where I'd come to revive, to wash out my clothes. This quiet landscape, the tundra under my thighs at this moment, indifferent as it might be to the presence of a squatter, is nevertheless heartening. It responds to my inquiring hand, to my scrutinizing eye, my flâneur's search for nothing in particular—or it will, if I demonstrate by my gestures a degree of respect, a capacity for wonder. This, at any rate, is the belief that guides me here, a belief that the physical land—a broadly encompassing term—is sentient and responsive, as informed by its own memory as it is by the weather, and offering within the obvious, the tenuous.

I've come back to the High Arctic to learn more than I was able to grasp on earlier trips, to recall things I'd forgotten, and to experience again the templates here that go deeper than I'm able to comprehend.

On this particular morning in late July I'm glancing up from a sympathetic novel about Claude Monet, *Light,* by Eva Figes, to take in my surroundings. According to orthodox geographers, I'm reading on one of the planet's outermost terrestrial edges, the flanks of a bare-boned island situated at the mouth of Alexandra Fjord; the fjord is positioned at the bottom of Buchanan Bay; the bay is a major inlet on the northeast coast of Ellesmere Island, the northernmost of the Queen Elizabeth Islands, an archipelago north and east of the North American mainland. From this island, called Skraeling, 660 nautical miles from the north geographic pole, I can see the ice wall of West Greenland standing like a palisade on the eastern horizon.

No one lives on Skraeling now. Eight hundred years ago it was seasonally occupied by a Paleoeskimo people known today as Thule (TOOL-ee). No one knows what they called the island.

My head is positioned higher than my feet in this slightly inclined cleft in the ground I've settled into, a shallow notch on a rocky slope

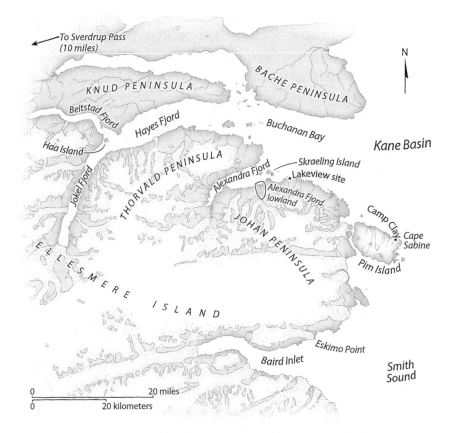

Johan Peninsula Area

tilted toward the water. I've gotten out of the way of a fitful breeze here to read Figes's *Light* and to annotate my notes about the Thule, whose abandoned houses and hearths I've come to Skraeling to see. The depression I'm in feels like a snug room in which to work for a while. Stretching outward from this place, in all directions, are the barren reaches of a polar desert. Extensive loose pack ice floats nearby in the bay. Beyond that, away to the east, lies the mostly frozen sea of Smith Sound. A lair like this makes it easy for me to concentrate, both on Ms. Figes's curiosity about Monet's *Nymphéas,* at this point in her narrative, and on the indefinite lives of the Thule, in whom I have a pressing interest.

I read and write for twenty minutes or so before I'm pulled up and out of my concentration toward movement in the land, and out into the land's bright colors. I watch for a while—the radiant light on high, orange cliffs opposite me, a flock of long-tailed ducks rifling by—and then return to my pages. A granite boulder supports my upper back. A cushion of tundra plants, composed mostly of dwarf willow and a type of wild rose called mountain avens, springs my hams above the frozen earth. I was here a good while this morning before I noticed light from the advancing sun starting to gleam on something within a thicket of willow stems beside me. I stared at the glint. I wormed my fingers down into the space. Debitage. The lithic debris a knapper creates when he's fashioning a stone tool.

Human spoor.

Once someone else had rested exactly here, with a different sort of work before him than my absorption in the behavior of light at Giverny and in academic speculation about the migration of people long gone.

The gleam of these gray chert flakes alters my frame of mind, as abruptly as if a stranger standing nearby in the stillness had suddenly cleared his throat. I set the book aside and reach for my binoculars, without needing to search for where they are. Suddenly I am hunting: What's out there? Just beyond my feet, the notch I'm lying in opens onto a stretch of dark seawater, a passage between Skraeling Island and the shore of Johan Peninsula opposite, a shore that forms the northern border of a thermal oasis called the Alexandra Fjord lowland. This park-like expanse of a few square miles is boxed in on two sides by talus slopes and high cliffs. At the far end, the double snout of a glacier abuts this haven for arctic plants and small creatures.

The water between me and the peninsula is almost entirely free of ice at this time of year. I sweep the dark water slowly, searching its surface methodically for the spectral white forms of beluga whales cruising just below. I look for a disturbance in the sunlight's sheen that might reveal where a ringed seal has come up briefly from its fish-hunting world to breathe before disappearing again. I parse the entire rippling plain of water bare-eyed, putting the binoculars quickly onto any anomaly.

I look about for common and king eiders, sea ducks I know nest on this island. They pass frequently, flying just off the water in long single-file lines. I watch for walrus, whose bellowing I heard an hour ago as I approached this spot from a campsite I share with three others. I watch this world from my sheltered niche unseen, as my shadow companion, the knapper, must have; but where I'm interested in the behavior and ecology of these animals, the look of them, the sensation of sudden contact with wild life, he would have also been interested in killing some of them, in eating and using parts of them to make his way in the world. If he hadn't been, if he wasn't alert to signs of their presence in the water while he sat here making a tool, if he didn't drain every bit of meaning from whatever came to his eye, his children would not eat. They wouldn't feast on eider eggs or consume the dark flank meat of ringed seals. He would not have the haunches and the backstrap meat of walrus and bearded seal to cache, his insurance against winter's throttling of all life in this place. He would not have oil from the beluga's blubber to light the interior of his half-sunken winter home, to warm it and take the hardness out of his frozen food stores.

It's different now. He's gone. His culture is a broken prayer wheel at your feet. Who can translate now what was once inscribed on its handle? His companions might conceivably have been the final expression on this island of Dorset culture, a people not destined to survive the arrival of the Thule, coming in from the west eight hundred years ago.

I can feel him here beside me, hear the plosive knapping of flakes from the mother stone, see him pausing to investigate the summer water, gazing into the volume of air above, studying the lush land across the way, where maybe he has seen wolf spiders hunting and Arctic blue butterflies feeding on highland saxifrage, and as a boy chased after flocks of purple sandpipers combing the wet meadows for food. Or maybe it was no Late Dorset man who nestled here, but instead a Thule person, whose people arrived in the vicinity better equipped for survival, with their walrus-skin boats, dogs of burden, and small sleds. To speculate productively about who left these flakes here, or what the knapper might have been fashioning, or whether men from both cultures might have chosen this same hiding place in which to work or from which to observe the movement of animals undetected, I will have to ask my companions. They reflect continually on the artifacts, the architecture, the remnants of human life they've found here on Skraeling, a place of seemingly insurmountable challenges for humans as summer gave way to winter hundreds of years ago. Neither Dorset nor Thule had much more to ensure their survival than animal skins, bits of plant fiber, animal bones, sinew, walrus ivory, animal fat, rocks suited to building, the occasional piece of driftwood. The archeologists I'm with know the many uses to which these things were put, understand the ingenuity behind the crafting, and view it all as evidence of the human determination to thrive, not to perish.

Missing from the scraps of skin clothing we have found here, from our sorting trays of bone implements and stone tools, from the broken harpoon tips and stone lamps that have turned up, is any indisputable evidence of how either Dorset or Thule people cooperated with each other here to ensure survival, of their approaches to parenting, of their psychological arrangements with the spiritual character of the land, their ceremonial life, their enduring narratives. This tantalizing evidence of what once had allowed these people to prevail in the face of great difficulty isn't waiting to be found. It wasn't destroyed, stolen, or lost. It's evaporated. We who wish to understand them no longer have it to work with.

I'D ARRIVED ON Ellesmere Island a week before. I flew in from Resolute, an Inuit settlement on Cornwallis Island, four hundred miles south and west of Alexandra Fjord. (At the time, Resolute enjoyed twice-weekly scheduled air service to Yellowknife, a thousand miles to the south, the capital of the Northwest Territories and the town from which I'd reached Resolute.) The plane that brought me the rest of the way here, a chartered Twin Otter, put down on the lowland across the way, beside a decommissioned Royal Canadian Mounted Police (RCMP) post. The only passenger on that flight, I'd shared the cramped interior with steel drums of aviation gas and helicopter fuel, boxes of fresh vegetables, bundles of shovels and ice chisels, packets of mail, and pallets of spare parts—an omnibus resupply mission, assembled to service several outcamps on the same day.

The Otter was chartered by Canada's Polar Continental Shelf Program (PCSP), which deploys scientists across the Canadian High Arctic every summer to conduct research. Alexandra Fjord was its first stop that morning. An archeologist from the University of Calgary, Peter Schledermann, had invited me to join him and two colleagues during the final phase of their field season on Johan Peninsula and Skraeling Island.

The landing strip at Alexandra Fjord is of a type locally called "make it or die," which may be something of an overstatement, but the strip *is* dramatically short and, midway along, it crosses a ridge. Once the plane touches down, you lose sight of the farther half of the runway. I've landed in more challenging places, but always savor the relief that comes with making a good landing in *any* hard place, especially in a plane loaded to its weight limit and carrying six drums of flammable fuel.

We touched down in calm air, under clear skies. If the weather had been dicey—cross winds, fog, sleet—the pilot might have passed up the landing and I would have ended up sometime later that day back in Resolute, waiting to hitch a ride on the next flight to Alexandra Fjord, which would be days later.

THE CANADIAN GOVERNMENT opened the RCMP post at Alexandra Fjord on August 8, 1953. It was meant to replace one set up across Buchanan Bay fjord on nearby Bache Peninsula, in 1926. Several nations back then, seeking mineral rights in this remote and rarely patrolled corner of Earth, had persisted in making vague references to "prior claims" in this part of the Queen Elizabeth Islands. Canada wanted to send a message to them by establishing a permanent official presence here on its international border. The resident mounties serving at Alexandra Fjord after 1953 were also there to inform Greenlandic Eskimos around Smith Sound—Inughuit—that they could no longer cross the forty miles of sea ice separating western Greenland from Ellesmere Island for most of the year to hunt, though they had been doing this since long before there was a Canada.

In 1962 the personnel at Alexandra Fjord were redeployed to Grise Fjord, a new RCMP post at an Inuit settlement on the south coast of Ellesmere Island. On the day I arrived, the still well-maintained RCMP buildings at Alexandra Fjord were occupied instead by a painter named Eli Bornstein and a photographer named Hans Dommasch. (PCSP had taken over the old RCMP post, which now occasionally supports the work of artists and writers billeted there.) Eli and Hans helped unload the freight meant for this camp and for the Schledermann camp. We all three waved goodbye as the Twin Otter hurtled off the strip and banked away to the north—and the great breadth of silence around us was restored.

Eli and Hans were delighted to see the Figes book I'd brought, which had just been published, and I was greatly relieved finally to be free of the hurly-burly of Resolute. Diesel generators hammering away all night, trucks and heavy equipment raising clouds of dust from the unsealed roads, air fouled with the odor of combusted fuel. A world of clocks and maintenance schedules, where people either showed up "on time" to dine in the PCSP galley or didn't eat, and where the offices one had to call at for weather reports, equipment requisitions, logistical support, and revamped aircraft schedules opened and closed precisely on the hour, in a place where the summer sun never set, half the population worked odd hours, and the weather followed a schedule of its own.

Eli made tea for himself and me and Hans went to bed. He'd been out for hours the "night" before the plane arrived, photographing icebergs grounded in Buchanan Bay. I stored my duffel and backpack in a corner of the dayroom, shared a few stories with Eli, and decided to go for a walk. In doing so, I was ignoring RCMP directives posted on the wall, warning visitors never to travel alone and to always carry a gun to protect against polar bears. I did pack food, water, survival gear, a radio, a first-aid kit, etc., so was not completely irresponsible.

Eli said he'd radio Peter, who was camped just a few miles to the east on Johan Peninsula, and let him know I was here and ready to join him in a few days when he moved his camp to Skraeling Island.

The RCMP post—a single-story headquarters building, flanked by four or five outbuildings, each painted white with red trim—is built on an old beach strand, a dozen or so feet above the fjord's high tide line.

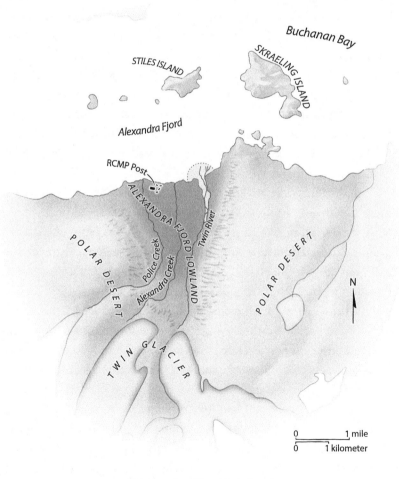

Alexandra Fjord Lowland

To the east, ocherous cliffs rise abruptly from an open plain to form a mesa about 2,200 feet high; to the south, Twin Glacier's two lobes debouch onto the lowland's gently sloping valley. To the west, vertical cliffs of gray gneiss and granite, less severely pitched, rise to meet the xeric landscape of a polar desert. The fjord lies to the north, with Skraeling about two miles offshore to the northeast. Next to Skraeling is a smaller island, Stiles, known locally as the Sphinx for its resemblance to the famous stone figure at Giza. The Sphinx: a lion with a human head. An Egyptian monument so named by Europeans for its resemblance to a winged monster in Greek mythology, one that lived in Thebes and was outwitted by Oedipus. The appellation seems bizarre here, someone's effort to diminish and subjugate the unknown, like putting a party hat on a dog.

From the cobble shore in front of the RCMP buildings, with my back to the great basin that cradles the biological oasis of the lowland, I face dozens of square miles of high, spare land on the other side of the ice-strewn water: the mesa-like Thorvald Peninsula on my left, and on the right, emerging from behind it and farther off, the larger Bache Peninsula. One feels the extent of empty land beyond the opaque shield of the water, and senses the threshold of one's own disappearance.

I turn inland, a small figure walking south into the U-shaped sanctum of the lowland. To reach the foot of the east lobe of Twin Glacier, which I plan to climb, I cross Alexandra Creek on a makeshift bridge and continue on until I reach the west bank of the Twin River. At virtually every moment the entirety of the valley is visible to me, a man standing there like a sparrow on the floor of a cathedral with its roof gone.

The air feels heated and smells of stimulated plant life, of rot and pollination. Given the desolation of the surrounding cliffs and the empty expanse of polar desert beyond, the soft wash of redolent air across my face is stimulating to the point of being erotic. The sensation surges through my body as I bend down to examine tiny flowers of purple saxifrage, to feel the knotty bark of thousand-year-old mountain avens shrubs, stems no thicker than my little finger. The hum of a polar bumblebee (*iguptaq,* in Inuktitut) hovering here, the fugitive odor of sap, the mobbing of blossoms by insects, is not unexpected, but the intensity of it brings me to a halt every few minutes.

I stop frequently to examine the evidence of lives lived out—the white rib of an Arctic hare, a dead collared lemming, still warm to the touch. No apparent wound, no clot of dark blood at the ear, the nose, the lips. Old age, perhaps. In soft soil at the edge of a sedge meadow, where wet ground grades into heath, I see the fresh, elongated toe marks of red knots and Baird's sandpipers, all now hunting thirty feet in front of me, a distance they continue to keep as I approach.

When I reach the river—I should have asked Eli about this—I find the water too deep for my calf-high rubber boots. I stuff the boots, my socks, and my trousers into my pack and step into the rushing meltwater, hoping to negotiate the rocky, uneven bottom thirty or forty feet to the opposite bank barefoot without falling. When I step out of the icy water on the other side, I begin to bark, doubling over like a wounded animal—the body's outrage—before putting my clothes back on.

At the foot of the glacier I squat down in order to hear more loudly the percolation of meltwater from the glacier's cold lip, the hissing bursts of air as small pockets within the ice release their stores of ancient atmosphere. With my head tilted this close, I can feel the glacier's frigid exhalation against my cheekbones. For a few moments the density of the silence in the valley exists in concert with the continuous sound of this huge object's meltdown.

After climbing several hundred feet up the incline of the glacier's tongue, I step over a cascading stream of water only a few inches deep, turquoise against the white ice, a white so intense I can't continue looking at it. This rill, five or six inches wide, runs so swiftly through the banked turns of its track that the ribbon of water seems almost to turn over on itself, like a Möbius strip. I climb higher, following the flow of meltwater until it diverges into a maze of rivulets, the headwaters of the rill. In little more than a month from now, all will again be quiet here. And the silence will extend from the glacier out across the whitened lowland and out over the fjord's thickening sea ice.

On my meander back to the north I follow the river. In places where one environment (the mesic tundra) meets another (the flowing river), what biologists call an ecotone exists. Some biological events, like evolutionary change in a particular species of animal, are more likely to be apparent here, where two different environments overlap. Halfway across the lowland, I bear off north and east toward the base of a scree slope at the bottom of the cliffs. I leave my rubber boots sitting on a prominent boulder, an erratic left in a colder era by the glacier behind me, and switch to a pair of light boots. The rubber boots will be hard to miss on the way back, if I remember to look for them.

The scramble to the top of the cliffs takes less time than I thought it would—up a slope of loose talus, across a bay of rock rubble, then about six hundred feet of easy clambering to reach the tableland. I'd expected an arduous climb, but had not considered how having been immersed in a sublime landscape like the one below would invigorate my enthusiasm to explore.

Crossing the lowland from the glacier to the foot of the cliffs, I'd stopped to examine nearly everything that caught my eye. I frequently felt the urge to glass the plain around me with my binoculars, looking for some hint—hardly a chance of this, but it seemed unmindful not to search—some sign of muskoxen or of the diminutive Peary caribou. Setting off again, I sometimes threw my arms up involuntarily, an expression of incomprehension that the world here in the middle of a polar desert could be so intensely alive, so elegant. The emotions born of what I was taking in, all the filaments of creation, peaked in feelings of tenderness toward everything here, a vulnerability to this life.

To my way of thinking the lowland was beautiful.

MANY YEARS BEFORE I traveled to the Alexandra lowland, I asked a professor of mine in graduate school, Barre Toelken, about a Navajo ceremony called Beautyway and about the Navajo concept of *hózhǫ*. *Hózhǫ,* a complex idea, is often loosely translated into English as "beauty," but the word refers as well to a state of harmony that pervades the world, and to a general state that in English means "to be in good health." Barre, who lived for a while on the Navajo reservation near Blanding, Utah, in the 1950s, and who still had family there, directed me to a book called *Navajo Blessingway Singer: The Autobiography of Frank Mitchell, 1881–1967,* and he put me in touch with an anthropologist friend, Gary Witherspoon, fluent in Navajo.

In my understanding, which is imperfect, Beautyway rites are conducted over a period of several days by a medicine person called a singer, in the home (usually a traditional hogan) of a "patient," with his or her family present. The patient is referred to as "the one sung over" and is conceived of as someone who has "deteriorated" or is otherwise in a state of spiritual imperfection. The Navajo way to view this state of deterioration or incomplete integration with the world is to regard it as normal, a condition that develops over time in every person. (The gradual, inevitable loss of coherence in a complexly organized Navajo system might usefully be compared with entropy in the second law of thermodynamics, as Clausius defined it.)

Restoring a person to a state of "beauty" requires that the singer "make it incumbent upon the universe" to re-create in the *patient* those conditions in the natural world that signify—for Navajo people— coherence or harmony. The rites of Beautyway honor those conditions, and this state of accord is represented for them by a series of sand paintings, which the singer creates and in the middle of which the patient finally sits.

The singer's intention, roughly put, is to make the one sung over "beautiful" again. A central tenet of Navajo philosophy, embodied in Beautyway ceremony, is faith in the processes of renewal, expressed as *Sq̨' ah naaghái, bik'eh hózhǫ́.* Navajo is notoriously difficult to translate into English, but these words, Witherspoon says, refer to the restoration of beauty through ceremony, "according to which conditions all around are blessed or made beautiful." Years later, Witherspoon refined this translation for me by referring to "the infinite repetition of the cycles of life[,] according to which there is beauty, harmony, and health everywhere."

The idea that "beauty" refers to a high level of coherence existing everlastingly in the world, and that beauty can be renewed in us through reintegrating ourselves with a world over which we have no control, has appealed to me ever since I became aware of this Navajo ceremony, a formal expression of that idea.

As I walked the Alexandra lowland, I was brought to a heightened sense of the "beauty" within it, a particular integration of color, line, proportionality, sound, smell, and texture. I was aware of its effect on me, and of how my vulnerability to it enhanced a feeling of health in me, of being in harmony with the world that existed outside my own thoughts and beyond my understanding.

FROM THE TOP of the cliffs, lying prone on the ground, steadying my binoculars on a rock to separate them from the slight trembling in my fingers (caused by blood coursing through me after the exertion of the climb), I saw gray and white bergs drifting in Kane Basin to the east, against a backdrop of Greenland's tidewater glaciers. That the rays of light reflecting from the sides of the bergs should remain parallel over a distance of forty-some miles, producing a sharply defined image of their contours, was as astonishing to me as the things themselves. Many of them were larger than the five square miles of lowland I had just crossed.

Before I began searching for a safe way down from the clifftops, I walked out to the northern rim of the tableland to get an unobstructed view of Skraeling Island. It appeared a thousand feet below me as a crinkled mass of browns, gray-greens, and tans, stark there two miles away in the dark water. I appreciated for the first time that it was actually a tombolo, two landmasses connected by a narrow isthmus.

Using a map of the island already in my head, I searched the surface for archeological sites whose names I'd become familiar with: Clinch Ridge, Ghost, Aivik, Oldsquaw. Most of these old Thule and Dorset campsites are situated near the island's coastline. A few are substantial enough to stand out clearly. Directly in front of me, for example, on the southwestern side of the smaller of Skraeling's two hills, was a cluster of twenty-three Thule winter houses, semi-subterranean homes with whale-rib-bone-supported sod roofs, all now dismantled. Taken together, the sites on the island represented the entire history of human life in the High Arctic, beginning with a few Independence I camps about four thousand years old and carrying up through camps once occupied by several other so-called Arctic Small Tool traditions (ASTt), identified as Pre-Dorset, Early Dorset, and Late Dorset, and including, finally, Thule sites from about eight hundred years ago. The remains of some modern Inuit camps were there as well, going back several hundred years.

Schledermann and his colleagues believe the seventeen-acre island was unoccupied for several hundred or more years at a time during those four millennia, most recently from about 500 BC to 700 AD, the period of time between two Paleoeskimo cultural phases, Early and Late Dorset. Whether humans retreated from here because of climate change or simply no longer passed this way, no one knows. However one reckons it, across a span of about four thousand years, one is confronted with the failures (and periodic inflorescences) of an extremely small group of people, several thousand human beings at most, pioneers who traveled here from what one day would be called Alaska, some 1,500 land miles to the southwest.

When I think of the Thule, I imagine a people hungry for movement, a people capable of making consistently good decisions when meeting with extreme circumstances, a people whose extended families perished, here and there, from just a little bit of bad luck—not finding enough food, say—but who were, overall, a resourceful and tireless people. A people, too, who no doubt lived in darkness of more than one sort, who gave darkness the respect it deserved but not more than that. I came to appreciate them in the years before I visited Skraeling and more so afterward, humans whose elders understood not only what to be afraid of but which of the things that appeared to be frightening could safely be ignored.

Everyone I know who's dug up the material culture of a vanished people somewhere on Earth longs for a conversation with the subjects of their inquiry, with the cave painters at Chauvet in the Ardèche Valley in southeastern France, with the Clovis hunters at Blackwater Draw in New Mexico, with the Semites at Ur, or with the Thule. If I could speak with the Thule, I would want to know what they found beautiful, and in what, precisely, had they placed their enduring faith.

BEFORE I STEPPED AWAY from the edge of the cliffs to begin my descent, I scanned the thousand feet or so of heights still above me for gyrfalcons and the lowland below, sidelit now by a sun in the northwestern sky. I isolated in my binoculars stretches of meadow and salt marsh, of heath and freshwater ponds. Nearly every animal living here is too small to pick out, except for the birds: small flocks of snow buntings and northern wheatears, flocks of glaucous gulls and black guillemots, one common raven, and a few long-tailed and parasitic jaegers. In the visual fabric of the tundra, however, I could make out patterns of subtle bright color, and see the sparkling demarcations of the creeks and their parent river, the gleam of the ponds. I could appreciate both the plein air transparency of this world and the reflections of light from its surfaces—water, leaves, rock shields, damp ground. The sheets of generalized color—beryl green, turquoise blue, lavender gray—comprised many distinct colors, no one of which could I separate out entirely with the binoculars. I could find no yellows, though I knew they were there. Some of these colors hung unsaturated in vegetation at the edges of melt ponds, where the water reflected these colors of the vegetation back at the vegetation, and colors on the bottoms of the ponds shone through the transparent water onto the undersides of leaves.

The French Impressionists, among whom Monet was so prominent, knew all this about reflected and enhanced light, and their work prompted many of us to think about it, about pigment and the world of incident and ricocheting light, and what was absent and present in the convergence of the two. And they taught some of us, probably without intending to, to see better. I can't help but think that the Thule also marked the subtlety of the colors I was seeing that afternoon, that like me they would have noticed the glistening surface of the Twin River as they waded across it, and thought that what gave the water life was not solely its movement but the reflections on its surface of wind-stirred plants and of birds passing overhead.

BACK AT THE RCMP BUILDINGS I found Eli and Hans discussing the contents of the upcoming issue of a Canadian journal Eli edited, *The*

Structurist. It was to be devoted to the subject of transparency and reflection, one reason they'd both been eager to read Figes's work. They were commissioning articles on reflection and transparency in music, architecture, and poetry, in addition to painting and photography. The issue was to include a portfolio of Hans's photographs of icebergs. I liked their enthusiasm. Their subject was not their own artistry but the complex business of finding meaning in the world, and of the creation and purpose of art.

As they carried on, I began preparing our dinner, wondering idly if there might be room in their journal for speculation about Dorset and Thule thoughts on all this, particularly on the absence of light, on darkness. Like Hans and Eli, I was so enthralled with the behavior of light around us that I'd failed to consider right away its inseparable twin, the midwinter night. Like the permafrost out there beneath a few inches of life-giving soil, it was invisible on this intoxicating summer afternoon in the hyperborean north.

One more reason, then, to inquire among the Paleoeskimos. What did humans do here when darkness banished the light and you were left with only your imagination, your small ball of saturated moss burning on the surface of a pool of seal oil in a stone lamp, your caches of meat bunkered by rocks too heavy for Arctic foxes to pry away but not so big a polar bear couldn't flip them aside?

Was the darkness, too, good?

THE FOLLOWING MORNING I left early to explore the west side of the lowland, where Police Creek carries meltwater from another glacier across the lowland and into Alexandra Fjord, and where there are extensive patches of grassy salt marsh. I began at the water's edge. To the east, the shoreline runs out against the base of the cliffs I'd climbed to see that part of coastal Greenland called Inglefield; to the west it continues on for some way, farther down into Alexandra Fjord, marking the northern edge of Johan Peninsula.

When I first began to travel far from home, I tended to pick up and keep things I found on the ground, whatever I wanted. By most people's lights, including my own, these were innocuous objects—seashells, feathers, a water-polished agate, the sun-bleached skull of a small mammal. Whatever my hosts or guides offered me, I was also pleased to accept, but these things I took from the land began to weigh on me over the years. I began to feel like a thief, though it was difficult to understand from whom I might be stealing. It was the deliberate disturbance of a place not my own, the ethics of that intrusion, that started tapping at my forehead.

Once, traveling with two archeologists on the North Rim of the Grand Canyon, in a place difficult to access except by helicopter, I came upon an Ancestral Puebloan site that had not been disturbed in the roughly six centuries since those people had walked away from it. It would have been easy to slip one of the small, stunningly decorated clay pots sitting there into my pack. There were dozens of them lying about. I thought about it. Not acting on the impulse came down to respect for the originators, long dead, and respect for the professional calling of my companions. And I suppose self-respect, not to mention the trust extended to me by these archeologists I was with.

The morning after we finished a survey and inventory of the site, I mentioned to one of them that I wanted him to know that I'd not taken anything.

"Yes," he said. "I noticed."

What to take, what to leave in the world's barely visited places—it's not a clear line. I worry about the propriety of such casual acts in a culture like mine, so keen on notions of private property, trespass, and ownership. On this second day of my exploratory hike around Alexandra lowland, I walked up on the remains of a small whale. It was almost completely buried beneath a blanket of viridian-green moss—the bright green color in a darker patch of green sedges is what drew me. The undisturbed bones of the left front flipper, the carpals and phalanges below the wrist of a mammalian forelimb, lay splayed in the moss like a human hand. The slow decomposition of the whale's remains—had it died at sea and been cast up here? Had it been butchered for food on what was once a beach, centuries ago?—seemed one of Earth's slowest ceremonies. My responsibility was simply to observe and move on, not to probe the ground for a tusk, which, if this was a male narwhal and not a beluga, might be there. So I went on. I did not mark the place and have remained quiet about it—until now.

When I returned late that afternoon, I asked Eli and Hans over tea what they knew of the Dorset, a culture strongly identified with a tradition of wood and ivory carvings of human faces that many believe reflect fear, or were meant to provoke terror. Eli, familiar with Dorset carvings, said he and Hans had chosen to come to Alexandra Fjord partly for the wealth of Paleoeskimo sites scattered along the shorelines of peninsulas here; and had chosen the place they were in for the comfort and shelter the RCMP buildings offered, an indoor place in a very outdoor place; and for the closeness of the buildings to the sea, where Hans could work with the translucency of stranded icebergs and with the broad palette of their whites, the whites of church linen, of ivory and alabaster.

He'd heard of Skraeling, he said, because it was the famous place where Peter Schledermann had discovered so many Viking artifacts— a ship's rivet, a piece of woolen cloth, an iron knife blade, a carpenter's planing block. Eli said he didn't know whether Norse people visited Skraeling from their outposts on the southern coast of West Greenland or whether those items had simply been traded this far north, and so come into the hands of Thule people as long ago as the twelfth century.

I told him this was a question I, too, had, and that I would ask Peter for both of us.

The discovery of Norse artifacts on Skraeling in the 1970s prompted heated international debate about sovereign rights in the Canadian High Arctic. Today, global warming has increased anxiety over the issue of sovereign ownership, with sea ice melting more extensively each year, making it easier to reach potentially rich oil and gas fields. Canadian Inuit and Greenlandic Inughuit are ill at ease with any Scandinavian suggestion that a Norse presence in Greenland and in the Canadian High Arctic, at roughly the same time that their own Thule ancestors were arriving from northwestern North America, could raise legitimate doubts about the primacy of Eskimo claims to the area, the question of later European colonization aside. Like Palestinian claims to lands in the Middle East or conflicting claims in Kashmir or to the Spratly Archipelago in the South China Sea, the question of ownership triggers anxiety and indignation in many, even among people who have no role to play in such disputes. Whenever a people emigrate in search of arable land, freshwater, or material wealth, or in order to enhance their political power or escape persecution, other groups, indigenous and nativist, step forward to interrogate them or to thwart them.

With the sea ice melting, these claims and counterclaims remain unsettled.

That evening a Canadian military helicopter flew into our camp to refuel, using drums the Twin Otter had dropped off. A civilian mapping crew was aboard, conducting the last few legs of aerial research needed to complete a set of 1:50,000 maps of Canada's High Arctic islands. I was planning to meet up with Peter and his two field companions two days hence at the RCMP post, but the pilots said if I could throw my gear together quickly, they'd run me four miles east up the coast to the spot where Peter's group was currently camped. I said sure, thanks, and a few minutes later climbed aboard the massive Chinook helicopter, hoping an impromptu arrival wouldn't complicate Peter's plan for pulling this Lakeview camp out two days later—with support from a much smaller PCSP helicopter—in preparation for making the final move of the season to Skraeling.

Peter was warmly welcoming. An extra person with little gear wouldn't necessitate an additional (expensive) helicopter trip. (Had that been the case, I would have been able to hike back out to the RCMP post, using a route I'd picked out while sitting on the open tail ramp of the Chinook's cargo bay on the way there.)

The Lakeview site on Johan Peninsula consisted largely of Arctic Small Tool tradition features, about thirty of them, along with some Thule features—tent rings, stone hearths and caches, and refuse middens. The evening I arrived, Peter, Karen McCullough, and Eric Damkjar were working on a dwelling from which they had raised a number of artifacts, among them some complete microblades about two inches long (the sort of small tool for which this Paleoeskimo tradition is named), a few projectile points, and a single side blade, all fashioned from pieces of medium-gray chert. Importantly, they had also been able to lift from a hearth in this dwelling a small bit of charred Arctic willow. It would later be radiocarbon dated at about 3,940 years BP (before the present), or about 2000 BCE (before the common era), placing this shelter at the very beginning of human occupation in the High Arctic and situating it in the period of several hundred years most archeologists refer to as Independence I. (This particular dwelling consisted of a barely discernible oval of gravel, approximately twelve feet by fifteen feet, enclosing a living area which included a round hearth constructed on exposed bedrock at the shelter's center. An inattentive passerby might easily have missed this meager evidence entirely, seeing it as nothing more than a patch of gravel and a few stray stones.)

Over dinner Peter mentioned another ASTt site, a little farther up the coast, that he, Karen, and Eric had also examined that summer. I went to have a look at it after the dishes were done. The scatter of stones here was easier for me to read as a former dwelling, nestled as the feature was in a narrow slot between two rock outcroppings. A square hearth box of small stone slabs set on edge stood out, as did several patches of gravel and sand that these particular Independence I people had built up in order to level the ground for sleeping.

I would learn in the weeks ahead that "structures" like these, along with their contents, are difficult to be definitive about. Though a region like this has had very, very few people passing through, possibly disturbing what earlier occupants constructed or left behind, those few, along with frost heaving, burrowing creatures, and weathering, have combined to slightly rearrange the places where people once slept and ate before moving on. Thule people, for example, used stones from ASTt sites to build structures of their own. And the rise and fall of sea levels over time—and the isostatic rebound of landmasses once the heavy burden of glacial ice was gone—can give a misleading impression of how far above sea level the dwelling once stood.

I walked south from this site—Peter told me it had been a summer site, that it was too exposed to have been suitable as a winter campsite—toward a gently rising upland where a bright green line, prominent in the low angle of light from a midnight sun, marked the upper reaches of a small stream passing near the archeologists' camp. Still intoxicated by images of the fecundity and dense living textures of the Alexandra lowland, and ignoring, again, the contrasting barrenness of the polar desert that surrounds the lowland and is so extensive here on the peninsula, I headed straight for the stream bank. The mats of emerald green moss on either side of the shallow water were so thick and sturdy, I took off my boots and socks and walked on them barefoot back to camp.

Encountering abandoned ancient human dwellings in Europe and Africa, I don't recall ever feeling a sense of tragedy. The former residents are usually too far off in time, and what might have been the prevailing circumstances for them in those places are too obscure to invoke. But it was not unusual for me to be overcome with feelings of melancholy before these ASTt sites. A few stones, expertly arranged to blunt the force of the wind, a hearth holding the charred remains of willow twigs thinner than a pencil. Sensing the tenuousness of such an existence, I often felt empathy with the anonymous and long-gone residents.

Walking the streamside that evening, I remembered for some reason a girl I'd known as a seven- or eight-year-old boy in California. Cerebral palsy had so restricted her ability to move about that her parents had confined her to a fenced yard at their home, to protect her. She lurched spasmodically in whatever direction she wished to go and could not fix her eyes on any one thing for more than a few seconds. Neighborhood children would stand at the foot of her gated driveway, gyrating and jerking, to mock her. For some reason she and I became friends. We mostly sat together side by side on the edge of a porch in her backyard. Sometimes I brought her things to look at. Her speech was all but unintelligible.

One afternoon the girl, Laura, managed to open the gate at the foot of her driveway. She began walking along the edge of the road, heading for my house. The street had no sidewalks, and the margin of oleander bushes, pepper trees, and eucalyptus was close on both sides. She veered in front of an oncoming car, as though suddenly losing her balance. The impact killed her.

I was inconsolable. It was the first time anyone I knew had died. A door of some sort closed. Or perhaps it opened.

Occasionally, I presume, this kind of childhood memory, the brutal, unregarding nature of everyday life, must come crashing through

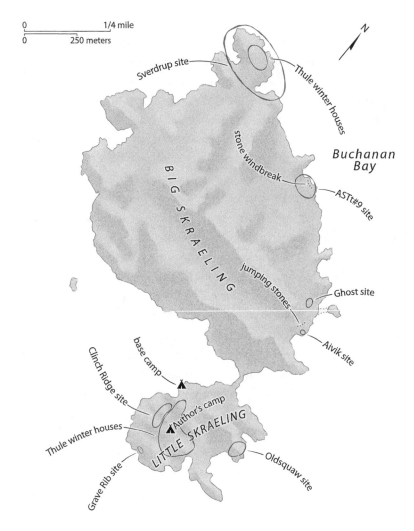

0 ————————— 1/4 mile
0 ————————— 250 meters

N

Sverdrup site

Thule winter houses

Buchanan Bay

BIG SKRAELING

stone windbreak

ASTt#9 site

jumping stones

Ghost site

Aivik site

base camp

Clinch Ridge site

Author's camp

LITTLE SKRAELING

Thule winter houses

Grave Rib site

Oldsquaw site

Skraeling Island

without warning for people in incongruous circumstances, as it did for me that cool summer evening walking barefoot alongside that stream. I recalled the innocence of those childhood days.

I wondered whether Dorset and Thule women had ever walked barefoot here on summer days with their children.

————

TWO DAYS AFTER I arrived at the Lakeview site, a helicopter ferried the four of us and our gear across the ice to Skraeling Island.

Peter has set up a camp in the same spot here most every summer for more than a decade. It was pitched on the edge of a small bay that, along with a dry isthmus, separated Little Skraeling Island from Big Skraeling Island. Peter, a handsome Dane, a person intently focused on his work and well known for his unprecedented finds of Norse artifacts in the High Arctic, set about erecting our cook tent with Karen, who was soon to become the new editor of *Arctic,* a primary journal of scientific research in the far north. Eric, a young Danish graduate student in archeology, and I moved a stove, a collapsible table, and other gear into the tent. He and the others then set up their personal tents close to the cook tent. With Peter's approval, I took my sleeping bag and other gear several hundred yards off to the east, to a stretch of higher ground where a large cluster of excavated Thule winter houses stood. I pitched my small dark blue tent beside a *karigi,* a semisubterranean Thule social and ceremonial structure that, without its roof, now sat exposed to the sky. My companions assured me, when I asked, that the floor of the karigi had been thoroughly scoured for artifacts the season before, and that it therefore was all right for me to enter this round structure and to sit on the stone benches abutting the low wall defining its perimeter.

Like the excavated karigi it was set alongside (before its roof was removed), the blue tent and its waterproof fly were protection against wind, rain, and heavy mists (and, one summer day a few weeks later, snow). Its walls also kept mosquitoes at bay and reduced the intensity of sunlight at night, making it easier to sleep.

An invitation to join a field party like this one comes with the understanding that you will participate in the work, arrive with appropriate clothing and gear, and be familiar with what your colleagues and others have written concerning the research being undertaken. On my side, I felt I needed to work with people in the field in order to understand their experience and points of view. I needed to know the feel, the smell, the sound of their work, to have those memories in my physical body. By the time I got to Skraeling, I had been with enough field parties for people to sense that I'd no interest in invading anyone's privacy or in exposing anyone's foibles or failures. The larger subject—in this case Arctic archeology—was for me always more inter-esting. This meant having to work around people's occasional outbursts of exasperation or ego with each other—or me. This is not infrequently the story with field parties conducting research in remote places, where weather, accidents, monotony, and close quarters can produce stress.

I wanted people to understand that I was not here to report on such things.

I thought it best at Skraeling to volunteer right away to be the regular dishwasher, to assure the others that I wanted to be helpful and didn't want to interfere in or disrupt their work, which was one reason I set my tent apart. With most field parties, anticipating tensions and preparing for them is never much of a problem; good humor begets a lot of joking at each other's expense, everyone works hard, and a sense of equity prevails. Often the indifference and imposing authority of a remote landscape by itself undermines the ordinary human tendency to become petty or unnecessarily dictatorial.

With this field party, however, the atmosphere was sometimes nota-bly tense. It's harder to join a field group partway through their season, after they've established a working rhythm with one another. The group by then shares too much recent history that doesn't include you, which was the case at Skraeling. Also, the three of them had just been visited by a Canadian television crew, whose presumption and insensitivity had made a bad impression. Additionally, there was a rumor that a group of tourists, the first in this area, was going to arrive later that summer. They were chartering in to Alexandra Fjord in Twin Otters and bring-ing inflatable boats with them so they could cross over to Skraeling from the mainland. My guess was that Peter was highly conflicted about the way his work—especially finding the Norse artifacts—had drawn popular attention, and about how this remote landscape, which so clearly fed his spirit, had become a destination for wealthy tourists and importuning television crews.

Peter, something of an introvert, was always gracious with me. He knew my written work and went out of his way to express his respect; but it was my own writing about the Arctic that, too, had cost him some of his privacy. This was another reason that I set my tent apart, to give him, every day, the undisturbed room he seemed to enjoy, a physical and temporal space where he was free to speculate without the threat of being summarized by someone, let alone confronted. The tent apart, of course, was good for me, too. It was a way for me to gently emphasize that I was—and needed to remain—apart, that I was a writer, not an archeologist, not a fan.

PETER HAD ME WORK alongside the three of them most days, carefully lifting up sections of matted plant life that had grown up over the millennia to obscure ASTt features. Most of the time I used my own pocketknife and a pair of tweezers to separate root threads and extract my finds—flakes of ivory and chert, globules of marine mammal fat, bits of worked skin, burins, and other small tools, whole and broken. I marked each find on a chart, the grid of which matched a grid of red twine that had been staked out tautly over the ground there to establish a pattern of identical squares. Other days, Peter urged me to just wander over Skraeling, to examine other ASTt sites and those of the Thule. He told me what to look for at each site, what not to miss.

On the northwest coast of the island was a second large cluster of Thule winter houses. The difference between those nineteen and the group of twenty-three where I was camped is that only a few of the former had had their collapsed sod roofs removed so that they might be thoroughly inspected. The rest stood undisturbed.

I needed little encouragement to begin crisscrossing Skraeling with a sketch map Karen made for me. At some of the sites she and Peter had singled out, I paused to make drawings. This handwork provoked questions for me about the way the interiors of these dwellings were arranged and about the direction in which the structure was oriented. As had been the case on the afternoon I'd spent in the cleft, reading Eva Figes, the luxury of sustained attention at these sites brought insight.

Three or four times during my weeks on Skraeling I climbed the six-hundred-foot heights of Big Skraeling Island and descended into the Sverdrup site, the much less disturbed village of Thule winter houses. The cultural debris still inside most of the dwellings—wood carvings, clothing, harpoon tips, hanks of skin rope, food—had lain frozen for centuries, protected from weathering and scavenging animals by the collapse of the buildings' roofs. The few dwellings that had been opened up, including the settlement's karigi, were eerily tidy and barren pits, swept clean of everything of archeological interest. In spite of these excavations, the enclave overall appears never to have been unquieted. To arrive here is like emerging from dense forest to find a group of people asleep in a clearing, set within a close surround of hills. Whenever I visited there, my eye would hold the slumbering houses against the whiteness of pack ice jammed against the nearby shore. I'd watch hoary redpolls, snow buntings, and Lapland longspurs flitting through the scene, landing on the sod roofs and calling sharply. These were the progeny of separate lines of Arctic life, all of them older here than the human line.

WHENEVER I CAMP for a while in a place this remote, I notice again the sensations of isolation. Logic tells you that all other humans except those in your party are far away. (Shortly after I left them, Hans and Eli departed the RCMP post, their field season concluded.) Most everything one associates with ordinary life—the noise of machinery like automobiles and leaf blowers, the anemic color of artificial lighting, the beckoning of electronic alarms, the sight and odor of waste, the monotony of queues, confinement in small work spaces—is far off. On some days the only sounds I heard between breakfast and supper were the voices of birds and the sharp explosions—like pistol shots—of stranded sheets of sea ice fracturing along the shore as the tide went out. The whine of insects. The snort and harrumph of walruses. The ping and click of rain on my parka hood. Sometimes silence so filled the air here I thought I could hear it—that it, too, had timbre and pitch. As I approached a site where my companions were at work, the clank of a steel trowel against a cobble and the murmur of voices became audible. The purl of sandy soil being poured through a sorting screen and falling into a plastic bin.

These sounds, tiny pebbles in the great basin of indigenous silence, led me to appreciate the way one century nestled within another here, the verticality of time in this place.

I remember studying a sequence of thirty-nine drawings as a boy.

They were included in a book called *Cosmic View* by a Dutch educator named Kees Boeke. In the first drawing, a girl, perhaps ten years old, wearing a long patterned skirt and a dark sweater, is sitting outdoors on a collapsible lawn chair. The view we have of her is from above and slightly oblique. She is holding a large white cat in her lap and seems bemused. In the next drawing, the scale of representation has increased, from 1:10 to 1:100. We see she is sitting next to a couple of parked cars and alongside part of a baleen whale, lying on its right side. The third drawing, at a scale of 1:1,000, reveals that she's sitting with these things in a courtyard at a school. The twenty-three drawings that follow this one carry us to increasingly greater heights above the girl and her cat, until, at a scale of $1:10^{26}$, the view of her is so intergalactic it seems to be located in the realm of fantasy. The twenty-seventh in Boeke's sequence of thirty-nine returns us to the girl and her cat at the scale of 1:10. From here we begin a journey in the other direction. First, at a scale of 1:1, we see a mosquito on the web of skin between her right thumb and forefinger. In the thirteenth and final image, at a scale of $1:10^{-13}$, we're inside a sodium atom, within a salt crystal on the girl's skin.

This simple exercise in scaling returns to me often. The images suggest tremendous depth and breadth in worlds one or more off from a world scaled to humans—a wasp's sense of the extent of the Alexandra lowland, say, or a view of Skraeling from a plane flying overhead en route to Moscow from Seattle, as contrasted with my view of a king eider nesting in a hearth box here. But Boeke's drawings encourage more than just thoughts about scale and point of view. They stimulate, at least for me, thoughts about the difference between my universe and that of the Thule. My umwelt and theirs, or mine and the umwelt of the wasp.

The text Boeke wrote to accompany the drawings is a treatise on barriers and limitations, which points, for me, to a certain conclusion. When a *boundary* in the known world—say, a geographical one for Thule people migrating eastward from Alaska, moving farther into an inhospitable world than anyone had ever gone—becomes instead a beckoning *horizon,* the leading edge of a farther destination, then a world one has never known becomes an integral part of one's new universe. Memory and imagination come into play. The unknown future calls out to the present and to the remembered past, and in that moment of expansion, the imagined future seems attainable.

In the autumn of 2008, the English artist Richard Long walked from Carnac, in Brittany, to the particle physics laboratory at CERN, outside Geneva, a distance of 603 miles. Carnac is the site of several thousand stone monuments erected in the last centuries of Europe's Neolithic period, most of them tall upright stones called menhirs. Long's walk, titled *Megalithic to Subatomic,* traces a journey similar to the one Boeke takes us on. Like much art of the past sixty years, as painting and sculpture moved out of the studio, Long's walk offers us a perspective on human existence prompted by questions that arose after Hiroshima and Nagasaki and the widespread deployment of nuclear weapons, questions mostly about the likelihood of human survival. The growing depletion of the world's natural resources, the desperation behind human diasporas, and the largely unaddressed problem of global climate change have propelled much of modern art even further out of the studio. The accumulation of such threats to human survival suggests the existence of an apocalyptic barrier where once, not so long ago, our way ahead looked almost clear. Our question now is, What lies beyond that barrier? Or more important, what is *calling* to us from beyond that barrier? We already know what is *pushing* us into the future.

Some contemporary art, art that is not about itself or about the artist, offers perspectives that conceivably could release us from the daily tyranny of depressing news, from the meretriciousness of decisions that commit us to the inevitability of environmental catastrophe. All great art tends to draw us out of ourselves. Through the imagination and skill of the artist, it reintroduces us to our surroundings, revitalizing them and revealing interstices there, potential points of entry for our imaginations.

THE THULE, as much as Peter, Karen, and Eric, are my constant companions here. When I recall a story I read months before in *Nature,* speculating about the Sixth Extinction, or a report that appeared in *JAMA: The Journal of the American Medical Association,* about rising cancer rates in first-world countries, I think about the indomitable, resourceful Thule.

In the austral fall of 1987 I was traveling through Namibia with a few people. We slept in the desert, here and there as we went along. We came south to Kalahari Gemsbok National Park (now Kgalagadi Transfrontier Park) in South Africa. One morning there I spotted a pale chanting goshawk in the top of a dead tree. This particular accipiter hunts other birds, as well as reptiles and small mammals. Like all avian predators of its type, the goshawk's hunting success depends on depth perception. The bird had its back to me as I approached. I imagined it gazing intensely at an expanse of savannah grass before it, searching for a creature upon which to swoop. As I drew closer, the bird rotated its head and stared down at me. Its right eye had been torn out of its socket. The hole was rimmed with blood-matted feathers.

It turned back to its survey of the savannah, ignoring me.

Often, when I want to give up, I think of that bird. How many other such severely wounded birds are there in the world, still hunting?

On another occasion, while working with a small field party a few miles from the South Pole—we were gathering samples from a snow pit to further document global climate change—I was given a tour of a scientific project under way at Amundsen-Scott South Pole Station, a research facility and planetary monitoring station established by the United States in 1957. This particular project was designed to locate sources of high-energy cosmic rays, and was part of an ongoing search for dark matter and dark energy in the universe. Astrophysicists theorize that dark matter and dark energy, difficult to detect directly, constitute as much as 95 percent of the mass of the universe; and they believe their presence can be inferred from data being collected by the Antarctic Muon and Neutrino Detector Array (AMANDA) at the South Pole. Most astrophysicists believe that our own and other galaxies are bathed in dark energy. The galaxies we see, then, are like tiny fish suspended in a vast ocean of unlit water.

The AMANDA detection device consists of a massive grid of phototubes buried hundreds of feet deep in the polar ice cap, all of them pointed toward the center of Earth. The tubes register the presence of muons, high-energy subatomic particles that have entered the opposite end of Earth (the North Pole) as neutrinos. In an experimental environment that is otherwise free of radioactivity, and dark, the muons emit a detectable particle called a Cherenkov photon. This evidence confirming the presence of muons registers on a bank of computers in a windowless room above the ice.

I loved the intellectual hunger behind this AMANDA experiment, the collaboration here between experimental and theoretical physicists, especially since I had encountered it (by accident) at a remote outpost like the South Pole.

What else was humanity straining to know in that same moment? Were research scientists elsewhere bearing down hard to understand how the biological fate of *Homo sapiens* will be affected by the extinction in some places of upwards of 60 percent of the populations of flying insects, including pollinators?

What would a Thule *isumataq,* a storyteller, a person who "creates the atmosphere in which wisdom reveals itself," have to say about birds that can't hunt, out hunting? About the critical importance of invisible matter. About a snowshoe hare sitting bewildered in its white coat on a forest floor of brown leaves, winter's not yet having arrived. (A second meaning of *isumataq* is camp leader, a person, in the case of the Thule, guiding a few families through an inhospitable environment to safety by knowing when and where to go. The isumataq is also perceived as a kind of allegorist.)

THE DESIRE TO KNOW MORE, to fashion ever more sophisticated systems of detection and measurement, is a desire not merely to know but to be prepared for the unknown. It is a quest, therefore, without an end. The half-blind goshawk and the recordings of the AMANDA probes came back to me at the Thule sites because here, too, on a different scale, were signs of change as the Thule drove out, absorbed, or perhaps never even encountered the Late Dorset. Evidence of the continuously unfolding universe, in a setting where it's emphatically clear that nature, the larger category that includes human nature and human history, is not static. It is design without an end. Its rubric is adaptation and change. Its imperative is adapt or die.

Our own imperative as modern social primates might be something else. Cooperate with one another or die.

Once the world quietens for hours on end, as it does here on the tundra, I find anxious questions about what kind of future humanity will make for itself as insistent as a thunderclap. One day I was working alongside Peter, inside the perimeter of a Late Dorset dwelling. "Probably used only briefly," he said of the place as he turned over the gravel we were kneeling on stone by stone with a small trowel. Here were two men, trying to tease out a few pieces of the puzzle of human survival, invention, and adaptability. I didn't say anything to Peter in that moment about the question of human survival—it's not always good for the visitor to attempt to make a point. But I felt the peculiar urgency of our particular task, to know the character and the fate of the Thule and Dorset who had gone before us.

When we finished dinner that night, after I'd done the dishes and we were all having last tea, hunched in our parkas outside the cook tent, I asked Peter what he thought we were looking for on Skraeling. Is the prize here archeological scholarship, or is it something we haven't spoken of?

NESTLED IN THE DECLIVITY in the tundra that day with the knapper's debitage, reading about Monet and listening for walrus, I recalled the bee-buzzed hours I'd spent crossing the lowland several weeks before. In my childhood I had known the kind of euphoria I'd felt that day, when the opportunity arose for me to go outside, to leave the domestic rooms, even the rooms where the doors were left open and the open windows were large. I would one day come to understand this impulse I'd felt as a child—to depart—in other ways, depending on the nature of the situation; but what I wanted most of all back then was to pass through the walls of the house. Setting Figes's book aside for a moment, I courted the memory of that day I'd walked on the lowland. The memory vibrated for me, like a struck tuning fork.

As I engaged the memory further, my fingertips were again winding through the willow stems and again I found the chert flakes. Sifting them, I thought how few archeologists who examine flakes like these and who write about them can fashion a stone tool. What do we miss, as a mostly indoor culture, making a few short summertime forays into Earth's remote country to inspect places where our ancestors once found a path, a way to live, but not, ourselves, possessing any of their fundamental skills? Not having cut meat with a stone, nor gone a week with only the soft edges of our skin clothing to sustain us, how well can we intuit the purpose of an arrangement of stones our ancestors made? Peter sometimes makes rhetorical observations something like this. How can indoor people understand outdoor people, with nothing but the intellect to work with, with no predisposition to inquire about or make use of what the body knows, what the foot has learned about balance, having easily crossed one terrain but having trouble with another?

The four of us are all studious individuals, squatting here before an Arctic Small Tool puzzle; but you could not call our focused intensity intimacy with the Thule or the Dorset. We're like blacksmiths, shaping bits of found scrap iron with our hammers, waiting for something we recognize to emerge.

We speculate continually.

I consider telling Peter about the pale chanting goshawk. He might enjoy the indomitability of the bird, this evidence of determined Life, so evident here on Skraeling. I wonder, too, if I understand the AMANDA experiment well enough to convey my enthusiasm about this kind of research question and my high regard for science, though archeology is actually one of the humanities. I could try to convey my euphoric experience hiking in the Arctic oasis across the water from our camp, but he might find my perceptions too abstruse.

PART OF THE PLEASURE of working alongside Peter is his general circumspection, his caution about what he can know; but he is also a searcher who wants to exercise a degree of control over what he discovers, who sometimes wants, in the way of many academics, to own the definition of what he has discovered, even more than he wants to own the thing itself. He is open-minded but of course opinionated about a subject he has given his life to, and he is not always forthcoming. We're both, however, intensely curious. We converse easily. I think it's because neither of us seeks to prevail, and anyway, our subjects are often too unruly, too full of guesses to allow for any final words.

I finished the Figes book that afternoon in the cleft, pushed a few belongings into my day pack, and descended a slope that leads to the shoreline, pausing at several ruins as I followed a long route back to our base camp. For the most part, the sites I passed were Early Dorset, maybe 2,500 years old. Peter thinks some of them might be Transitional, however, meaning they were built by people who were neither late Pre-Dorset nor Early Dorset but a distinct tradition somewhere between the two, which some archeologists call Independence II and others call Transitional. (This particular tradition shows signs of having been influenced by a contemporaneous Greenlandic culture of the time called Saqqaq.) A roughly circumscribed area that I stopped to examine—doing this feels like watching an animal that doesn't know you're there—included several Transitional and Early Dorset and Thule sites together. The entire complex is named the Grave Rib site, after an upright weathered rib bone, possibly a walrus's, which someone had once driven into a Thule burial mound here.

ONE MORNING a helicopter arrives in camp, the same helicopter and pilot that had moved Schledermann's camp here from the Lakeview site on Johan Peninsula. The four of us have packed our gear and are ready to depart for Haa Island, situated between the entrances to Beitstad and Jokel Fjords, twenty-two miles to the west. As we're boarding the helicopter, Peter tells me there is not enough fuel to safely ferry five people out to the site and back. He regrets, he says, having to say I will not be able to go with them. It's his call, but I know the range of this helicopter, fully loaded, in calm weather at sea level, and suspect he has some other reason for not wanting me to come along, which he doesn't wish to make clear.

I'm irked, watching them fly off, but soon decide, with everyone gone for the day, that it's a good time for me to bathe. I heat a small basin of freshwater and take it down to the edge of the fjord, where I strip, wash, and then wade into the cold water to rinse off.

I'm into fresh clothes and have hung my wet laundry out on a line to dry when I hear the helicopter returning. The pilot is close enough for me to read his hand signals. I understand he's motioning me to approach the helicopter as he lands. He props open the side door with his foot to shout over the noise of the rotors that he's sorry Peter bumped me from the flight. Do I want to fly with him now on an errand he has? He's spotted a beach-cast narwhal with a large tusk on the north shore of Pim Island, about twenty miles to the east, and wants to retrieve the tusk. Yes, yes, I say, and run to grab my day pack and survival gear.

We land upwind of the carcass, which is bloated and rank. Polar bear bait for sure, and I'm not comfortable standing around without a rifle while the pilot chops the tusk out of the whale's head with a hatchet. It's eighty-two inches long. He lashes it to one of the helicopter's skids and thirty minutes later we're back in camp. He drops me off, crosses the water to the RCMP post, tops off his fuel tanks at the barrel farm, unloads the tusk, and heads west to a scientific camp at Sverdrup Pass. He's carrying some resupply to two women there studying Arctic hares. Their camp is about thirty miles west of Haa Island. He'll pick Peter and the others up on his way back.

After making lunch, I set off to find a stone fox trap, located along a route Peter has drawn up for me in one of my notebooks. Thule people trapped both polar bears and Arctic foxes in these simple stone enclosures. A sliding stone slab, tripped by the animal's tugging on a bait line, dropped down to pen the animal in a constricted space, immobilizing it. Thule pursued this lethal strategy with bears and foxes for two reasons—for the raw materials the animals offered them and to keep the foxes, especially, from disturbing Thule meat caches. Both types of Thule trap can be found in the Bache-Thorvald-Johan peninsular region around Buchanan Bay. Peter refers to this part of Ellesmere as "the crossroads to Greenland," because it was from here that different pulses of human migration made the crossing eastward across Smith Sound to the northwest shore of Greenland.

The trap Peter described to me was on the southern coast of Big Skraeling Island. Knowing my companions would not be back until suppertime, I decided to take a less direct route to the trap than the one Peter had marked out for me along the shore. I wanted to look at a set of "jumping stones" early occupants had set up for a game like hopscotch, and to see how the stones at another campsite had been assembled to create a windbreak. I'd thought frequently about Thule people's cleverness with stones. Their work doesn't show the same skill with fitting stones together tightly that their Inca contemporaries had, but they were able to move massive rocks, sometimes over considerable distances, to establish foundation walls for their houses or in order to construct bear traps. They fitted "flagstone" floors together; made benches, meat platforms, and hearth boxes of stone; and carved stout but elegant soapstone lamps that held seal and whale oil for cooking, heat, and light.

SKRAELING IS NOT a large island, but it has a lot of topographic relief, so it has many nooks, clefts, and elevations where anything could have happened or be happening. From the start of my walk, then, I was alert for signs, and conscious of not letting my mind drift off somewhere else.

I tried to get out of myself, to enter the country.

When I was young and just beginning to travel with indigenous people, I imagined that they saw more and heard more than I did, that they were overall simply more aware than I was. They were, and they did see and hear more than I did. The absence of spoken conversation whenever I was traveling with them, however, should have provided me with a clue about why this might be true; it didn't, not for a while. It's this: when an observer doesn't immediately turn what his senses convey to him into language, into the vocabulary and syntactical framework we all employ when trying to define our experiences, there's a much greater opportunity for minor details, which might at first seem unimportant, to remain alive in the foreground of an impression, where later they might deepen the meaning of an experience.

If my companions and I, for example, hiking the taiga encountered a grizzly bear feeding on a caribou carcass, I would tend to focus almost entirely on the bear. My companions would focus on the part of the world of which, at that moment, the bear was only a fragment. The bear in this case might be compared with a bonfire, a kind of incandescence that throws light on everything around it. My companions would glance off into the outer reaches of that light, then look back to the fire, back and forth. They would repeatedly situate the smaller thing within the larger thing, back and forth. As they noticed trace odors in the air, or listened for birdsong or the sound of brittle brush rattling, they in effect extended the moment of encounter with the bear backward and forward in time. Their framework for the phenomenon, one that I might later shorten to just "meeting the bear," was more voluminous than mine; and where my temporal boundaries for the event would normally consist of little more than the moments of the encounter with the bear, theirs included the time before we arrived, as well as the time after we left. For me, the bear was a noun, the subject of a sentence; for them, it was a verb, the gerund *bearing.*

Over the years I absorbed two lessons from indigenous people about how to be more fully present in an encounter with a wild animal. First, I needed to understand that I was entering the event as it was *unfolding*. It started before I arrived and would continue unfolding after I departed. Second—let's say we didn't disturb the grizzly bear as he fed, but only took in what he or she was doing and then slipped away—the event itself could not be completely defined by referring solely to the physical geography around us in those moments. For example, I might not recall something we'd all seen a half hour before, a caribou hoofprint in soft ground at the edge of a creek, say; but my companions would remember that. And a while after our encounter with the bear, say a half mile farther on, they would notice something else—a few grizzly-bear guard hairs snagged in scales of tree bark—and they would relate it to some detail they'd observed during those moments when we were watching the bear. The event I was cataloging in my mind as "encounter with a tundra grizzly," they were experiencing as an immersion in the current of a river. They were swimming in it, feeling its pull, noting the temperature of the water, the back eddies, and where the side streams entered. My approach, in contrast, was mostly to take note of objects in the scene—the bear, the caribou, the tundra vegetation. A series of dots that I would try to make sense of by connecting them all with a single rigid line. My friends, in contrast, had situated themselves within a *dynamic* event. Also, unlike me, they felt no immediate need to resolve it into meaning. Their approach was to let it continue to unfold. To notice everything and to let whatever significance was there emerge in its own time.

The lesson in those experiences was not just for me to pay closer attention to what was going on around me, if I hoped to have a deeper understanding of the event, but to remain in a state of suspended mental analysis while observing all that was happening, resisting the urge to define or summarize. To step away from the familiar compulsion to understand. Further, I had to incorporate a quintessential characteristic of the way indigenous people observe: they pay more attention to *patterns* in what they encounter than to isolated objects. When they saw the bear, they right away began searching for a pattern that was resolving itself before them as "a bear feeding on a carcass." They began gathering various pieces together that might later self-assemble into an event larger than "a bear feeding." These unintegrated pieces they took in as we traveled—the nature of the sonic landscape that permeated this particular physical landscape; the presence or absence of wind, and the direction from which it was coming or had shifted; a piece of speckled eggshell under a tree; leaves missing from the stems of a species of brush; a hole freshly dug in the ground—might individually convey very little. Allowed to slowly resolve into a pattern, however, they might become revelatory. They might illuminate the land further.

As much as I believed I was fully present in the physical worlds through which I was traveling over the years, I understood over time that I was not. More often I was only *thinking* about the place I was in. Initially awed by an event, the screech of a gray fox in the night woods, say, or the breaching of a large whale, I too often moved straight to analysis. On occasion I would become so wedded to my thoughts, to some cascade of ideas, that I actually lost touch with the details that my body was *still gathering* from a place. The ear heard the song of a vesper sparrow, and then heard the song again, and knew that the second time it was a *different* vesper sparrow singing. The mind, pleased with itself for identifying those notes as the song of a vesper sparrow, was too preoccupied with its summary to notice what the ear was still offering. The mind was making no use of the body's ability to be discerning about sounds. And so the mind's knowledge of the place remained superficial.

BEFORE I RECOGNIZED it for what it was, I was nearly on top of the windbreak Peter had directed me to at Skraeling ASTt site number 9. The wall was almost three feet high and about fourteen feet long, built between the shore, close by, and what would have been the rear wall of what some Arctic archeologists call a Transitional dwelling. An uncomplicated thing, meant to break the force and chill of the wind. It seemed dormant and resolute, like a volcano.

Archeologists have removed most of the artifacts they've uncovered on Skraeling, but the aura of the presence of early occupants remains as strong at some sites as the feeling of shared desire and fate that comes to me from handling their tools and carvings, the durable evidence of their occupancy. No dwelling seems irrelevant or redundant. Each one speaks of a few specific people who've left behind something that clings to their hearth boxes (each stone placed just so), their stone pavements, the circles of anchor stones for their portable tents. When the collapsed roof of a Thule winter house is removed and its interior is exposed for the first time to continuous solar radiation, the floor begins to reek of the animals the Thule occupants once hunted and ate—small whales, bearded seals, ringed seals, walruses, geese, fish, Arctic hares. It's possible to pick up and eat scraps of animal flesh off the floor (which some find palatable, if not tasty). The size of the entrance tunnels to the winter houses and the dimensions of their interiors hint at the physical size of the people. In a hearth box you might find a handful of "boiling stones," smooth cobbles heated up in the fire's embers and then dropped into a skin bag filled with some type of cold soup.

All across Skraeling the message for me was the same for all of the traditions: tenacity and utility. These were eminently practical human beings. It's far more difficult—impossible, really—to imagine accurately the ceremonies that assuaged their grief or raised their hopes, or the episodes in their lives that brought on convulsive laughter. It's hard to picture children. (Elsewhere, not here, researchers have found structures they characterize as "Thule doll houses.") It's difficult to grasp how truly daunting places like this could be, with months of darkness, the bone-splitting cold, and the sudden, inexplicable disappearances of populations of food animals when before they had always been there.

SOMETIME IN THE EARLY 1850S, two charismatic Inuit men, Oqe and Qitdlarssuaq (also called Qillaq), led about forty other Inuit northward from a settlement at Pond's Bay (now Pond Inlet), in northern Baffin Island. They crossed fifty miles of spring ice at the mouth of Lancaster Sound, to arrive on the shores of Devon Island. Some historians, perhaps to dramatize where no additional drama is really required, believe the men and their followers were running from numerous blood feuds at Pond's Bay; but it could also have been that these two men had learned from European whalers of the existence of Inughuit—Polar Eskimos—living in northwest Greenland. Familiar as they were with the brisk trade in walrus and narwhal ivory that went on at Pond's Bay, they might have hoped to make contact with their cultural cousins in Greenland, to see if they might prove to be a new source of ivory. Or perhaps, like their Thule ancestors, they had no plan but the excursion itself, a journey framed by someone's vision.

Over the next four or five years Qitdlarssuaq's group continued north, finally reaching the central coast of Ellesmere Island, probably in the vicinity of Johan Peninsula. The group by then is smaller. Oqe is thought to have turned back for Baffin Island with some people, but there is no record of their ever having gotten back. Qitdlarssuaq led his followers across the sea ice in Smith Sound and there, in the vicinity of present-day Etah, they met the Inughuit. To their amazement, these Eskimos lacked three tools the Inuit regarded as indispensable: the bow and arrow, the kayak, and the leister spear (*kakivak*), an implement designed to both impale and secure a fish. Anthropologists speculate that either famine or epidemic disease might have wiped out so many people in northwestern Greenland that the few who were left didn't know how to make these things anymore. So they devised a life without them.

Qitdlarssuaq's group stayed in Greenland about seven years, passing on their technical knowledge and intermarrying with the Inughuit. And then most or all of them decided to return to Pond's Bay. The homeward journey proved a disaster. Qitdlarssuaq died soon after the group recrossed Smith Sound. Those who remained split into two groups and wintered in two separate places. Only one of those places, Makinson Inlet near Cape Faraday, provided sufficient food. The following winter the Makinson Inlet group overwintered there again, along with survivors from the other camp. During that winter, however, the marine mammals around Makinson Inlet migrated elsewhere and starvation set in. Two men, Minik and Maktaq, began to kill weakened people and eat them. Five people, including two children, escaped to the north and were able to kill and retrieve a seal, which got them through the winter. The following spring these five returned to Greenland. Nothing is known of what happened to the others.

The comfortable isolation my companions and I enjoyed on Skraeling—with cookstoves, plentiful food, helicopter transits, moderate summer temperatures, and regular radio communications with PCSP in Resolute (which warned us of approaching storms)—insulated us from the harsh and unrelenting capriciousness of weather in the High Arctic, a reality the Thule, their predecessors, and their progeny of course could not avoid.

THE FOX TRAP, an oblong box with a square cross section about eight inches wide and twenty inches deep, had not been disturbed in the hundreds of years since it had been set. Peering between two stone slabs that formed one corner of the trap, I saw the skeleton of a fox. Its snout rested on the mandible of a ringed seal fetus—the bait. The leather thong that had once suspended the sliding door had disintegrated, but I could see how this slab had been rigged to slide home.

A fatal moment for the fox. For me, a kind of shrine.

Years in the future, walking through densely populated areas laid to waste by natural disaster, like Banda Aceh in Sumatra after the 2004 Boxing Day tsunami, or being driven down shell-cratered streets in Kabul, I had settings in which to consider how Death, that forever famished and indifferent visitor, is more apparent in some places than others. What I felt before the skeleton of the fox—killed but never put to use—was not sadness, not tragedy, but renewed awareness of the ineluctable horror of life, of which my own culture seems sometimes weirdly innocent. In 1998 a boy in my community in Oregon killed his parents and then shot and killed two students at his school before he was subdued. Predictably, residents asked, "How could this have happened here?" though over the past thirty years incidents like this have detonated randomly, and regularly, in dozens of American communities, and they have been followed by explanations that were unsatisfying and incomplete. One does not find "evil" in these events, one finds desperation and pain, the merely human. At Makinson Inlet, in the face of starvation, one finds cannibals but, too, an unknown person inspiring enough, skilled enough, to get two other adults and two children away, clear of the horror.

In the modern era, witnessing social, economic, and physical breakdown in traditional villages in Africa or rural Australia or in the barrios, the favelas, the ghettos, or the townships of major cities, I've come to believe that the root cause of this breakdown has nothing to do with the absence of "civilization" or the presence of "evil" but has almost entirely to do with the unremitting presence of political repression, poverty, racism, and living lives of servitude. The problem of ensuring human survival in these places, let alone providing for a human efflorescence, is staggeringly large. The situation cries out to be completely reimagined, or as some traditional people say, "It needs to be redreamed."

WHEN I HAD SET my tent up next to the karigi, it was with the idea, presumptuous of course, that the Thule ghosts there would enter my dreams. That I remember, they never did. My dreams were about travel corridors in the land I was in, pedestrian narratives about the small drama of my own life, and demi-visions turning on the familiar axes of allegiance and betrayal, of fulfillment and longing.

If I had wanted to ask the Thule, and I did, how they managed the darkness that befell them, whether it was from murder or starvation, or only nightfall when it came for good in the autumn, and what they placed their faith in, I also wanted to know about the shape of their dreams, however incomprehensible they might be to me. In the long night of winter, did their dreams become longer, more elaborate? Was there such a thing as a summer dream? Was the dark a teacher or was it an oppressor?

In 1949 Eigil Knuth, an independent Danish archeologist, from an aristocratic family and with a background in art and architecture, found the wood frame of a large skin boat, a Thule *umiaq,* near Kap Eiler Rasmussen, in northeastern Greenland. The frame was splayed open under a thin layer of snow, about five hundred yards from the shore of the Wandel Sea, an embayment of the Arctic Ocean. A few of its parts were very likely of Norse origin—several nails and a piece of oak. Disintegration and scavengers like foxes had reduced the walrus-skin hull to scraps. Amid the frame members, Knuth and his colleagues found some pieces of firewood and a few bone implements. In the surrounding area they also located several Thule culture objects—hare snares and a toy sledge. Knuth's conjecture was that people had simply walked away from the umiaq and left all this behind.

Archeologists now speculate that this Thule group arrived at Kap Eiler Rasmussen some time in the early 1400s, near the end of a warming period in the High Arctic, having paddled around the northern end of Greenland and put in here as winter was coming on. The site, known as Kølnæs, is the easternmost reach of Peary Land, the northernmost of the High Arctic oases, a place that compares with the lowland at Alexandra Fjord or with Truelove Lowland on Devon Island. From here, north, east, and south, there is nothing but ocean. Human beings—or these people at least—could go no farther.

Among the records of humanity's most extreme peregrinations, this one has, for me, become the most prominent. It marks the end of a stupendous journey. The crafting of the boat and the nature of the tools abandoned here evince great technical skill and a considerable capacity for innovation. The cache of firewood in the boat speaks to a certain measure of wealth, if the term is understood to mean more than is needed for survival. These people were whale and muskoxen hunters, fully capable individuals. And then something happened. Perhaps the onset of cooler conditions froze their avenue of retreat, making the boat of no further use the following summer. And having exhausted the local supply of food—or maybe, as would happen at Makinson Inlet, after marine mammals had left the area—the travelers had moved inland with their pack dogs, carrying only absolute essentials and hoping to find muskoxen. They had to cross Peary Land to the west of them and then navigate rugged country between there and the northwest coast of Greenland, where their home country was and where they might find marine mammals again and feed themselves.

I want to think they found muskoxen and made it the five or six hundred miles back to a place where their relatives were amazed, but not stunned, to see them again.

A replica of their thirty-five-foot umiaq was put on display at the Viking Ship Museum in Roskilde, outside Copenhagen, in 1980. One day I hope to see it, to feel the way the walrus hide comes tightly over the gunnels and is bound to the frame with bearded seal thongs. Few handmade but well-designed objects like this can be understood quickly, but several hours spent alone with one of them can make a kind of conversation possible. One does not necessarily learn the "truth" of the thing, but an image of the imagination that created it might emerge.

The umiaq found at Kølnæs surfaces in my mind frequently as a symbol of both Thule ingenuity and resourcefulness. What a remarkable notion, what courage and self-confidence to abandon both it and part of the tool kit that made the physical world manageable for them, to leave it all there on the beach and head inland. To bivouac, to sleep and move on, to search for the distant movement or the dark dot that might mean food. Once winter came down like a sheet of iron and food eluded them and the living became even harder, did their dreams change? Moonlight and starlight reflecting off the snow might have made it possible, when the skies were clear, to navigate; but any search for food animals in the winter gloaming was problematic. *Unnuiijuq,* the Inuit call this ordeal, "looking for food in the dark."

Anthropologists and archeologists I've asked speculate that in winter darkness Thule people slept for hours on end. Conceivably these long slumbers opened up large dreamscapes, which might have functioned for them like the cycles of myths they listened to in difficult times from an isumataq or an *angakkuq* (shaman). In the modern era we're less familiar with such epic dreamscapes (and of course far less attentive in our secular lives to the myths upon which Western culture itself is founded). In the West, industrialization brought with it a new prescription for rest—eight hours of uninterrupted sleep; and that put an end, for most working people, to the natural rhythms of human dreams. Our dreams are now regularly truncated by "the need to get up and get going," a timing dictated for daily life by clocks. What Shakespeare in his plays called "second sleep" meant the sleep that came after a period of wakefulness following "first sleep." During that interval, sleepmates spoke to each other about the imagery of their dreams. In so doing, they maintained intimacy with a way of seeing the world that waned with the rise of rationalism.

The challenge in addressing the utility of our dreams is not whether to reject them outright in an effort to privilege the sort of logical truth the rational mind offers us. It's to picture a conversation between imagination and intellect, one that might produce an advantageous vision, one the intellect itself cannot discern and which the imagination alone is not able to create.

Lying by the karigi on Skraeling some nights, I wondered whether Thule dreamers, like the women and men who abandoned their umiaq at Kølnæs, rediscovered the thread of their resolve in the narratives of their winter dreams, both the intention and the means by which to make another sort of life.

In my idle thoughts before sleep, I more often thought, however, of the Thule stone bear traps.

I HAD THE CHANCE once to visit the ruins of a penal colony built by the French, in 1852, in Les Îles du Salut, a tiny archipelago eight miles off the tropical coast of French Guiana. It was erected primarily to isolate political prisoners like Alfred Dreyfus from French society, but also served to rid France of its "undesirables"—criminals, political objectors, the mentally impaired, and the destitute. Most were confined on Île du Diable, the Devil's Island, a location made infamous by Henri Charrière in his novel *Papillon,* and by a movie of the same name.

I've long been drawn to prisons, both as symbols of punishment and as monuments to injustice (as well as justice). Here is where we house wrongdoers, however that category might be defined in any particular age. Prison buildings serve a legitimate social purpose, but they also stand, metaphorically, for a kind of social confinement that many who live in restrictive or repressive societies experience every day. In a culture in which invasions of privacy are routine or where human rights violations are common, people not actually living in prisons nevertheless sense that they are. They feel ill treated by indifferent people in positions of power and believe that their ordinary movements and activities are unreasonably restricted by bureaucrats. They feel violated by security cameras, by drones, and by the incessant mining of their smartphones, computers, laptops, and other electronic devices for the personal information that creates Big Data, warehouses of particularized fact that in turn make surreptitious intrusions into our private lives possible. And they feel, further, that their voices of ethical objection directed at government and businesses are repeatedly ignored by ideologues and by the entrenched corrupt in power.

Having inspected several abandoned colonial penal colonies and spent time visiting a couple of modern prisons, I've come to understand more clearly the somewhat facile metaphor of feeling imprisoned in a free society. I've also thought about how people living *outside* a prison environment, people who are denied certain of their basic freedoms, nevertheless believe they are living in a state of unfettered freedom.

One would like to think that in an enlightened time prisons would exist largely to rehabilitate those criminals who can be rehabilitated, to isolate the violent and psychopathic from the rest of society, and to punish those guilty of serious crimes. But that is not always the case. Among the incarcerated in many countries, including my own, are many who would be better off cared for in mental institutions, as well as first-time offenders, charged with only minor crimes, who are left to fend for themselves with professional criminals and predatory gangs. Moreover, in most countries allocations for the education of prisoners—the single most effective way to reduce the rate of recidivism—are paltry to nonexistent, in both government-run and for-profit prisons.

Prisons, then, can be understood to represent some of society's deepest unresolved problems: racism, the cultural capacity to inflict unnecessary cruelty on others, and ambivalence about how to deal with sociopathic personalities in public life. Prison buildings themselves eas-ily provoke questions for me about whom society really wants to be rid of, and about how various political and religious forces within society combine to impose harsh judgment here, frequently on those without influential advocates or competent legal counsel. If wisdom and equity played a larger role in these matters, justice generally would look very different from what currently is in place. Prisons remind us of more than the inevitability of terrible human failure, of social intolerance, political totalitarianism, and the seemingly ineradicable presence of injustice. Importantly, they also remind us of how easily one's admirable instinct to be empathetic can perish when faced with the realities of daily life in prison. If prisons are awful places—violent, mind-numbing, and unsafe—then, obviously, prison reform is called for; if people are imprisoned who shouldn't be there, then social reform is called for.

Prisons seem to me, and perhaps to others, like canaries in the pro-verbial mine shaft. In a free society, we must always ask, Who exactly is to be placed in them? And how is someone to look after their own salvation in such places? And do prisons symbolize, even in free soci-eties, malignant intolerance—the inability of judges and others with discretionary power, for example, to empathize? To create a better social order, one must accept what prisons reveal about the full spectrum of human nature (the incorrigible behavior of the career criminal or the psychopath's incapacity to empathize) and disown the naïve belief that those incarcerated represent a class of people that poses a great threat to social stability. A greater threat, in my view, are those who deny or ignore the reasons for refugee diasporas, the collapsing populations of wild animals, and the neurosis of consumerism, all in an effort to ensure their own financial well-being.

MOST OF THE ROOFS in the one-story whitewashed cellblocks on Île du Diable have caved in, and the cells themselves, along with the corridors connecting them, are heavily overgrown with tropical vegeta-tion. Wild fig trees have entirely sealed off the entrances to some cells, and their massive trunks block a number of passageways. Even with sunlight pouring in and the jungle actively reclaiming the site, however, one can easily picture the bleakness that once characterized this compound—the regimentation of daily life, the petty acts of humiliation by the guards, the prisoners' despair, and the accommodation of pathological perversions on the part of both prisoners and guards. Dreyfus, wrongly convicted of treason (partly because he was a Jew), was eventually exonerated; many others sent here for punishment, however, had done little more than incite the wrath and indignation of the powerful. The illegitimacy of their convictions, frequently around issues of religious belief, social class, and political philosophy, eventually led to widespread condemnation of these particular prisons. Most of them were closed or torn down before the end of the nineteenth century.

The impulse behind the establishment of the colonial prison, however, the impulse of nations to rid themselves of criminals, idlers, and those who criticize government, did not end with the termination of colonialism. The too-aggressive pursuit of criminal justice easily led—and still leads, in some totalitarian regimes—to the realization that the prison compound offers a solution for neutralizing political foes, a place where such foes can be silenced and effectively buried.

Opposing opinions about how to design a just prison system—whether to punish, on the one hand, or to rehabilitate and forgive on the other—are difficult for most people to balance. What I found moving through the derelict buildings on Devil's Island and sitting awhile in some of the cells, looking at the evidence around me, was great sadness at the way we behave toward one another. How wretched it must have been, having indifferent strangers designating precisely which spaces you would be permitted to occupy, strangers who also had the authority to arrange the order of your every waking minute, and to be made to accept, every day, the denial of your own instincts for self-preservation.

Whoever ended up in these colonial prisons was treated like a faulty appliance—dropped off at a repair shop, struck with a ballpeen hammer to make it work properly, and thrown back on a shelf at night.

I sat in the empty cells on Île du Diable in the same way I sat in front of the empty dwellings of the Thule, wondering where the path to safety lies in our time. Wondering about the fate of those who, uneasy, are increasingly raising their voices. Wondering, considering the many apparent threats we can see on the horizon, whether what is to emerge for us is an unimaginable darkness of social disorder and ecological disaster or the fully imagined landscape of a second, a very different, Enlightenment.

THE WORK ON Skraeling had a pleasing order to it. We proceeded each morning to a partially worked site and took up again the task of collecting, accessioning, and speculating. We sifted the thin soil searching for anything relevant—a flake of gray chert or dark argillite, a shriveled thread of sinew—and worked the entire ground near us methodically, measuring, cataloging, making sketch maps. The work never seemed tedious. Birds flew over, calling. Cirrus clouds scudded west on a prevailing wind. Our concentration was focused so narrowly that we hardly spoke to each other, unless it was to further the task at hand or to share an occasional insight. If we were far from camp, we'd each eat the lunch we'd brought and I'd go to work on my notes, or ask the others about what it was they were doing, and what images they'd formed of the people whose dwellings we were so respectfully, I thought, taking apart.

Occasionally I saw something I wanted to pocket, but I never acted on the impulse. One of Peter's great fears, he tells me, is that artifacts from four thousand years of human history are scattered everywhere across the ground here, and the only thing that keeps people from walking off with one of the more spectacular objects—an ivory carving, a spearpoint—is the great difficulty of getting here and the basic honesty of those who do come.

In many of the archeological sites I've visited, the overriding apprehension—after anxiety over inadvertently damaging the site or its contents—is the same: thievery. When thieves arrive, it is not just the objects themselves that are lost. The continuity of the record of human occupation is destroyed, and the sense of who we are and where we might be headed is compromised. It's codexes pulled from a library's shelves to fuel an itinerant's campfire.

It is discomforting to consider thievery in such a remote and pristine setting. But anyone who works in archeology today knows incidental and professional thievery thrive in every environment where the question is not *Where have we come from?* but, *What is it worth?*

A VISITOR TO Les Îles du Salut today is meant to appreciate the contrast between the barred cells on Île du Diable, with their thick walls,
iron doors, and porthole windows, and the patrician appointments of
the prison commandant's manse, across the water on Île Royale. The
commandant's quarters enjoy unobstructed views of the sea in every
direction. Fine linens are here, a silver coffee service, closets with many
changes of clothing. The screen doors in this spacious dwelling are
lightly sprung, and its mahogany shelves are laden with memorabilia
and curios. The visitor is reassured: the innocent must always be protected against the machinations of the malcontent, the evildoer. One
might as easily leave here, however, with another impression: the insidi-
ous nature of social oppression, and the many ways in which those who
dwell in the manses continue to create such places, securing in the cells
opposite those who threaten the social order that those in charge prefer.

What resonates for me in this place, and other places like it, is not so
much the past. The unavoidable and dangerous question for the traveler
here is: Who now occupies the commandant's house? Who prefers that
you do what you are told? Who is it that calls for you to be silent? These
questions go back a dozen millennia for us, to the rise of now-obscure
people like the Abu Hureyra, in Syria in 13,500 BCE, and to the people
of Karaca Dağ in Turkey a thousand years later, and after that to the era
of the first cities and the rise of complex societies. The difference now
is that so much of what we depended on back then—wood, fish, freshwater, arable land—is now much diminished. The cities have increased
in number and size. And the fear of living a life of real impoverishment
in the future is spoken of in every café.

EMPATHY FOR EACH OTHER'S predicaments, it seems to me, is the
starting point for any system of justice in our time. The caution
here—as the prior of a monastery once put it to me—is to understand
that justice without liturgy is barbarism, and liturgy without justice
is sentimentality. I took him to mean that to pursue justice outside
an ethical framework (the Bible, the Qur'an, the Constitution of the

United States) would be intolerable in a society that enshrines its ethics; and to imagine that evil is not a force in the organization of human societies is to remain unenlightened.

I was able to ask archbishop emeritus Desmond Tutu once about prisons, how he viewed them in the context of the disintegration of apartheid. He said something I thought oddly in accord with what the prior had said to me. With an end to that murderous and racist regime in South Africa, he said, you had two choices as you addressed the challenge of rebuilding: pursue justice at the expense of peace, or pursue peace at the expense of justice. The answer he and his colleagues found was Truth and Reconciliation, a legal process that established a middle ground, the very hardest ground to establish and then to hold. In the Truth and Reconciliation Commission hearings, those who had been harmed were asked to describe what was done to them, and those who had caused the harm were asked to tell the truth about what they had done. Both spoke in the same courtroom at the same session, seated in front of each other. The result of these hearings, said the archbishop, was reconciliation. Having the ones who were harmed describe what was done to them and the ones who harmed them acknowledge in detail what they'd done, and explain why they had caused harm, ensured both a measure of justice and a measure of peace. Those in the courtroom charged with passing judgment on the accused sought to locate and encourage within themselves a capacity for empathy.

Only the very worst offenders were sent to prison.

ABANDONED COLONIAL PRISONS like Devil's Island are teachers in my mind, monuments to the immorality behind imperial thinking and to the unilateral power of those rulers who enforce compliance with their wishes. The effort behind colonial imprisonment was neither to rehabilitate, to socialize, or to educate. It was to *punish,* relentlessly and heartlessly, and to thereby create, it was hoped, loyal citizens, those who would become the servants of empire. The system was driven by resentment and pettiness and it was enabled by social indifference (and, too, by the political impotence of those few who objected).

The threat of its full resurrection, in my mind, is never far off.

PETER AND THE OTHERS returned from Haa Island late that afternoon, pleased with what they'd found—some very large Thule dwellings,
which they would later describe as the last major Thule encampment in
the "crossroads to Greenland" region of Ellesmere Island. Along with
these dwellings, they discovered artifacts that told them these particular
Thule were hunting inland for caribou and muskoxen, probably in and
around Sverdrup Pass. The four of us began discussing how unusual
this evidence was. Thule are commonly described as "*whale* hunters" or
"marine mammal hunters." Peter also wondered whether Thule people
like these might have occasionally built their winter villages near the
site of a beach-cast bowhead whale. It would have provided them with
enough protein and building material to get twenty or thirty people
easily through a long winter. The carrion might also have drawn in
polar bears for them to hunt.

That evening in my tent I listened to classical music on a small
battery-powered cassette tape player, the music rising up through my
pillow. This was my usual habit on such trips, and this time I'd chosen,
with no originality, the Finnish composer Jean Sibelius, whose melancholy music fits the stony barrens and stretches of tundra here. I'd
also brought some of Bach's chamber music, the sonatas and partitas
for unaccompanied violin, and a few of the Beethoven symphonies. I
find the mood in Sibelius's *The Swan of Tuonela* and the evocation of
the forested landscapes of Finland in his *Tapiola* transporting. Tuonela,
the land of Death in Finnish myth, is encompassed by a river of black
water upon which a white swan floats, singing. Bach's Partita for Violin
no. 2 includes many variations on a single theme, suggesting the infinity
of meaning that might be drawn from something that, initially, seems
quite simple.

That particular evening I was listening to the Beethoven Fifth Symphony and drifting through the events of my day—looking inside the
fox trap, catching the image of a perfect reflection of the lowland cliffs
in the flat water of Alexandra Fjord, feeling the heft and spring of the
narwhal tusk the pilot had retrieved. Recalling my emotions in some
of these scenes, I envisioned the bow moving rapidly over the strings

of a cello. The musical phrases increased the intensity of the events I was recalling, opening them up to an unexpected degree. The music sharpened my comprehension of what had happened that day.

Once when I was talking with the American composer John Luther Adams about the way music might pry something ineluctable out of a particular landscape, and about how certain geographies can increase the intensity of certain pieces of music, we realized that the enthusiasm we each felt for certain landscape paintings was stimulated by qualities in the paintings similar to those that composers use in creating successful music. In music, purity refers to a lack of diffusion, while in painting, purity is a measure of saturation. Harmony and timbre compare in painting with the hues of a color. Loudness compares with visual brightness. The totality of patterns present in a piece of music—melody, harmony, timbre, rhythm, texture—the landscape painter achieves with the purity, amplitude, and hue of her colors. Adams calls these patterns in music and painting "ecologies." In many of Adams's twentieth-century compositions, these ecologies limn the subtleties of specific landscapes, including the sonic components and patterns of those places. Landscape painting, more obviously, does the same, creating realistic or abstract interpretations of moments—or even years—in such places. Astonishing to me is that it is possible to find in the "totality of patterns" of a particular piece of music a unity that compares with the unified biological, geological, and geographical ecologies of a landscape, many of which specifics a composer might actually be unaware of.

The music I listened to those nights on Skraeling brought the land in closer, deepened it to my senses. The music conjured, to my way of thinking, the prior residents. It beckoned to the Thule hunters and their families to become known.

IN THE MORNING the helicopter came again, and the four of us and the pilot flew east together, tracing the northern boundary of Johan Peninsula. Along the way we put down to examine two Thule polar bear traps, built, curiously, only about twenty feet apart on a stone shelf some five feet above the water. The traps were constructed of stone slabs, braced and reinforced with boulders. The low-ceilinged tunnels inside both traps were about eleven feet long. In order to reach the bait at the far end of the tunnel, a bear was forced to inch forward on its belly. Tugging the bait loosened a wedge and the door slab slammed down. Stretched prone in this position, the bear lacked leverage. It couldn't use its great strength to break free.

If the bear was not already dead when the hunter arrived, the hunter removed a few stones at the head of this mechanism and drove a spear into the bear to kill it. I thought about the bear, about *pihuqahtaq*, "the one who walks," after we left. Inuit hunters in Nunavut today say of the polar bear, *nanuq*, "he's the one most like us." A polar bear stands erect like a person. They still-hunt seals, a technique requiring terrific patience. They construct snow houses for shelter and for birthing. They move seasonally, as Inuit did in the old days. Polar bears are considered the most skilled of hunters, the angakkuq's most powerful helper, and the most intelligent of all animals. The bear is symbolically complex for Inuit people, seen as a mediator between the sea and the land, as one who moves easily between the human and the nonhuman world. They are thought to live in villages where they appear to each other as humans. Inuit dead are given water removed from the body of a bear, if it is available, to help them on their journey.

The scene in which the Thule hunter comes upon the trapped bear, the living hunter encountering the dead hunter, is as eschatologically complex, probably, as anything in the Thule world. (A branch of theology, eschatology deals with death, destiny, and the final journey of the soul.) I wanted to listen to it spoken of. The inert eight-hundred-year-old trap was a reminder that in this extreme locale, *survival* was the dilemma every hunter had to deal with.

WE LANDED AT Cape Sabine on Pim Island, at the eastern tip of Johan Peninsula, putting down amid a jumble of rocks and boulders in a trough between two rocky ridges. In the summer of 1884, the survivors of the Lady Franklin Bay Expedition (1881–1884) awaited rescue here. It was the second winter of waiting with their leader, Lt. Adolphus Greely.

Approaching the ruins of this starvation camp in silence, we could see that very little had been disturbed over the past century. Here a piece of worn-out clothing, there half a dozen rusting barrel hoops, the staves having been burned to provide warmth. Peter tells me that whenever he comes here, he brings an offering of food, remembering the men, especially those whose bodies were buried here.

Peter went ahead to the graveyard with his gifts while the rest of us separated in the chilly air and began walking slowly around the shabby bivouac barracks.

The place is called Camp Clay.

When the Second International Polar Conference met in Bern, Switzerland, in 1880, the United States agreed, for its part, to establish a research outpost along the northern reach of Ellesmere Island's east coast. In the late summer of 1881, Greely and his party were put ashore at Discovery Harbor, Lady Franklin Bay, where they erected Fort Conger, their winter quarters. Throughout the winter of 1881–82 and the following summer, Greely, his twenty-two men, and their two Inughuit dog handlers, hunters, and survival experts, named Thorlip Christiansen and Jens Edward, conducted field surveys and laboratory research. When the rescue ship that was to have come for them in August of 1882 didn't arrive, they were forced to overwinter again. They resumed their field and laboratory work, hoping for rescue late in the summer of 1883. On August 9, running dangerously short of supplies at Fort Conger and knowing the seemingly capricious pack ice might be holding up the rescue vessel, Greely ordered his men to abandon the camp. His plan was to advance southward along the Ellesmere coast, watching for any sign of a food cache that might have been left behind the previous year by the vessel sent to rescue them. They kept a sharp eye out, too, for a second rescue vessel, one they fully expected to encounter in the weeks to come.

The party of twenty-five men (they'd left twenty-three sled dogs behind at Fort Conger), carrying a sixty-day supply of food packed into a launch and three boats, found no food caches along the way and saw no sign of a ship. By mid-September Greely had given up hope of being rescued that year. The party put ashore at Eskimo Point on Johan Peninsula, about twelve miles south of Pim Island, where they managed to reconfigure several old Thule winter houses to create winter shelters. The Inughuit hunters, who were described by the party's second-in-command as "worth their weight in gold," killed a 600-pound bearded seal, which greatly improved their overwintering prospects. Some weeks later one of the Inughuit, Jens Edward, drowned while hunting seals to secure more food for the months ahead.

One of the search parties Greely sent out from Eskimo Point that autumn discovered a small food cache at Cape Sabine on Pim Island. Greely decided to move the entire party there, despite the strenuous effort they'd already made to create a winter quarters at Eskimo Point, and despite the fact that it would have been far easier to move the food at Cape Sabine to Eskimo Point than to move the boats and the rest of their equipment and supplies north to Cape Sabine. The men were openly critical of Greely, and Greely realized his men were beginning to doubt him. It was the start of a chaotic time.

At Cape Sabine the men built a low stone wall in the shape of a square and inverted the launch over it. The boats' oars served as rafters and the sails were sewn together to make a ceiling. The whole of it was roofed over with a layer of sod. They chinked the walls with clots of earth, with socks and other bits of cloth. (The supplies that had been left for them at Cape Sabine in the fall of 1882 were so inadequate they seemed an insult. The rescue efforts of both 1882 and 1883 were poorly organized and then badly managed.) Of the twenty-five men who went ashore at Eskimo Point on the last day of September 1883, only seven survived long enough to be rescued, and of those, only three seemed ever to fully recover. Most of the men starved to death. One committed suicide and another was executed for stealing food in a camp plagued by theft and mistrust.

I TURNED AWAY from the party's now-roofless shelter, with its empty tin cans and abandoned scraps of clothing, and began walking toward "Cemetery Ridge," where Peter had been and where I would find, I knew, the ashes of the party's last fire, still undisturbed. The bare ground around the hut walls had been trampled flat a hundred years ago. Denuded back then, the ground had never recovered. It was muddy now from summer melting in the upper level of permafrost.

On the ridge, a glacial moraine of coarse gravel, I found a series of parallel depressions. The first graves, empty now. (The bodies of five of the dead buried there were washed out to sea before two rescue ships, the *Bear* and the *Thetis,* finally arrived to take aboard the living and the bodies of the exhumed.) One survivor, Corporal Joseph Elison, who'd lost his feet and seven fingers to frostbite and weighed only 78 pounds, died en route to Godhavn, Greenland. There the body of the other Inughuit man, Thorlip Christiansen, was taken ashore and turned over to residents. The body of Jens Edward was never recovered.

I was already familiar with many of the details of the overly optimistic Greely expedition, and with the failings and ineptitude of some of his men. I also knew about the lack of integrity and the absence of courage that had characterized the first two rescue missions, which were the indirect cause of the eighteen deaths at Cape Sabine. I could not in that moment on the ridge, however, find the frame of mind in which to assign blame to anyone. I felt only sorrow. That I could not feel any admiration, though, shamed me. By this point in my own life, I knew how easily things could go bad for a small party in the Arctic, how Death could draw near, even in the long days of summer; but there was at Camp Clay an element of vainglory, of willful ignorance, that undid my instinct for compassion, disrespectful of the dead as this is to say. Greely's party, like Sir John Franklin's party exploring here forty years before Greely, had no clear idea of what they were getting themselves into, or of how flawed their idea of establishing a "supply line" of small food caches in country like this was, how far short of perfection their survival skills were. The two Inughuit had served the expedition well and faithfully when they might easily have put themselves first and walked away. Still, Greely wrote that these Eskimos were "unable to appreciate the objectives" of his expedition. Most of Greely's party, too, condescended to the Inughuit's knowledge of how to survive in these circumstances, feeling it beneath them to inquire among such people, to adopt their strategies or their methods.

If there is a human tragedy at Cape Sabine, these notions of racial superiority are a prominent part of it.

I found myself greatly conflicted, staring at the depressions in the gravel ridge and slowly walking the whole of this "God forsaken" site. I picked up a strip of shirt cloth with a button still attached, the broken frames of a pair of glasses. I set them both back down. I climbed over a jumble of red granite boulders and ascended the larger of the two ridges that cradled the camp and watched my friends below, walking slowly with their heads bowed, unsure of how to behave, how to regard the evidence.

We were glad, each one of us, that we had not been called upon in our lives to endure what went on here.

TEN MONTHS BEFORE I arrived at Cape Sabine I stood on the tidal flats of the River Dart, in Devon, in conversation with the polar explorer Wally Herbert. He and three companions had arrived at the North Pole with their dog teams on April 6, 1969, probably the first people to do so, the enduring claims of partisan supporters of Admiral Robert Peary's claim to have gotten there in 1909 notwithstanding. Herbert would later write of Peary's "self-destructive craving for fame" and expose in a book called *The Noose of Laurels*, in relatively gentle and compassionate words, the dishonesty behind Peary's claim.

I asked Herbert how he imagined the conversation might go if he were to find himself in a room with Peary, Robert Falcon Scott, Vilhjalmur Stefansson, Roald Amundsen, and the other polar explorers of a hundred years ago, if it was only them in the room, no reporters, no one who had not been to the extreme outer edge of human endurance in that environment.

"If there were no others, just us?"

"Yes."

"We would be respectful. Compassionate, perhaps even solicitous about each other's health. It wouldn't be necessary to say much."

All of these men, Peary most all, have been criticized as vain and self-promoting, all but the Norwegian explorer and Nobel Peace laureate, Fridtjof Nansen.

After Herbert answered my question, he went back to watching a flock of sheep grazing in a lush pasture with stone fences on the far side of the river. The tide was going out and the mud flats around us had become more extensive. On a steep upland behind Herbert, I could make out through a copse of trees three large estate buildings. This had once been the home of an English explorer and master mariner both of us admired, John Davis (1550?–1605).

"I had a talk with Peary about this once," Herbert volunteered.

In 1985, he went on, he'd traveled to Washington, D.C., at the invitation of the National Geographic Society, a longtime champion of Peary and of his claim. They wanted Herbert to write a definitive article about Peary's quest for the North Pole for *National Geographic*. With the society vouching for him, the Peary family allowed Herbert to examine Peary's personal diary from the spring of 1909, the record of his approach to the pole, which until then they had declined to make available to historians.

While he worked on the article at the society's headquarters in Washington, Herbert told me, he occasionally went to visit Peary's grave, across the Potomac River in Arlington National Cemetery, in Virginia. He'd sit on a bench opposite Peary's catafalque. One day, after he'd finished the lunch he usually brought along, he climbed the steps of Peary's monument and placed his hands palms down on a stone slab directly above the coffin.

"I said, 'Why did you lie? You know I've been there, and you know that I know you lied. Why?'"

Herbert told me that at that moment Peary's coffin became visible to him through the stone. It began to rise and seemed to shed water as it did. He was able to make out Peary's face. His eyes were open and he was staring at Herbert, but he didn't speak. Herbert repeated the question. Why had he lied? Peary continued to stare, and then the coffin began to sink. Water washed over it and Peary's face became distorted. The coffin stopped and began to rise again, until Herbert could see the face clearly once more. Peary looked at him without expression, Herbert told me, and then said, "Be kind."

Earlier that day, in his studio, Herbert had shown me his Xerox copy of Peary's 1909 diary. Using maps of the regular patterns of sea ice movement around the pole for reference, he took me page by page through Peary's diary, pointing out faint pencil marks in the margins where, he said, Peary had calculated the numbers he would need—the geographical positions and the distances in miles-made-good over the moving ice. Peary then went back and inserted these numbers in spaces he'd previously left blank while writing his daily entry. What he'd done, Herbert explained, was to determine the figures he would have had to have in order to support the claim that he had actually gotten to the pole.

The historical issue, Herbert told me, was not really whether Peary had reached the pole. The real question was why he had lied about reaching it. It was a few hours after this, sitting on rocks on the tidal flats of the river, that Herbert described the talk he'd had with Peary that day at the cemetery.

Herbert later sent me a limited edition print of one of his paintings, which has hung ever since in my studio. It shows ten dogs approaching a landfall over the sea ice. They're arrayed in a fan hitch, pulling a sled over crusted snow. It's clear they've been traveling nonstop for some time—the dogs' hitch lines are tangled from their crisscrossing behind each other for miles. It's also clear from tracks in the snow that three other sleds have recently come this way, though these sleds are not visible in the distance. The painting is from the point of view of the last sled driver—Herbert—and the place in the background is the north coast of Spitsbergen, in the Svalbard Archipelago, the first sighting of land for Herbert and his companions in more than a year.

Two of the ten dogs are looking back over their shoulders at the driver. They, too, seem to understand what they'd accomplished.

THE EVENING I WAS planning to listen to the Beethoven Ninth I abruptly changed my mind about how to go about it. I took the tape player and my day pack, left the tent, and headed across the island's isthmus for the Sverdrup site.

Much of classical music encourages, for me, thoughts of a physical or metaphorical landscape somewhere. I have to pause and recall the meaning of some of the technical terms—adagio, toccata, fugue—but it's a relatively easy and straightforward task to distinguish the four movements of a Beethoven symphony, and to remember from one year to the next (sometimes) the structure of a symphony like the Ninth, an homage to the brotherhood of mankind, inspired by Friedrich Schiller's poem "Ode to Joy." The moment in the symphony I most frequently recall is when the baritone's voice is heard for the first time, in the fourth movement. Emotions raised by the music alone are suddenly invigorated by the sound of a human voice. The abstractions of the orchestra—the musical tones—are joined in that moment by literal words, language sung by the soloists and the chorus. The effect is so profound that some listening intently during a live performance momentarily lose their composure. You can sometimes hear it in the audience's quick inhalations.

For as far back as I can remember, a feeling of affection toward what the poet Adam Zagajewski calls "the mutilated world" has welled up in me when I've listened to the Ninth Symphony, to Mahler's Second Symphony, to Bach's *Passion According to St. John,* or to the contemporary music of the Estonian composer Arvo Pärt. In my experience, a change in the quality of light falling on a hillside or a single choreographed movement by a ballerina might as easily release in someone else similar feelings of tenderness toward the wounded world, and feed the hope that these wounds might somehow be healed.

My faith in the capacity of certain works of art to break down ingrained prejudice, to undermine cynicism, to open a calloused heart, was strong at this particular point in my life on Skraeling Island, but fragile.

Leaving the tent, I had no inkling of the size of the error I was about to make.

I crossed the saddle of the ridge on Big Skraeling and descended into the old Thule camp. I wanted to listen to the Beethoven Ninth in the presence of the ghosts I felt were here. During my days on the island, looking at objects the Thule had made, I'd come to admire them, in that uninformed way one might feel admiration for a grandparent, having had an unanticipated but deeply illuminating conversation with her or him. I could put aside the violence and ethical failures of which I knew the Thule, like every other people, were capable, their lack (in the Western mind) of ambition or of any focus on the sort of progress that my own culture values so highly.

It is the case, I think, that it's what is decent, brilliant, and wise in a people that now we most need to know more about, and need to share with each other, not the banal evidence of their miscalculations or the supposed absence in them of the kind of sophistication we imagine ourselves to be in exclusive possession of.

Crossing the island, I thought it was the *light*, not the darkness, in ourselves that best characterizes our efforts, and that it is this very light that we're most in danger today of not recalling.

The Ninth Symphony, with its "titanic emotions," its faith in the Divine, its unrestrained evocation of selfless generosity, is an expression of belief in the brotherhood of all humans. I wanted to sit in the karigi in the Sverdrup camp, with the Thule ghosts occupying those stone benches, and play this music for them. Even if it proved to be only cacophony in their ears, tedious and grating, I wanted to offer it. A gesture of respect toward those who had repeatedly faced extreme difficulty and found a way through. I imagine that their triumph was akin to that of my own ancestors in Europe, Magdalenian Cro-Magnons, or of Afghan agriculturalists in that war-torn country today, continuing to plant and harvest and feed their families in the river valleys of the Hindu Kush.

I stood up in the middle of the karigi and spoke quietly to the stone benches. I said where I was from, what my culture valued, and I recalled some of the worthy things my culture had done. Just a few sentences. I said I admired them, admired their success, and that this music I was about to play was what someone in our culture had created, and that for almost two hundred years my own people had considered it one of our greatest works of art.

I set the small tape player upright on the stone floor and pushed PLAY. The tinny, dimensionless sound unfurled in the cool air. I imagined in that moment that the sod houses around me were a herd of bison, turned in for the night on a grassy common. But with the robust assertions of the early chords of the First Movement, I began to sense my mistake. What I was doing seemed suddenly so incredibly ignorant that the brilliance of the music couldn't suppress a sense of humiliation that began to rise in me.

I let the First Movement play out, Beethoven's sketch of humanity's struggle and triumph. I wanted to believe that what I was hearing was, in Richard Wagner's words, "a struggle conceived in the greatest grandeur of the soul contending for happiness against the oppression of that inimical power that places itself between us and the joys [that are offered to us]." But in that moment those thoughts seemed irrelevant. What I had done was not simply ignorant. It was evidence of an arrogance I was mortified to discover in myself.

I turned the tape player off and stood there on the flagstone floor facing the empty benches for a few moments. I put the machine in my pocket. Nothing to be said, really. I desperately wanted to find words that would reestablish some common ground with the Thule ghosts. Instead, I apologized for my intrusion, thanked the audience I imagined sitting there for their tolerance, and retreated. I walked backward until I passed through the entrance to the karigi and stepped into the world outside.

Making my return over the ridge, I recalled how often in the past I had resisted the impulse to share those aspects of my own culture that I admired—our capacity for generosity, for example; our willingness to respond in an emergency—with members of a culture that had experienced the brutal force of colonial intrusion. I knew the only right gift to offer people in these situations is to listen, to be attentive. In those circumstances, giving in to the urge to say something is often only self-indulgent or self-serving. The voice of my culture has already been heard, repeatedly. Loudly. It would have been better that night to have sat at the entrance to the karigi in silence and let the hours pass.

I knew this, but in my earnestness, I forgot.

As I descended the south slope of the heights on Big Skraeling, my thoughts were jumbled. It was easy for me to understand how I had betrayed myself, that pride was what was behind it. What I couldn't understand was the sadness pressing down on me, a sadness so deep it pushed aside the feelings of self-recrimination. Was it because I had to accept in myself the mistakes that can occur even when one has nothing but good intentions? Was it that my faith in the centrality in all cultures of "the beautiful," in the primal importance of the numinous dimensions of life, and in the possibility of defusing the tensions in human societies around race and cultural differences was childish?

This terrible moment triggered a crisis of self-confidence.

SOME YEARS AFTER this experience at the Sverdrup site, a mutual friend brought Arvo Pärt and me together at a house on the coast of Oregon, a few miles from Cape Foulweather. The Estonian composer and his family were renting the house for part of the summer, and our friend was hoping Pärt and I might embark on a project together. At the time, Pärt was trying to finish a large-scale piece called *Adam's Lament*. Working on it, he told me, was a continuing struggle. He said he "did not have enough tears" to return to work on it just then.

Pärt's music is austere and contemplative. Human suffering and Divine consolation are prominent in his compositions, and the resolution he sometimes finds is majestic. Our conversation, predictably, touched on compassion and despair in our personal lives and work. He described the political oppression he'd grown up with in the Soviet Union, a psychic wound that still had not healed, he said, though he and his family had recently moved to Berlin. We talked about the social responsibility of the artist in developed countries, where some people are impatient with intellectual complexity and opposed to the mixing of cultures, and where the intelligentsia were often suspicious of "beauty."

At one point I tried to explain to him the effect a short composition of his, *Cantus in Memory of Benjamin Britten*, had had on me. I told him about my experience at the Thule site, the collapse of self-confidence that had followed, and how difficult it had been to regain my composure. The source of my grief, I felt, was mysterious, but in the seven minutes it took me to listen to his *Cantus* for the first time, I said, my anxiety around this event gave way to peace.

I saw puzzlement in Arvo's face. His English was not as good as his wife's. Her English was excellent and she was helping us out. We were standing on a balcony overlooking the beach and the ocean beyond. Nora took hold of the front of my shirt and of the lapel of her husband's vest and began gently rocking us back and forth. She began to cry, whispering, "Yes, yes, yes," telling her husband he had been understood, telling me that what her husband composes can reassemble a person.

———

THE AFTERNOON OF our last full day on Skraeling, I packed up nearly all of my gear so I wouldn't delay our departure the following morning. I only have to roll up my sleeping bag, collapse my tent, and shove a few things into my duffel. The weather is damp, misty. I put on a pair of waterproof wind pants and pull on a waterproof anorak over a wool shirt and down vest. The wool pants I'm wearing have been repaired in three or four places. I once burned a hole in one of the socks I'm wearing, by putting it too close to the cookstove to dry. I'd had to stitch up a spot on the back of the vest where a sled dog had bitten me. The working history of these pieces of well-worn clothing is part of what makes them comfortable. The repairs are reminders to be careful.

I help sort and pack what remains of our food and protectively wrap artifacts from the sites. Peter and I make a last run in our inflatable Zodiac to the mouth of the Twin River to fill water jugs. It's been on my mind to ask Peter why he thinks that the form of everyday tools either doesn't change or changes only in minor ways over long periods of time, throughout Dorset history, for example. In Paleolithic caves in Cantabrian Spain or the Dordogne Valley in France, the style of Cro-Magnon painting doesn't change much either, even over thousands of years. Why? It's not the right time for this question, though. We're having trouble hearing each other over the sound of the outboard engine, and Peter is trying to concentrate on navigating through loose pack ice.

I never follow up on this with him. There's too much to do on our last day, too many details to remember; and Peter wants to get in one last session of concerted effort at one of the ASTt sites before the day ends. While we're filling the water casks, however, a related cultural question starts to take shape in my mind. We've long known that biological evolution explains the development of myriad life-forms over time. Evidence for the geographical radiation of physical forms, based on a relatively few body plans, is especially prominent in fossil records from the time of the great biological extinctions at the end of the Permian and Cretaceous periods and, before that, perhaps at the close of the Precambrian eon. Scientists, however, have only recently begun to raise questions about the *psychological* evolution of any single species,

because the evidence for it is not preserved anywhere and meaningful psychological evolution is generally not expected to have occurred
except in the hominin line. Up until the end of the Pleistocene epoch,
about 12,000 years ago, behaviorally modern *Homo sapiens* (as opposed
to anatomically modern *Homo sapiens*) remains virtually unchanged for
more than 200,000 years, although *Homo sapiens* has of course by then
developed various cultures far more complex than those of any of his
hominin predecessors or relatives. Until nearly the end of the Pleistocene, *Homo sapiens,* like all other animals, had evolved almost entirely
in response to the forcing pressures of his *physical* environment. Today,
the rate of change in his *cultural* environment, especially the rate of
change in the evolution of technologies like information processing,
electronic communication, and artificial intelligence, has become so
great that some in the older generations of the most technologically
advanced societies have lost a fundamental rapport with people only a
few generations removed from them. They process and evaluate information too differently.

   Homo sapiens is now probably evolving more quickly in response to
changes in his cultural environment than in response to changes in his
physical environment.

   To put this in other terms, perhaps in terms that are too simplified,
questions about the history of *Homo sapiens*'s physical evolution have
nearly been eclipsed in importance in the twenty-first century by questions concerning the evolution of the human mind. And by questions
about how the *cultural* world of human beings—their art, economies,
technologies, forms of government, and social organization—might
reflect or influence human psychological evolution. While changes
in the physical environment seem poised to have a significant impact
on the physical evolution of *Homo sapiens* in the relatively near future
through selective pressures exerted, for example, by global warming
and exposure to industrial toxins and viral disease, the pressure being
exerted on *Homo sapiens* by changes in his cultural environment is
much more difficult to track or even discern. The urban cultural environment of major cities might soon become too electronically complicated for many older or rural people to effectively manage, and *Homo
sapiens* might not have the ability to prevent a kind of chain reaction

that will leave a significant part of the human population without the psychological resources to cope with the challenge of a physical environment suddenly hostile to human survival.

This of course is an apocalyptic vision, and also one based on an imperfect understanding of complex threats that remain poorly defined. It also ignores the capacity of human beings both to imagine and to build another environment, different from the one they find themselves in. But the setting on Skraeling Island—human survival with limited resources in a demanding environment, achieved by societies of practical and resourceful people—urges one to consider those unsettling moments in which an accomplished, professional person in a field like law or medicine becomes flummoxed by a world of electronic communication and information storage in which terms like "access," "lost," "authoritative," "verify," and "private" have come to mean something different from what they once meant. The usual dismissive reaction to this worry, to imply that there is nothing to be concerned about, one just gets a younger person to engineer the necessary exchanges, fails to recognize that the older person is potentially a reservoir of knowledge that will disappear if he or she is not one of those directly managing the nuances of an electronic exchange. And it fails to take into account the process of disintegration that has occurred historically in human societies where a dominant society effectively eliminates a smaller culture not through genocide but by simply ignoring its elders' objections and focusing instead on converting its children. By inducing, or forcing, children to speak the dominant society's language, and by installing them in civil service situations, where they often become indispensable as translators and negotiators, the dominant culture effectively erases their birth culture.

DURING THE WEEKS I spent on Skraeling, it was the lifeways of the Thule and their predecessors, as I've said, that most preoccupied me, primarily the striking fact that they lived on the absolute outer edge of a geography that historically had rarely been inhabited by human beings, and that they nevertheless discovered, or invented, here, despite the cold, the dark, and widely scattered sources of food, the materials and ideas that worked for them. What they put their faith in, what images dominated their dreams, where love and beauty and tolerance fit in the thoughts of their elders, are things archeology cannot recover. These things might be intuited, however, in conversation with their progeny, the Inuit and Inughuit, or in conversation with elders among those cultures not so thoroughly separated from the world of the Thule as my own culture is.

In every culture in which I have encountered formal elders, the people who carry the history of what will work and what won't, I find them to be among the relatively few people in those cultures adept at thinking and working outside the constructs of their own metaphors and myths, while at the same time attending to the ways in which their history is compelling them to act. They know the difference between a world that is being imposed on them and the freedom to choose the life they want. What upends elders is the seductive attraction of that imposed world—the allure of material comfort and wealth, the advertiser's promise to satisfy every appetite. All of this they regard as venal. They see succumbing to it without questioning it, without resistance, as a desire to die.

THE LAST DAY before a group decamps from a remote place is usually a helter-skelter affair. So many tasks to address, so many last-second decisions to make before the scheduled arrival of a bush plane or a helicopter which can't be kept waiting long. I'm happy Peter has managed things in a way that gives us an additional few hours to work calmly at the Oldsquaw site. (*Clangula hyemalis,* once called oldsquaw, is now called long-tailed duck.) I like the orderliness of archeological work. Being familiar with the not-quite-boring techniques for effectively scouring the floor of someone else's former shelter makes me feel comfortable and useful here. The possibility that something unknown might be revealed is exciting, the possibility that one of us might unearth an object that has not been exposed to light for more than a thousand years, perhaps a small Dorset carving of a polar bear's head, cupped for all of us to look at in someone's dirty, wind-chafed hand, swollen and pink from the cold.

Before we hike up from camp to work the Oldsquaw site for the last time, I visit the sheltered spot on Little Skraeling I've found above the Grave Rib site, the knapper's station, where I finish a sketch map of where I've been on the island, so I can recall later where things were. And I search in vain for a small sheet of folded paper on which I've written the dimensions of the hut at Camp Clay, to stick in my Skraeling notebook. Was it a brass button from an officer's tunic that I'd seen in the mud there that day? (Later, Peter says yes.) Considering some of the dwellings I've examined here, I try to recall what I've read of the Eleusinian Mysteries, the ceremonies of death and renewal Greeks performed at Attica 2,500 years ago, during the same years Early Dorset people might have been enacting a similar ritual about the inevitability of death and their hope for the return of light. An enactment of the Dorset Mysteries.

I consider again my sense that the Thule created out of the barrenness of this land and its darkness something aesthetically pleasing, that they might have viewed their dwellings, in fact, as appealing according to principles similar to the old Japanese principles of beauty in spareness, of *shibui* and *yūgen*. Perhaps they knew how to turn the darkness and the cold and the absence of life inside out.

I'm neither the disinterested observer on this island nor the skilled archeologist, only someone taking notes and making drawings on the edge of minor and major mysteries. Seven years before I came to Skraeling, I was camped with a small group of Inuit hunters on the sea ice of Admiralty Inlet, at the northern end of Baffin Island. They were hunting narwhal. The ice camp had been set up next to open water on Lancaster Sound where narwhal were milling, waiting for the ice in Admiralty Inlet to break up. I was involved with the gutting and butchering of these small cetaceans. Eskimos, in my experience, are generally not comfortable with a white man writing things down in a setting like this, where wild animals have given over their lives to the hunters. Too often trouble follows, usually in the form of indictments from people in unrelated cultures who criticize the hunters as barbarians and condemn their way of life. For this reason I work up my notes only when I am alone in my tent, out of sight. Still, my hosts intuit my frame of mind, my way of life. They call me *naajavaarsuk,* their word for ivory gull.

Naajavaarsuk, a colonial seabird, is the only purely white seagull. But it wasn't the color of its feathers that caused the Inuit hunters to choose that name. When seagulls crowd a gut pile on the sea ice, left there by polar bears (or Inuit hunters), it's the larger gulls—Thayer's gulls, great black-backed gulls, and glaucous gulls—that monopolize the offal, muscling each other aside. The smaller ivory gull tends to stand on the perimeter of the action, darting in to snatch something when there's an opening. It was this way I had of actively participating, but then stepping back to observe what others were doing, that brought the first man to say naajavaarsuk.

PETER, KAREN, AND ERIC ARRIVED at the Oldsquaw site soon after I did. Weeks before, they'd set up a string grid that divided the flat ground around the site into squares about twenty centimeters on a side. The squares are numbered and they're keyed to numbered squares on sheets of paper. Anything found within a square is recorded on the matching square on that sheet. Like James Cook, with his lines of latitude and longitude, my companions had found a grid of squares helpful. The precision here implied that the coordinates themselves had a kind of authority.

What was learned here was mostly learned by taking dwellings like this Late Dorset structure apart, in order to speculate productively about how everything here might work. It's a laudable endeavor, but what can be said of a culture like mine, which can't leave what it finds alone? Are we not in some way just like the Thule, whose habit of dismantling a Late Dorset structure to make one of their own we sometimes rue?

My days on Skraeling Island had mostly been about two things: gathering knowledge and accumulating experience, the kind of experience that feeds, metaphorically, into other experiences, illuminating both of them if one is lucky. I was on the outside looking in here, like a director watching some other director's staging of a play like *Richard III,* but also like an actor *in* the play, looking out into the audience, into the face of the reader. Being here was an opportunity to share an experience, and then leave the reader to draw his or her own conclusions, against the backdrop of many other stories the reader has already heard about our human relatives and about those two perennial questions: Who are we and where are we headed? The narrative of our ancestors, this story in which sophisticated people supposedly replace "primitive" peoples, is far from being straightforward. It might not even be correct.

If no one really knows what the world is actually like, whether it's there in three dimensions or ten, whether the urges to love and forgive or to murder and abuse represent one road or two, whether the restlessness and appetites of the contemporary world are the first sign of an eclipse or not, then I'd be pleased to know what these mute Thule might have to say about it.

WHEN PETER EXCAVATED the karigi at the Sverdrup site, where I'd thought to inform the Thule about beauty, he discovered a Late Dorset figurine hidden beneath a flagstone, a wooden carving of a face. He and his colleagues have found related Dorset carvings in ivory— a needle case, a goose, an Arctic hare—all hidden beneath the floors of karigis by people who arrived after the Dorset. The wooden face from the Sverdrup karigi is an elongated visage, about two inches high. The mouth is open in what some believe is a scream. Small wood splinters have been forced into the face. Whether the scream is meant to induce terror or to register horror no one can say. But the Thule apparently saved these images of their predecessors to turn over in their hands and to consider in their own ceremonial houses.

When we finished for the day, after we had rounded up our tools and the others had started out for our base camp, I walked south to the edge of the Oldsquaw terrace where I could look down on another terrace below, which ran out to the sea. I was able to make out Thule burial cairns there, stone-lined food caches, the husks of excavated winter houses, and sets of long axial features (parallel lines of stones on either side of a central hearth, bisecting an oval living space), the latter characteristic of an ASTt site. I recalled, standing there, the last lines of Constantine Cavafy's poem "Waiting for the Barbarians":

> . . . *what's going to happen to us without barbarians?*
> *Those people were a kind of solution.*

I DESCEND TO the lower terrace, doff my pack, and sit an hour in a sheltered depression with my binoculars, watching the water. Mist settles on me and my rain gear gleams in soft light filtered through an overcast sky. I rise and shoulder the small gray pack with its books and pens, the bird guide, pilot bread, a thermos of water. I descend to the gravel beach and squat there like a primate, poking at the ground with a willow twig, turning over the sea-washed, ice-scoured rocks.

I walk past the dark blue tent, its entrance zipped shut against the mist and a rising east wind. I walk north and east to a pond where two eider ducklings paddle slowly, warily, over water corrugated by the wind.

Soon probably, I think, they'll learn to fly.

I sit awhile with the birds, until I realize their parents are circling, reluctant to land. I start walking toward the tents of my companions, obscured now by a slanting fall of white. (I'm reminded of that lovely phrase "snowflake Appaloosa.") I pass the place where I worked with them that afternoon. The string grid is gone, the Late Dorset site descending again into the anonymity in which it was first found. I am some yards past the first site before I turn back. I remove from my pack a package of pilot bread crackers. I place several of them on a flat stone that sits within the perimeter of the old dwelling. An offering, left behind in this Dorset narthex by a heathen.

The next day, on the opposite side of the island, a helicopter lands. It takes me and my companions away.

Puerto Ayora

Isla Santa Cruz

Archipiélago de Colón

Eastern Equatorial Pacific

00°44'36" S 90°18'32" W

t's too warm here, too humid, for me to sleep. Or I'm just not acclimated enough to this place yet. I've set the door ajar, opened a window opposite as wide as it will go to the night air, and tied the window curtains aside (which might otherwise have provided some privacy in this small room). I've done all this with the reasonable expectation that a cross breeze, an onshore wind, would materialize. Miles inland, air still rising from the old lava flows, like radiation from a slag heap at a foundry, should be providing the convection that would draw air off the dark ocean tonight, but it's not.

A film of sweat remains on my supine body.

This discomfort is only a minor distraction, however. In my hurricane mind, the churning of esoteric information goes on, thoughts I can't seem to organize well enough to create any point of stillness. Genetic variability, how that plays into speciation, is not a realm I've much familiarity with. Instead I have books and some correspondence with experts, an approach to learning that is one part gaping wonder, one part respectful critique. To gain a little more room in which to work out some of the biological problems, I've been reading the British cosmologist Stephen Hawking tonight on the unboundedness of the universe, the entity containing all other entities and still expanding, 13.8 billion years after it apparently initiated itself. Out there, so very far from here, I'm thinking one might lay out the genetic variables in the joinery of haploid cells expecting—given a blackboard that big—

to sort it all out. On my damp, narrow bed, I can only steeply arch my back and gaze upside down through the open window at the geography that preoccupies Hawking, the depthless black with its pulsing crystal dots. A region where there is no upside down. And where, I read once, there is no "nothing." Plasma physicists define the permeability of free space—the emptiest parts of the universe—as $4\pi \times 10^{-7}$ newtons/ampere². There are always atomic particles there, until you reach the edge of the universe and the province of true Nothing.

Stacked on a folding chair next to the bed are disquisitions on natural selection and genetic drift, refinements on the original sketches that Wallace, Darwin, and Mendel worked up, not that many years ago. Holding these loose pages down—in case the breeze does come—is a one-pound book about something else entirely, a popular history entitled *Satan Came to Eden*. In it one finds lust, self-delusion, and what some say is murder and others call pitiable death. The setting of this melodramatic Eden is the Galápagos Islands in the eastern equatorial Pacific, the place I'm lying in tonight being Puerto Ayora, on Isla Santa Cruz.[1] And Satan, as I understand the concept, is as complex an entity as any idea either Hawking or Darwin ever addressed, though probably active in the minds of a great many more people.

Farther from my bed is the room's second chair, on the seat of which are more books and several manila folders. On top of these is a folded letter from a friend, a man who came to this place in 1960 with a poorly planned and ill-fated expedition, hoping, along with his companions, to actually build an Eden here. They fell short of their goal.

Satan Came to Eden is a single minor work in the extensive library of books chronicling the cultural history of the Pacific Ocean. Like many of the others, it deals with the effect cultural illusions about this ocean have had on (mostly Western) explorers hopeful of discovering an Earthly Elysium. In the Pacific, no dragons waited in ambush along unexplored edges of the sea, as had historically been the case with the Atlantic. Here there were only teasing possibilities, cultures of, reportedly, voluptuous ease, made idyllic by "salubrious zephyrs" in places like Tahiti.

No Tahitian zephyrs are passing through my room this evening, but there is plenty here for me to read. Somewhere in it all, I know, is that geologically complex story of how the moon departed, leaving the crater outside my window that now cradles Magellan's "pacific" ocean.

THE FIRST HUMANS to put ashore in the Galápagos might have been Polynesian explorers, hundreds of years before Western cartographers began guessing where exactly to situate the larger islands and associated islets of this archipelago in Earth's newly discovered ocean. The isles lie on the equator, about six hundred miles off the coast of Ecuador, just south of an active tectonic boundary separating the Cocos and Nazca Plates. The Western discoverers were Spanish sailors and ecclesiastics, blown off course while sailing from Panama to Peru, in 1535, with Fray Tomás de Berlanga, the bishop of Panama, aboard. By the late sixteenth century, the Galápagos had become a center of operations for pirates and coastal raiders, as well as a regular watering station and depot for merchant vessels and ships of exploration. These ships frequently left goats and pigs behind, a form of insurance for shipwrecked sailors who might find these coasts. Also important to ships' crews were the islands' giant tortoises. Stored upside down in ships' holds and left inverted there, some lived as long as a year before being killed for fresh meat.

Shortly after Darwin's arrival in the islands aboard the naval barque HMS *Beagle,* in 1835 (Ecuador annexed the isles in 1832), the archipelago became a popular way station for American whalers. The *Plymouth,* out of Sag Harbor, New York, with the mestizo adventurer Ranald MacDonald aboard, anchored here briefly in 1845. The crew filled their water barrels and took on firewood, checked for any letters that might have been left for them in a barrel on Isla Santa María, and then deposited in that same barrel correspondence to be carried east by vessels bound for Atlantic ports. In 1845 whale ships outbound from here were mostly sailing for fresh whaling grounds in the Sea of Japan.

In his *The Encantadas, or Enchanted Isles,* Melville wrote that the Galápagos Islands present the sailor with a Plutonian sight. He compares the mountainous landscape, with its plains of volcanic rubble and its parched coasts, to the ruins of a burned-out prison colony. By not emphasizing fog-draped forests in the damp highlands of some of the larger islands, or the abundant (and docile) bird and marine life, and by demonizing the place as an Orphean haunt, Melville fixed the islands in the minds of many of his nineteenth-century readers as bleak and unwelcoming, and condemned the archipelago itself as a weird and exotic destination.

Melville used the imprecise Spanish word *encantada* to emphasize the feeling of mystery that emanated from the isles, a characteristic his fellow whalers often referred to when discussing the Galápagos. (Seamen also intended the term "Las Encantadas" to convey the fact that complicated currents streamed through the islands, making the archipelago difficult to navigate. Sailors maintained as well, even after the widespread use of chronometers made it possible to accurately determine longitude at sea, that the islands remained difficult to find.) Once Western scientists came to appreciate the complexity of the islands' terrestrial, avian, and marine life, and the resident animals' remarkable docility, *encantada* (when used by English-speaking people) came to mean something much less ominous—a bewitching place, more inspirational than funereal.

This once-elusory Pacific archipelago has become, today, more than a mere focal point for the study of island biogeography or adaptive radiation, two modern disciplines that have produced much of the evidence validating Darwin's and Wallace's idea of evolutionary descent by natural selection. The islands' unexpected preindustrial silence, easily and quickly accessible to a visitor to the islands' interiors or along its unpopulated coasts, is a rare and deeply satisfying experience for most tourists. Because of the presence of so many animals here indifferent to human life, and because of a peculiar "end of the road" undercurrent in Galapagean society, of people questing for a final meaning in life, a vision of the end, one might characterize this Pacific destination as *una tierra de los sueños,* a dreamscape. This remains true despite the fact that commercial ecotourism, illegal market hunting, government corruption, creationist dogma, and naïve schemes for amassing material wealth have been colliding here, dramatically, for decades.

The average visitor does not journey to the Archipiélago de Colón in search of scientific elucidation or of Pacific island history, but rather with hopes of finding that state of pure wonder so effortlessly attained here when standing, say, before a flock of ethereal flamingos, dozing

one-legged in a saline estuary, or before a lethargic cast of marine iguanas on a ridge of lava, creatures seemingly left over from the first days
of creation. People come here to clear themselves out, to rediscover an
inner core of tranquility in a place they believe to be uncorrupted.

On the dirt road that passes near the cabana in which I now lie, in
anticipation of an evening breeze, I once fell into conversation with a
local man about this feeling that so many visitors have of attaining here
a state of transcendent peace. Concerning the modern significance of
these islands, he had this reflection: *"La tierra puede transformar el alma
y lamente, y corazón de todos los hermanos."* ("It's possible for this place
to shift your soul, to ameliorate the pain of modern existence, to elevate
the heart of everyone who visits here.")

We'd begun speaking that afternoon, he and I, mostly in English,
about images of life and death in the islands. The intensely hued vermilion flycatcher of the high country, who will land on your head and
yank out a few hairs for its nest. The swallow-tailed gull, which will
fly backward against a headwind in order to settle more gently on its
nest; a bird, moreover, that has evolved to hunt successfully in darkness
far out at sea. A wounded fur seal, hauled out on shore rocks, a chunk
of shark-ripped flesh the size of a pineapple hanging from its flank.
A dying immature masked booby, hung up like a broken kite in the
branches of a palo santo tree, having misjudged a swirling wind and
now, as you come upon it, blinking its last looks.

The young masked booby dying in a palo santo tree, which I mentioned to this man, was not the only dying—or deceased—bird I came
upon the morning I visited their nesting grounds on Isla Genovesa.
There might have been twenty-five or thirty dead, immature masked
boobies. The opportunity to behold this scene of natural carnage, while
at the same time not losing track of the transporting beauty of orchids
in full bloom on the slopes of the islands' volcanoes, was one more thing
that Las Encantadas offered me as a guest: the possibility of holding
both images together in the same moment.

I GOT UP in the dark, determined to make something good out of my
sleepless night. I turned on the lamp by the side of the bed, a 40-watt

incandescent bulb, its weak illumination further muted by a crude lampshade fashioned from the translucent skin of an *Opuntia* cactus. Two geckos shot across the red concrete floor and sprinted up the white-washed wall opposite, hanging motionless there, tense as strung bows, while I pulled on a pair of shorts and a T-shirt. I closed the window, locked the door behind me, and set off for a breakwater of igneous boulders flanking Pelican Bay in front of the Hotel Galápagos's lodge.

Even at night, with its emphatic horizon missing, the Pacific is vast. Most of us perceive this ocean as a single undifferentiated entity, though Cook gave these waters to the cultural West with a distinct face, and not merely by discovering islands within what had once been a colossal emptiness—Norfolk Island, the New Hebrides, New Caledonia, the South Sandwiches—and confirming the location of islands like Easter Island, South Georgia, Tonga, the Hawaiian archipelago, and the northern and southern Marquesas, which might have been seen earlier by other mariners. He was the one who gave the Pacific a continuous surface. After Cook, scraps of the former Western Ocean—the coastal waters of the Americas, the Polynesian waters, the separate Philippine Sea—became, together, a piece of whole cloth. Cook anticipated, too, the widespread revelation, two hundred years later, of a unique Polynesian epistemology, a singular way of viewing the human-occupied world, one in which the primary frame of reference was not land surrounded by water but a mass of water containing widely scattered bits of land. The Polynesians, he intuited, were *ocean,* not land, dwellers.

Frequently, from the shores of Earth's five oceans and also from the bridges of ships, I've tried, hour upon hour, to understand the oceans not as waiting grounds, empty places waiting to be defined by an event, but as a type of consciousness. The modern oceans, evolved from the Panthalassic Ocean of the Paleozoic, from the Tethys of the Mesozoic, and a few other primal bodies of water, all of them moving on now toward something with yet another name, proceeding according to a clock different from the one I keep, also different from the clock that was ticking while thirteen species of Galapagean finch evolved from a single common ancestor, and different from the measure of geological time that applies to the evolution of the archipelago's shield volcanoes, rising from a single hot spot in Earth's mantle, a vent that periodically erupts through the Nazca Plate and is the explanation for the ancient heights and calderas we today call Isla Santa Cruz, Isla Sin Nombre, Isla Marchena, and so on.

Beneath a sky alive with universes, sitting here this May evening on a breakwater by the dark ocean, which sighs now and then in lazy susurrations against the rocks, I feel an urge to get closer to the water, to raise the Pacific out of its thinghood and into the personhood that Cook and others who sailed so very close to it knew. I picture the tracks of their voyages, a spider-web maze of lines drawn on a large map I keep at home, the routes that brought the Pacific out of obscurity for my ancestors: Magellan's crossing; the passages of the cultured English pirate William Dampier; the voyages of Tasman, Bering, Lapérouse, Roggeveen, Mendaña, Wallis, Heyerdahl, and Charles Chichester; of Spain's Manila galleons; and the tracks of the pioneering British research vessel *Challenger* in the nineteenth century. And of the Polynesian double-hulled exploration canoe *Hōkūleʻa* in the late twentieth century. Lines imposed by me on a surface that maintains no such records, and whose lack of roadways, water courses, and mountain spines compels the imagination to remedy the tracklessness, to make something up.

The Pacific's inconstant surface—what Shakespeare in *The Winter's Tale* called its "unpath'd waters"—was once a symbol of the unknowable. I sit before the unknowable then, imagining these years-long voyages. I try to see past what could have been the tedium of Sir Francis Charles Chichester's single-handed circumnavigation of the planet (he put in at only one port), recalling his harrowing doubling of Cape Horn. I imagine the giant squid, hunted a mile deep by a sperm whale, and Don Walsh sitting on the bottom of the Vitiaz Deep. I think of *Halobates,* riding the Northern Equatorial Current.

If I were to scribe a line on a map of the Pacific this evening, straight away to the northwest, it would not cut a shoreline but for Galápagos's for 6,098 miles, not until it came to the Aleutian Islands. If I were to draw another line straight south, it would not encounter a coast until it met the wall of the Abbot Ice Shelf in Antarctica, 4,993 miles distant. If I looked to my left and imagined the far-off Bay of Panama, and then to my right and envisioned the Philippine Sea, the span would be more than 10,000 miles. The Pacific is twice the size of the Atlantic, a comparison perhaps too incomprehensible to convey meaning. If in a cartoon, Mount Everest were placed on the floor of the Mariana Trench south of Guam, its peak would fall 6,800 feet short of the surface of the Pacific. If one were truly to comprehend the size of the thing, one would be halfway to imagining God.

I've sat still so long here, and my pupils are so dilated, I can make out three brown pelicans dozing on the water, not twenty feet away on Pelican Bay. Sometimes in the presence of such apparent innocence—these birds who are oblivious just now to all that is hidden and potentially threatening in the lightless world we share—I recall that line Conrad gives Kurtz in *Heart of Darkness,* when he tries to engage Marlow's imagination around the unaddressed barbaric nature of the jungle reality Marlow has stumbled into: "The horror! The horror!"

I am attuned to this. The seeming innocence of the birds and the starvation camp at Cape Sabine. A lyrical afternoon walking across the Alexandra lowland and the bodies of the luckless stacked like industrial kindling beside the burial pits at Banda Aceh.

The Bay of Panama. It lies below the indecipherable horizon there in the east. I can imagine it, picture Vasco Núñez de Balboa, standing on his historic (though today still unidentified) peak in Darien, toward the end of September 1513. He's ordered his Indian guides and his soldiers to remain behind, a few hundred feet below the top of this mount. As conquistadores went, Balboa was not as ruthless, not as vain, as hungry for silver as most. But I can see him standing there with the dog, his formidable *perro de presa,* Berganza. A soldier dog, trained by conquistadores to chase down Indians and tear them to shreds. Perros de presa were raised to be symbols of Spanish virility and prowess. Broad-chested, with short hair and a high, wide forehead above small eyes. The muzzle short, the mouth wide, the canines long. They stood to a man's knee. They were bred from bull baiters, canids used in the corrida dramas, dogs that created such a spectacle of gore and mayhem they were eventually replaced by human banderilleros wielding barbed picks.

It does not take much in the way of travel or the reading of histories to turn up the barbaric perros de presa nearly every culture has arrayed against those it hates, or those whose possessions it desires. When thinking of the conquistadores, who maintained the trappings of civility while they loosed their dogs to savage people, I think about the bankers in Amsterdam who first underwrote a nascent Portuguese slave trade out of West Africa, as immoral an enterprise as anything the Mongol pariah Timur Lenk ever imagined. And once those slave-based economies began to falter, with the rise of capitalized industrialism early in the nineteenth century, what of the English bankers, who took this for-profit commerce in human beings from Dutch financiers and the Portuguese, but were able to distance themselves later from Britain's history of rapacious behavior?

The moral oblivion of the slave trade. The piracy and ransacking of Spanish villages by Drake and the other West Country mariners. These things do not seem at all immediate in the modern world, nor any longer even relevant. Indeed, to recall them and to express outrage, regret, or sorrow is regarded by some as unworldly, as if conquistadores like Pizarro and perros de presa like Berganza were part of the West's uncivilized past, largely gone, or an unfortunate aspect of the human desire to possess, to exercise control. Most people do not wish to hear about what the historian David Stannard calls "the worst demographic disaster in the history of the world," the elimination of the Indian populations of the Americas.

For schoolgirls in northern Nigeria trying to run from Boko Haram raiders laughing at their panic, for impoverished Christians in South Sudan trampled by Janjaweed cavalry, for a family blown piecemeal across a city square by one of al-Assad's barrel bombs, the sixteenth century is now.

Back in my room in Hotel Galápagos, laid out on a ceremonial cloth with a few other things, is the eight-real piece salvaged from the *Nuestra Señora*. It reminds me not to forget the ease with which sixteenth-century Spaniards succumbed to the temptation to exploit the native peoples of the Americas, because their deaths did not matter, and, for them, there was no accountability. The coin fills me with wonder and dread whenever I pick it up. It represents the spirit of Leopold of Belgium, centuries later and closer to my own time, holed up in his country estate in Laeken, while his functionaries bled the Congo basin of everything marketable, working to death, murdering, or otherwise doing away with ten million Africans. It reminds me of the soldier-thug Joseph-Désiré Mobutu, who collaborated with Belgian intelligence and the Central Intelligence Agency in Washington in 1961 to assassinate Patrice Lumumba, Congo's first democratically chosen prime minister. Four years later Mobutu, with American support, staged a military coup in Congo, a country he would rename Zaire and rule as a dictator for thirty years, enforcing policies as indifferent to human suffering and misery as Saddam Hussein's, and, as Mobutu Sese Seko, amassing a personal fortune of some four billion dollars.

It is easy to misremember the Mobutus, the Batistas, and the others; or perhaps it is not possible to recall them at all clearly if the goal is to facilitate Western-style progress and its twin sister, profit, if the agreement is not to focus on the past but to lament these regrettable and uncommon aberrations, and then to move forward. What the coin tells me, though, nine-tenths of an ounce of Mexican silver, is that it is dangerous to believe the past is behind us, that a remedy for barbarism has been found. Is it not, in fact, barbarism that sits well dressed and well spoken today in a corporate boardroom in Frankfurt, Shanghai, or Delhi, as far from human suffering as the bombardier flying back to Tinian Island aboard the *Enola Gay*? Or is barbarism a term reserved instead only for those flying planes into the World Trade Center towers?

History tells us that with every great empire comes great barbarism, that the two are inseparable, so that to diminish barbarism you must dismantle the empires. This forces the question of what, really, civilization brings to people that they did not already have. And why is civilization so hard on the people who turn it down?

The inequity, the memory of it, upends my thoughts. I accept its inevitability, but cannot accept the scope of expression we permit it.

WHEN I STAND UP, the pelicans are startled. They paddle slowly away. No breeze yet, but it's cooler now. I'm curious about the town, Puerto Ayora, this late at night, and head somewhat aimlessly in that direction, over the small bridge that crosses the head of Pelican Bay. Two pariah dogs pause in the road to watch me pass. Neither domesticated nor feral, these are the town's scavenger animals. They make their living within the ambit of Puerto Ayora but are associated with no house. They do not venture out into the countryside, where they fear the feral dogs that travel there in small packs, subsistence hunters of iguanas and feral pigs, dogs not to be put off their hunts by a human with a treat or a handout.

From the bridge I move into town, past shuttered houses and shops. Nothing seems to stir here. I move like a specter through the streets. Laundry hangs limp in the humid air to dry. Children's toys lie inert on the ground. Eventually I find my way back to the water, at the end of a private dock at the Hotel Delfín. Looking across Academy Bay from here, I can just make out the Hotel Galápagos, though not the cabana beyond it where I am staying. The few lights that still burned on the hotel grounds earlier are now dark, the hotel's generator having been shut down at midnight.

A crescent moon, late to rise, silhouettes about thirty sailing and motor yachts and several sport and commercial fishing vessels anchored in Academy Bay. I once interviewed a man, Dennis Puleston, at his home on Long Island. He'd sailed to the Galápagos (and all around the Pacific) in the 1930s, while in his twenties. Later he wrote a book about his adventures, *Blue Water Vagabond*. In his eighties when he spoke to me, he struggled to translate his experience, speaking of it as if trying to reach me from another world, a world without GPS, without onboard radar to penetrate darkness and fogbanks, a person guided solely by his magnetic compass, his charts, by a feeling for the wind and the sea, and the look of a windward horizon. The Galápagos he knew, I understood him to say, is no longer there.

IN THE 1960s—I'll tell this story in its shortest form—a group of anthropologists, academics who disputed the idea that Polynesians reached the Hawaiian Islands, the Marquesas, the Societies, and other South Pacific archipelagos by accident, set about trying to prove that Polynesian navigators, departing their homelands in more densely islanded Micronesia, knew exactly what they were doing. The scholars intuited that Polynesian navigators had found their way south to New

Zealand, as far east as Rapa Nui (Easter Island), and as far north as Hawai'i by employing a sophisticated awareness of currents and wave and wind patterns; by noting the direction in which transiting birds such as black noddies and fairy terns were flying, early in the morning or late in the afternoon; and by using rising and setting stars to set and hold a course.

What the scholars most needed to know, in order to confirm their ideas and proceed with their work, was what the navigational techniques used by the Polynesians actually were and how they were employed; and what a Polynesian voyaging canoe—thought by them to be a double-hulled, catamaran-style vessel—looked like. Length, beam, draft, architecture of the hulls, design of the steering oar, and how its masts were rigged.

The anthropologists, maritime historians, and other researchers involved in the project eventually earned the respect of native Hawaiians, who joined them as colleagues. Working alongside each other, they settled on the rough dimensions of a practical double-hulled, two-masted vessel; on the canoe's rigging and the shape of its sails; and on the design of the sailing platform and other details. Then on the island of Satawal, in the Caroline Islands, they located a man still familiar with the techniques of traditional Micronesian navigation. A group of native Hawaiians apprenticed themselves to the Caroline islander, Mau Piailug, and began learning how to read wave patterns, cloud color and shape, ocean currents, changes in water depth, prevailing winds, and the presence of freshwater lenses sitting on the ocean's surface, the discharge of nearby rivers. These variables, however, were only part of what composed the dynamic system of open-water navigation that Polynesians had historically relied on. The other part was an ability to read the stars throughout the solar year, to monitor their changing positions as the hours, the days, and the seasons passed. The discipline needed to learn and then recall these movements provided the navigator with a "star compass," a mental construct by which he could set and hold a course. Polynesian navigation was fundamentally different from the system that grew up in the West, where a sextant, paper charts, a magnetic compass, and ships' rutters (logs) provided the most reliable guidance. The latter depended on instruments, which could be lost overboard, and on paper, which was perishable; and it was a system that was most accurate in stop-time. The Polynesian system was held in the mind, where it could not wash overboard or be misplaced; and its time frame was dynamic, constructed for use aboard a *moving* vessel.

Pulling together all the information they'd gathered, the group built and, on March 8, 1975, launched from a beach on Maui, the *Hōkūleʻa,* a traditional 62-foot double-hulled Polynesian voyaging vessel. The crew of young Hawaiians who'd trained under Piailug, who was sailing with them, navigated it unerringly across 2,500 miles of open ocean to Tahiti, using only traditional methods of navigation.[2]

A year after the *Hōkūleʻa*'s initial voyage, an archeologist named Yosihiko Sinoto found some parts of an 800-year-old voyaging canoe preserved in a saltwater swamp on the island of Huahine, in French Polynesia. The vessel had apparently been driven ashore and crushed by a tsunami. His examination of the debris—of its 18-foot-long steering sweep, for example—tended to confirm that the design group in Hawaiʻi had done an exceptional job of matching the *Hōkūleʻa*'s design to the design of the original prototype.

Thirty years later, when I visited the swamp where Mr. Sinoto made his discovery with him, the two of us were invited to join several others for dinner. There we were introduced to one of the *Hōkūleʻa*'s navigators and to some of the vessel's crew at that time. The crewmen prevailed on their navigator, a young Cook Islander, to show us the tattoo they'd chipped in to get him in Tahiti, after he'd proven himself. The modest navigator was reluctant but eventually pulled his T-shirt up over his head and turned his back to us. From the nape of his neck to his coccyx, the major stars of the Southern Hemisphere were inked across his back in their familiar relationships to each other. Riding the star pattern was a brightly colored marine iguana, the tip of its tail resting at the spot where a man's tail might have been, its body crossing at an angle over his spine, and its head swiveled to stare out at the viewer from the base of the man's neck.

Later in the evening, the tattoo now hidden again under his shirt, the young man explained to us what it meant to be able to navigate the *Hōkūleʻa*. At the urging of a few anthropologists, he said, some of his people had set about rediscovering a way to navigate without conventional Western instruments, across a seemingly empty wilderness of intimidating space. This came at a time, he told us, when dominant cultures around the world had begun to worry that despite their scientific and technological sophistication and their large reserves of material wealth, they were losing their way. They seemed to traditional people like cultures trapped aboard a rudderless ship, sailing very fast over a deceptively calm ocean.

"Once we too, Polynesians, felt lost as a people," he said. "Now we have something to offer others, a way to regain confidence."

IT WAS LATE when I returned to my room. Waiting here for me was John C. Beaglehole's reverent biography of Captain James Cook, which I was two or three hundred pages into; and my notes from William Beebe's *Galápagos: World's End,* which I wanted to review, now that I had returned to the islands for a third time. Beebe's 1924 book had turned many people's attention toward this Pacific outpost, and a few readers, believing the untrammeled archipelago was the paradise that might finally be their sanctuary, looked for a way to get there.

I'd arrived in Galápagos by myself, a week ahead of a group of fourteen people, mostly strangers, whom I was to join on a tour of the islands. I came early in order to have some time in the evenings just to read, after daytime strolls. But tonight the reading would have to wait. The heat and humidity that had propelled me from my room had abated some, and I'd sufficiently worn myself out now to be able to fall asleep at this late hour. Had the electric power still been on, the blast of light from the room's overhead bulb would have sent the geckos scurrying, but it wasn't and they were not alarmed by the beam of my pocket flashlight. I opened the window by the bed, propped the door open with a chair, wove a length of rope across the doorway to keep the pariah dogs out, and slid quickly sideways into sleep.

MY REGULAR ROUTINE here in the morning is inquiry. Every day I consider how fortunate I am to be free to wander in Galápagos. Not everyone gets to come. Pay attention to small things I tell myself. Look closely at what are clearly *not* the answers to some of your questions. Do not presume that later you'll be able to read about something you've witnessed today.

Even if I've gotten to bed late, I'm usually up at first light, which comes without warning on the equator. Just a few minutes, really, between full night and full day. The hotel generator starts up at six, breakfast is served at six-thirty. I usually go for a walk around Puerto Ayora afterward and watch people come to terms with the day—the unloading of foodstuffs at the market (fruit and vegetables from fincas in the highlands), children escorted to school, someone shouting at a recalcitrant internal combustion engine.

Some mornings I stop at the shipyard to see how repairs are going on a fishing boat that's got a rotting keel. Or I walk up to the Charles Darwin Research Station to use the library. One morning I borrow a small rowboat and cross Academy Bay to keep an appointment with Karl Angermeyer, whose family has been in Puerto Ayora for many decades. When we speak, he offers me some particulars of Galapagean history not in the books I've read, going over the particulars of scandalous events in the 1930s on Isla Santa María, for example, or relating a bit of folklore about Darwin that I know not to be true. As a paterfamilias in Galápagos, Mr. Angermeyer resembles other people I've spoken with in villages in South America or Asia who are simply trying to keep track of what is important to remember, a worldwide effort in small settlements to keep one's people from slipping off into caricature or oblivion.

Another morning I catch a ride with someone headed for the airport. My aim is to visit a resident in the highlands I'd met on an earlier trip, a man I'd taken an immediate liking to, Steve Divine. I hadn't been able to walk the extent of his finca in the highlands back then. Now I could. The narrow road north from Puerto Ayora to the airport takes a traveler to the north shore of Isla Santa Cruz, where a narrow seawater breach—a natural canal—separates it from Isla Baltra, where the airstrip is. The road first ascends into agricultural areas in the highlands, passing through the small settlements of Bellavista and Santa Rosa before descending again to a great plain of dark a'a and pahoehoe lava, a land hostile to nearly every seed and spore that falls upon it.

My acquaintance drops me off at the gate to Steve's farm. When we ran into each other a few days before in Puerto Ayora, Steve invited me up for a walk in the scalesia forest around his place. He's very well informed about the biology and ecology of the islands, someone with an attractive attitude of allegiance to the country all around him. As I've pursued my reading about the archipelago, both the popular literature and the scientific literature, and become more conversant with its basic geography and natural history, my conversations with Steve have offered helpful clarification and surprising connections, the type of insights only a residency, a true apprenticeship, can offer. He calls the palo santo tree, which conserves moisture by shedding its leaves in the dry season, a "hard to die" tree. And it was he who told me that on the islets of Plaza Sur and Plaza Norte, off the eastern shore of Isla Santa Cruz, I would be able to see swallow-tailed gulls flying backward against a headwind to land gracefully on their nests.

Steve is articulate about—and insightful regarding—smoldering resentments in the islands, most of these due to tensions between the staff of Galápagos National Park (about 97 percent of the land in Galápagos and virtually all the archipelago's near-shore waters fall under the jurisdiction of the park) and a group of settlers in the islands who feel the park's boundaries shut them out. They want their hunting and fishing activities to extend into the park. At the time I spoke with him, these resentments were particularly strong in the settlement of Puerto Villamil, on Isla Isabela, where residents had deliberately started forest fires inside the park's boundaries and had established illegal commercial fisheries in near-shore waters. Steve could see both sides, but had no respect for the villagers' violent way of registering their complaints. Too often in the past, when settlers have entered the park to hunt feral cattle or cut down large matazarno trees for lumber, or when they've defiantly ignored catch limits and harvesting seasons for lobsters, sea cucumbers, and other marine life, the shouting and insults voiced at town meetings have turned into fistfights. Once, villagers stormed the Charles Darwin Research Station, the symbol of scientific research in the islands and the strong international interest in the islands' protection, smashing windows and destroying years of scientific records. On other occasions they've deliberately killed tortoises living in the park. (The image of the giant tortoise, the park's iconic animal, appears on the park's logo.)

The root of this disagreement is class resentment. A relatively small, well-educated, conservation-oriented group of caretakers finds itself in direct conflict with a much larger group of working-class fishermen and subsistence farmers, mostly Ecuadorian nationals, who until they reached the islands were largely unaware of the international movement to conserve the islands' ecology. Its ideals are almost incomprehensible to them. Like the American colonists who arrived at Isla San Cristóbal aboard the *Western Trader* in 1960, these Ecuadorian nationals arrived in Galápagos with unrealistic ideas about an archipelago that has very little freshwater or arable land, and only minimal municipal services. Basic commodities like flour, cooking oil, and paper products are expensive and in limited supply; the infrastructure to support health care is primitive; there are few paying jobs; and thousands of well-to-do tourists, innocent of the economic and social complexities here, arrive every week. The visitors' money is eagerly sought, but many residents consider their presence a nuisance.

In almost every public meeting place in Puerto Ayora, you can hear the discussions about unemployment, class privilege, disputed jurisdiction around park boundaries, and schemes to circumvent or solve these problems. The fundamental disagreement among island residents is over the relative importance of the biological integrity of the islands. The disagreement is exacerbated by differences of opinion over the place of conservation in a human community with high unemployment, and it is further complicated by disagreement over the need for economic growth and development in the islands. Too, some Ecuadorian nationals deeply resent the investments international conservation groups are making to preserve the park.

For many years park custodians have argued, unsuccessfully, for a limit on the number of visitors to the island. (Most all visitors sleep aboard tour boats and take their meals there. They may come ashore, sometimes just briefly, only at a restricted number of sites in the islands, where guides try to ensure they don't wander off the established trails, disturb animals, take souvenirs, or discard trash.) When I came to the Galápagos for the first time in 1986, the annual limit, set by the Ecuadorian government, had just been raised from 18,000 to 25,000. The actual number of visitors that year was 32,000.[3]

Part of what motivates Ecuadorian nationals to move to Galápagos can be found in Ecuadorian folklore, the widespread idea in mainland Ecuador that anyone can make a fortune from "tourism" on "the Ecuadorian frontier." The fortunes to be made here, however, are mostly made by business acquaintances of politicians in Quito, who decide who will and who won't get a license to operate a tour boat in Galápagos.

The deeper one digs into the phenomenon of Galápagos, the more one finds the kind of thievery and injustice that infect ordinary life wherever in the world economic opportunity and political malfeasance drive "progress." My sympathies in the Galápagos, I found, lie both with the misled Ecuadorians who arrive here on government-subsidized flights from the mainland with a defective dream, and with members of the park's staff, an underfunded and underpaid group of dedicated scientists and conservationists trying to control illegal hunting, fishing, and timber theft and to mitigate the impact of more visitors than they believe the park can support. And my affection lies with immigrants like Steve, who are sympathetic to both sides of the conservation issue and who try to encourage in visitors the same degree of wonder they themselves feel about protecting the islands from economic exploitation. Some visitors to the islands have considerable economic and political power in their home countries, and people like Steve hope they will bring those strengths to bear on behalf of the archipelago.

ONE MORNING, on a previous visit, when Steve and I were having coffee on his porch, a friend came by to catch up on local news with him. While they spoke, I began making notes on the conversation Steve and I had been having. My attention, however, soon shifted from the page in my notebook to a patch of darkness developing on the surface of the ocean. It was taking shape in the middle of a vast sheet of incandescent light. The ocean, some miles away and far below us, was visible over the canopy of a forest sloping away from Steve's house. A huge vertical mass of gypsum-white cloud was sliding over the sun, dimming the polished silver surface of the water. My glance took in the great sweep of Isla Santa Cruz's southern coastal lowlands, where Puerto Ayora was, the treetops nearby, and finally an open grassland that abutted Steve's compound. Here beside me was a pond, on which a pair of white-cheeked pintail ducks were feeding. My eyes then settled on the glitter of an orb weaver's web, ballooning in a light wind. The sunlight was so strong, and the air it moved through so clear, that even at a distance of several feet I could see perfectly the tiny distinctive spines on the body of a yellow-and-black star spider.

Steve's friend departs. We sit in silence. The weather, despite the threat of a storm rising on the southern horizon, is conducive to well-being. The coffee excellent. Neither of us tries to voice a thought about the state of our ease.

I WANTED TO GET BACK to the hotel in time for lunch, so I said good-bye and set off down the road, hoping for a lift.

When the overloaded airport bus comes into view, swaying drunk-enly on its ruined shock absorbers, I can see that the luggage rack on the roof is already crowded with people. I find a place on the rungs of a ladder mounted on the rear bumper. By the time we reach the village, I'm covered with dust. It's caked on my skin because of the heat and humidity. I ask the hotel's proprietor, Jack Nelson, whether I can take a brief shower, my second of the day, freshwater being at a premium. Sure, he says—and he'll set some lunch aside, including fresh vegetables from Steve's finca, delivered the night before.

The interwoven nature of quotidian life in Puerto Ayora—Jack's sister, Christy Gallardo, runs the town library—makes me feel comfort-ably situated here. But wary.

Returning to my room, I give the manzanillo tree in the courtyard a wide berth. Its milky sap causes skin to blister and itch, like a brush with poison ivy. My room, cabana numero cinco, I had noticed the day I arrived, is across the road from the cemetery. Because volcanic soil is so difficult to excavate, most of the graves there are sarcophagi, with minimal decoration. The whitewashed concrete coffin boxes rest squarely on the hard ground. A few are elevated on pier blocks.

Always a caution, a cemetery.

On that first visit to Galápagos in 1986, I was so transfixed by the range and extent of bird and animal life on the islands, and by the inshore life that overwhelmed me as a snorkeler—by the seeming miracle of it all—that I missed initially how thoroughly and intimately life and death are mixed here. Paddling through mangrove thickets along the coast or pushing through heavy vegetation in the highlands, one is acutely aware of the living: small birds twitter and flit constantly in the understory of the scalesia forests. In sheltered lagoons, visitors float on transparent water above pods of spotted eagle rays undulating slowly past below. On coastal shores, flocks of ruddy turnstones and sanderlings probe the beach in search of food, along with black-necked stilts, oystercatchers, and yellow-crowned night-herons, and sally lightfoot crabs make short scampers across the surface of the water. On clinker plains in the lowlands, where there is little vegetation, death is a more striking part of the tapestry of life. My initial encounter with life's insistent companion, as I mentioned to the colonist that day, came on Isla Genovesa, the remnant of a volcano's collapsed rim, an island less than a fiftieth the size of Santa Cruz.

We had arrived at Genovesa on an overcast day. The motor yacht anchors in Bahía Darwin and the small group of us make our way up the Prince Philip's Steps trail to a lava plain that forms the island's flat summit. A strong onshore wind sweeps through the chirping of birds nesting among bare, mute-spattered rocks—thousands of red-footed boobies, masked boobies, great frigatebirds, and wedge-rumped storm petrels. The gaunt plain is strewn with sun-bleached palo santo twigs, with the castings of short-eared owls, and with the skeletons of large birds, stripped of their flesh by Galápagos hawks. The air is ripe with the odor of fish, which adult birds are disgorging for their chicks at hundreds of nests.

Clots of entwined cinder-colored marine iguanas, each the image of the stygian imps some sailors once thought they were, stare across this hysterical landscape, still as gargoyles. Mockingbirds snatch booby chicks, kill and eat them only a yard or two from the patch of pea gravel where an impassive parent sits. Short-eared owls emerge from crevices in the lava to savage the young of petrels. The wind sends frigatebirds, surprised by a sudden change in its force or direction, cartwheeling across the ground that surrounds their waist-high nests in clumps of saltbush.

The skeletons of birds deceived by the wind hang like auguries in the limbs of trees. Masked booby chicks, gorged on fish to the point of stupefaction, and lacking the muscle tone to bring themselves erect, lie sprawled on rocks beneath the trees. The larger booby chick in a nest occasionally kills the smaller one.

Above it all maneuver adult birds, just in from foraging in the ocean, dazzling in their acrobatic management of the shifting wind. The marine iguanas drop from the rocks into the same rich waters, like seagoing lizards. Tenacious cacti and muyuyo and lantana shrubs have found footing in interstices in the lava rubble and are thriving. The extent of death here burnishes life, and the vivacity of the living diminishes the tyranny of death.

I recall here a musical term: motet. Suspended over all that I saw was a thick cloud of birdcalls, raucous to some ears. A cacophony. A motet is "a vocal composition in polyphonic style on a text of some sort."

What text is this, here on Isla Genovesa, with the exuberant peeps of satisfied hatchlings and the squawks of those whose lives are ending?

SOME EVENINGS I skip dinner at the hotel and eat in one of the cafés in Puerto Ayora, hoping I might meet someone from one of the other settlements in the islands, Puerto Baquerizo Moreno on San Cristóbal or Santo Tomás on Isabela, which I have not been able to visit yet. I can also practice my miserable Spanish and try to get the run of Ecuadorian Spanish, different to many ears from Cuban Spanish or Argentine Spanish. One night I meet one of the park's guides for langostinos (prawns). The following morning, he tells me, he's taking a small group of volunteers by boat around to Tortuga Bay on the south coast of the island, a stretch of beach where endangered Pacific green turtles regularly lay their eggs. The group might see females coming ashore or even burying their eggs; but his main intention, after walking everyone through a lesson in the ecology and conservation of green turtles, is to have the volunteers clear the beach's wrack lines of jetsam, the garbage thrown from passing ships.

I'm welcome to join them, he says.

We do not see any turtles that morning but locate a number of pits where females have recently laid their eggs and covered them with sand. According to the guide, each pit shelters hundreds of eggs. The hatchlings will emerge at night, when darkness gives them a better chance of crossing the exposed beach below the high tide line. Once in the water, though they'll still face other predators, they'll be safe from mockingbirds, Galápagos hawks, and ghost crabs.

We leave the beach by boat at midmorning, carrying several dozen bright yellow bags filled with trash and feeling very good about our work.

That night I decide to return to the beach using an overland trail. I hope to see hatchlings emerging. The two-mile route leads to the beach over rocky ground and is flanked by dense shoulder-high brush. The air is muggy, but I can make out the trail easily enough in the light of a waxing moon. I don't need to turn on my flashlight. (In the half dark, a flashlight beam would only make the visible world smaller.)

Just as I step onto the beach, I catch movement at the water's edge. Mockingbirds. In the minutes before I arrived, a group of hatchlings had apparently tried to cross to the water. There might have been a scene here then as brutal, on a smaller scale, as tyrannosaurs attacking a scurrying herd of herbivorous ankylosaurs in an open grassland. There are no hatchlings now on the beach. A few moments and then the mockingbirds are gone. I find the pit the hatchlings came from above the high tide line and sit down on the sand nearby, where I have a view of open ground to the left and right. An hour passes. The absence of light and color make the beach even more still.

I'm adrift in memories of dark nights spent watching for animals in other places when I see, moving past my hip, a disk darker than the sand and half the size of my palm. I rise quickly and scan the beach near it methodically with my flashlight. It seems to be alone. The line of its advance is a determined strike for the water. I walk alongside it. No birds are aloft, but I soon see a ghost crab, then another, scuttling toward us. I fend them off with my foot, but they're as determined as the young turtle, circling behind me, charging in. I consider carrying the hatchling to the water, but this feels immediately like crossing a line—too much interference. How is one ever to measure these things? I stay with the skirmish, protecting the turtle until it's safely away in the surf.

I've seen what I'd come to see at Tortuga Bay and now turn back for Puerto Ayora, feeling suddenly very tired. A long day, and I've made it even longer with the hike out here. From time to time a cloud crosses the face of the moon, and it becomes so dark I have to feel ahead with my toes for secure footing. I don't want to turn the light on. At some point a large animal thunders past me. A dog. I hear its panting as it passes. The ticking of its nails on the rocks suddenly stops up ahead. It's looking back at me. I sense something else coming up behind me. Gooseflesh runs up my back and I swing the flashlight that way. No eye shine. An empty back trail. I shut the light off, to keep the outer dark from getting darker, and continue on. I don't hear the dog ahead anymore and am sure he's gone.

Images of feral dogs begin to crowd in. I hear a swish of brush and am certain the noise is behind me. Then the ticking of claws again. I leave the light on now and begin to trot, looking back over my shoulder and up ahead into the beam of light, eager now to reach the streets ahead. When I see the silhouette of the first house against the starry sky I realize I was much closer to the village than I'd thought, and slow to a walk. The dogs overtake me before I can react—four of them, calico-patterned, race by on the left. Ten yards ahead they break sideways into thick brush. I stand stock-still until I hear nothing again.

What was I thinking, running like that? That I could outrun what I imagined was there?

They were toying with me.

Two pariah dogs asleep at the edge of the village do not lift their heads from the dirt as I pass. Back in the room, sitting in a wicker chair, I try to separate the world of threat the hatchling faced from the world of vulnerability I had just reacquainted myself with.

A FEW DAYS AFTER my encounter with the dogs, I have lunch at the Hotel Galápagos with Bruce Barnett, a biologist studying feral dogs in Galápagos at that time. Many types of dogs go feral in the islands—terriers, spaniels, German shepherds, bulldogs, retrievers, Great Danes. All are phenotypic expressions of the same genotype—*Canis familiaris*. In the wild, most of these phenotypes disappear. The range of canid expression, in other words, shrinks. Depending on where and how they live in the islands, feral dogs occupying the same area begin to resemble one another after only a few generations. In a region of lava flows and sparse vegetation—above Bahía Isabel on Isla Isabela, for example—where feral dogs hunt mostly marine iguanas, Barnett found that the dogs have the same look despite their different ancestries: long-legged, bat-eared, short-furred, bare-bellied animals, adapted to the hot, sun-blasted lava fields over which they travel.

Barnett's research appealed to me. It had none of the cachet that attaches to studying the systematics of Galapagean finches. It was, instead, focused on the morphology and social behavior of an animal that, strictly speaking, doesn't belong here, and it was research related to the everyday lives of human beings, from whom the dogs had fled or who had abandoned them.

In addition to studying feral dogs, Barnett was involved in a more contentious issue, the eradication of populations of feral goats, pigs, cattle, rabbits, cats, and donkeys, each of which has had a major ecological impact on every island on which they've been found. (Feral dogs, pigs, goats, and cats were already established in the islands by the time Darwin arrived in 1835, though their populations were relatively small. Exotic plants, too, had already established themselves, from seeds inadvertently brought ashore on the shoes and clothing of ships' crews. And by then, too, upwards of 10,000 tortoises had probably been taken off the islands to provision sailing ships. Later, scientists would continue removing even more of them for their collections. [I saw the hollow carapace of the only giant tortoise ever found alive on Isla Fernandina on a research shelf alongside the carapaces of other Galapagean tortoises at the California Academy of Sciences museum, in San Francisco, in 1994. It had been collected in 1905–1906. Along with the other carapaces neatly arrayed around it, the carapace from Fernandina had the appearance of a book from which all the pages had been torn, leaving only the covers.])

Over the years, in an effort to preserve the general outline of the flora and fauna of the ecosystems Darwin and other scientists first described, multiple eradication campaigns have been mounted, some of which were large-scale military-style operations. Professional hunters from New Zealand, for example, using helicopter gun platforms and coordinating their attack plan with sharpshooters on the ground using Judas goats as lures, removed nearly 100,000 goats from a single island in one year.

While the effort to maintain a "pristine" Galapagean ecosystem is in one sense worth praising—left unchecked, sizable populations of exotic plants and animals can radically upset an ecosystem comprised of species that have been evolving alongside only one another for thousands of years—attempts to distinguish between "indigenous" and "exotic" species are problematic. One view holds that ancestors of the finches, the tortoises, and other iconic Galapagean creatures were themselves "colonists," while the Norway rats, the pigs, and the rest who arrived with the sailors were merely "invasives." But drawing the line today between those that belong and those that should be exterminated is tricky, politically as well as biologically.

What most angers ardent conservationists on Galápagos is the imprint of mankind here—the palm trees, the mosquitoes, the havoc wreaked on the islands' vegetation by goats, the carcasses of sharks washed up on island beaches, their fins cut off by commercial hunters to supply Asian markets. As far as practical, conservationists would like to see all the feral animal populations removed. No more pigs rooting out green turtle eggs, no more cats plundering Hawaiian petrel nesting burrows, and no more indigenous mammals—five species of rice rat, for example—lost to the black rats that came ashore from ships.

No one would argue—probably—that people abandon the islands; nor does anyone—probably—want to accommodate the wild pigs and goats. It's the middle ground that no one can seem to find. Moving darkly through the discussion of what belongs here and what doesn't—for example, which are the preferred plants and which should be eradicated?—is the age-old disagreement in human societies concerning human immigrants. The echoes of racist rhetoric, nativist prejudice, and economic self-interest in these sometimes volatile conversations on Galápagos are unmistakable.

MY WEEK AT Hotel Galápagos is nearly at an end. The people I'm scheduled to travel with for the next ten days aboard the *Beagle III* will arrive tomorrow at Baltra. I'll be over there to meet them at the airport and to board our motor yacht. I'm packing my bags and dive gear and making some notes—Barnett has given me copies of some of his technical papers—when I hear a distinctive rap on my door. Jack, the proprietor. He's come to say that there are two men in the lobby who wish to see me. They're local residents and want to know whether I'm available to write up the story of their many trips to Thailand together.

I meet the men in the hotel lobby. I'm uncertain about their reasons for wanting to meet but want to do Jack a favor, as he has done many for me. We shake hands and move to seats in the restaurant. No one here at midmorning. Jack serves us coffee—on the house—and they launch into their proposal.

The men are about twenty years older than I am, two bumptious, overweight Americans, very enthusiastic about Thailand as a tourist destination. I'm not the person for their project, I tell them right away, but to be polite I ask about Bangkok. Have they seen its floating market, the Wat Phra Kaew complex? Have they visited Chiang Mai? It's quickly apparent that none of this is of interest to them. Their interest is in sex, especially with adolescent girls.

I push back from the table, say politely that I am not able to accommodate them, and leave the table without shaking hands.

As I walk out of the restaurant I catch Jack's eye in the kitchen, cutting up tomatoes for lunch. To be civil, I gesture back with a shrug, when what I want to do is confront him.

The darker part of the real world is always just around the corner.

Back in my cabana, trying to defuse my anger and refute what I'd witnessed, I remembered a story that put me back on track.

A friend of mine, an ethnomusicologist, had gotten funding to study percussion ensembles in West Africa. After a week in one village, when he felt comfortable enough with his host family to ask the question, he inquired about ritual scarring. He wanted to know in particular about a man whose scars were in the shape of crescent moons, randomly spaced on his arms and back. His host said the man had molested a girl, and all the men in the village had bitten him.

WHEN I FIRST CAME face-to-face with extensive schools of brilliantly colored tropical fish in Galápagos, in shallow water at Gordon Rocks near the Plaza islets, I shouted so forcefully with excitement that I spit out my snorkel and choked. The fish that caused this outburst were blue-eyed damsels, dozens of them moving through the gin-clear water, together with schools of dusky sergeant majors and yellowtail surgeonfish. Blue-eyed damsels are darkish fish with yellow lips and tails and bright blue eye rings, about five inches long and compressed vertically. What made me shout was more than the vividness of their colors, though these hues were intensified in the sun-shot water by a glycerin-like mucus that covers their scales. It was the sheer *number* of them, all turning away from me in unison, as though they comprised a single organism. And it was how perfectly placed they seemed to be in the world, in that particular spot, in the moment that I found them.

On subsequent dives on that trip—growing accustomed now to the use of the snorkel, diving deeper, learning how to stay down longer— I could not turn away from the marvel of these fish—wrasses, parrot fishes, Moorish idols, butterfly fish, the damsels, triggerfishes, needle-fishes, stargazers, halfbeaks. Back on the boat I pored over Godfrey Merlen's *A Field Guide to the Fishes of Galápagos*, the fish chapters in Roger Perry's *Galapagos: Key Environments*, and dive notes and drawings that the guides gave me. The popular names for the fishes—grunts, sleepers, knifejaws, boxfishes, triggerfishes, and the rest—were captivating: guineafowl puffer, tinsel squirrelfish, clown razorfish, rainbow scorpionfish, Sheepshead mickey.

It was some years after that first exposure to Galapagean fish that I returned to the islands as a scuba diver with the equipment and skills needed to descend into deeper waters and to stay down longer. It meant having a better opportunity to see whitetip reef sharks and Galápagos sharks, and being able to inspect reef life in a more leisurely way. One day on this trip six of us dove on a small seamount called Roca Redonda, hoping to find moving "walls" of hammerhead sharks there, not an uncommon sight in water to the north of Isla Isabela. When I rolled backward off the gunnel of the panga in my scuba gear and stabilized myself in the water, I saw several scalloped hammerheads about fifty feet below me, moving slowly past the vertical wall of Roca Redonda in water through which I could see no bottom. As we descended, the hammerheads drifted off into the gloom. When our group leveled out at about eighty feet and looked straight up, we saw the sharks gathered together again about thirty feet above us. Sixty or seventy of them, some close to twelve feet long. They moved lugubriously in an open pattern, suggesting a lattice, backlit by the sky.

The French use a phrase to describe the scuba diver, *l'homme sans poids,* a weightless person, a play on the alliterative French expression *poissons sans poids,* the beguiling weightlessness of a fish in water. Weightlessness allows us to ascend the walls of Roca Redonda slowly, inch by inch, inspecting tiny stone ledges on which small animals and plants play out their lives in shadows thrown by the passing sharks. Suspended as each of us is, we can swim along the wall "upside down" and study the underside of the surface water above, see bursts of wind hitting it, wavelets forming and collapsing on it. Somewhere shallower than here you could watch the bottom passing below you and see on the sand there the movement of clouds overhead. To possess this ability to go left or right, forward or backward, up or down, all at the same time, to be released from the constraints of gravity, gives you a frame of reference like a tethered astronaut's. This is what Icarus wanted. With this perspective, the third dimension that birds and fish move through effortlessly is yours for a few moments. And then the nature of their lives opens up more fully for you. The hammerheads move past us like swans milling on a city park pond. When we lose sight of them again, they've formed up in a shape like a leaning wall, a kind of close vertical stacking that birds don't readily employ because it would compromise the lift they need to fly.

The wall of them slides into the distance like a shoji screen.

I could be wrong, but the hammerheads seem hardly to have noticed us. At the time I encountered them, years ago now, they had yet to be harvested in great numbers by Galapagean fishermen, solely for the commercial value of their fins. The upwelling of nutrients around Roca Redonda maintains schools of medium-size fish feeding in the area—jacks, basses, and groupers—which in turn (once) fed the numerous hammerheads.

Back on the boat that day, we were toweling off when someone spotted a school of common bottlenose dolphins about half a mile away. It had been only a few days since fifteen of us had been in near-shore waters with masks, fins, and snorkels cavorting (as we imagined it) with this same species of dolphin. They'd breach all around us, bolt off, then come racing back. Trying to engage with them quickly wore most of us out. The dolphins approached closely but never close enough to be touched. When we saw dolphins again, only a couple of people expressed an interest in getting back in the water with them.

I left our yacht with two others and a boat driver in a panga.

This pod of dolphins was less interested in us than the other dolphins had been, but when we got close, we rolled out of the panga anyway. Maybe they would come to us. The water was very deep, water open-ocean sailors call blue water, because of its cobalt hues. What appeared beneath me was not dolphins but something I would not ever have expected to see. Thirty feet below was a female sei whale, a dark steel-gray form about forty-five feet long, nursing a single calf. In a moment like this, the mind recovers equilibrium before the heart can reclaim its normal rhythm. It pieces together the light and the shade, the lines of the forms in the water, to make a coherent image.

I breathed slowly through my snorkel, floating motionless, watching the creatures until their undulations took them into indistinction. I hoped my two companions, who had dropped off the opposite side of the boat, had seen the whales. It turned out they hadn't. Underneath them when they first looked down had been a young sperm whale.

Back on the boat, how could we relate the details of what had happened without distressing those who'd chosen not to go?

———

A COUPLE OF NIGHTS LATER, the *Beagle III* was motoring southwest across Bahía Isabel, an embayment on the west coast of Isabela. I went to the bow after supper to watch the overcast night. The western headland of Punta Cristóbal ahead was barely readable, an opaque black inseparable from the opaque black of the sea, but distinct against the slate black of the sky. A light breeze crossed the bow.

To the southwest, toward the open Pacific, I saw something in the blackness I could not immediately understand, a straight whitish-turquoise line aimed at the starboard bow and continuing to extend toward it as I watched. The line widened as it drew closer and then suddenly hooked at the bow like the letter *j*. A dolphin, now riding the pressure wave just ahead of the boat. He or she was the first of six, all arriving within minutes of each other—from behind us, from up ahead, from the port side—each trailing a pale turquoise line: bioluminescence, from plankton excited by the dolphin's movement through the water. With the six of them riding the boat's pressure wave in parallel, they lit up an oval of water in which each animal was a distinct ghostly silhouette. I could hear their plosive breathing and smell their rank breath.

The most intense illumination, a radiant creamy white, surrounded the dolphins' heads and the leading edges of their pectoral fins when they extended them slightly to steer. The light was less intense along their flanks but didn't start to fade until it was fifteen or twenty feet behind their dorsal fins and flukes (the dolphins were seven or eight feet long). They rose in curvets from the water, streaked ahead, crossed beneath one another and veered away from the boat, constantly changing their alignment. Fish burst away from them, triggering additional blooms of bioluminescence. From moment to moment I couldn't tell whether the original six were still with us or whether some had gone and others had arrived. The air thickened with their mewling squeaks. By my watch, the dolphins were on the bow for an hour and a half.

When the last one departed, the sea went dark again until, minutes later, a patch of bioluminescence more than fifty feet across suddenly opened in front of the boat. I fully expected the *Beagle III* to plunge in. I gripped the boat's steel railing and fought off feelings of vertigo. We crossed through the light and the moment of panic faded. I set my eyes again on Punta Cristóbal, which, despite the long interlude with the dolphins, did not appear to be any closer.

Even with nothing to be seen, I continued to stand on the bow, waiting to see anything I could. Eventually I saw a flock of swallow-tailed gulls, the night hunters, passing to the west with the slow wingbeats of egrets. They are a kind of bird the *Hōkūle'a* navigators look for just before dawn, when they are reliably headed back to land and to their nests.

When I left the bow to go below, it occurred to me that the large patch of bioluminescence could have been caused by a whale suddenly halting its effort to surface until we'd passed.

EVERY COMMERCIAL AND PRIVATE boat tour in the Galápagos—some commercial tours are only a few days long, others last a couple of weeks—must follow the itinerary given to them by national park officials. This ensures that no tour group shows up at the same place, at the same time, as another. (Passengers may go ashore only at one of the park's approved landing sites, of which there are about sixty. In order for visitors to see as much of the variety of Galápagos as possible, the itinerary assumes that most tour boats will be in transit during the night, while passengers sleep. That way, visitors often awaken to find themselves already anchored at the place they'll disembark to after breakfast.)

Evenings at sea aboard the *Beagle III* at this time of year are mostly warm, and some of us forgo the bunks below to sleep out on the decks, where the Milky Way is particularly dense with stars on a clear night. On the morning I'm thinking about, our itinerary has us anchored at Isla Rábida. The engines have been shut down for a while; a few Galápagos doves are asleep on the ship's railing. No movement, no gas or electric light is visible ashore. The silence here is a presence, like the air inside a large rotunda. It's finally rent by the sound of a knife blade stopped by a cutting board—the cook, preparing cantaloupe for breakfast.

We're at Rábida to visit a colony of greater flamingos living at a salt lagoon. They feed in the shallow water on a type of crustacean that makes their feathers pink, and the females lay their single white eggs on small heaps of mud on a salt pan nearby. They're among the wariest of the islands' birds, and having heard the boat's engine earlier, they are arrayed on the far shore of the lagoon when we arrive via a short trail. Something about the arrangement of space here and the scheme of simple colors—the broad turquoise lagoon, the deep blue of a high-pressure sky, the long pink line of distant flamingos in front of a wall of green mangrove trees—induces silence in the eight or ten of us standing there. We move like tiptoeing parents in the bedroom of a sleeping child. Outside of the soft puff of an intermittent breeze against my ears, the only sound here is the bleating of the birds, a sound like the honking of geese. The overall impression is that the birds are pink, but with my binoculars I can separate out individual feathers: salmon, vermilion, scarlet, coral.

We're downwind of the birds and the intermittent breeze brings their floating feathers to us over the water—breast and nape feathers, coverts and scapulars—floating curves, turned up at each end like a child's drawing of a canoe. The stately hesitation of flamingos feeding and the trembling of hundreds of feathers on water lapping the lagoon's edge create an opening that was not there when we arrived. Vulnerability and a feeling of friendship—with the birds and among the people I'm traveling with.

We return to the boat carrying that eloquent silence.

WE'D SPENT THE last part of one afternoon around Isla San Salvador swimming with fur seals and hiking up to an old saltworks. We were en route to Isla Genovesa when I saw something unusual on the water inside San Salvador's Buccaneer Cove, near Cabo Cowan. The sun was setting and its last low rays were reflecting on splashes on the surface water, about a thousand yards away. At first I thought these were the plunge dives of blue-footed boobies, which feed near shore and dive straight into the water, like pelicans. The pattern of the splashes, however, never changed. Not boobies, then, but animals struggling in the water. Caught in a net. Galápagos sea lions (*Zalophus californianus wollebaeki*).

Our guide, Orlando Falco, prevailed on the captain—who was

reluctant to get involved—to change course. Orlando knew that Gala-pagean fishermen had started using sea lion carcasses as bait to catch sharks, and that Asian factory ships had been calling at villages like Puerto Villamil that year, offering to buy shark fins and supplying fish-ermen with nets. The entire business—factory ships putting in at the villages, netting the sea lions, killing the sharks, selling their fins—was illegal, but the park had neither the funds nor the personnel to stop it.

When we got closer, we could see that about fifteen sea lions were tangled in the net and drowning. Some were bound up together in the mesh, each one fighting the others to reach the surface to breathe, before being driven back under by their companions fighting for air. A crewman lowered a boat for us. Working from one side of the panga, Orlando and I began cutting the animals out of the net with our dive knives. Three other people, counterbalancing the boat on the other side, were holding flashlights for us to see. The most desperate of the sea lions were trying to clamber aboard the boat. The boat driver was using an oar blade to keep them from biting the two of us or turning the boat over. It took about forty minutes to free them all. As well as we could determine in the dark, all but one of the fifteen were able to swim away.

In the melee, Orlando and I had cut each other's hands and arms, and when we got back aboard the *Beagle III,* we saw that our shins were black and blue from banging against the panga's gunnel. Strangely, each sea lion seemed to understand at some point what we were trying to do. As I moved to cut the last few strands of green line from around a sea lion's head, it stopped fighting me and trying to bite. It rested calmly in the water.

Two years later, when I returned to Galápagos, I saw the carcasses of thirty or forty finless sharks cast up like driftwood on a dozen beaches. The fishermen's practice then was to throw them overboard and leave them to die after cutting their fins off.

AS WE CRUISED through the islands and went ashore one place and another, I began to recognize patterns of color and the presence of shapes and forms that, taken together, suggested this setting and no other. On the northeast coast of Alaska is a place called—in the Iñupiaq language of the Iñupiat there—*Naalagiagvik,* "the place where you go to listen." The name refers to the practice of a particular Iñupiaq shaman who visited this area regularly to listen to the voices of animals and to voices not audible to others, like those of her ancestors. From this ensemble she built up the guiding stories her people steered by, the stories that gave them a direction in life and kept them from harm.

Inspired by this act, and by the enormous metaphor it was, the composer John Luther Adams, an Alaska resident then, designed an installation for a room at the Museum of the North, at the University of Alaska at Fairbanks, called "The Place Where You Go to Listen." A continuous stream of precisely modulated electronic sounds flows from a suite of speakers here, tones created by the dynamics of Earth itself. Seismographic, geomagnetic, and meteorological data from outstations across Alaska feed into computers in Fairbanks, which using Adams's algorithms create a "whole cloth" of intricately woven tonal values. The viewer/listener at the installation sits on a bench facing a set of five glass panels. The panels change color according to the season of the year, the weather, and the time of day. The pattern of changing color reinforces the sensation of being present to a real landscape, one made richer by the translation of Earth's constant micro movements and phenomena like the northern lights into the patterns of a singular sonic landscape.

For many years John and I have exchanged ideas about the nature of music, language, and the natural world, and how combining them in various ways can tell us where we are, literally and figuratively. Prompted by John, I've tried, wherever I've gone, to stay attentive to the sounds that emerge from an environment, believing with him that each place offers a unique pattern of sound, arrangements that change over time, depending on the season, the temperature, the humidity, the strength of the wind, and the hour of day in that locale. I've long been attracted, too, to the work of other artists who have a gift for patternmaking—painters, choreographers, composers—and who respond to the world that lies outside human control. Each singles out components from what's available—tones, hues, movements, progressions. Brought together successfully, these components offer us art so well integrated, so seamless, we call it beautiful, in the way the particle physicist's singularity or the Greek philosopher's *Theosophos* can be said to be beautiful.

When I look at a landscape as a traveler, if I'm diligent, I can sometimes make out an inherent visual pattern, a sort of topology of lines and colors, into which movements fit—a bird gliding downwind across a brindled cliff face, above a dark ocean, say, or a range of treeless hills shadowed by a passing storm. Within the pattern, the separate pieces—sound, color, movement—all inform one another. In the end, whatever the components, it seems impossible to separate them again into single elements. They fit together so well that an analysis of exactly what is going on, rather than an uncritical appreciation of the phenomenon, is likely only to distance one from the scene.

Recalling my conversations with artists about the guesswork associated with creating a pleasing pattern frequently enables me to see better what is before me, see it not so much as art but as essence—the essence of a place, for example. In Galápagos it was—to pick something at random, seen from a passing boat just offshore of an island—a white ramulose horizontal strip of leafless palo santo trees, barely separated from a parallel strip of white beach below by a matte black line of lava, and then a dark ocean below that, holding a bright white line of surf against the lesser white of the beach, all of this surmounted by a pointillistic forest of scalesia trees on a slope above, with the dark ocean weakly reflecting the land and sky.

The image did not have to have meaning. This was only the *presence* of the place, in the middle of a March afternoon on the equator.

ON ONE OF the last days of our excursion aboard the *Beagle III,* our group landed at the foot of a trail on the east coast of Isabela. The trail here led up to the rim of el volcán Alcedo, a climb of about 3,700 feet. I was eager to climb Alcedo for several reasons. The largest undisturbed population of giant tortoises in Galápagos lives inside its crater, and the daylong hike to the rim takes one through all of the islands' vegetation zones, from the coastal plant communities up through the transitional dry zone of the scalesia forest to the wet-zone communities and the pampas around the crater. In addition, because the climb is physically taxing and requires making an overnight camp in the wet zone, few visitors attempt it. It's one of the least-visited approved sites in the archipelago. It still has the look of the Galápagos Darwin found.

If the giant tortoise is the archipelago's most iconic animal, the shield volcano is its most iconic landform. Molten lava, periodically bursting through the Nazca Plate from a hot spot in Earth's mantle, flows out across the ocean floor and cools, creating a shield that slopes away gently on all sides. Each succeeding lava flow adds another layer of molten rock to the shield—more height, more breadth—until the flows breach the surface of the ocean. The flows continue, but at this point what began as an underwater volcano is now a volcanic island. As the Nazca Plate advances eastward, the stationary hot spot bursts through at another site in the plate, eventually creating another island, the islands eventually forming an archipelago. (The oldest island in Galápagos, Plaza Sur, is to the east of Santa Cruz. Among the youngest is Fernandina, far to the west. The hot spot sits today between Fernandina and Isabela, close to La Cumbre, which is an active volcano.)

Once a Galapagean volcano becomes extinct, like some of those on Isabela, its vent begins slowly to collapse, forming a dry crater. (The ancient disintegrating crater on Genovesa has collapsed so completely that it's flooded today by the ocean.) The most spectacular craters in the islands are those of the still-active volcanoes on Isabela like Cerro Azul, and of course the crater of La Cumbre, which is filled with glowing lava, not tree ferns, bromeliads, and liverworts.

Earlier, on a visit to Puerto Villamil, our guide had hired a man with a stake-side truck to take us up to the rim of el volcán Sierra Negra, at 4,890 feet. From there we'd have a spectacular view of Santa Cruz to the east and, to the northwest, of Fernandina. Ten of us were standing on the truck bed, leaning against its wood railings, as the vehicle left Villamil and began laboring up a steep grade in first gear. Suddenly the engine quit. The driver quickly put his foot on the brake. He then informed us that because he had no starter motor, he was going to have to jump-start the truck by rolling backward down the dirt road and then easing the clutch in. Which he did. A few minutes later the engine quit again, just as he was coming through a curving section of the road. As the truck began to roll backward through the curve, it became clear that for some reason the driver wasn't able to engage the clutch—and that he had lost his brakes. A few of us made moves to jump off the truck, but we were too late. It was gathering speed and fishtailing. Twice it rose up on one side, threatening to flip over, before rolling to a stop in a barrow ditch by the side of the road.

We decided to walk the rest of the way.

The view from Sierra Negra that afternoon was theatrical, a vista encompassing nearly a thousand square miles in which not so much as the wake of a boat upset the illusion of the extent of space around us. It spread outward from droplets of dew on blades of grass on the crater rim to heavy surf breaking on the beaches of Santa María, fifty miles to the south.

On our way back to Puerto Villamil, ever-trusting, we accepted a ride in the tip bin of a brand-new Hino dump truck. Bright yellow. We passed the aging Daihatsu stake-side that we'd started out in, still tilted precariously in the barrow ditch. I felt a pang of sympathy for the driver, standing there with hands in his pockets, facing expensive repairs, and with no sign of help on the way.

I've been in situations like this before, a near miss and you walk away with a good story; but memories of this potentially fatal mishap remained unsettled in my mind for hours. The driver had accepted a fee without mentioning the marginal condition of his truck. We were lucky it hadn't flipped when the clutch and then the brakes failed. We were miles from help in Villamil, and even farther from competent medical help on Santa Cruz—if we could raise anyone there on our shipboard radio. And yet here was this middle-aged man trying to make ends meet.

What was he going to say to his family about the money our guide asked him to give back?

IN THE CRATER of Alcedo we would encounter giant tortoises on their own terms and, if we stayed long enough, see something of their way in the world—eating, resting, mating, sleeping, moving with determination on their elephantine legs, wallowing in drip pools beneath scalesia trees festooned with epiphytes, gazing into the distance with pensive faces, reckoning events impossible for us to imagine. The carapaces of the oldest of these are encrusted with lichens and used by hawks and vermilion flycatchers as observation posts.

We spent a full day inside the crater, fogged in and chilly. Those without hats sported helmets of beaded moisture on their hair. The melancholy atmosphere encouraged a kind of brooding in me, about the finless sharks, the poverty in Villamil, and war zones I'd seen. The reaction of the tortoises to our presence seemed to unfold in slow motion. They were like wizened sentinels, waiting for us to pass on. Theirs were intensely local lives.

The floor of the crater was pocked with volcanic vents encrusted with canary-yellow sulfur deposits. (We came upon a lone feral donkey here, remnant of a herd that once carried panniers of sulfur down to the coast for shipment to the mainland, another early effort, like the one at the salt mine on San Salvador, to develop an export economy in Galápagos. The animal was in good flesh, alert and wary, in great contrast to the bone-thin, scarred, limping horses standing abject and catatonic in the dirt streets of Villamil.)

When we broke camp in the morning, I lashed a large volcanic cinder to my pack. I'd read that some creatures in the Galápagos might have originally dispersed through the islands by riding volcanic cinders carried by the archipelago's strong currents and pushed by the wind. I had a hard time imagining a rock floating in salt water, and curiosity led me to break the park rule about leaving everything in the park undisturbed. The cinder was about three feet in circumference but it weighed little more than a pound. I carried it down from the crater rim to the edge of the water and flung it in. It sank with the impact but quickly bobbed to the surface, where it floated with about half its mass clear of the sea.

VILLAMIL. THE SETTLEMENT was like a fishhook in my mind. It's easy to make the mistake when traveling abroad of finding only the good or only the bad in a place, easy to miss how complicated the weave of bad with good is. If archetypal goodness in Galápagos is represented by the idealistic park ranger who will not accept a bribe, archetypal evil is represented by those residents of Villamil who started the forest fires on Isabela, killed the sharks for their fins, and harassed and terrorized the family of a park ranger who tried to live there, driving him and his family back to Santa Cruz.

The day the group of us returned from Sierra Negra, our guide told us that in light of the accident with the truck and his having to ask the driver to return the fee, it might be good to act in a friendly way in Villamil, to buy a few sodas and trinkets from vendors. We can take the time to do that, he said. Another plan came to mind right away. Could I find someone to escort me to the ruins of a penal colony west of the village?

There are a few images in a poem by the American poet Robinson Jeffers called "Apology for Bad Dreams" that rise up regularly in my mind. Jeffers writes in the poem about the contrast between beauty and violence. He suggests that the world cannot be understood without accepting both. He says that it is not good to forget, in the pursuit of virtue, the degree to which immorality defines our condition and our history. In an essay called "In Defence of the Word," the Uruguayan writer Eduardo Galeano refers to the writer as "the servant of memory," both his or her own memory and the memory that his people have of what has been done to them. Galeano is arguing, essentially, that a writer who lies ceases to be a writer; and that a writer is obligated to resist complacency and to remember the things the ruling classes hope will be forgotten. He is thinking, I believe, of writers like Rian Malan in South Africa or Bei Dao in China.

Many years ago I attended a trial in Delmas, South Africa, during the last years of apartheid. Delmas is about forty miles southeast of Pretoria. The federal government chose Delmas as a venue to make it more difficult for foreign reporters based in Johannesburg or Pretoria to attend. Nineteen black men were being tried for sedition. The charges were trumped-up distortions of the truth, but the court found all nineteen guilty, and sentenced them to death. (With the collapse of the apartheid government several years later, their appealed convictions were reversed and the men were all set free.) A friend in Johannesburg

suggested I attend the trial before I left on a long trip into the Namibian outback. He said I would get a feeling for the racist government's
psychopathic indifference to human rights and to truth.

Having attended that trial in March of 1987, I never afterward
entered the outback in the same frame of mind. I carried with me the
faces of those nineteen men and the spectacle of bigotry that unfolded
in that federal courtroom. (I also wondered more often, in those days
after Delmas, about my own state of moral oblivion, the indifference in
myself that left me blind to injustice in places far from home.)

It's not difficult to locate the ruins of penal colonies set up on the
edges of empires. The British sent their undesirables first to North
America, until the American Revolution forced them to ship people to
Australia, during the era of transportation. The French used Île du Diable and New Caledonia, the Spanish sent theirs to Ushuaia, in Tierra
del Fuego, and the Portuguese, who began the practice of using colonial
prisons, built one in Madeira. The question for me is not only who is
justly or unjustly condemned, or whom a nation tries to rid itself of,
but the capacity of nations to indifferently expunge human life, like
Portugal under Salazar, or Panama under Noriega, or Indonesia under
Suharto.

Jeffers would have argued, I think, that it is foolish to believe one
can actually eliminate the brutality men are capable of; but one can
reduce the level and the extent of cruelty. And Galeano would have
argued, I believe, that not only must the horror of these places not be
forgotten but any effort to suppress these stories must be exposed and
discussed openly, if democracies are to function.

If you never pass through Nagasaki, if you never see the ruins of
POW camps in North Vietnam, if you don't walk the Shoshone slaughter ground at Bear River, Idaho, it's easier to believe that these things
are merely historical, or to believe that the era of death camps, penal
colonies, and raids on American Indian encampments no longer exists,
that these places are now only symbolically important. It is to put forth
the idea that autonomous drug cartels in Mexico can be brought to heel
by a strong government.

Our guide said no to my going to the prison grounds outside

Villamil. "You won't learn anything out there," he said. "It's all overgrown."

WHEN WILLIAM BEEBE, a biologist and explorer affiliated with the New York Zoological Society, published *Galápagos: World's End* in 1924, he lit up in the minds of many of his readers a vision of Elysium. People inferred from his romanticized account that anyone with the means to get there would be able to pursue a self-sufficient life of tropical leisure. Planning a trip to Galápagos aboard a tramp steamer became a fad, for a while. Several American owners of private yachts refitted their vessels to support scientific exploration in the archipelago, and an unknown number of Europeans set sail for the islands with the hope of leading lives free of bourgeois convention and of establishing small-scale export businesses. Few of these ventures came to anything, but among them were some that burdened the archipelago with stories of tragedy, mystery, and pathos. It's the rare visitor today who leaves the islands without having heard something of the gossip and speculation concerning the people involved in those misadventures. The best known of these stories is famous all out of proportion to its banal content, but decade after decade it seems to grab the attention of visitors who want to know "what really happened" on Santa María in the 1930s.

A man named Friedrich Ritter, a German doctor with a practice in holistic medicine and an enthusiastic proponent of Nietzschean ideas about male supremacy, arrived at Isla Santa María in 1929 with his companion, Dore Strauch, a German woman with multiple sclerosis who had initially been his patient. They were intent on building an idyllic haven for themselves on the island. Each had walked out on an unhappy marriage in Berlin, and the exotic nature of their new home, together with the bohemian trappings of their relationship, made them the titillating subject of numerous articles in popular European magazines.

Late in the summer of 1932, a more down-to-earth German couple from Cologne, Heinz and Margret Wittmer, arrived on Santa María, and they, too, began homesteading, not far from Friedrich and Dore. Within a couple of months four more people arrived in the area, a ménage à trois and an Ecuadorian man the other three had hired to do their chores. The group was led by a flamboyant self-styled "baroness" who held in her thrall the two men she was involved with. The Wittmers had a respectful but cool relationship with Dore and Friedrich, but the two couples were united in being disapproving of the "Baroness Eloise von Wagner Bosquet" and her two lovers, Rudolf Lorenz and Robert Philippson. The baroness, one suspects, was ultimately to blame for much of the petty thievery and intrigue that came to characterize the small community; and when relationships with their respective lovers began to deteriorate, both Dore Strauch and Rudolf Lorenz began to seek sympathy and comfort at the Wittmers' farm.

One day the baroness informed the Wittmers that she and Robert were leaving Santa María. They were sailing for Tahiti. No one actually sees them leave, but neither is ever heard from again. At the same time, Rudolf arranges for a ride on a boat bound for Guayaquil, on the Ecuadorian coast. Months later, he and the boat's captain are found dead on a beach on Isla Marchena, nearly a hundred miles to the north. Friedrich dies on Santa María, claiming to have been poisoned by Dore. Dore returns to Germany. The Wittmers stay on.

Until her death in 2000, visitors who stopped at the Wittmer compound, at Black Beach on Santa María, could socialize with Margret Wittmer. They could sip a glass of her fermented-orange "wine" while she signed their copies of *Floreana,* her version of what had happened on the island in the thirties. A short, stout woman, Mrs. Wittmer had about her a strange air of self-importance. She seemed to possess a permanent sense of irony about life, believing that most of the people who come thousands of miles to behold this famous Darwinian shrine to biological evolution, the Galápagos archipelago, were actually far more eager to speak with her about the baroness.

WHEN THE *BEAGLE III* left Black Beach, after my second encounter with Mrs. Wittmer (we'd met earlier, on my first trip to Galápagos), we doubled the coast to the north and anchored at Bahía del Correo. The famous post office barrel here, refurbished many times, is still used by tourists (though they're no longer able to have their postcards hand-canceled by Mrs. Wittmer). In a mostly dependable but somewhat haphazard way, mail leaving Bahía del Correo today, bearing the correct amount of Ecuadorian postage, will eventually get to its destination.

The setting for the decorated barrel, a dusty clearing set back from the beach, has the faux charm of tourist kitsch with its many hand-lettered signs affixed to posts in the ground (*San Francisco 3452 miles*), but it's good fun to post a letter or a card and nearly everyone does. Our guide tells us how the barrel once served whalers and others far from home, and his words are as poignant as they are historical. He speaks about the tenuousness of this kind of communication in the nineteenth century, the likelihood that a letter would be destroyed in a shipwreck or that a question written out with great anxiety and earnest thought ("Will you marry me?") might go unanswered for several years. The hopefulness of it all, the desire to be known and heard when so far from home, to be remembered, gave all of us an insight into the vastness of the commercial world nineteenth-century whalers lived in, the contingency of their emotional lives.

From Bahía del Correo we have a long run to Academy Bay. One of the guests has prevailed on the captain to be allowed to take the helm for part of the crossing. The seas are relatively calm, the weather is good, and there are no hazards between here and there. The captain graciously turns over the wheel. When he does, I ask him if he has time to talk. Yes, he says. We sit on the aft deck beneath an awning. The cook brings us dark Ecuadorian coffee, and I ask Captain Eugénio Moreno about orientation and navigation. How does he know where he is going?

The captain is a private person. He rarely engages with passengers beyond a few polite sentences, but he is not reticent. I got to know him on another trip and learned that he can be very forthcoming and that he also has a good sense of humor. Once, when we were making a fast crossing between two islands together in a panga, a school of flying fish shot over the bow. All but one veered off and that one hit me hard in the chest, knocking me backward off the thwart I was sitting on. The captain laughed so hard he could hardly steer the boat. I laughed, too, a reaction that confirmed our good acquaintance.

Captain Moreno didn't say anything about the importance of magnetic compasses (though he was trusting the gentleman upstairs to hold a heading of something like 018° magnetic for the next thirty minutes). Instead he followed up on an earlier conversation we'd had about the *Hōkūle'a.* He was very interested in what the *Hōkūle'a*'s navigators were doing. He told me that a mainland person would automatically be suspicious of such traditional non-Western techniques, but that they made excellent sense to an island person like himself. He told me that after so many years in Galápagos he had a feeling about where to go when the seas got rough and how to get there. It wasn't in the charts, that information. It wasn't determined by bottom soundings with sonar or by weather faxes. I told him that some of the Polynesian navigators were able to lie down in the trough of a canoe and sense by the way the water knocked against the hull where they were in the currents. And that a blind navigator once worked in Micronesia. He didn't doubt it.

With his last sip of coffee the captain looked past me at the line of the ship's wake, which he had to lower his head to see clearly, underneath the keel of the panga hanging from davits in the stern. He raised his eyebrows quickly and said he would see me later. When I turned to look at the wake, I saw that its shape was serpentine. To steer the course he was asked to keep, the helmsman had been overcorrecting constantly to maintain his heading.

I sat alone at the table after the captain went to the bridge, experimenting with my own sense of orientation. Where was I? Riding very comfortably over the Pacific, in glorious weather, shaded from the sun, enjoying the breeze, glancing at seabirds as they passed. Soon lunch would be served: a crisp salad, fresh fruit, fresh fish. Later I would go back to the Gabriel García Márquez novel I was enjoying. I would also talk with the guide about the biology and ecology of the red-billed tropicbirds circling the *Beagle III* just then.

I seemed to want for nothing.

But where was *here* in that moment? The nexus of *here,* a place like the mail barrel's, a pinpoint that links a speaker to a listener, the *here* to some *there* across an intervening distance—where was this particular *here* located? I will try to tell you, from my deck chair in the stern of the *Beagle III,* with the captain now at the helm and three waved albatrosses in the distance, animated specks of white gliding across Wedgwood blue, headed south for their colony on Isla Española.

To the northwest of me at this moment, then, lies the Hawaiian archipelago. To the southwest is New Zealand. To the southeast, beyond Peru, lies Bolivia. To the northeast, Panama. Four points of a compass rose, converging here, on the *Beagle III*. In Hawai'i, one finds the still-unresolved matter of a military takeover of these islands by the United States in 1893, in support of American sugar and pineapple growers who wanted the native Hawaiian monarchy deposed and a new and different apportioning of lands in Hawai'i. Native Hawaiian people, kanaka maoli, continue to bring their story to the attention of the world. In *Aloha Betrayed: Native Hawaiian Resistance to American Colonialism,* the author, Noenoe Silva, quotes the African historian Ngũgĩ wa Thiong'o, speaking to the core of the native Hawaiian problem today: "[T]he biggest weapon wielded . . . by imperialism . . . is the cultural bomb. The effect of a cultural bomb is to annihilate a people's belief in their names, in their languages, in their environments, in their heritage of struggle, in their unity, in their capacities and ultimately in themselves."

In southern Bolivia, outside the city of Potosí, is a mountain called Cerro Rico. The Inca ruler Huayna Capac, after learning from his men that the mountain had warned them with a terrifying bellow to stay away, had named the mountain *potojsi,* "a great thundering noise."

Over the centuries, indigenous Quechua miners took untold tons of silver from Cerro Rico for their masters. Now mostly all they mine there is tin. As many as 8 million Quechua workers have been killed by cave-ins, by accidents with explosives and smelters, and by silicosis, an occupational disease. The Quechua say they used to believe that the mountain was sacred, that it was the home of a personified energy that gave Quechua people life. After the Spanish forced them into making their living as enslaved miners, the personality of the mountain began to shift in the Quechua imagination. Today it is seen as something quite different, a Beelzebub. On El Día de los Compadres, an annual festival day for Quechua men, the miners descend deep into the mines to pay homage to a Quechuan god named El Tío. They explain that the benevolent life-giving force that once resided here has now abandoned the mountain, leaving it to El Tío, a hellion whom they represent with life-size papier-mâché constructions as a spike-eared human with red horns, a large phallus, and a conquistador's goatee.

The American anthropologist June Nash quotes a Quechua man on this perversion of sacred life-giving energy. It has been reduced to a papier-mâché model, festooned with streamers of colored paper called *serpentinas*. Its mouth is stuffed with coca leaves and surrounded by gifts of alcohol and tobacco. Today, the miner told Nash, "[w]e eat the mountain, and it eats us."

In Christchurch, New Zealand, five Native American skulls were once kept on display at the Canterbury Museum, that of an elderly Lakota woman, that of an Arapahoe man, and those of three Salish Indians. The skulls were sold to the museum in 1875 for seven dollars, apparently by an American paleontologist named Othniel Charles Marsh. At that time, the heads of Native American people were routinely gathered up at the site of a massacre as souvenirs. Native American repatriation committees are working today to locate the appropriated heads of their ancestors and to bring them home for burial. When the late Wiyot painter, sculptor, printmaker, and carver Rick Bartow, an Oregon resident, was invited to Christchurch by Maori artists in 1994, his hosts asked him if he knew about the skulls in the Canterbury Museum. He didn't. The Maori said they were anxious about the situation but didn't know how to approach the museum. Bartow didn't either, but he asked someone on the museum staff for permission to improvise a ceremony and to contact members of a repatriation committee in Umatilla, Oregon.

On the day of the ceremony a museum curator placed the five skulls on a table and stood away. (This was in the hours before the museum opened to the public.) Maori women sang Bartow into the room. He entered shoeless, holding cedar branches he'd gathered in a nearby park and carrying tobacco and water. He was weeping. He recalled later that his tears came partly from his knowledge of human tragedy (such as he had seen while working in hospitals in Vietnam), from his own feelings of unworthiness in that moment, and from the relief he was feeling with the prospect of repatriation. He lit the cedar branches and cleaned the room with their smoke. He bathed the skulls and placed an offering of tobacco in front of them. He picked the skulls up one at a time and cradled them, brushing the forehead with his hand while he wept. He spoke words he does not remember now. Finally, with the skulls all sitting on the table again, he asked them to tell him what they needed. He said he was there to bring them home.

When I spoke with him later, he declined to repeat what they said.

After the ceremony he walked to the River Avon, in the park where he had gathered cedar branches and, praying, placed the remains of the ceremony in the moving water.

EARLY IN HIS NAVAL CAREER, then-lieutenant Robert Peary developed the idea that in order to succeed in life, he had to find some consequential project or enterprise to associate himself with, and then ensure that his name would be the first mentioned when people referred to the project or event. When military and commercial interests in France and the United States began to plan the building of the Panama Canal, Peary put his name in as a candidate to represent the U.S. Navy. He was selected, but he saw eventually that there were too many others—industrialists, politicians, military officers—involved. He wouldn't be able to establish himself as the project's visionary. There were too many decisions to be made that he wouldn't be able to control.

He resigned his position and shifted his aspiration to the conquest of the North Pole.

To sail through the Panama Canal today, in an era when robots build cars and space probes have entered the sphere of the Oort Cloud that surrounds the solar system, is to be moved to silence by the sheer scale of engineering, by the audacity behind constructing this shortcut to the Pacific. Peary's instinct about the project was correct. To this day, in the minds of most who see it, something of the miraculous still clings to this series of locks. But no single name.

I made the passage through the canal once from the Caribbean side, aboard an icebreaking research vessel headed for the Weddell Sea, in Antarctica. When we entered the Gatun Locks at the north entrance, Russell Bouziga, the Cajun captain of the *Nathaniel B. Palmer,* ordered the bridge cleared of all the ship's supernumeraries, including a small party of scientists headed for Antarctica. (This was the *Palmer*'s first voyage. When it entered the Weddell Sea a few weeks later it would be the first vessel to do so in winter since Shackleton's *Endurance* was crushed there in 1915.) I was leaving the bridge with the others when the captain told me to stay where I was. (A couple of weeks before this, during the ship's sea trials in the Gulf of Mexico, I volunteered to attend to an urgent problem for him. It involved diving to inspect the *Palmer*'s inoperable bow thruster. The professional diver on the ship's crew had gone ashore for the day and my offer to help [I was a certified diver], not knowing what would be required of me, made an impression on the captain during a time when his crew was overworked trying to meet inspection deadlines.) Bouziga, a Vietnam combat veteran, told me the reason he asked those who were not members of his crew to leave the bridge during the Panama passage was because so many laborers had died building the canal. Most were now buried below us in unmarked graves. It was his feeling that only workingmen should be on the bridge for the crossing. For the hours it took us to reach the Pacific, Bouziga asked for complete silence.

BEHIND ME, now that Captain Moreno had taken the helm, the wake of the *Beagle III* was straight as a runway.

And my *here* was, again, here, awash in equatorial sunshine and soft air, riding the untroubled waters of El Canal de Santa Cruz, two magnificent frigatebirds moving west on deep, slow wingbeats, and the *Beagle* bound for a supper of fresh Bacalao (grouper) and langostinos in Puerto Ayora. The episodes that memory had brought forth, of my conversations with the native Hawaiian patriot Noa Emmett Aluli, of purchasing the silver real in Christiansted, of the time Rick told me about the Maori invitation to ease the universal grief of traditional people, and of the honor silence Russell has asked for, to acknowledge the work ordinary laborers had done in their anonymous way, were now situated in the undisturbed present, this moment here, where the world seems more benign and the opportunity to forgive, to accept, floods the heart.

To recall trouble does not necessarily mean to dwell on what once had happened. The recollections also bring with them the relief that perspective offers.

AT THE CORE of Darwin's idea about evolution is a very simple observation: every living thing has parents.[4] The germ cells of each parent carry sets of genes that, though generally the same from one generation to the next, hold out multiple possibilities for the offspring of any given set of parents. The nature of this event, what will actually come of combining these two packets of information, is what scientists call stochastic (i.e., a result that can be approached statistically but that cannot be predicted, i.e., random).

With the aid of sophisticated tools of observation, geneticists today can identify whole swaths of genetic material in a particular genome—the human genome, for example—and say with some confidence which genes influence which trait in the ontogenic development of a member of a new generation. They can say, in other words, which genes will influence the color of a baby's eyes and what color they're likely to be. What they are not able to predict, however, despite their knowing the color of both parents' eyes, is what the baby will actually look like or—a much more complicated question—how the baby will behave. They can speak with some precision about the first part of the event (the creation of a new person, the beginning of its ontogeny, or growth) but not the second (who the person will be). No one can know what, in fact, is to come.

Change in the genome of a species over time and the relative quickness with which genetic mutations can start to alter the familiar (and similar) appearance of members of a species are not phenomena easy to characterize. Together with the speed and direction of genetic change taking place because of selective pressures being exerted by an animal's changing environment (changes in its climate, catastrophic geological events, chronic pollution of its environment, changes in the species it shares an environment with), random genetic change creates a landscape of quicksilver movement for an evolutionary biologist, paleoanthropologists, and other scientists trying to guess where any particular species might be headed, including human beings.

We know that as one generation succeeds another within a given species, the species changes, usually incrementally but occasionally dramatically. (A species is not so much a permanent thing as a point on the developmental line of that thing through time.) We know that genetic mutation interacts in some way with a species' environment such that some randomly generated changes in the nucleotides of a gene are reinforced, in order, generally, to perpetuate fitness in a species. But there's a great difference between the fate of an individual of a particular species and the fate of that species itself. Parents who have great hopes for their own children in an environmentally compromised world may therefore also be parents who despair over the fate of mankind.

The reason radical changes in the human environment caused by the Industrial Revolution are so anxiety-producing is not that they keep us from predicting a benign future for ourselves. We've never been able to do that anyway. It's that, apparently, major changes in *Homo sapiens*'s physical environment are occurring with what scientists believe is unprecedented speed. However well individual people might manage in the face of these changes in the decades ahead, the future of the *species* remains as open a question as it was for all the other hominins we're related to, none of whom, it's important to note, are still with us.

It is characteristic of our age that the urge to commit to the eschatology of a particular organized religion, and the intense critiques of all of organized religions' eschatologies, are both driven by the same conviction: no one knows where we're headed. We know only that we will change over time, and that our long history as a relatively stable species (200,000 years or so for anatomically modern man) is no guarantee that these changes will take place slowly, especially in the Anthropocene.

Darwin receives most of the credit for the way modern people imagine themselves changing through time as a species and for people's ability to conceive of their phylogenetic ancestors—from australopithecines and several species in the genus *Homo* through to, probably, *Homo heidelbergensis*—as being both different from and similar to themselves. In reading *The Voyage of the Beagle* as a young man, I was struck by two things that I felt shaped the way Darwin came to understand biological evolution. One was the effect of his years-long experience aboard the *Beagle*. During his time at sea in this two-masted barque, in weathers calm and calamitous, he developed, I thought, a more informed sense of the sheer size of Earth, which he'd had little inkling of as a boy growing up in Shropshire. Each day he spent offshore was one more day in which he stood at the center of almost unlimited space, an expanse of water and sky defined only by the continuous line of the horizon. Every hour at sea, becalmed in the Atlantic's doldrums or scudding before a quartering wind, the breadth of the ocean he saw and the great reach of the inverted bowl of the sky were in stark contrast to the world Darwin was familiar with as a passenger aboard HMS *Beagle,* little more than a mote of dust in the vastness between continents. This contrast, day in and day out, between the unknown ocean and the familiar ship, I felt, encouraged Darwin to develop a similar figurative vastness around the idea of evolution just then beginning to mature in his imagination.

The extent of pelagic space surrounding the *Beagle* compelled Darwin, I believe, to take what he thought he knew about change in an entirely new direction. The *Beagle* offered him the comfort and reassurance of a known cultural world: each leech line, halyard, clew line, and sheet on the *Beagle* was meant to do something specific in a complicated (though not complex) system, and he leaned toward complexity. Each book in the personal library Darwin took aboard the *Beagle* at Devonport in December 1831 addressed a topic he was already more or less familiar with. His table conversation with Robert FitzRoy, a zealous Christian fundamentalist and the *Beagle*'s autocratic, mercurial, and very empirically minded master, was predictably orthodox, class sensitive, and reactionary. I have to imagine that Darwin's glances out the open door of FitzRoy's cabin, as the inscrutable and protean ocean rolled past each day during their meals together, spoke volumes.

It struck me as indicative, too, that the subject of the first paper Darwin wrote that was based on his experiences aboard the *Beagle* concerned the nature of dust blown westward from North Africa on a harmattan wind, dust he found covering the ship's main deck one morning. Far out to sea, he discovered indisputable evidence of the impermanence of the enduring world.

The second thing I feel significantly shaped his ideas about biological evolution early on were the three volumes of Charles Lyell's *Principles of Geology.* The ramifications suggested by Lyell's central idea—that Earth was far older than was commonly believed—created a kind of fever in Darwin's mind. (Darwin received the first volume as a gift from FitzRoy, on their departure. The second and third volumes reached him later in Montevideo and at Port Stanley, respectively, in the Falkland Islands.) In the 1830s, geologists could be roughly separated into two camps. Uniformitarians maintained that Earth had changed only gradually through time; catastrophists argued that changes evident on the surface of Earth had occurred suddenly. Each tended to stress the characteristics of physical geology that most strongly supported their views. Uniformitarians pointed to the gradual buildup of sediments on the floors of lakes that give rise to sedimentary rocks, such as sandstone and shale. Catastrophists brought in volcanic eruption and unconformities in layers of rock. What Lyell, a uniformitarian, introduced was a temporal framework within which to reconsider either one of these positions. What he placed before his (for the most part) religiously conservative Christian colleagues was a span of time over which geological processes had been at work that was immense, vast beyond their reckoning. Archbishop Ussher's six thousand years would not begin to cover it.

What Lyell offered Darwin was the second part of the context in which to consider biological evolution. With the enormity of space surrounding him almost every day, and with Lyell, who gave him an enormity of time, Darwin was able to see biological evolution as a very long road branching off and unfurling through time, that phenomenon more profound historically and infinitely more complex to him than the impressive but still incidental English machine that he was sailing aboard.

I can imagine, too, that Darwin sensed yet another stimulus for revising his nascent ideas about biological change through time. It came during the thirty-five days he spent meandering though the islands of Galápagos in 1835 while FitzRoy was making marine soundings. While islands might offer the traveler firm footing and certain concrete realities to deal with—species of plants, layers of sedimentary rock, catchments of freshwater—they are also each circumscribed by a shoreline. In an archipelago, similar islands are almost always visible nearby, across

a watery surface that is forever in motion and *not* so easily characterized by empirical measurements. And to reach any of these other places
requires some sort of assistance—a ship or a boat. Darwin must have
noticed that certain birds and sea mammals—sea lions and fur seals,
for example—actually moved easily among the islands. Later, when he
understood for the first time how finches on the various islands differed
from one another, he must have considered again what might explain
the *thing* that held all this biology together. What, in other words, was
the nature of a biological archipelago—a number of distinctly different
but similar things (islands), each individual biome characterized by its
own menagerie of life-ecologies but all these biomes subtly related. He
must have seen in the connection between biology and geology here, in
this microcosm, the adumbration of something quite new.

DARWIN CREATED a massive scientific and cultural disturbance in the
West with the publication of *On the Origin of Species*. Theologians
viewed his ideas about humankind as a most serious threat to orthodoxy
and social order. This Shropshire gentleman, they believed, was arguing that biological change followed no preset course and that it had no
purpose other than biological fitness in the moment. Moreover, it was
change without a destination. In essence, they saw, he was saying not
only that *Homo sapiens* was *not* headed in any particular direction as a
species (i.e., *Homo sapiens* was changing over time but not "improving")
but also that humanity had no end point in "perfection." Further, he
was implying there was no hierarchical arrangement of Earthly species,
at the apex of which *Homo sapiens* was to be found. According to one
of his most astute biographers, Janet Browne, what most frightened
Darwin about his own ideas was not the threat they presented to fun-
damentalists among the theologians of his day, but that he was implying
there was little difference, ultimately, between people and animals. In
other words, that humans were not set apart by having "a soul."

Darwin's assertions were shattering to consider because they suggested for some the ultimate meaninglessness of human life. Resistance
to his ideas of adaptation and change remains as strong in some quarters

today as it was in England in the 1870s. (One of the most peculiar pockets of resistance to the theory of evolution happens to be in Galápagos, where fundamentalist Christians have sought successfully to become certified as official guides. They want to offer visitors alternative ideas about evolution and to emphasize the "special creation" of human life.) Darwin strived to restrict the discussion of his ideas to the province of biology, but of course he was unsuccessful. Atheists, agnostics, political revolutionaries, existentialists, and eventually social Darwinists—all commandeered his insights and took them off in directions Darwin himself probably would never have headed. Darwin was a cogitator, not an agitator, someone better imagined as a philosopher than a provocateur. He was, in fact, the herald of a kind of thinking that wouldn't come to the fore in another discipline—particle physics—for another forty years. Like Copernicus before him, and like Freud and Jung after him, he changed fundamentally the way we imagine ourselves in the world.

At the time Darwin was writing, scientists trying to explain the natural world concentrated their research in only two areas, chemistry and physics. Biology was viewed primarily as a descriptive science, a sort of stepsister to the other two types of inquiry, and further, one largely in the hands of gentlemen naturalists, people with nothing more serious on their minds than the twittering of birds and the blooming of roses. What Darwin did was to put biology on a level with physics and chemistry as a path of inquiry into the nature of the natural world. His work in systematics and his theoretical work in evolution took biology into areas of unpredictability and indeterminacy, which physics in fact had been aimed toward since the time of Democritus, and which it finally attained with the development of relativity theory and quantum mechanics. Darwin anticipated Heisenberg's famous insight, that the kind of indeterminacy that characterizes quantum theory is present in all natural systems. He wasn't implying that the human effort to achieve, say, moral progress without a map wasn't possible. He was saying it wasn't possible to make such a map in the first place.

Though inclined as a scientist toward the immutability of laws—physics and chemistry had laws: Newton's first law of motion, say, or Boyle's laws about the behavior of gases—Darwin was actually giving us reason to consider that biology had no laws. Instead, it had situations, like evolution or parthenogenesis or mitosis. People wonder sometimes whether Darwin might have written in a different way about biological evolution if he'd had Mendel's "laws" of inheritance to work with. The answer to this is that if Mendel had preceded Darwin, we likely would never have gotten the Darwin who created such an impact. Mendel's observations about genetic inheritance were not about predictability, as Mendel had hoped. They were about probability, and probability was awaiting a mathematics that would make change more comprehensible.

When I was seventeen, traveling though Europe with a group of my prep school classmates, I remember emerging from the Uffizi Gallery in Florence one afternoon with my head spinning. The artwork had made a terrific impression on my adolescent and febrile imagination. Trying to walk off the excitement, I ended up on the Ponte Vecchio, staring down at the Arno River. I had a book in my shoulder bag which I'd been trying unsuccessfully to read for several weeks. I was in over my head with it. It was called *The Orientation of Animals: Kineses, Taxes, and Compass Reactions*. The authors were an entomologist named Gottfried Fraenkel and a zoologist from Sri Lanka, Donald Gunn. Their subject was the way animals, mostly insects, orient themselves in the environments in which they live. Given their needs—to feed and to survive long enough to reproduce—and the quality of their environments, and given a particular moment in time—the temperature and humidity, the season, the angle of the sun—how did they orient themselves in the physical world? The subject was deeply fascinating to me, that animals could be understood in terms of how they adjusted themselves daily in a changeable environment in order to satisfy some utilitarian need. Or simply to please themselves.

I stood at one of the Ponte Vecchio's portals between two shops, struggling with the meaning of some of the technical terms the authors were using, and, not for the last time, closed the book, distracted by the muddy surface of the rain-swollen Arno. And wondering where, in a world of autonomous insects evolving in some South American jungle at that moment, Botticelli's *The Birth of Venus* might fit, that painting having left me in a state of wonder as deep as the one Messrs. Fraenkel and Gunn had taken me to.

LATE ONE AFTERNOON, our guide aboard the *Beagle III,* Orlando Falco, offered to take a few of us swimming at Bartolomé, a small island off the east coast of San Salvador. I brought my fins and my mask along, though Orlando said there wouldn't be much in the way of tropical fish there. The bottom was composed mostly of old lava flows around a few pockets of sand, but there was no coral. The water was so clear I could read the textures of ropy flows of pahoehoe lava twenty feet below the panga. After everyone else had jumped overboard, Orlando took me a little farther offshore, into deeper water, and motioned for me to drop overboard there.

I rolled out of the panga and kicked hard for the dark bottom I saw below. The bottom that came into focus, however, was not a continuation of lava flows from the shore of Bartolomé—it was a huge school of orange-eyed mullet. Before I could halt my descent, the schooling fish parted, rising up around me in the form of a hollow cylinder. As I continued downward, the fish below me parted to reveal a white sandy bottom at about thirty-five feet. When I turned over to look back up at the fish from below, I saw that the elongated school stretched off more than a hundred feet in both directions. The lowest layer of this lens was about five feet off the bottom. The mullet were swimming in tight synchrony, veering and milling.

Thousands of them moved in unison above me, like a single thunderhead.

When I needed to ascend I put my hands together over my head like a springboard diver, kicked, and started moving up through them. When I glanced down, I saw the white bottom wink out beneath me and slowed my rise. Wherever I extended my hands now, the fish moved gracefully aside. When I pulled in my legs and hugged them to my chest, the fish came in closer, and for a few moments I was entirely surrounded. When the last layer of fish divided above me I saw the white bottom of the panga through about ten feet of water.

That minute and a half with the orange-eyed mullet was an experience my body as well as my mind continued to remember. Here, for me, was the edge of the miraculous. In every corner of the world there was such resplendent life, unexpected, integrated, anonymous.

ON THE LAST DAY of our ten-day excursion, the *Beagle III* put in at Puerto Ayora. We had a dinner of sea bass, rice and beans, and a fresh vegetable salad. The evening air was beginning to cool. We said good night to one another and headed for our rooms at the Hotel Galápagos, me to cabana numero cinco to begin sorting out my gear, to shower and pack. In the morning we'd all board a bus for the overland trip to Baltra and the flight out to Quito.

I wasn't ready to go. I picked at my belongings, indecisive about what to put where. My mind was surging in several directions at once, trying to align Darwin's catalytic history here with environmental degradation today in the islands, and with the future of possibilities for *Homo sapiens*. Somewhere the Australian philosopher Val Plumwood has written that humanity's task now is to "resituate non-humans in the ethical and to resituate humans in the ecological." Having an ecological—rather than a solely political or economic—view of *Homo sapiens* and knowing that the physical environment exerts a selective pressure on the human genome lead to a straightforward observation: to care for the environment is to care for the self. To run roughshod over the environment is to subscribe to the belief that humans are free to remain indifferent to their physical environment, that natural selection doesn't apply to them. That humanity's biological future lies, instead, I suppose, not with natural selection but with genetic engineering, with the edited genomes of CRISPR (Clustered Regularly Interspaced Short Palindromic Repeats) babies. Designer children.

From the small window in my bathroom, after my shower and after I'd turned out the bathroom light, I could see the palisades of the night standing on the plain of the Pacific. What would Cook and Darwin have said to each other about the grids each man had used to navigate? How might they have defined that word *archipelago*? And what might Darwin have offered us if he'd sailed with Cook and not with the fundamentalist FitzRoy? And what might either of them have said to the crew of the *Hōkūleʻa*, after being apprised of the Pacific islanders' trust in "navigator birds" to help shape their course?

I'D BROUGHT ALONG to the Galápagos a three-by-four-and-a-half-foot navigational chart published by the U.S. Defense Mapping Agency Hydrographic Center, in 1978, which incorporated a sequence of fifteen revisions. Based on British Admiralty chart 1375 (which is based in part on FitzRoy's sounding in the Galápagos in 1835), it was entitled "Archipiélago de Colón" and rendered in a scale of 1:600,000, meaning an inch here equaled about 9.5 nautical miles. The islands were represented in shades of gray, and a series of concentric lines revealed the topography of each one. The waters between them and the ocean surrounding the archipelago were white and hatched with lines of latitude and longitude. A navigational grid of thin green and maroon lines, based on magnetic compass bearings, was imposed over the other lines. Random stretches of water were stippled with numbers designating depths and with tiny, feathered arrows showing the direction of currents. Numbers atop the arrows indicated the average speed of these currents.

I wanted to fix this third journey to Galápagos in my memory. I was using the chart as a framework in which to envision the events of the trip and to establish the sequence in which they had occurred. I noted each of our landfalls: the dive at Roca Redonda, where it was not possible to step ashore; the route of our climb up the eastern flank of Alcedo, above el Canal de Isabela; the crossing of Bahía Isabel, where we'd seen the dolphins slipcased in phosphorescence. It was an exercise for me in recollecting and imprinting.

As I walked around the room, folding clothes, putting notes in manila folders, cleaning the residue of ocean air from the lenses of my binoculars, I would step over to the map spread out on the bed and fix on it from memory the place where something else I'd just recalled happened. It was mildly disconcerting to me that Bahía del Correo was actually north and east of Margret Wittmer's Black Beach, not to the south of her holdings, as I had pictured it. And Isla Darwin and Isla

Wolf were much farther north of the main islands of the archipelago than I had imagined them to be.

When I had everything ready, I prepared to roll the map up and slip it into its protective case. As I brought the large sheet of paper across the bedside lamp I was suddenly reading the archipelago as if it were lit from beneath. The weak lightbulb was the volcanic hot spot under the Nazca Plate, venting through La Cumbre and Cerro Azul.

I removed the cactus-skin shade from the lamp to make the hot spot more intense. And then suddenly I saw the situation in reverse—the bulb was the sun and I was looking at it from the ocean floor, not seeing the islands in an overview as the peaks of a cluster of volcanic mountains but seeing them from below as floating objects on the surface of the ocean, like volcanic cinders. To enhance this illusion, I rotated the chart, swapping top for bottom. What was the "upside down" of this place now? And if that was sunlight I saw dimly through the surface of the ocean, backlighting the islands but revealing no detail, was it now all right that *west* was on my right and not my left? If I rotated the map end for end to put the west back on my left, the "bottom" of the archipelago was now apparently its "top."

What would be the correct orientation for a navigator at the start of a voyage through Galápagos? What alternatives might there be for a mariner steering here by conventions other than port and starboard, east and west? What would the navigator in the *Hōkūle'a* suggest here?

In a corner of the navigational chart I saw the following in purple letters: "WARNING: The prudent mariner will not rely on any single aid to navigation."

Like Cook, Darwin wasn't navigating in Galápagos with an existing map. He was making a map.

Jackal Camp

Turkwel River Basin

Western Lake Turkana Uplands

Eastern Equatorial Africa

3°06'08" N 35°53'18" E

W̲e've come down from the north on the dirt road from Lokwakangole, five Kamba men and myself, riding in a couple of long-wheelbase Land Rovers. One of the Rovers, outfitted as a utility vehicle with a cargo bed behind the cab, carries a bulky load of camp gear under a dusty green tarp. The others are all older than I am, in their fifties. Here in Lodwar each of them shakes hands with a young Turkana man who's been waiting for us. A tall, shoeless fellow in gray shorts, he wears a dark blue short-sleeve shirt with vertical stripes of small yellow triangles. His forehead, cheeks, and chin are neatly ribbed with small cicatrices.

The Turkana man helps two of us, Onyango and Nzube, untarp the load in the utility vehicle and the three of them roll two 45-gallon plastic water barrels over to the town well to fill them. There is a polite exchange among them in Swahili, the Kamba men asking the Turkana man about the quality of the water before they start pumping. The head Kamba man, Kamoya, accompanied by the other two in our group, leaves to visit small shops scattered along the disjointed, uneven lanes of the village, and I follow. Kamoya is looking for flashlight batteries, millet, a single roll of black-and-white film, and cigarettes. I pause before a shop window to study the reflection of my face. Rivulets of sweat cut through the dust caked there. It glistens on the bosses of my cheekbones and pools at the base of my neck, filling the shallow depression between my clavicles.

The village simmers in the afternoon heat. I adjust my wide-brimmed hat. In every direction I turn, someone is looking at me.

The proprietor of a store where Kamoya purchases a 25-pound sack of millet asks where the four of us are from. Wambua and Bernard Ngeneo drift out the door with their purchases without answering. Kamoya remains behind to respond to the question. I dawdle, pretending to search for aspirin on a shelf of medicaments, but I am listening for Kamoya's answer.

Kamoya says they are Kamba, from the south of Kenya, and that they've been working for a white man up north at a place called Nariokotome. Does he know it? About fifty miles north of Lokwakangole? Oh, he knows of it, yes. It's the famous place where Richard Leakey found the skeleton of a boy who lived there very long ago, the man says. Actually, the man who discovered the 1.53-million-year-old *Homo ergaster* skeleton known as the "Nariokotome boy" is standing right in front of him, though Kamoya doesn't clarify this point.

And what are you doing here? the man inquires, gesturing with his head toward the door the other two have just walked out. Kamoya says we're going to be camping south of the Turkwel River for a few weeks, to the east of Lodwar. We're going to be looking for rocks, to see how old they are. The country Kamoya refers to, the man is aware, is mostly scrubland, overgrazed by Turkana livestock—goats, camels, sheep, and donkeys. He reacts as if Kamoya is deliberately leaving something important out.

Though he himself does not smoke, Kamoya buys a pack of cigarettes to mollify the proprietor, thanks him, and leaves.

I leave without having bought anything.

The others come into view up ahead at a petrol station. Kamoya's longtime friend Nzube Mutiwa is waiting in the driver's seat of the utility vehicle. Onyango Abuje and Christopher, the young Turkana, are topping off the diesel tanks in the Land Rovers. They've already filled the water barrels, rolled them up onto the ute's cargo tray, retarped the load against the clouds of road dust they expect to face, and are now checking the tire pressures all around.

I ask Kamoya, driving the other vehicle out of Lodwar across the short concrete bridge that spans the Turkwel River, about the conversation in the store, which had been conducted in Swahili. Why had he stayed behind to answer the nosy man's questions, instead of leaving with the others? And didn't it seem to him, too, that the proprietor was a bit suspicious? Kamoya says to the first question, "You don't want to be rude someplace where people don't know you." He answers the second question with a flick of his left hand, a gesture of indifference.

We cross the river on the dirt road that goes south to Lokichar, but almost immediately turn left onto another road. We're headed east toward the western shore of Lake Turkana, about 35 miles away. This particular road will bear off to the south before it reaches the lakeshore, however, and afterward pass to the west of the Nachorugwai Desert. Lake Turkana, known informally during the British occupation as the Jade Sea, was also once called Lake Rudolf. It's relatively narrow and fills a depression about 149 miles long, part of the East African section of the Great Rift Valley. It has no outlet.

Our vehicles move steadily east. Nzube remains a ways behind us, to keep clear of the worst of the billows of dust, which hang nearly motionless in the still air above the road. Some miles east of Lodwar, we turn left at an unmarked place, gear down, and proceed north in four-wheel drive across rough tilted terrain. We negotiate the soft silt of wadi courses, push through stretches of thick brush, and cross small, hard patches of water-rounded gibber stone, finally coming to a halt before the rampart wall of a gallery forest, a line of old acacia trees marking the west bank of the Kerio River. Ten days later, when our water barrels are empty, we will dig in this dry riverbed for water to quench our thirst, to wash ourselves and our dishes, and to wash our clothing, of which the Kamba men have very little.[1]

Kamoya directs the setting up of camp—the cook tent here, the sleeping tarp there, a worktable over in that spot under the trees, firewood in this spot. A clothesline is to be strung between those two trees. I look for ways to be of use. The others, moving quickly and efficiently through a familiar routine, politely accommodate my efforts. Wambua goes off to fix a flat tire. Onyango and Bernard walk off a ways to dig a latrine.

After the campsite has been squared away, Christopher brings glasses of lemonade from the cook tent and the six of us sit together on folding chairs in the shade of the tall acacias. Kamoya describes the country around us in English, occasionally in Swahili or Kamba. He wants to get everyone oriented. Three miles to the north, the Turkwel River. Directly to the south, the Napedet Hills. Off to the west, the Loima Hills. A thinly vegetated country of lava tuff and imbricated cobbles, of mud-flows, sand depressions, and exposed gravel surrounds the camp. We will scour this landscape methodically, day after day, looking for fossil evidence of the ancestors of *Homo sapiens*—the ramus of a hominid jaw, an astragalus from the foot of an early human, a fragment from a pelvic girdle. Looking for any hint of human ancestors, of australopithecines, or early "maybes" in the human evolutionary family, like the hominid *Ardipithecus ramidus,* emerging from the rock and earth around us. These would be the deep and the very deep ancestors of all of us sitting beneath the acacias, not to mention the ancestors of Helen of Troy, of the seventh-century Chinese poet Du Fu, of the builders of Tenochti-tlán, of Neanderthals buried at Shanidar in the Fertile Crescent, also of the three Turkana men now striding swiftly toward our camp from the west. Their haste causes their multicolored cloaks to flare like wings in the heated air. One grips a *vimbu,* a Turkana fighting stick.

I had been helping Christopher gather and store bundles of firewood before we sat down for lemonade and to listen to Kamoya's descriptions. I was studying birds in the acacia trees around us after Kamoya spoke, comparing them with images in a bird guide I had, and was making notes in a spiral notebook as the Turkana men approached. I was the last of the six of us to notice the men, after Wambua signaled to Kamoya by lifting his chin sharply, and Kamoya turned in his camp chair to see what had caught Wambua's attention.

I put my notebook aside and look around apprehensively at the others.

In their left hands the three Turkana men carry their head stools, *kitis;* in their right hands two of them carry *mkwajus,* short walking sticks. They cross the camp's perimeter line and quickly seat themselves a short distance from us, near an acacia.

They have the indignant air of men who have been slighted.

The tallest of them, wearing a thin wool cloak with wide vertical stripes of red separated from wide stripes of dark blue by narrow yellow stripes, begins to address Christopher, who is peeling potatoes in the shade of the trees. He jabs his vimbu repeatedly into the ground for emphasis. His hair is impeccably groomed: a chignon at the back of his head, a foreknot above his brow, and a skullcap of cornrows, the whole of it meticulously plastered with red and gray mud. A single ostrich feather is spiked vertically in the chignon. Steel bracelets on his wrists clink brightly in the dry air when he gesticulates. Thin metal hoops sway and jerk at his earlobes.

While the man continues to direct his tirade against Christopher in Turkana, Kamoya begins to approach. He lifts his chair, moves it ten feet closer to the Turkana men, then sits down and waits. He does this several times until he is seated directly in front of the three of them, at which point he waves Christopher away, back to his cook tent. Kamoya begins speaking quietly, deferentially, in Swahili. The other man renews his berating, now in Swahili. Kamoya doesn't interrupt him. Nzube translates for me. The men feel insulted. No one has asked their permission to camp here. But his anger and suspicion go deeper than this. The Turkana believe we are here to look for commercial gemstones and geodes. No, says Kamoya, we're looking only for old bones, and we're doing this work with the support of the government of Kenya. Kamoya, of course, is implying that he does not need these men's permission, but he is respectfully receiving their objections. He listens patiently while the other man explains the tenets of traditional hereditary land ownership among the Turkana people, which, he finally comprehends, is not going to change Kamoya's mind.

One senses he and his ancestors have been losing this argument for more than a hundred years now.

The other two Turkana men are younger, less emotionally involved. Only one of them wears steel and leather bracelets, steel ear hoops, and a sheathed wrist knife. His cloak is orange with tiny yellow decorative stitches. He has a wooden plug in his lower lip. His hair, too, is plastered flat and done up in a traditional way. He turns away occasionally during the conversation to laugh at the absurdity of Kamoya's statements. The third man wears a cotton bedsheet for a cloak, dyed purple and brown in a sort of paisley pattern. Like the other two, he wears open leather sandals and his face is ritually scarred, but his hair is cropped close to his skull and he is balding.

Kamoya's courteous manner, his unperturbed composure and tact, have effectively defused the situation; but this is not an injustice that can be rectified. It can only, for a time, be mitigated. I've witnessed this sort of confrontation between legal and traditional owners in a variety of places—in a Warlpiri Aboriginal village in Australia's Northern Territory; in a farming community on the Ghorband River in northern Afghanistan; and in the High Arctic Inuit town of Pangnirtung, on Baffin Island. If everything that comprises a particular stretch of land—its still and moving waters, its trees and animals, its weather and footpaths and its rocks—is viewed as a community to which human beings, too, belong, it can neither be sold nor owned, even by the people who feel they belong to it. What the Turkana man is saying, I believe, to the Kamba man is, "Why didn't you knock when you came into my house? Why didn't you explain what you wanted before you walked into my home?" And what Kamoya is saying to him, not without some reluctance, is that it is not necessary to ask. That's no longer the way it's done.

What these awkward and painful encounters come down to now, in all the old colonial places, is whose authority can most effectively be enforced. The older Turkana man knows he is going to have to concede. Short of armed revolution, pointless and suicidal, he has no other choice. But he does not want to compromise his dignity or to lose self-respect by being silent. He does not want to sever his connection with a metaphysical world that comes to him daily in his dreams and in his waking hours, as it has for his ancestors for millennia. So he has come and complained and taught the younger men how to do it, hoping they will listen and take it in and continue. And he has gone away bitter, as have so many before him in the other colonial places.

I ask Kamoya what was really going through his head during the exchange with the Turkana men. He understands, he says, that he has the authority to speak as he did, operating as he is under the auspices of the National Museums of Kenya. The presence of a white man, he tells me, added additional weight to his authority. And the material wealth on display—the Land Rovers, the gear spread around the camp, the presence of the Turkana servant—added even more. But what he really thought about the ethics behind it all, in a countryside twisted into a new shape by colonization, I never knew. My guess was that he didn't particularly like sending the Turkana men on their way.

The only ethics I really needed to probe in this situation, anyway, were my own. What were my own reasons for not asking permission? For not having knocked?

WITH THE DEPARTURE of the Turkana representatives, Onyango and Ngeneo look for a shady spot in the acacia gallery to take a nap, the hottest part of the day having arrived. Wambua goes back to patching an inner tube and Kamoya and Nzube settle into a game of checkers, using bottle caps filled with black or white plaster. Christopher stands in the front door of the cook tent, drying a roasting pan and watching the Turkana men dissolving into the distant scrubland. I see he has carefully trimmed the branches of a toothbrush shrub in front of his cook tent, leaving behind short, rigid pegs on which he has hung up cups and cooking utensils.

I return to the bird guide and my notebook. As Kamoya spoke to us about the surrounding landscape, I came to understand that most everyone had been here before. The work we'll do in the days ahead continues a survey Kamoya began in these late Miocene deposits several years earlier. (I had to glance at a laminated reference card stuck in the pocket of my notebook to remind myself, as Kamoya spoke, that the Miocene epoch preceded the Pliocene epoch in the Tertiary period.)

Off and on, as I recall Kamoya's presentation to us and jot down recollected details, I'm scanning the acacias for birds I can hear but, unfamiliar with their calls and songs, can't identify without the bird guide. (Wherever I travel, I purchase a guide to local birds. The book quickly becomes a vade mecum, a reference to regional life, the paragraphs of which, thankfully, lack the tendentious prose and political commentary of many travel guides.) With the help, then, of Williams and Arlott's *Birds of East Africa,* I'm able to identify Cape rooks in the trees, dark crows with slender bills and long neck feathers; white-headed buffalo weavers, a heavy-billed, red-rumped bird, a seed-eater; mourning doves, with their pale pink breasts and carmine eye rings, uttering a familiar guttural murmur; and superb starlings, small, plump birds with rufous-chestnut breasts, a narrow white breast band and, depending on the angle of the sun, metallic blue or metallic green wings. There are thirty-four species of starling arrayed in several genera living in East Africa; thirteen species of dove; and forty-nine different weaverbirds, including the gray-headed social, the strange, the compact, the Somali yellow-backed, and the northern masked.

The Cape rook is the lone East African rook, one of only two rooks, in fact, to be found in all of Africa.

For a while I am lost in the fine points of the descriptive paragraphs that differentiate among the forty-nine species of weaver; but I cannot maintain the diligent frame of mind required. I glance up to watch the distant Turkana until they are absorbed, like the last fragments from an explosion.

BEFORE I FLEW UP to Lokwakangole from Nairobi to join Richard at Nariokotome, also the British paleoanthropologist Alan Walker, and Kamoya and his crew, I had lunch alone one day at the New Stanley Hotel, where I was staying. I asked the waiter where he was from. Lodwar, he said. Oh, I said, I'm headed that way myself. Did he have any recommendations? No, he answered, but then, as an up-and-coming waiter at a fine hotel in the big city, and perhaps sensing I was a little naïve, he added, "Those people up there are primitive, but very okay."

I could understand better now what he'd meant. One more group of isolated, underinformed, decent people, these particular Turkana were still confounded by the effrontery of invaders of one kind or another—missionaries, shills for commercial ventures, petroleum geologists—but compelled to face the consequences of not being strong enough or smart enough or ruthless enough to effectively resist. Their losses seem unjust and cruel to many people living today in the countries that originally colonized places like northern Kenya—but inevitable. Until now, this has been how Western civilization has made its way, securing resources and creating the lebensraum it wants, however that might be morally defended or achieved.

However one might philosophize or rationalize around the injustice, the departing Turkana were left with one more wound to their epistemology, the gall of their powerlessness, and the knowledge that we had, in effect, the permission of God to look into the origins of man here, that the nobility of our task, like the economic imperative behind the seismic trucks thumping the desert to the north of us in search of oil deposits, was but one more of the civilized world's trump cards.

As evening comes on—quickly, with no lingering dusk at 3° northern latitude—flocks of doves roosting in the acacias offer a plaintive chorus and I regain a sense of equilibrium again around these old issues of colonial injustice. Their voices dampen my frustrations, embrace my confusion, and I become once more a student of the inquiry into man's origins that has brought me here, work for which I have enthusiasm and respect. What mostly brings me out of my funk is the image of Kamoya before the irate Turkana, defusing violence, but not demanding obeisance in the face of his ultimate authority.

I close the bird book and go with that image into the evening.

MY FIRST TRIP to Africa, to the southwestern quarter of the continent—Namibia, Botswana, South Africa, Zambia, Zimbabwe—presented me almost daily with what I thought of as archetypal imagery. In Harare one morning, the old Salisbury of Southern Rhodesia before independence, I became lost trying to find a bakery the doorman at my hotel had recommended for breakfast. I found it eventually, but not before being swept this way and that through the streets of Harare by throngs of people headed for work—courteous, smiling, seemingly uncalculating people, dressed in bright fabrics. Lively rivers of people enthusiastic about life and among whom I was welcome. Under a full moon one night in Zambia, I walked alone out onto a splinter of land that fronted Victoria Falls, on the Zambezi River. I had on flip-flops and wore only a pair of shorts. The mist boiling up like smoke from the plunging cataract completely soaked me. When I turned to go, putting the full moon behind me, the only moonbow I've ever seen appeared ahead of me, suspended in the night air.

I don't know where to start with the range of African animals I felt so privileged to see on that first trip. I spent an hour in the company of two black rhinos, a critically endangered mammal, a creature of astonishing size, more reticent than its less endangered cousin, the southern white rhino. The time I had with them, watching from the roof of a Land Rover, felt like the final hour with a revered grandmother before she was called across to the other side. My grandchildren will likely never have the opportunity to see a black rhino except in a zoo. In the Kalahari Desert, in western Botswana, I occasionally encountered small herds of oryx, a large, exceedingly wary antelope with long lance-like horns. Oryxes seem to me robust, capable, and dignified as they cross great stretches of waterless and unvegetated land, as if they took water and fodder from the air. In northern Namibia, near Etosha Pan, studying what was left of a Burchell's zebra carcass from a blind—a spotted hyena kill—I saw five species of vulture feeding together: Cape, white-headed, Egyptian, lappet-faced, and white-backed. An ecological Venn diagram working the carcass.

Encounters with dozens of species of wild animals, in as many different settings on that first African trip, fixed that part of the continent in my mind as Edenic, despite what I knew to be the extensive destruction of wildlife taking place there, to sustain local markets for bushmeat and the market abroad for substances used in traditional Chinese medicine; an Eden, despite my having attended the apartheid-era trial in Delmas, despite my having moved through war-battered villages along the Angolan border in Namibia, where South West Africa People's Organisation guerrillas were, at the time, fighting the South African army, illegally occupying Namibia and forcing thousands of Namibian families to pay the price for their illegal incursion.

My encounters with elephants, African hunting dogs, springbok, Kori bustards, warthogs, impala, lions, ostriches, giraffes, and the others always provoked in me the same two emotions: wonder and gratitude. I felt profoundly lucky to be able to see such things with my own eyes, in landscapes that had no closing hours, no fence lines, no cultivated fields, no built structures. These encounters were so biologically and metaphorically rich that they seemed to have no bottom. The experiences of that first trip did not provide an antidote to what I'd been exposed to in that courtroom in Delmas, or blur my memories of the faces of the famine-stressed children I'd seen in the Caprivi Strip in Namibia; they kept despair over the fate of humanity at bay. Each one—the free animal in its domain, the condemned men in a South African court—intensified for me the authority of the other one.

I was immensely glad to be back in Africa again, where, by most accounts, everything began for us.

THE ONGOING ACADEMIC and popular debates about human origins—about exactly where the line of human descent lies among hominoids in the human evolutionary family—captivated my imagination as a young man, so much so that I boldly wrote the well-known paleoanthropologist Louis Leakey when I was nineteen, asking if I could work as a camp boy at Olduvai Gorge in Tanzania (Tanganyika at that time), naïvely and self-importantly wishing to make myself part of his and his wife Mary's efforts to learn more about where we came from. The Leakeys had brought the search for the origins of mankind into people's living rooms in the early 1960s with a series of spectacular finds at Olduvai: the fossil bones of *Homo habilis, Homo erectus,* and a robust australopithecine. (Kamoya and Nzube were both working with the Leakeys at Olduvai at the time I wrote to Louis.)

Louis and I were not able to put a plan together that would bring me to East Africa during the time they were working there, but many years later, still keenly interested in human origins, I wrote to his son Richard, by then himself a well-known paleoanthropologist, especially because of work he'd conducted in northern Kenya around Lake Turkana. I asked to visit with him at his sites at Nariokotome and Koobi Fora. Richard wrote back to say yes, please come, and extended an invitation to visit him first in Nairobi, where I could view the collection of hominin skulls at the National Museums. Then, he wrote, we'd travel together up to Nariokotome. When I inquired about spending some time in the field, if possible, actually searching for hominid fossils with Kamoya Kimeu and his colleagues, Richard arranged for me to go with them to Nakirai, a camp east of Lodwar where Kamoya had been working for a while.

(Nakirai was the Turkana designation for that place situated among acacia trees along the Kerio River where we were to camp. Kamoya told me it means "the place of the jackals.")

What interested me about Nakirai—and Nariokotome and also Koobi Fora on the east side of Lake Turkana—was the certainty that seeing them would add greatly to my sense of what the search for human origins looked and felt like. Most of the academic discussion about human origins tends to speculate on how *Homo sapiens* came to be, based on relatively scant (and somewhat problematic) fossil evidence. Too often, it seemed, these earnest disagreements drifted toward tedious and pedantic sparring. The chance to work alongside people like Kamoya, searching for the residue of ancestors tens of thousands of generations removed from us, a chance to experience the physical place where a large portion of the evidence has come from, promised something richer than what I had been gleaning over the years from the pages of *Science* and *Nature*.

In the spring of 1984, during a visit to New York to see my ailing stepfather, I saw an exhibit at the American Museum of Natural History featuring many of the most famous hominin fossils. (Some owners, fearing loss or damage to these irreplaceable objects, had sent replicas instead.) The show was called "Ancestors: Four Million Years of Humanity." The physical evidence used in support of any of several different routes humans might have followed from their australopithecine ancestors to the fire pits at, say, a Cro-Magnon shelter like Gönnersdorf in the Rhine Valley made a profound impression on me. Afterward I walked for miles across Manhattan, in an expanded state of awareness, knowing the immense length of the line of our descent, from a protohuman existence some five or six million years ago to the present moment, which found humanity in the forms I now saw, cultural sophisticates navigating Manhattan's streets impatiently at dusk. For the first time, I think, under the spell of the show, I saw them as the last surviving members of the hominin family.[2]

I walked across Central Park to the Upper East Side and then headed downtown on foot, passing the apartment buildings of the very-well-off and, at the southern edge of Midtown, passing through the Murray Hill section of the city, where I'd spent my teenage years. Eventually I found myself south of Houston Street, in SoHo, a part of the city famous at the time for its thriving community of artists and galleries.

While I walked I tried to picture the gap between the individuals I took note of on the street and the individuals in the museum, the fossil skulls, of course, of *particular* individuals, put on display millions of years after they had died. What would our ancestors look like to us if we viewed them from a vantage point in time other than our own? What if it were possible to stand beside a *Homo habilis* father while he shaved meat from the femur of a gazelle for a meal 2.2 million years ago, in what is today the Rift Valley in Tanzania? What if you could watch a young *Homo erectus* woman cracking nutshells beside her sister in what is now Hebei Province in China, 600,000 years ago? What if it were summer in the hills of Cantabria, in what is now northwestern Spain, 41,000 years ago, and you could lean against a tree and watch while a *Homo sapiens* girl approached a *Homo neanderthalensis* boy? What if you were standing in a Natufian settlement in what would one day become Jordan, at the dawn of "the age of civilized man" 11,000 years ago, watching a child inexpertly thread a bone needle while her mother rolled her eyes at the attempt? What if you had more than this one moment of your own life from which to see, seated next to a table on a sidewalk at a restaurant in SoHo, where a woman is toasting her husband with a glass of Chablis on the occasion of his just having sold a painting to the Guggenheim Museum?

What if the perspective you could imagine for yourself, the foundation for your ethics and your politics, was not the condescending *now* of right now?

WHEN I WAS YOUNG and wanted so much to see the world, my mother gave me an atlas, *Hammond's Illustrated Library World Atlas,* published in 1948. Overlaying some of the pages with tracing paper and using colored pencils, I lined out the journeys I wanted one day to make. Down the Yangtze from Chongqing to Shanghai. Across the Great Victoria Desert in Australia. From Panama across the Darién Gap to Tierra del Fuego. Decades later, I took the old atlas down off a shelf. Its maps were now much dated—French West Africa and Belgian Congo, gone; Yugoslavia, dismantled; Ceylon had become Sri Lanka and Siam, Thailand; the Ellice Islands, Tuvalu. Several loose sheets of tracing paper fell free as I paged through the book. I'd completely forgotten about them. As I retrieved and stared at these tracings, I realized I'd managed to make many of these journeys in the intervening forty years. In that moment in front of the bookshelf I suddenly felt the familiar sensations of wonder and gratitude, emotions I'd known for years now as a traveler, but this time they rose up around the daydreams of an eight-year-old who wanted to go but didn't know how.

Maps held me in thrall when I was young. They combined, in a single two-dimensional space, both the broad all-encompassing reality that a great journey makes possible and the particularity of those places along the way that comprise the journey. To behold the map is to imagine, in the same instant, both the arc of the journey and the moments that will make it up. This, to make a bit of a leap, is part of the genius for me behind Monet's impressionistic representations at Giverny. The unfocused colors of these sketch-like images mimic the sketchiness of one's general recollection of movement through a particular geography, while at the same time the painting consists of myriad discrete dabs of color. One of Monet's contemporaries, Camille Pissarro, painted panoramas of Parisian streets that work before the eye in the same way maps do: you appreciate the entirety of the area and, simultaneously, its discrete components.

Travel, to my mind, can become a type of mapmaking. As a young man I read Darwin as much for the epic journey he describes in *The Voyage of the Beagle* as for his thoughts on how living organisms change through time while proceeding in no predetermined direction. I later read James Cook because I was attracted to and admired his disciplined effort to travel deliberately, and also for the way traveling had taught him about the world. I read the biographies and autobiography of Ranald MacDonald because his quest for identity made him a different sort of traveler than explorers like Cook or Tasman or Matthew Flinders.

Traveling, despite the technological innovations that have brought cultural homogenization to much of the world, helps the curious and attentive itinerant understand how deep the notion goes that one place is never actually like another. Traveling encourages the revision of received wisdoms and the shedding of prejudices. It turns the mind toward a consideration of context and releases it from the dictatorship of absolute truths about humanity. It helps one understand that all people do not want to be on the same road. They prefer to be on their own road.

Darwin taught that, like the panda or the thresher shark, *Homo sapiens* is an animal without a destination, and like all other animals is known only in its present form, a transitional form, even if that form, like the coelacanth's, is stable for a long period of time. Modern humans are part of a continuum that stretches from *Homo heidelbergensis,* some 250,000 years ago, to a descendant standing up ahead of us, invisible in the ether of the Anthropocene.

No other creature, as far as we know, is as focused on identity and destiny as *Homo sapiens.* The desire to have special meaning in the world is one of the reasons I think humans search so diligently for the bones of their ancestors. The mere fact that they existed and are somehow related to us is reassuring to an animal that tends to view itself today as a barely tethered balloon reacting to the rising winds of accelerating cultural change. Many of us are not adapting smoothly to a rapidly changing world, especially psychologically. Our ancestors offer us historical meaning, but they give us no indication of the future. And what is true for us is true for every other animal: no matter our impressive history, every day we advance figuratively into evolutionary darkness. And, because we are inescapably biological, we have no protection against extinction.

I ARRIVED AT Nakirai carrying as little prejudice about the origins of man, I hoped, as was possible. (I trusted my experience there would have its own logic, that I need only be attentive and prepared to participate.) Over the years, camped with researchers in the field in different situations, I've found it helpful to maintain a curious instead of a skeptical frame of mind, at least initially. The world, I think, is mysterious at a fundamental level; and we're free to engage with it at any depth we wish, free to listen to any serious intelligence trying to make (usually limited) sense of the world's mysteries—the pi meson, the nutritional needs of a black rhino, the psychology of the Thule. And we're free to be enthusiastic about any human effort to comprehend humanity's place in the world, no matter who does the comprehending. The quest for a state of equipoise, of balance in the face of the paradoxes and contradictions that come with chaotic cultural forces, seems to be more valued in times of great stress than the overconfident pronouncements of authoritarian professionals, who too often seem only to encourage or suppress panic or fear in their listeners.

It's good to know where you come from, so that you do not live as though you're lost, someone wearing a mask of confidence but feeling no measure of assurance—about anything.

IF THERE IS a trick to a search for the meaning of a particular fossil bone, it's that in order to perceive what no one else has seen there, it's necessary to overcome one's preconceptions, to shift one's point of view, to give up orthodoxy. The capacity to do this is what, for me, not to belabor the point, constitutes part of the genius of Cook. He sought to navigate where others had been satisfied merely to sail. And it is part of what distinguishes Darwin as well, a man who stepped away from more than one kind of orthodoxy.

One emerging view of *Homo sapiens* among evolutionary biologists is that he has built a trap for himself by clinging to certain orthodoxies in a time of environmental emergency. A belief in cultural progress, for example, or in the propriety of a social animal's quest for individual material wealth is what has led people into the trap, or so goes the thinking. To cause the trap to implode, to disintegrate, humanity has to learn to navigate using a reckoning fundamentally different from the one it's long placed its faith in.

A promising first step to take in dealing with this trap might be to bring together wisdom keepers from traditions around the world whose philosophies for survival developed around the same uncertainty of a future that Darwin suggested lies embedded in everything biological. Such wisdom keepers would be people who are able to function well in the upheaval of any century. Their faith does not lie solely with pursuing technological innovation as an approach to solving humanity's most pressing problems. Their solutions lie with a profound change in what humans most value.

THE DRIVE FROM Nariokotome to Nakirai through the village of Lodwar, setting up our camp, and dealing with the Turkana men brought us nearly to the end of the day. Christopher made supper, and after we ate, I asked Kamoya if he had a few minutes. We'd gotten to know each other the week before at Nariokotome, but that was in a camp being run by someone else. This was his camp. I wanted to understand how things might go best here. Kamoya was both fraternal and avuncular where I was concerned, and I understood I had plenty to learn. In a certain sense, I really didn't know where I was, and we were about to plunge into a time—mostly the early Pliocene—that was floating free for me, conceptually, and was therefore nebulous.

The work itself was complex, a combination of abstract awareness and empirical proficiency. In a couple of days Wambua would find a late Miocene hominid tooth, a dark bluish-gray object with a weak sheen to it, indistinguishable to most people from a random piece of polished quartz. It was a spectacular and important find, there being a gap of some one million years in the late Miocene fossil record for hominids. It is still hardly comprehensible to me at this point that an animal dies, that scavengers scatter its bones, that some of them end up in a river where they are eventually buried under layers of sediment; and that over millennia minerals in the sediment slowly replace the organic molecules in the bone, creating a fossil. Some six million years after a primate dies of a heart attack or is killed by a predator, a man in khaki shorts and a blue-gray short-sleeve shirt bends over it with a stick. He nudges undistinguished bits of gravel aside to fully reveal one of the animal's lower left premolars.

Wambua was a person who knew where and how to look, and who often was aware right away of the significance of what he saw.

EACH MORNING we head out at about six-thirty in the Land Rovers, leaving Christopher behind to mind the camp. Each day we search another large, undemarcated tract of open land on foot. Most often we line ourselves out six abreast, about twenty feet apart. When we encounter dry water courses with eroding banks, we concentrate as a group on these banks, because fossils frequently emerge here. Like the unperturbed drifts of the Grevy's zebras we sometimes see grazing around us, our movements conform to the contours of the land. I adjust my pace to the movement of the other five, searching my assigned ground, but I look up regularly to mark the heading and pace of my companions. I'm usually working off somewhere to Kamoya's right. We walk in silence, and I'm conscious of not breaking Kamoya's concentration every time I have a question or come upon something that seems interesting. But if anything looks promising I alert him.

I carry my notebook, a water bottle, a small tape measure, a pocket-knife, a compass, my binoculars, a first-aid kit, sunscreen, and an extra pair of sunglasses in a small day pack. The others carry nothing but versions of the same type of stick I use, made from the peeled branch of a toothbrush shrub. The stick is about two feet long and straight, but it's been cut in such a way as to leave a short spur, the base of another branch, resting across the palm of one's hand. The distal end of the stick has been sharpened to a blunt point. Onyango shows me how to make one, and with beginner's luck, it turns out very well. Wambua offers to trade his stick, which is not quite so straight, for mine. I agree, believing the favor will be repaid some day.

Each man has his own technique for searching effectively. Onyango moves slowly, hesitating, with the deliberation of a heron stalking fish in shallow water. He's the only one in camp besides myself who reads regularly. Ngeneo moves with impatience, swiping at the ground with his stick and looking away distractedly. He is always the first up in the morning after Christopher and often whistles while he's searching. Kamoya, a stout man with a large head, moves confidently and calmly, like a bull among heifers, the bridge of his bifocals usually riding far forward on his nose. He looks up from the ground and out around him in a regular rhythm, alternating with his intense concentration on the ground immediately in front of him. Wambua moves slowly, like Onyango, but with his hands clasped behind his back, pausing briefly to study a single patch of ground and then another, and sometimes bringing his stick around to poke sharply, as though he meant to wake up a stone. Nzube, a small-boned, diminutive man with a high fore-head, covers more ground than anyone. His eyes jump very quickly from spot to spot.

During our days together, each of these men will make a significant find, which suggests that no single approach in the search for fossils is superior, or that almost any method will lead to success. The indis-pensable element in each man's scrutiny of the ground is his ability to recognize what is important, to differentiate quickly between what is significant and what is merely interesting. In the first few days working alongside Kamoya, I struggle to develop the right search images, to separate the metal from the dross. But I've done this before, search-ing for meteorites amid other similar-looking rocks in Antarctica, and looking for bits of grain in sandy soil at Ancestral Puebloan sites, in the American Southwest. It takes a few days of intense concentration, and then I, too, am able to recognize what the others are looking for.

Depending on what we discover at any particular place, we might come back again the next day or move on to another area. How much time we spend at a particular place depends on the geological age and richness of the fossil beds there. An astonishing aspect of the men's work, to me, is that most of them can recall precisely where they found anything of note. One day I found what I knew by then was part of a hippopotamus skull. Unless we found a hominid fossil close by, I knew it would be unlikely that anyone could raise the funds necessary to excavate this hippo, but I was surprised when I called Kamoya over to see it that he seemed to take only cursory notice. Three or four days later, however, when we were working nearby again, he walked straight over to the hippo fossils from several hundred yards away. The fossil bone fragment I'd found wasn't apparent to the eye on that level plain of ground, dotted with brush, until you were standing right next to it. I couldn't identify the markers he was using to navigate successfully to this spot, over what, to me, seemed undistinguished ground.

He said he wanted to examine the hippo more closely.

On another day, when the heat of the afternoon sun was driving us off a couple of acres of dark cobbles, and I would have said we were still several miles from the vehicles, Nzube led us toward a rise, over barren ground as anonymous to my eye as the surface of the ocean. When we topped the rise, I saw our vehicles just below—"cars," they always called them—with canvas bags of cool water hanging from the side mirrors.

Once I got better at recognizing fossils, I felt more comfortable accompanying the others on these searches, and Kamoya began to share more details with me about what was going on around us at any particular place. He'd describe its geological history, saying, maybe, that four million years ago this or that place was a swamp or a savannah. Soon I could identify carnivore teeth, crocodile skulls, upper and lower turtle plastrons, bovid mandibles, fish bones, and slender strings of tiny rocks that were in fact the intact vertebral columns of small rodents. One day Kamoya came over to me with a molar. "Hominin," he said, but with so little enthusiasm I was puzzled. "Maybe a couple of hundred years old," he continued. He walked back to his search line and dropped the tooth where he'd found it. I brought him a bone from a large fish once, a lightly constructed plate with radial struts. It looked something like a mammalian scapula. I left my stick behind, standing up at the find place. The bone had mineralized in such a way that the iron oxides in it gleamed in tiny spots of color across its surface, iridescent hyacinth blue, mauve, indigo, the pale purple of lilacs, and the dark purple of eggplant.

Kamoya examined it with appreciation and then slipped it into my shirt pocket. "Lots of fish around here," he said.

On another occasion Kamoya brought over a giraffe tooth to show me. While I turned it over in my fingers, trying to absorb its distinguishing diagnostic characteristics, I caught the slightest movement of Kamoya's chin, indicating something in the distance. I tried to take in whatever it was with a sidelong glance—two Turkana boys following us, a third of a mile away, trying to stay out of sight.

WE RETURNED TO Nakirai, the Jackal camp, about one each afternoon for lunch, which Christopher would have ready for us. He got up to make breakfast in the dark and we ate it quickly, during the first few minutes after sunrise, before hundreds of flies were shaken out of their night stupor by the warming air and drawn to the moisture in our food.

Christopher slept in the cook tent; the rest of us slept in the open on thin mattresses, side by side on a large green tarp. Christopher never wore shoes in camp; the others went barefoot as well, but put on sandals before we left for the day. I switched from flip-flops to a stouter pair of shoes before leaving camp, and put on sunscreen. One morning Wambua stared at a nonexistent watch on his wrist with a look of exasperation as I delayed everyone by applying sunscreen, which they of course didn't need. We laughed.

During the hottest hours of the day, after lunch, we'd drift apart, the six of us, most to take a nap. I'd catch up on my notes for the day, go back to reading Chinua Achebe's *Things Fall Apart* or Thomas Pakenham's *The Scramble for Africa*. Kamoya might be on the radiotelephone with someone at the museum in Nairobi or be playing checkers with Nzube or Onyango. Wambua, a muscular, broad-chested man with a thin mustache, who squinted and spoke rapidly and passionately about things, might be lying supine on his bedding, smoking a cigarette. He, Nzube, and Ngeneo are the only smokers. At dawn, Wambua smokes a cigarette quietly before rising from his sleeping pad.

LATE MOST AFTERNOONS Turkana families approached our camp, walking in out of the arid lands where they live and graze their livestock. They squat at the periphery, watching us closely as we go about our business. If we approach them, they begin to complain in Turkana of illness, of their need for transportation to Lodwar. Someone back at one of the *bomas* has been bitten by a puff adder (*akipoon,* in Turkana) and needs attention. A baby, perhaps malarial, is held out to Nzube, as though he were capable of magic. They ask for shoes, shirts, pants. Twenty or thirty of them are still there, silent, staring quietly at us from the dark when we go to sleep.

Kamoya is always gentle with the Turkana people, whose curiosity and desires are constantly before us, but he's firm in his dealings with them. One evening he explains to me how he manages his compassion, the impulse to be generous. In fact, we have very little extra of anything in our camp. We have a few five-pound bags of millet, which Kamoya sometimes trades for a goat to slaughter. He drives very hard bargains, and there are always displays of disbelief, outrage, disgust, and bafflement on the other side during negotiations.

One night I notice a young woman massaging her forehead in a way that makes me think she has a headache. She is wearing traditional clothing and a garrote of interwoven bead necklaces, three or four inches high, as well as several brass arm rings. She's sitting on a log with some other women. Kamoya, when I ask him, says it's all right to offer her some of my aspirin. I bring Christopher with me to interpret. She says yes, she has a headache, but when I offer her two tablets, an older woman seated next to her says something to Christopher. He turns to me. "She says, 'We prefer to lack it.'" What she means, I understand, is that people like myself tend to view people like her as "lacking" certain things, but that sometimes they actually prefer to lack them. What I regarded as an act of charity is, for her, an opportunity to decline those things that might lead to compromises in her life.

As it happens, the morning after we're told that someone's father has been bitten by a puff adder is actually the morning Kamoya has chosen for us to go to Lodwar to replenish our supplies. First we drive both Land Rovers to a Turkana settlement where the older man is reported to be near death. As we approach the man's *awi napolon* (the thornbush-encircled residence of a prominent male), we see the man who supposedly needs to get to the hospital right away sitting outside, waiting for us in his finest clothes, holding his *mkwaju*.

"Nobody at death's door here," Kamoya says as we pull up.

The man simply wants to see Lodwar, where he's never been.

Both vehicles quickly fill with women, men, and children. Several more cling to the spare tire mounted on the rear door and clamber onto the roof, straddling the tanks of extra fuel secured up there in the roof rack with another spare tire.

It takes a long while to reach Lodwar. When we arrive, Kamoya explains to everyone the time and place to regroup for the return trip. Everyone scatters. Not all of them return to the rendezvous at the appointed hour, though others replace them. On the return trip to Nakirai we drop some people off and pick others up, including the man who was not bitten by a puff adder. Disgusted with Lodwar, he had started walking home soon after he got there. Some of the people who hitched a ride with us had never been inside a motor vehicle. They marvel at how the side windows slide open, at how the door handles work.

People standing on far-off rises in the land with their goats stare at us as we pass. People inside the vehicle, sitting in the middle seats, shout with all their strength to hail them. The herds of goats break like waves over elevations in the scabbed land.

OCCASIONALLY WHEN I can't sleep I take my binoculars out and gaze into the wilderness of constellations above, the silver path of the "milk road" we pass beneath as the night advances. It seems to me that I can sometimes feel the planet rotating underneath me and I imagine sunlight falling hard on the other side of my darkness, falling on French Polynesia. When I was a boy, there were no other planets to be aware of outside the immediate companions of the sun. Now we know there are hundreds of planets out there, just in our own galaxy. Radio telescopes, which can see where we can't, have made images of them. When I was learning about Copernicus's universe, people thought all life depended on photons—incoming solar radiation—to survive. Today we know that tube worms and other life-forms living around thermal vents on the oceans' floors don't need sunlight in order to make their sugars. They require only sulfur. When I was in grade school, people thought most all life on Earth roamed the surface of the planet or swam in its waters. Today we know that the greater part of Earth's biomass lives underground. When I was learning to swim, people believed that the continents were stationary. Now children are taught that 120 million years ago South America and Africa sailed apart, and that 50 million years ago India plowed into Asia, creating the Himalayas.

In order to gather in the stars directly above me, I have to focus my binoculars out beyond a rete mirabile of thin, crisscrossing acacia limbs overhead, with their tiny, moisture-hoarding leaves. I locate the supernova 1987A in the Large Magellanic Cloud to orient myself, and then search for some of the less complicated southern constellations I'm getting to know—Triangulum Australe, a small constellation close to the south celestial pole, and Crux, the Southern Cross, the long axis of which points almost exactly to that spot, the south celestial pole. (In the Northern Hemisphere, a triple star system, Polaris, twinkles in the night sky one degree from the north celestial pole; in the southern sky, there is no comparable star to serve as a stationary marker.)

I can easily locate the Small Magellanic Cloud, 210,000 light-years away. Both "clouds" are apparent to the naked eye. They seem to be part of the Milky Way, but each is actually a galaxy entire unto itself. Taken together, the Magellanic Clouds represent a portal that opens onto the nearer realms of deep space. Astronomers include both, along with the Milky Way, in our Local Group, fifty-four or so galaxies, many of them dwarf galaxies, forming together a thin ellipsoid, distinctly separated from the galaxies of other local groups. In this ellipsoid, Andromeda stands at one end and the Milky Way at the other, about 2.5 million light-years apart. Our Local Group is one of about fifty such local groups arrayed relatively close to us. Astronomers organize all these local groups into superclusters. Each local group in a supercluster might contain anywhere from a dozen to a thousand galaxies. The number of superclusters in the known universe runs into the millions.

The calculations outstrip meaning.

To pull far back to something easier to imagine—our own galaxy, with only about one hundred billion stars in it—and focus again on the Southern (latinate) Cross, I can distinguish four terminal stars, one at each end of its two arms. The closest of the four to us is Gamma Crucis (Gacrux), only about 90 light-years away, while the farthest off, Beta Crucis (Mimosa), is about 353 light-years away. The Southern Cross (Crux), one of the smaller of the southern constellations, also frames most of a dark nebula called the Coalsack. And the Coalsack is visually adjacent to a loose swarm of about a hundred brilliant young stars collectively called the Jewel Box. Like every other constellation, Crux is a three-dimensional object with a two-dimensional identity.

If I were to dig down this evening into the rocky soil under my head and find the cranial cap of a Pliocene hominin, could I determine how this individual took in the stars? With only this remnant stony basin, part of a distant relative's head, could I learn what these stars provoked in her? Or him. I have to think it provoked nothing. It's too early in our history, they say, for such thoughts of the stars.

Some nights, when a slight breeze puts the limbs of the acacias into a gentle motion, the barely audible sound of this soughing mimics for me the hum of the stars, and the glitterings of these suns seem like the overtones that sometimes carry beyond the bowing of stringed instruments. On those nights I might try to force the ramulose arrangement of these convoluted branches and twigs into the as-yet-unsettled pattern of human evolution. The trunk of the tree represents the kingdom Animalia. Where major limbs branch off into various phyla, I follow the one that represents chordates, the animals with backbones, and from these the branch representing the class Mammalia, the mammals. From that branch another diverges, the one that represents the primates, starting some 55 to 65 million years ago. Here, among the Eocene prosimians and Oligocene anthropoids, among hominids like *Sivapithecus,* *Proconsul,* and the Old World monkeys, I might find a road leading to the primates who read books. We're all right picking this trail up about 30 million years ago with *Aegyptopithecus,* a possible ancestor living in the early Oligocene. In the mid-Miocene, fifteen million years ago, *Kenyapithecus* is a possible ancestor. The path from there to *Homo* is indistinct. Baffling, really.

A few not-for-sure hominids—*Sahelanthropus tchadensis, Orrorin tugenensis, Ardipithecus kadabba*—turn up in late Miocene deposits. Gracile (i.e., slender, lithe) australopithecines are present by the early Pliocene and one of them, possibly *Australopithecus afarensis,* might be a direct ancestor. By now the ancestors of gorillas and chimpanzees are well along on their own paths. Gorillas diverge from the human line about 11 million years ago, chimpanzees about 7.7 million years ago. At this juncture, one might hope to imagine a continuous line, as one species of hominin evolves into another, and then that one evolves into yet another, but this kind of thinking is fundamentally flawed. The actual evolution of *Homo sapiens* and every other animal along the branches of this imaginary tree is dazzlingly complicated.

The conceptual problem is easy to visualize. Most of us are familiar with the popular branching diagrams found in textbooks that represent evolution with neat lines of continuous descent, where the lines divide and subdivide like this:

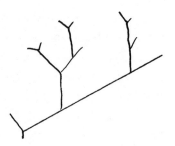

The difficulty with this diagram is that evolution doesn't proceed in this manner. It works more like this:

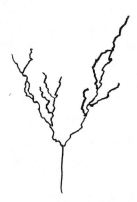

Lots of dead ends and genetic variations, together with some inter-breeding. Some lines of descent run closely parallel for long periods of time before terminating or branching off in a noticeable way. An additional complication is that evolution in the animal kingdom actually looks a bit more like this:

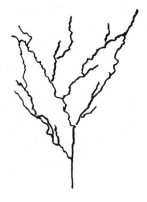

as, for example, when *Homo sapiens* and *Homo neanderthalensis,* whose lines diverge about 450,000 years ago, interbreed in Europe or western Asia 400,000 years later, producing a "hybrid" human cohort that soon dies out but which leaves virtually all non-African humans with a small percentage of Neanderthal genes.

Succinctly put, it's hard to say precisely where any individual human being came from. We all seem to have descended from relatively small, isolated human populations, many of which might have interbred at some point and some of which conserved the genes that give them a distinct look. Walk down the streets of Montreal or Singapore or Istanbul and you can see how varied the species *H. sapiens* is, even though from a certain perspective we all look pretty much alike. Paleoanthropologists like to point out that individual humans in the taxon *Homo sapiens* are more closely related to chimpanzees in the taxon *Pan troglodytes* than sheep are to goats. (Humans and chimpanzees share more than 98 percent of their genes.)

STARING INTO THE distant treetops, looping and swirling in the night breeze, and hoping for more of the order of life than nature offers, I wish the wind would drop. I'd like the branches to stand utterly still so I might discern in all that jumble a single continuous line to *H. sapiens.* But, figuratively speaking, the breeze never lets up. Even if it did, it would be easy to miss any single branch obscured by another, or to perceive two branchlets emerging from the same branch when in fact each is growing from a separate branch. Or to make sense of any instance of inosculation, of one branch growing *into* another, which has occurred in the history of *H. sapiens.*[3]

The path of human evolution is not the completely hopeless muddle it might seem to be, however. The problem is that the tree is a misleading metaphor, one, and two, there are too few human fossils extant for anyone to be definitive about what preceded us in the hominin line. These conceptual and empirical problems, although very real, can be set aside temporarily, however, in order to say that we do have a fairly good idea about who our recent ancestors were.

Most paleoanthropologists generally agree that roughly 11 million years ago, in the mid-Miocene, a single primate gave rise to two separate lines of development. One—both are still ongoing—is represented today by the gorilla (*Gorilla gorilla*). The other line, sometime in the late Miocene, came to be represented by a primate whose own genetic line eventually diverged into two separate lines of expression. One line, that of primates ancestral to chimpanzees, led to the chimpanzee and bonobo (*Pan paniscus*) of our time. The other line produced the hominins, some of which are ancestral to humans. These ancestors all eventually disappeared as distinct species, evolving into something else or becoming extinct, with some of them living alongside other hominin species ancestral to humans for tens of thousands of years before succumbing to the selective pressures of ecological upheaval. (It appears, for example, that *Homo ergaster,* sometimes referred to as "African *H. erectus,*" *H. habilis,* and the robust australopithecine *Paranthropus boisei* lived together in some parts of Africa for this long.)

The first group of primates clearly ancestral to *H. sapiens* are the gracile australopithecines. Among them are *Australopithecus afarensis, A. africanus,* and *A. sediba.* They begin appearing in the fossil record in the early Pliocene, 4 to 5 million years ago, and several survive into the late Pliocene. One of them, possibly *A. afarensis,* is the likely progenitor of *H. habilis.* And *H. habilis* might conceivably be ancestral to *H. erectus.* Or *H. erectus* could be a descendant of a lineage yet to be discovered.

However he came to be, *H. habilis* is making stone tools by about 2.6 million years ago. *H. erectus* shows up about 1.89 million years ago, first as *H. ergaster* in Africa and later as *H. erectus* in eastern Asia. (A descendant of *H. ergaster, H. heidelbergensis,* is the probable progenitor of both *H. neanderthalensis* and *H. sapiens.*) *H. erectus* might have survived in eastern Asia until about 100,000 years ago, and *H. floresiensis* might be among its descendants. *Homo floresiensis* survives until about 54,000 years ago in southeast Asia.

Homo ergaster, who has fairly strong support among paleoanthropologists as *H. sapiens*'s earliest direct ancestor, is often the starting point today for a consideration of the one characteristic that finally radically differentiates *H. sapiens* from all other hominins: culture. (Bipedalism and a significant enough increase in brain size over australopithecines had been in place a long time among humanity's close relatives.) By about 200,000 years ago or so, anatomically modern man is hunting and gathering and living a social life in Africa. *H. neanderthalensis,* perhaps descended from a different population of *H. heidelbergensis* than the African one, is doing the same in western Asia and Europe.

IT'S IMPORTANT TO NOTE here that each of these hominins appears in the fossil record at a time closely associated with dramatic climate change. Such ecological changes—in the amount of moisture available, for example, which in certain areas favors the survival of extensive forests or the spread of grasslands—might favor one hominin species over another; or these ecological changes might accelerate the evolution of a new hominin species, one better suited to the new climatic conditions.

With the arrival of behaviorally modern, or cognitively modern, *H. sapiens* about 55,000 years ago, culture begins to play a role in the evolution of man as important, eventually, as environment. In this sense, *H. sapiens* makes himself exceptional among all other animals in the late Pleistocene by becoming as strong a force himself in the process of evolution by natural selection as meiosis.[4]

The tendency of some to exaggerate our own importance as a species in the great theater of life on Earth is a sign of hubris. A more biologically informed or enlightened, and certainly secular, point of view is that man is better off viewing himself as a flawed rather than a potentially omnipotent creature, an animal with no more of a guaranteed future than any other animal. This perspective, some argue, that we are not the be-all and end-all, might eventually lead to better politics and to the development of more equitable social and economic systems worldwide. Still, *H. sapiens*—i.e., culturally advanced man—*is* exceptional. The provocative question is, Where will his exceptionalism take him?

Like all other creatures, biological man is evolving in response to both natural and anthropogenic selective pressures such as deforestation and ocean acidification, the latter of which will dramatically affect its supply of protein in the near future. Other selective pressures, like that from global climate change, are powerful enough to render certain of humanity's responses to them, like technological innovation, irrelevant. Further, the culturally generated selective pressures now affecting human evolution have, since the start of the Industrial Revolution, become so significant they've caused the extinction of hundreds of other species and triggered the Sixth Extinction of biological life. Those same anthropogenic forces, operating alongside familiar natural forces, are now shaping the evolution of all Earthly life.

The alarming situation here for humanity is that *H. sapiens,* though it has asserted itself as the dominant species on Earth, is at the same time the potential victim of its domination over virtually all Earth's ecosystems. If *H. sapiens* were to become extinct, the event would simply be regarded as evolution continuing to unfold, a biological future for *life* but not one that any longer included humanity.

It is worth noting as well that *H. sapiens*'s anatomical evolution, as measured today by genomics, has begun to accelerate. Among some evolutionists, this situation foreshadows speciation.

HAVING PUT FORTH this generalized and somewhat conjectural sketch of human origins, I can now situate, in a more meaningful way, what Kamoya and his colleagues were interested in finding during the time I spent with them around Nakirai. They were looking in particular for late Miocene/early Pliocene fossils in the larger hominid family, but they also hoped to find, for example, fossils in the developmental lines of the gracile and robust australopithecines. (In deposits this old, they were not likely to find the fossils of any species in the genus *Homo.*)

The interpretation of hominid fossils is a relatively circumscribed pursuit. The research consists almost entirely of examining and reexamining a relatively small collection of fossils. Most interpretations of the evidence represent responses to straightforward and long-established questions about anatomy and phylogeny. Each new fossil amplifies or refines what we know, and of course the physical evidence itself has tremendous authority. It is in the allied nascent field of evolutionary psychology, however, that the opportunity today for stunning insights into the evolution of human beings now seems to be much greater, because of recent advances in neuroscience. And it is also here that speculation, formal theorizing, and laboratory research are likely to create a broader and more informed public debate than human evolutionary biology has had to face in the past from the very large number of people who do not believe in hominin evolution.

The more provocative interpretations of man's history now no longer revolve around whether *Australopithecus sediba* or *A. afarensis* is directly ancestral to man but around the question of what happened to a single, relatively small group of *H. sapiens* living in the general vicinity of what is today Djibouti, in the Horn of Africa, about 55,000 years ago. A number of scientists lean toward the view that what occurred was a small change in the structure of the human brain, a minor encephalic event that would nevertheless prove to be extraordinarily adaptive, and which probably accounts for the sudden appearance at this time of stunningly complex human cultures. The Middle Paleolithic, which began about 120,000 years before with *H. sapiens* making stone tools far more sophisticated than *H. erectus*'s tools, ends here, and the Upper Paleolithic begins. Keeping track of human origins now requires classifying *H. sapiens* not by changes in his anatomy but by the development of increasingly rich and diverse regional cultures.

This singular group of people, who might once have called what is now the Danakil Desert in Ethiopia home, was clearly a different kind of people. Today they are called behaviorally modern or cognitively modern people to distinguish them from all other humans living at the time. They soon crossed the nearby strait of Bāb al-Mandab, which connects the southern end of the Red Sea with the Gulf of Aden, and entered the Saudi Arabian peninsula. From there they migrated north and east, replacing *H. neanderthalensis* in western Asia and Europe and replacing the descendants of *H. erectus* in southern Asia; and then populating Australia by crossing the nascent Timor and Arafura Seas that separated, at the time, Sunda (the contiguous landmass of Indonesia) from Sahul (Australia and New Guinea).

In the millennia to come they would populate Micronesia and then Polynesia, reaching as far across the Pacific in voyaging canoes as the west coast of South America. Having developed extremely effective techniques for hunting large mammals, suitable clothing, and portable shelters, they would move north into Siberia, eventually crossing the Bering land bridge and spreading through the Americas, into the islands of the Caribbean, and as far south as Tierra del Fuego. The trajectory and the speed with which people moved and adapted to nearly every sort of Earthly environment and became encultured in their places, considering the pace of *Homo*'s early history in Africa, are staggering.

UP UNTIL ABOUT 55,000 years ago the genus *Homo* represented only a single thread in the elaborate and incomprehensibly large tapestry of biological life. A predator and a scavenger, as well as a species of prey, especially for large cats, he was a highly social primate whose altricial young required an unusual amount of attention during their first two or three formative years. He was in no way the dominant animal in his own range. Wherever he lived, in relatively small, isolated populations it is thought, he competed with other animals for food and also for water when it was scarce. Through long periods of global cooling and warming, he adapted and endured, like other animals.

A disinterested observer, following the development of *H. sapiens* from the time he becomes clearly distinct 200,000 or so years ago until the advent of behaviorally modern man, might have marveled at *Homo*'s use of fire—making it, transporting it, and using it to prepare his food. He might have admired his stone and bone tools, and the way he made use of other materials, like animal skins and wood. *Homo* might not have seemed any more remarkable to an observer, however, than some of the other animals he lived among in Africa. He possessed no capari-son that could compare with the rococo plumage of many birds. He was less intimidating than a rhino, less dangerous than a mamba, less exotic than a giraffe, less nimble than a guenon. He would have stood out primarily as a maker of things, and as the single surviving species of bipedal hominid. Compared with a chimp, he had greater dexterity, and he might have been as persistent a cursorial hunter as an African hunting dog, and have been notable, also, for the things he carried with him from one place to another. In addition to fire and his children, this included his tools, his hunting implements, and perhaps containers of water. He would have drawn an observer's attention but not com-manded it.

And yet there was something there.

Homo sapiens's small, scattered populations might well have seemed of little consequence in the larger tableau of Africa's savannah wildlife, but a thoughtful observer of anatomically modern man 100,000 years ago would have marked the potential in such things as *Homo*'s attention to patterns of social order, the nature of his intense curiosity, and the adumbration of a quality no other animal seemed to possess, which one day would be called intelligence, an ability to assemble things—fiber, the passing hours, sounds—into complex patterns that would one day be called weaving, calendars, language, logistics, and art. It would have been an eerie thing to comprehend, as it is eerie for us today to find in the eyes of a chimp the glimmer of something that for a moment seems human, a look that says, "I know."

WHILE I WAS WALKING the semiarid lands around Nakirai, searching for traces of our hominid ancestors with Kamoya and the others, it sometimes occurred to me that I'd unconsciously situated myself in a sort of interstice, a middle ground from which I was peering out. It lay between what creatures like Kamoya and I and the others had evolved *from* and, looking in the other direction, at what we'd become. The pivot point for me, the place in my mind with such a powerful before and after, lay with that nameless group of people in the Afar region of Ethiopia 55,000 years ago. Without intending to, they separated themselves from the galaxy of African wildlife and emerged as something else, not yet the founders of civilization but no longer truly wild. These were the first creatures to shimmer with intentionality.

I looked back along a narrowing corridor into a far-off haze that obscured a few species of australopithecines. And then, like a man who stares to his left for a long while and then turns to look to the right, I saw something like a dispersion of fireflies, just a few at first, then swarms, and then the rippling explosion that becomes culture and then high culture in the hands of its inventor, modern man. People are disembarking on the shores of northern Australia; Mousterian Neanderthals are giving way swiftly to full-blown, Magdalenian humans; the first Natufian cities are crystallizing in eastern Anatolia. The Hittites, the Phoenicians, the emperors of the legendary Xia dynasty in China, and the line of the Pharaonic kings in Egypt, the Aztec empire—all bloom. Parisian salons support erudite disquisitions on philosophy, the world wars leave millions dead on five continents, successful heart transplant surgery is performed for the first time, in Cape Town, and all the rest of invention, modification, improvement, and domination carries down to these six middle-aged men in shorts, on foot in the Nakaisieken Desert in Kenya, with their highly evolved theaters of space and time, in which they live and think, in which they contemplate meaning and ultimate meaning.

We are so small in the desert, and the range of human personalities so great, the different sorts of intelligence extant in the panorama of the many still-distinct human cultures so large, the greater or lesser capacity among individual humans to think clearly or to imagine what isn't so obvious, the many distinctions between what is real and what isn't, according to different systems of metaphysics that . . . the possibilities in all this are so extensive that to gather it all under one name, *Homo sapiens,* borders on absurdity.

Walking the desert every day, I feel no compunction about imagining australopithecines or early *Homo.* No sense of ethics or morality seems to come into play. I feel no stake in whatever they were. They are like objects to me. After that group leaves the Afar region 55,000 years ago, however, I find I cannot think of them as objects. They are more like relatives, like harbingers, people with whom I share a fate.

The australopithecines send a message forward in time with no ominous note in it, no hidden threat. The message we read from the 1,800 generations of humanity that became historical following, possibly, a slight change in the structure of the human brain, a story about cultural achievement and human brilliance impossible adequately to honor, seems to carry within its heart, in contrast, a warning.

SOME NEUROLOGISTS, in an effort to have us more fully appreciate and understand the varieties of mind that can and have emerged from a brain as complex as ours, speak of certain neurological disorders such as Tourette's syndrome, Parkinsonism, catatonia, and manic depression as "psychological conditions" rather than disorders. What these conditions have in common is unusual perceptions about the rate at which time passes. Those "afflicted" with these conditions perceive the amount of time an event takes to unfold as being either greater or lesser than is the norm for *H. sapiens*. What this implies is that some minds might be better adapted than others to dealing effectively with those parts of the contemporary social environment that are characterized by rapid rates of expansion and change, like the environment created by information technology. Some minds thrive here; others founder. (Environments created by information technology have a significant impact, according to some, on the rapidly shifting dynamics of human social organization, affecting the way we relate to one another. This in turn helps to shape the expression of certain human emotions and impulses, such as generosity and aggression.)

In addition to varieties of temporal scale, one assumes that there are "disorders" of spatial scale, such as agoraphobia; and that having one of these "disorders" might conceivably either constrain or improve the opportunities for envisioning a viable human future. Being able to imagine alternative temporal and spatial frameworks in which to implement a more benign human future—being able, even temporarily, to eliminate the sort of tyranny that the press of time or the limits of space can induce, producing despair instead of hope—seems to be a crucial part of conceiving of a future for *H. sapiens* that is not dystopian.

In speculating about a human future today, one is compelled to consider the role of natural selection more broadly. In addition to evolving in a physical environment of global climate disruption and unprecedented population growth (unprecedented for a large terrestrial mammal, one that now occupies multiple ecological niches from which thousands of other organisms have been displaced), *H. sapiens* is now also evolving in response to an increasingly pervasive cultural environment. A question that quickly arises is: To what degree do man's built environment and his cultural environment exercise a selective pressure on, for example, temporal and spatial "disorders" such as manic depression and agoraphobia; on such mental conditions as autism, narcissistic personality disorder, and psychopathy, all characterized by a lack of empathy; and on the continued existence of such characteristically human behaviors as altruism and aggression?

Speculation along these lines can easily produce considerations that are chilling. From the point of view of an evolutionary psychologist, it is a relatively straightforward matter to posit that a significant number of human beings do not have the ability to cope readily with the cultural environment their species has created (contributing, many psychologists and psychiatrists believe, to the rapid growth in the general population of anxiety disorders that require pharmaceutical management). Further, even evolutionary biologists now emphasize that by both actively and passively contributing to the ongoing development of an environment that is chemically toxic, and by continuing to devise systems of information exchange that elude the grasp of some portion of the human population, *H. sapiens* now faces historically unprecedented selective pressures that might strongly influence the evolution of *H. sapiens* over a relatively short period of time.

Where humanity will be in one hundred years is no longer solely a question of global warming, the disrupted ecologies of viruses like Ebola, and genetic mutation caused by exposure to synthetic chemicals. In the short term, the percentage of the human population that can cope most successfully with change in the cultural environment (but also, importantly, without pharmaceutical support) might be playing the more critical role.

A not unwarranted, though perhaps extreme, reference for the consequences of rapid genetic change along the path that led to fully modern man is the relationship between *H. sapiens* and *H. neanderthalensis* during the millennium or so the two occupied the Middle East together, some 50,000 years ago. Descendants of the small population of *H. sapiens* that migrated north and east from Bāb al-Mandab five or six hundred centuries ago simply overwhelmed *H. neanderthalensis*. Despite interbreeding successfully with *H. sapiens* and perhaps learning from them (and leaving behind a distinct and poignant archeological layer in Europe in the late Middle Paleolithic identified as Châtelperronian), *H. neanderthalensis* fades to the point of extinction, like an untended fire.

Five hundred thousand years ago, when *H. sapiens* and *H. neanderthalensis* began to diverge on the evolutionary path that led to behaviorally modern *H. sapiens*, they might have continued to look very much alike, for tens of thousands of years. When they came face-to-face in Europe again—and in the Near and Middle East—what made them different was not so much the way they looked but the radically different levels of complexity in their cultures. Their speciation, in other words, was more a matter of different *psychologies* than different morphologies. For a while—no one is sure for how long—they coexisted in Europe, occupying separate but adjacent territories, until *H. neanderthalensis* made his last camps, possibly in the vicinity of the Cape of Gibraltar, and then disappeared.

The difference between what is today the flickering hint of speciation on the horizon for *H. sapiens* and an event we can look back on, the survival of *H. sapiens* and the eclipse of *H. neanderthalensis,* is that with any future divergence in *Homo,* geography might not play the strong role it traditionally has. Two increasingly different groups of *H. sapiens,* one with a high degree of technological competence, the other far less able to manage psychologically in this realm, might come to represent distinct populations not because they are separated by geographic space, once a requirement for speciation, but because they are divided by electronic space: they will have ceased to communicate effectively with each other. The psychological space between them might rapidly become too great to bridge, leaving both groups isolated on either side of a chasm, and neither group in a superior position.

Of course such a scenario might never develop. Viral pandemics, nuclear war, crumbling national infrastructures, economic catastrophe, genetic mutation as a result of exposure to toxic substances—any of this might take *Homo* in some other direction. It is not possible to say anything definitive here, except perhaps that dramatic change in the near future seems to be in the offing, and if the species is to achieve its aspirations for justice, reduced suffering, and transcendent life, and if it is to prevent the triumph of machinery that it so clearly fears, an unprecedented level of imagination is required.

WHATEVER HAPPENED to us in northeastern Africa long ago, it's important to understand that what set behaviorally modern man apart so dramatically from other populations of *H. sapiens,* and from Neanderthals, was his ability to recognize and manage various forms of complexity, including social complexity. A widely held view about the enlargement of the frontal lobes in *Homo* is that they enabled *Homo* to far surpass earlier hominins in developing and maintaining extensive social relationships, in creating kinship systems that were apparently both stronger and more effective for maintaining feeding and breeding strategies, which ensured the survival of a sufficient number of offspring.

In short, behaviorally modern man was probably more adept, more capable, and better organized than anatomically modern man, or any other species of *Homo* he might have encountered.

A second important point about behaviorally modern man—hereafter referred to simply as *Homo sapiens*—is that he continued to evolve. As he dispersed into a wide range of climatic environments, an impressive array of phenotypes developed from the human genome. In other words, the genetic material available to *H. sapiens* provided the foundation for a diverse but not particularly variable group of human forms. Climate, diet, and physical environment exerted selective pressure, and as a result, some groups developed lighter skin, others were taller or had thicker hair shafts or better resistance to specific diseases, developed wet as opposed to dry ear wax, or became adapted to life at higher altitudes. Genetic evidence—geneticists say 2,465 human genes, about 13 percent of the total in the human genome, have been actively shaped by *recent* evolution in *Homo*—suggests that man adapted quickly and extensively as the species dispersed and took up life in impressively different habitats—the North American Arctic, the Kalahari Desert, the Amazon rain forest, the islands of Micronesia.

Two keys to understanding the origins of modern man—the development of language and the emergence of ceremony—present researchers with virtually no durable evidence to contemplate, but it's widely believed that both developed, perhaps gradually, over the past 50,000 years. And both point to an increasingly complex social life for *H. sapiens*. Today, the careful use of language—sincere, thoughtful, respectful—and participation in ceremony still create an atmosphere of powerful social cohesion when human beings come together. And ceremony also functions as an antidote to loneliness.

World history is full of inspiring charismatic figures—Muhammad, the dissenter Jesus Christ, Jeanne d'Arc, Mahatma Gandhi, Albert Schweitzer, Dorothy Day, José Martí, Martin Luther King Jr., Wangari Maathai—but historians of social change often point out that meaningful social change, the kind of change that improves the conditions in which people live, comes about through the work of *many* people. A charismatic figure might galvanize change and stand as its historical representative, but human beings are social animals. They take care of one another through continuous social interaction. The popular notion that in bad times heroes show up is an enduring literary device, but it is wiser for a population in difficult straits to effect a means of courteous and respectful social exchange—conversation and ceremony—than to wait for a hero to speak. I emphasize this because I've so often been struck by the difference between a society that believes wisdom is part of the fabric of a *community,* and that it is best represented in the words and actions of particular people (elders), and a society that believes wisdom is only to be found in certain people. The difference for a community would be the difference between choosing to act heroically as a group or waiting for a hero to act.

The human effort to listen to each other is, for me, one of the most remarkable of all human capacities, though, compared with commentary about, say, the origins of art in human culture, hardly a word is ever said about the human capacity to listen to another person. I bring this up because if the creation and maintenance of effective social networks, a particularly striking human attribute, is necessary to protect individuals against threats to this species' health, then the ability to listen carefully to one another becomes critical.

In looking back on our origins, we might easily fall prey to two misconceptions. First, that *H. sapiens* evolved toward perfection (as opposed to simply changing in response to changes in its environment); and second, that whatever might have been lost from one millennium to the next as modern man evolved is something that we are well rid of. The idea of "improvement" in a species over time has no footing in evolutionary theory. And it could be that something *H. sapiens* "lost" on his way to modernity, perhaps a willingness to cooperate closely with others on a daily basis, is something he can reclaim because he has, unlike any other animal, a historical imagination and a knack for innovation.

A key to maintaining large, effective social networks is having the ability to comprehend what someone else is thinking and, importantly, being able to understand that *whatever* someone else is thinking, it might be different from what you yourself are thinking in the same situation. It's not until a child is four or so that she or he can grasp that someone else who sees the same world they do sees it differently. Evolutionary psychologists, in an attempt to describe levels of increasingly more complex human awareness, call this achieving the second level of intentionality. Levels of intentionality are arranged hierarchically within a framework called theory of mind, where the first level of intentionality is an awareness only of one's own perception of reality and the belief that all others see the world this way.

Awareness at the third level of intentionality would be the ability to grasp how someone else is interpreting the thoughts of a third person in a group conversation. Evolutionary psychologists assume that most adults are able to achieve a fourth level of intentionality. Some can operate at the fifth level of intentionality; a few may be able to achieve a sixth or even seventh level of intentionality.

Here's one way to understand levels of intentionality:

First level: I think this about X. Doesn't everybody?

Second level: I think this about X, but understand you think something different about X.

Third level: I think this about X. I know you see X differently, and I understand that Jane thinks yet something else about X, different from what you think or I think. But, I also understand you're not aware that Jane interprets what I just said to you differently. She's not hearing what you're hearing when I speak to you.

Fourth level: I think this about X. I know you and Jane each have different views about X, and I also know you can't grasp what she's thinking about what I just said, that you did not hear it the same way she did. If Richard had been here, I know he would not accept my assessment of your misinterpretation of what Jane is actually thinking about the conversation you and I are having now at this table.

The ability to operate at high levels of intentionality in social situations is crucial for achieving high levels of cooperation in social situations. The ability to understand how others perceive a situation amounts to a kind of empathy. It can lead to an amelioration of tensions in a group faced with a problem. Or, of course, it can lead to the manipulation of others in the group. At higher levels of intentionality, one can find great empathy, great compassion, and a great capacity for cooperation—or just the opposite.

What most of us notice in small social settings, I think, is not the ability of another person truly to empathize with someone else's point of view, but the *inability* of some people to do this. They cannot entertain another point of view without fearing the loss of their own, or they're simply less capable of being empathetic. Autistics and psychopaths, though some in either group might be highly intelligent, have a limited ability to empathize, to move into higher levels of intentionality and to be attentive there to the needs, the fears, and the hopes of other people. They quickly become impatient with a world not organized according to their preferences.

The relevance of these thoughts about empathy, higher levels of intentionality, and social cooperation to the search for human origins lies with the convergence in our time of two extremely powerful forces. One is ecological, our ability essentially to bypass nearly every natural control on the increasing size of our population. Only a catastrophic viral outbreak or widespread nuclear warfare now threatens man's ability to exploit all of Earth's ecosystems in order to secure energy, sustenance, and economic profit. (Unfortunately, continued success here will likely produce diminishing psychological rewards, and will move humanity closer to an existence not dissimilar to the life of a machine.)

The second powerful force is the accelerating rates of change in the man-made world, when considered in the light of recent research on the biochemical, morphological, and histological landscape of the human brain. The social and cultural development of *H. sapiens* over the past 55,000 years has been relatively swift. The rate of change in the *man-made* environment today is so great, in contrast, that the idea that one generation is meant to teach the next generation how to manage has begun to seem quaint. Also, evolutionary psychologists question whether all human beings have the same ability to adapt to such rapid change. Some people, then, might seem marked for marginalization if human societies are to continue to pursue higher levels of efficiency or adapt easily to increasing amounts of social control. We're reluctant to address publicly the inability of certain people to "keep up" for fear of being regarded as bigoted, intolerant, chauvinistic, or xenophobic. The failure to speak in defense of marginalized or persecuted groups is a constant in recent human history. It is also an act of cowardice that asks to be reckoned with before mankind faces truly staggering shortages of energy, freshwater, and food.

If *H. sapiens*'s future is threatened by environmental factors, both natural and anthropogenic, and if the ability of many people to cope with the complexity of the man-made environment is compromised, and if the need for cooperation seems great, how are we to tone down the voices of nationalism, or of those in support of profiteering, or religious fanaticism, racial superiority, or cultural exceptionalism? If economic viability trumps human health in systems of governance, and if personal rights trump community obligations at almost every turn, what sort of future can we expect never to see?

THIS LONG SPECULATION about the fate of modern man is a simplified, perhaps somewhat simplistic, overview of a problem not exclusive to any single nation or people or style of governance. All people, every culture, every country, now face the same problematic future. To reconsider human destiny—and in so doing, to leave behind adolescent dreams of material wealth, and the quest for greater economic or military power, which already guide too much national policy—requires reassessing the biological reality that constrains *H. sapiens*. It requires "resituating man in an ecological reality." It requires addressing the inutility—the biological cost to the ecosystems that sustain him—of much of mankind's vaunted technology. Whether the world we've made is not a good one for our progeny—asking ourselves about the specific identity of the horsemen gathering on our horizon and what measures we need to take to protect ourselves—requires a highly unusual kind of discourse, a worldwide conversation in which the voices of government and those with an economic stake in any particular outcome are asked, I think, to listen, not speak. The conversation has to be fearlessly honest, informed, courageous, and deferential, one not guided by concepts that now seem both outdated and dangerous—the primacy of the nation-state, for example; the inevitability of large-scale capitalism; the unilateral authority of any religious vision; the urge to collapse all mystery into one meaning, one codification, one destiny.

When I've passed through different troubled parts of the world and sought local advice—on Indian reservations in the United States, at Banda Aceh in northern Sumatra after the Boxing Day tsunami in 2004, in Western Australia during the heady days of mining ceaselessly for iron ore (financed by the Chinese)—I've seen the same pattern of coping with disaster. Deferential local cooperation. This suggests to me that for many people in difficult circumstances, the notion of needing help from a centralized authority, especially one living at a remove from the problem, and the notion of fully protecting certain types of economic progress are not much on people's minds. What I see consistently in these situations is the emergence of individuals who embody that culture's sense of competence into positions of authority. They are its wellspring of calmness. They do not disappear with defeat or after setbacks. They do not require reassurance in their commitments to such abstractions as justice and reverence. In traditional villages they're called the elders, the people who carry the knowledge of what works, who have the ability to organize chaos into meaning, and who can point recovery in a good direction. Some anthropologists believe that the presence of elders is as important as any technological advancement or material advantage in ensuring that human life continues.

I've not traveled enough, read enough, spoken to enough people to know, but this observation feels almost eerily correct to me. At the heart of the generalized complaint in every advanced or overdeveloped country about the tenor of modern life is the idea that those in political and economic control are self-serving and insincere in their promise to be just and respectful. I sat down once at my desk and wrote out the qualities I observed in elders I'd met in different cultures, nearly all of them unknown to one another. Elders take life more seriously. Their feelings toward all life around them are more tender, their capacity for empathy greater. They're more accessible than other adults, able to engage in a conversation with a child that does not patronize or infantilize the child, but instead confirms the child in his or her sense of wonder. Finally, the elder is willing to disappear into the fabric of ordinary life. Elders are looking neither for an audience nor for confirmation. They know who they are, and the people around them know who they are. They do not need to tell you who they are.

To this list I would add one more thing. Elders are more often listeners than speakers. And when they speak, they can talk for a long while without using the word *I*.

Living in one of the most highly advanced of human cultures, I often wonder, What have modern cultures done with these people? In our search for heroes to admire, did we just run them over? Were we suspicious about the humility, the absence of self-promotion, the lack of impressive material wealth and other signs of conventional success? Or were we afraid they would tell us a story we didn't want to hear? That they would suggest things we didn't want to do?

SOME NIGHTS WHEN I awakened in darkness in Nakirai, lying supine with Ngeneo on my right and Kamoya on my left, I'd quietly slide out from under my bedsheet and go for a walk in the moonlight. Most often it was the memory of the intensity of some moment I had found myself in that sent me off into the dark. I might have been staring into the Jewel Box with my binoculars and wondering if these stars were old enough to have planets yet. Or trying to imagine what some set of fossils once looked like as a fleshed-out living animal. I wouldn't walk far, and I swept the ground continually with my flashlight for snakes. The guidebook I was using warned that red spitting cobras (*emun lokimol* in Turkana), black mambas (*emun lokipurat*), and northeast African carpet vipers might be out hunting at night in this locale.

The thought of an encounter with a poisonous snake, by day or night, was never far from my mind. A few days before we arrived at Nakirai I was out with Kamoya, Wambua, and Nzube, looking for fossils in a wadi south of Richard Leakey's camp at Nariokotome. We were carefully searching the walls of the cutbanks. The wadi was about twenty feet wide and the banks were about four feet high. Suddenly Nzube and Wambua, walking about thirty yards ahead of us, came running hard around a bend toward Kamoya and me, shouting in Swahili *"Ikuuwa!"* "Mamba!" said Kamoya, yanking me toward him as we bounded up a notch in the cutbank with the others. My left biceps still in Kamoya's grip, I turned to watch the mamba pass. It was about eight feet long, a uniform olive-gray color, moving swiftly through the middle layer of twiggy brush growing on the opposite cutbank. An otherwise understated guide to the poisonous snakes of Africa says the black mamba attacks "with unbelievable speed and ferocity" and that it often lifts its head high in an effort to strike a person in the chest.

Years before this, on the upper Boro River in northern Botswana, during an afternoon nap which the guides urged our small group to take because of the terrific heat that day, I walked away from the others and ranged around a stand of acacias, looking for anything interesting. When I knelt down to peer into the entrance to an aardvark den, one of the guides came running, waving me off and shouting, "Get away! Get away!" When it's this hot, he said, mambas seek out the cool air in aardvark dens.

So I step carefully. Most nights I move off no more than a dozen yards from my companions. I want to listen to the Nakirai night, apart from the breathing of the men, to hear the crickets and the barely audible seethe of the acacia twigs brushing against one another. I listen for the alarm calls of birds who suddenly find a predator, a snake or striped polecat, in their nests.

I look at my companions asleep beneath the acacias. I get along well with everyone but Ngeneo, who often seems out of sorts, even sullen, when he's not engaged in some task with the others. And Wambua, who keeps up a gruff countenance. Of us all, Christopher aside, Wambua is the one most likely to be keeping his own counsel.

I know this rhythm of camp life "out bush" from other experiences. One day moves easily into the next at a camp like this one, while a small number of people pursue empirical evidence that will support some idea or theory. The momentary annoyance of biting flies, sunburn, unfamiliar food (boiled goat intestine one morning for breakfast at Nakirai), compromised hygiene, minor wounds, or a lost notebook fades before the intense pleasure of looking for the things you search for every day and sometimes find. The nature of these things and the impossibility of ever fully grasping what they are are intoxicating to feel. You do not want to reduce the mystery they represent with terse or restrictive language, for fear that their ineffable essence will then slip away from you.

IN THE LATE PALEOZOIC, 300 million years ago, Africa was embedded in the center of Earth's single supercontinent, Pangea, and Pangea was surrounded by Earth's superocean, the Panthalassic. At that time, "Africa" included three other landmasses. Later they would separate from it and become Madagascar, the Arabian Peninsula, and a swatch of land that today stretches north from Jordan to the Bosphorus, the strait that connects the Black Sea to the Mediterranean and that separates Anatolian from European Turkey. About 160 million years ago, Pangea split apart to become the supercontinents of Laurasia, in the north, and Gondwanaland, to the south. In a scenario now familiar to many schoolchildren, Gondwanaland soon disintegrated, with South America moving to the west, Australia and Antarctica moving away to the south, and the subcontinents of India and Iran drifting to the north. By 90 million years ago Africa finally stood apart, separated from South America by the Atlantic Ocean, from Europe by the Tethys Seaway (the proto-Mediterranean), from India by the Tethys Ocean itself, and from Madagascar by the Mozambique Channel.

Several million years ago the Great Rift Valley was continuing to open in East Africa, creating a geography that would become the major focus of research into humanity's origins. The section of the African Rift Valley farthest to the west, running roughly north and south, embraces a series of lakes, from Lake Albert (or Mobutu) in the north to Lake Nyasa (or Malawi) in the south. The other major section of the African Rift, forming, together with the western section, the African half of

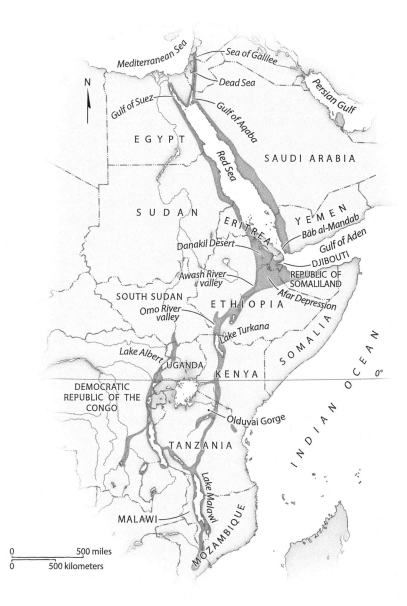

N

Mediterranean Sea
Sea of Galilee
Dead Sea
Persian Gulf
Gulf of Suez
Gulf of Aqaba
EGYPT
SAUDI ARABIA
Red Sea
SUDAN
ERITREA
YEMEN
Bâb al-Mandab
Danakil Desert
Gulf of Aden
DJIBOUTI
Awash River valley
REPUBLIC OF SOMALILAND
SOUTH SUDAN
Afar Depression
ETHIOPIA
Omo River valley
Lake Turkana
Lake Albert
SOMALIA
UGANDA
KENYA
0°
DEMOCRATIC REPUBLIC OF THE CONGO
INDIAN OCEAN
Olduvai Gorge
TANZANIA
Lake Malawi
MALAWI
MOZAMBIQUE

0   500 miles
0   500 kilometers

Great Rift Valley

the Great Rift Valley, runs northeast and southwest. It, too, includes a line of lakes, the largest of which is Lake Turkana. The other major section of the Great Rift Valley begins in western Syria. It includes the Dead Sea and the Gulf of Aqaba, and becomes the trough that holds the Red Sea. At the southern end of the Red Sea, a sequence of lava flows formed on the African continent an area now called the Afar Depression. It includes the Danakil Desert and the modern nations of Djibouti, Eritrea, Ethiopia, and the (internationally unrecognized) Republic of Somaliland. It also marks, as I've said, the area in Africa from which most paleohistorians believe modern man departed the continent, crossing the gap at Bāb al-Mandab, sometimes translated as the Strait of Grief or the Strait of Tears or Sorrow.

Three million years ago, australopithecines lived throughout eastern and southern Africa, and sites in the southern portion of the Great Rift Valley have proven so far to be the greatest repository of their fossil bones and those of some of their progeny. The first of these hominin fossil sites to draw widespread interest was Olduvai Gorge in Tanzania, where Louis and Mary Leakey discovered "Zinjanthropus," a robust australopithecine now called *Paranthropus boisei*. Subsequently, their son Richard and his wife, Meave Leakey, developed hominin fossil sites farther north on either side of Lake Turkana. Still farther north in that valley, the American paleoanthropologists Donald Johanson and Tim White developed sites in the Awash River valley, in the Afar Depression, among them Hadar, where the fossil bones of the gracile australopithecine *Australopithecus afarensis* (popularly known as Lucy) were found, as well as sites in the lower Omo River valley which yielded the bones of both *Paranthropus boisei* and *Paranthropus aethiopicus*.

In recent years paleoanthropologists have made strenuous efforts to involve geologists in their excavations in the northern section of Africa's Rift Valley in order to establish a more detailed and accurate framework for dating hominin fossils. Once certain geological strata were identified and dated, researchers were able to focus their attention on various well-defined layers of Pliocene and Pleistocene deposits with the hope of finding in them both australopithecine fossils and those of hominins in the genus *Homo*.

It was research like this, into the geology of the African part of the Great Rift Valley, that gave Kamoya and his crew confidence that the land they were now searching might produce fossils in the hominin line. It was a search that could have been successfully conducted nowhere else on Earth. If it were pursued in geological layers of a similar age in Australia or Siberia or North America, nothing would turn up. Hominins didn't live in those places at those times. It is only here in this geography, employing a group of men scouring the ground with the instincts of professional trackers, that scientists have the greatest chance of success.

We all knew—the others better than I, of course—which contours in the land were the best to follow, and how to sweep and probe the surface of the land visually. We knew the age of the deposits we were searching, so we had ideas of what we might come upon. In these layers of ancient lake and river deposits, eroded and sorted over time by flows of water, we hoped for one or two crucial pieces of the jigsaw puzzle of hominid evolution that goes back 8 to 10 million years, 95 percent of which remains to be found.

EACH MAN'S SEARCH for fossils shifted periodically during the day from moments of intense scrutiny to moments of distraction. If one of the Turkana men who had confronted Kamoya or one of their young proxies was shadowing us, it's possible that he might have noticed a hominid femur one of us had actually missed. But, being Turkana, he might leave it be and never mention it to anyone, because it was not important to Turkana people. He would be concentrating, for the most part, only on what it was that we were occasionally picking up and *not* putting back on the ground. They would look at the places where any one of us had stopped and reached down for something, and they would try to figure out what it might have been that had caught our attention.

I wondered later how these days might have gone had I been traveling instead with five Turkana historians and working hard to understand the trustworthy matrices with which they walked this land.

Whenever I glanced over at Kamoya, I was struck by the distinctive rhythm of his scrutiny. He would glance up, look away over the arid plain that fanned out before him to the horizon and then look back at the ground at his feet. To keep alert, to better inform himself, he regularly changed his spatial reference. This didn't mean he would find more fossils than the other men—there was too much chance in the thing for that. But it did mean that he was continually cultivating a comprehensive sense of what we were trying to do—develop a framework, a spatial and temporal matrix, for the big questions each of us, in his own way, carried: Who are we? Where have we come from? Where are we going?

Philosophers continually rephrase these questions and, along with the rest of us, speculate on how to answer them. One doesn't have to be a philosopher, however, to appreciate how hard such questions push at us today or to want to refine the way the questions are posed. I've followed men like Kamoya for years, up sand rivers in Australia's Tanami Desert and across the ice of the northern Bering Sea. I've tried to learn from these people. They are acutely sensitive to the shape of the unbounded spaces they're moving through. They have an awareness of the nested character of the temporal framework that contains them, like a stack of graduated bowls, each set within the next larger one—the time of day, the particular day in a lunar cycle or solar year, all of that situated within a cultural epoch. I enjoy their company partly because they know in any given moment, as I do not, precisely where they are. It gives the best of them an almost preternatural poise. In the vast expanse of the multifaceted unknown, they are certain of their location.

ONE DAY WHEN the weather promised to be a few degrees hotter than usual and Kamoya planned to explore an area where the stony ground radiated heat intensely, he suggested I spend the morning in camp. He had a sheepish look about him when he made this suggestion, and I suspected he knew I knew that. He had, in fact, some sensitive business to attend to, and it was actually Richard's wish that I not be involved that day. I resented being told to stand clear, but understood it.

A few years before this, Kamoya had found the skull of a Miocene ape in a dry riverbed nearby, at a place they named Kalodirr. Richard and Meave Leakey later discovered several more genera of Miocene apes in these deposits. As the richness of these potentially highly important deposits became more apparent, Kamoya took on the task of trying to improve the boundaries of the search area at Kalodirr. The field trip that Richard had graciously arranged for me to join included some work in this sector, and I assumed he was anxious that word about the richness of the site might inadvertently get out if I went there.

The field of paleoanthropology, particularly for those who have a stake in one or another version of *Homo*'s origins, is characterized by unusual levels of suspicion and jealousy. Its practitioners are often guarded and proprietary, especially when it comes to unpublished data and fossils that have not been fully described in the scientific literature. They don't want to feed speculation before their own formal positions have been set out.

While one might be tempted to belittle such self-interest (or be amused by the arrogance that sometimes goes along with holding such a position), there's good reason for guardedness. The discoverers of important hominid fossils enjoy a kind of notoriety and fame far beyond the reach of most academics, and their ability successfully to pursue such expensive, logistically complicated, and labor-intensive work depends heavily on securing grant money. And grant money most often goes not only to high-profile subjects like the search for human origins but also to people whose skill and success make them newsworthy. In short, for paleoanthropologists, securing grant money to continue their work is not all that different from running a successful business in a highly competitive market.

Kamoya needed to spend his day at a site Richard didn't want known. I was happy to stay behind, and Kamoya was grateful for my keeping up appearances by agreeing not to go out that day—in the same fierce noonday sun we'd been working in the day before.

In camp, I collected firewood with Christopher, worked on my notes, and washed a few clothes that had stiffened with salt residue from sweating. At midmorning Christopher brought me a cup of black tea served on a saucer, the way only Kamoya and I were served. (The other men just got the cup.)

It would be a while, I knew, before Christopher had to begin fixing lunch. I asked him if he would talk to me about his scars. He's eighteen, the first of us up every day and the last to bed. He's fluent in Turkana, Swahili, and English. The rise and flare of his large upper lip, his small ears, and his long skull give his face a distinctive cast. He enters into conversation easily with others, whenever his work permits. He works hard to keep our camp orderly and neat, but there is no ostentation in his movements.

The outer edge of each of his eye sockets, forward of his temples, is marked by a vertical set of three parallel cicatrices, all about three-quarters of an inch long. Similar sets of horizontal scars appear on his chin and on the boss of each cheekbone, and a sixth set of scars, these about twice as long, appear on his forehead above the bridge of his nose. These facial scars, ritually incised during a rite of passage when he was young, have a certain elegance to them, especially in comparison with the other quotidian scars he bears, all of them minor save for a vicious-looking healed cut on the outside of his left knee, from an operation.

When he left camp to gather firewood, Christopher wore a cheap pair of plastic and rubber sandals. In camp, where he went barefoot, I could see the soles of his feet were cracked and heavily calloused. Scars on his forearms and shins, on his knees, and on the backs of his hands were the signs of an engaged physical life. They offered a history of his body, and made his body somehow more authentic, more authoritative.

Christopher demurred when I asked for details about the ceremony that had given him his facial scars. It was not appropriate, he said, to discuss these things with a non-Turkana person. He spoke to me about the broken knee that had occasioned the operation, but more than this, he wanted to know anything about the United States. What was it like to live there? Did you see movie stars? Did everyone own a car? How hard would it be for him to travel there?

I might have said that it was inappropriate for a person from the United States to discuss life there with someone who had always lived in a place like Lodwar, because of the likelihood of being misunderstood. Instead, I told him about my home in Oregon, about the dense, towering forest there and the rain, the salmon and black bears, the long drive to town for groceries. I asked him to tell me about growing up in Lodwar. He did. After a while he excused himself and went into the cook tent to fix lunch.

After he left, I sat and listened closely to the calls of red-billed hornbills, a ubiquitous sound, it seemed, in this particular countryside; and to the twitter of Somali sparrows feeding in the crown of a few Borassus palms standing among the acacias. The birdcalls, in a certain way innocuous, opened the land up into three dimensions, creating a larger scape of reference than the small bolus of domestic space that I'd been sharing for an hour with Christopher.

ONE AFTERNOON WHEN I was staying at Nariokotome, Richard flew several of us across the lake in his small plane to visit Koobi Fora, a research camp he'd established years earlier in an area where he'd made some spectacular finds. It was now a permanent complex of buildings. He hoped it would one day develop into a resort of some sort for people who wanted to see this part of Kenya's Sibiloi National Park. Visitors to Sibiloi, a nature preserve of close to a thousand square miles that includes a crocodile sanctuary, can expect to see an impressive array of wildlife—cheetahs, reticulated giraffes, Grevy's zebras, golden jackals, leopards, several species of gazelle, and a large antelope called a topi. The park is undeveloped, remote even by African standards, and largely waterless. Periodically it serves as a hiding ground for Somali bandits. Except for those attending college classes at Koobi Fora, few people were visiting this area at the time I was there.

When we arrived, Richard indicated he'd prefer to slip away quietly from the dozen or so adoring graduate students living there. He suggested we take off on a short trip to the north in one of the short-wheelbase Land Rovers. This would give us a chance to walk over the land together and to talk. One of the things uppermost in his mind at that time—he had just returned from a lecture tour in the United States—was the degree to which attempts to refute or deny evolutionary theory were accommodated in America. He grew increasingly incredulous speaking to me about it as we drove along. He recalled several talks before large audiences at major universities in which it seemed to him the moderator was going out of his way to encourage creationists to offer their views during the question-and-answer period. At one point he became so exercised, recounting one of these scenes, that he nearly drove up an embankment flanking the dirt track we were following.

I wanted to ask him what he thought of the work of an assistant professor of anthropology at Boston University at that time, Misia Landau, but I didn't follow up. At the first mention of her name, Richard's knuckles turned white on the steering wheel. Instead, I posed a less inflammatory question, about *Homo ergaster* and the evolution of Asian *Homo erectus*.

When she was a graduate student at Yale, Landau wrote her PhD thesis on the way the story of man's origins is usually presented by paleoanthropologists. She argues that their writing is "characterized not by a set of fossils or theoretical principles but instead by a common underlying narrative structure." It was her opinion that it was this "deep narrative structure," not the fossils themselves, that accounts for the plausibility of the explanations in these narratives. She found this to be true, she said, both in articles written for professional journals and in the popularizations prepared by scientists for readers with no background in paleoanthropology.

Emphasizing the relative paucity of evidence upon which to base any reliable paleoanthropological narrative about the origins of man, Landau asserts that "the most characteristic feature of paleoanthropology" is its ambiguity. She argues further that the "form of content [of] the narrative of human evolution conforms to a traditional and explicitly literal model—the hero story."

In camp, on that day I stayed behind with Christopher, I reread a paper by Landau, which had appeared some years earlier in *American Scientist,* called "Human Evolution as Narrative." It's partly a scholarly attempt to describe two academic approaches to discussing how narratives work (structuralism and hermeneutics). She's careful to refer only indirectly to contemporary paleoanthropologists such as Richard Leakey and Donald Johanson, to avoid the charge that her argument is directed against them (the egotistical behavior of both men having drawn comment in professional circles). Instead, Landau sticks to critiquing earlier scientific writers on the development of *H. sapiens,* such as Darwin, Henry Fairfield Osborn, Grafton Elliot Smith, and Thomas

Henry Huxley. But her central thesis is that scientific writing, in particular writing about the origins of man, is not anywhere near as objective as its practitioners would like to believe, that in fact it's culturally influenced. And in the case of writing about human origins, scientists, she asserts, are writing a story with a known end in mind—the perfection of biological life in the form *H. sapiens*. These writers, she says, regard all of humanity's ancestors as "transitional" to modern man; and they treat these ancestors as evolutionary dead ends, as failures. Earlier hominids, evolutionarily complete in the moment and sublimely successful ecologically, are accorded no respect, writes Landau. They're seen as mere stepping-stones to *H. sapiens,* as creatures whose protohuman natures have no intrinsic worth.

Landau's argument, that paleoanthropologists are as human as the rest of us, that they should enjoy no professional immunity when it comes to trying to tell a credible story, is perhaps too insistent. The story they tell, of course, *is* a hero narrative, and it is in fact built on a relatively small body of evidence; but in the end, the scientists are only professionally curious people, striving to be rigorous and as touchy as any other group of professionals when it comes to having their stories critiqued by laypeople.

Sitting there at Nakirai with Christopher, rereading the Landau paper, and swiping away at flies landing on my face in search of moisture, I recalled the thrill of those weeks when I was first in touch with Louis Leakey, hoping to be offered a spot helping out at the Olduvai Gorge camp in the early 1960s. The first time I read Misia Landau's work I felt a similar exhilaration. Her topic was the lifelong subject of people like Joseph Campbell, author of *The Masks of God* and *The Hero with a Thousand Faces*. How, people like Campbell asked, do human beings put into words their ideas about the meaning of human life? How do they convey through art and religion (and for Landau, through science) their beliefs about the significance of human life? They do it partly by investing in certain transcultural stories, like the one about the adventures of a culture hero, which, after a period of trial and hardship, always ends in triumph.

Many people are familiar with some version of this story, whether the hero is Prometheus or Siddhartha Gautama or Superman. When we hear a story that approximates these stories of a culture hero, we become more comfortable as listeners. We're more prepared, as Landau says, to believe the story is true. But what if things have changed so drastically in the human world since the time of culture heroes in China, in India, and in the Mediterranean that we are no longer really comfortable with such a story? If we have only nostalgic affection for it? In an era of an exponentially expanding human population to feed and decadent wealth piling up in many countries alongside lethal poverty, the culture hero is perhaps no longer relevant, because the scale of the trouble is beyond him. What if what replaces the hero and the hero's perilous journey is the self-sustaining community? What if we are now at the end of questing for security as we have understood it, at the end of thousands of years of questing for peace and wisdom, in Xia China, in Periclean Greece, in nineteenth-century North America?

What if the horizons of greatest importance are now, instead, to be found *within* us? What if we need an entirely different kind of story to sustain us, Jung's journey or Thomas Merton's journey or even Aung San Suu Kyi's journey, instead of Aeneas's or Alexander the Great's?

ONE DAY MARY Leakey took me to lunch at the Muthaiga Country Club in Nairobi. Between the world wars this club functioned as the social headquarters of a dissolute community of British colonials and wealthy idlers, most of them living in an area they called Happy Valley, in the Kenya Highlands north of Nairobi. The club came to stand for all that was imperious and obdurate about the British occupation, including its patronizing and racist views of black Africans. The day I had lunch there, the club seemed cloaked in the assumptions of another era, the preserve of people who were entertained by the ineptitude and corruption of black Africans trying to figure out how to achieve and then manage political independence.

I'd grown to admire Mary Leakey for several reasons. She was a successful paleoanthropologist at a time when the profession accommodated very few women, and she'd pursued a pioneering study of African rock art as a mostly unheralded researcher. And she'd accomplished all this while working in the shadow of her self-regarding and lionized husband, Louis, who was not always discreet in his relationships with admiring women. She was an astute observer of human foibles and dubious about human virtue.

I enjoyed her candor and the enthusiasm with which she discussed what interested her. When the waiter arrived, she ordered for both of us. Afterward, we took our coffee outside on the veranda, where she enjoyed a cigar. In her fashion, she then held court, but with no trace of overbearing self-importance. We spoke mostly about prehistoric art, about her research into it in Tanzania, about the cave paintings at Altamira in northern Spain, and about the White Lady, a famous pictograph in a grotto in western Namibia, which I, too, happened to have seen. She made no effort to be definitive, to insist on the validity of her own interpretations. She seemed content to marvel and speculate from the security of a confident and well-informed intellect.

Out of the blue she surprised me by saying, "Well, has my son been rude? Has he shown contempt for your ideas?"

She seemed delighted to have posed the question, and eager to hear something in response that would make her laugh. I said no, that Richard had been welcoming, courteous, accommodating. He was confident about his own views, I said, and I thought he liked to spar. I also thought he might have been slightly suspicious about my reasons for wanting to visit Nariokotome and my wanting to accompany Kamoya and the others to Nakirai. But I understood all that, I said. (In a letter to Richard, before I came to see him, I'd said something about enjoying my conversations with Donald Johanson, an archrival of Richard's at the time.) But no, I thought Richard quite decent. Just a bit wary.

She gazed at me through the blue haze of her cigar smoke with a bemused look, as though I did not really understand how the world worked.

KAMOYA AND THE OTHERS came back into camp about one. When I asked if they'd found anything interesting, he said they hadn't, but that it had been a good day. "So," I asked, "none of *Turkanapithecus*'s pals turned up?" A short, bright laugh lit up his face. Sometimes I thought Kamoya was so full of laughter that any jostling of his state of mind would cause it to burst from him. I'd watched him sit by, composed and seemingly indifferent, while one white man upbraided another white man for his faith in the capacity of Kenyan blacks to accomplish anything of importance without the guidance of white people. Later, when Kamoya told this story to a group of black men, his retelling was punctuated with sudden bursts of laugher, the laughter provoked by the absurdity of the notion.

After lunch, and after Kamoya had had his game of checkers with Nzube, and after we'd all had a nap, and after Christopher had brought the two of us afternoon tea, Kamoya talked with me about his early years. He'd grown up in a much more colonial Kenya than the one he was living in now. Back then he made one pound sterling per month working at a dairy. One day he took some milk for himself and was caught. His employer offered him three options: three months of work with no pay; twenty-five strokes across the back with a beating stick; or six months in jail. He refused it all and that same night slipped away, returning to his homeland in central Kenya. He was twenty-three. In the months following, when he heard about a need for workers at Olduvai Gorge in Tanzania, where the Leakeys were looking for hominid fossils, he made up his mind to apply. The Leakeys had obtained funding from the National Geographic Society to continue their work, after Mary's discovery of a partial *Paranthropus boisei* skull, the spectacular find that made the Leakeys internationally famous and which boosted the search for mankind's ancestors into prominence. Once at Olduvai, Kamoya and seven other Wakamba men began digging in what they at first understood was a graveyard. Kamoya remembers that there was little water at Olduvai, but that they ate well, and that when he showed a knack for the work, Mary invited him to dig alongside her so she could teach him. Still, he told me, he remained uncertain about the nature of the project. "No one in school," he said, "had ever mentioned anything about this." He took note of the relative wealth and social status of the white people who came to visit Olduvai, and of how impressed they were with the Leakeys' work. Whatever it was that he was helping with, he finally decided, this would be a good job for him.

In 1964 Kamoya found—when he tells the story he uses the pronoun *we,* not *I*—an australopithecine mandible at Lake Natron. It was the first of many major finds by Kamoya, for which the National Geographic Society would award him the John Oliver La Gorce Medal in 1985, by which time Kamoya had found more hominid fossils than any other person.

He was glad he'd been selected for the job at Olduvai and glad, he told me, that he'd stayed with it, despite his early misgivings. Paleoanthropology had provided him with a very good life.

The last slant rays of the setting sun picked up a bird in one of the acacias and prised it from the camouflage of its perch. A dark chanting goshawk. A hunter. Kamoya indicated it with a gesture, raising his eyebrows and tilting his head. I told him the story of the pale chanting goshawk I'd seen in Kalahari Gemsbok National Park (now Kgalagadi Transfrontier Park), in South Africa/Botswana. Where, figuratively, the human eye might gather a single pixel of information in its search of unfamiliar country, these accipiters gathered ten pixels. I told Kamoya I wished I could experience that degree of acuity.

Kamoya wondered what the camps were like around Kalahari Gemsbok. The difference between here and there, I answered, was that there you had to secure your camp against troops of baboons, and there were also spotted hyenas to deal with, so you couldn't sleep outside on a tarp like we were doing. You had to sleep inside a tent and be sure to zip the flaps shut, or the hyenas would stick their heads in and bite you. And some nights you could hear the stomachs of elephants gurgling and churning, they were standing so close to the tents; but you didn't hear their footfalls. In the morning we'd sometimes see their spoor only a foot from the tent walls.

"Do you imagine," I asked Kamoya, "that we are comprehensible to the Turkana?" He shook his head no. I told him that I sometimes thought of us as a flock of birds moving through the Turkana universe, gleaning seeds the Turkana had no interest in. At other times, I said, I thought of us as a pack of hyenas, like the ones who came into a camp I was in in northern Botswana once and took a few things we needed. Without a gun, you could do nothing about them. You just put everything you wanted to keep in a safe place and got out of the hyenas' way.

I told Kamoya that I'd seen tribal people treated like pariahs when they approached camps I had been in, and that this sort of thing happened in cities as well. And that I was sometimes embarrassed by the attitudes of my own people, even though I knew that, occasionally, tribal people who turned up in remote camps were conniving. They were looking to steal things, and they seemed to enjoy how uncomfortable their begging made us. It was hard to know what to do. We condemn racism and say it's driven by ignorance and fear, but it's also a tool some people use to survive. Everywhere in the world I've been, I told Kamoya, you see people making distinctions just as vicious and unwarranted about someone's social or economic class, though this kind of dismissal of a person is not as widely condemned.

I told Kamoya I admired his tact and his empathy in dealing with the Turkana people. Maybe, like genocide or exploiting people for profit, racism is a failure of empathy, an inability to imagine more than one's own point of view, an effort to solve paradoxes rather than learning to live with them.

I didn't try to frame for Kamoya a thought that had been running through my mind, about the psychological struggle for equality that I'd seen among tribal or traditional people in so many places. Everywhere I'd been I'd watched indigenous people attempting to pull themselves through that jagged hole that would give them entry to the white world. For some, acculturation was a transformation necessary to survive, to eat, work, and raise a family.

I wondered what it might be, in the thoughts we were exchanging, that made Kamoya sense their complexity, that his feelings about racism, for example, were so loaded with the possibility of being misunderstood that he might quit thinking of which words to use and, like me, just stare into the side-lit world around us, the last bright minutes of violet light before the swift fall of equatorial night. As we rose from our camp chairs, two Abyssinian rollers streaked past, twenty feet over our heads. They are stout, large-headed birds with formidable hooked bills, radiant azure plumage, chestnut-colored backs, and cobalt blue wing tips. The outermost tail feathers on each side have developed into streamers. They trace the bird's passage through the air like Japanese calligraphy brushes.

––––––––

THE NIGHT AIR is still when we retire, completely without tension. Later, a soft wind comes up, blowing south from the Turkwel River. Past midnight a low-pressure cell somewhere farther to the south of us begins to draw air off the plain on the west side of the lake. As the breeze builds into a wind, it gathers mosquitoes off the river. It carries them southward over the desert, where they find little to attract them. They prefer the environs of the river, where animals come regularly to drink. Around three in the morning, thousands of them drop out of the wind and are upon us. We sit up quickly, alert, casting about frantically for insect repellent.

Six men in their undershorts in the pale moonlight, the hiss of the spray cans, hands slapping skin, the soft curses. Wet with chemicals, we return to our sheets. Wherever I touch my body I trigger an angry itch. In the morning, forever measuring things, I count more than sixty bites.

I've had only a few encounters with mosquitoes since coming up from Nairobi, but I've been taking an antimalarial called chloroquine regularly. No drug can prevent malaria, but a few moderate the symptoms. In the back of my mind is the thought that the generations of protozoa for which the mosquito serves as a host turn over so quickly that eventually the drugs serve only to weed out the susceptible, leaving behind the drug-resistant phenotypes of this protozoan to breed up, in my case, a population of chloroquine-resistant parasites. Then an alternative drug is called for. Before I left home, my travel doctor had told me I'd be safe around Lake Turkana with chloroquine. No chloroquine-resistant strains of the protozoan were known to be there, he said. Still, it was a bit scary, being surprised like that.

In the morning we're all scratching ourselves and shaking our heads.

We leave camp together at the usual hour, on what will be my last day with these men. My scheduled flight to Nairobi from Lokwakangole two days hence has been canceled. Kamoya says I should go up to Lodwar this evening, and then in the morning get a *matatu,* a small passenger van, which will take me to Kitale. From there I can get a bus south to Nairobi. He's been on the radiotelephone to arrange a room for me at the Turkwel Lodge. We'll drive up after tea, with Onyango and Nzube.

The day before, Nzube had gone off on his own to walk a section of land he and Kamoya thought promising, and this is where we're going this morning. The first place we sweep is a low hill, composed mostly of mixed gravels. We find part of a large turtle plastron exposed there. I find a few crocodile teeth, which look like gleaming agates in the fine rubble of a gravel outwash. There seem to be fish bones everywhere. The richness of Nzube's discovery is spellbinding. Everywhere I look across the dozen square yards immediately around me, I'm able to pick out the fossils of an animal I recognize.

Most of the surface of Earth, outside of Antarctica and the deep beds of the oceans, has been examined with discriminating sets of eyes, but the details have been assembled by cultures with dissimilar feelings about what's not worth remembering and what's worth knowing more about. What the Samburu in northern Kenya know about their place, what they can recall and enumerate and elucidate for a stranger, or Nunamiut Eskimos living in the Brooks Range in Alaska, or pastoral Rabari nomads in Rajasthan, Apuriná in the upper Amazon, Bedouin in Algeria, or Pintupi in the Gibson Desert in central Australia, represents an expanse of knowing born out of long and intimate contact. When scientific observers arrive to study a place, another layer of knowing comes into play. Turkana people no doubt know about the crocodile emerging here at my feet, but they've let it be. This morning, people with different ideas about what is to be valued or carried off to show others stand about, staring at an anonymous crocodile from the Miocene, and at the turtles and fish that once swam with it.

In these situations, I try to keep track of two important questions. Has anyone in the scientific group spoken to the local people about this? And what is the level of recent disturbance here? Anthropogenic disturbance. What has already been taken away from this spot, and what has been added to it in recent human lifetimes? This morning, the land all about us looks overgrazed, punished as it were, by herds and flocks of domestic animals, but otherwise not greatly disturbed, unless you are thinking about yet one more place in Africa that has been stripped of its wildlife. No stone ruins are here, no signs of agriculture. Still, most everywhere we have gone we've found the tracks and tread marks of oil exploration vehicles. The heavy vehicles break apart the thin hemispheres of fossil mammalian crania. The wind blows in the odd bit of paper, or perhaps a plastic bag that hangs up in the brush and eventually disintegrates in the sun. I've made little effort to learn to distinguish the spoor of zebra or of the relict populations of gazelles and antelopes here. I know how infrequently we cross their trails or see their droppings. Every few minutes, though, we find the spoor of camels, donkeys, sheep, and goats.

We spend about forty-five minutes at the crocodile site, picking up pieces of the fossil animals here, turning them over and over in our hands, fitting some of them together without comment.

From the crocodile site, the six of us fan out, signaling to each other occasionally, but we find nothing as impressive as that. Like casual strollers we wander through the bones of this Miocene menagerie, evidence of another time, days when there was water everywhere.

A historian interested in more than just these ancient creatures that so intrigue us, someone who wanted to offer up a comprehensive sense of the place to a person who did not have the means or the time free of obligations to come here, would want more than what the six of us might be able to offer. She would want someone who knew the plants, who could pick out their seeds amid the silt and gravel—someone who could recognize the pollinators, could sort through the shriveled acacia pods, the bleached beetle carapaces, the palm nut husks, the smallest feathers of the sandgrouse—and say what it had once meant and what it means now. It has always slightly amazed me how infrequently scientific expeditions make room for people who are highly conversant with the place but whose goals are not scientific, people who have no command of the technical vocabularies of science, who aren't as constrained as the logical positivists when it comes to a philosophy of being, who wear the wrong clothes, show up with the wrong color skin, or lack professional ambition.

We move over the gravel plain, circling back gradually to the place we've parked the cars. Some distance shy of that, I signal Kamoya for help. The point of my stick rests next to a fossil.

"Coprolith," he says. "Crocodile."

If you took it apart grain by grain, would you be able to discover what it had eaten in the hours before it left this behind?

At the cars we drink cool water and watch a flight of C-130s lumbering north over the Loima Hills to the west, ferrying supplies to southern Sudan, where in those days, blacks and Arabs, Christians and Muslims were at each other, lethally, in the old way.

CHRISTOPHER HAS BAKED a loaf of bread. We can smell it downwind of camp as we approach. Goat stew and rice. Dark tea. I will miss the hospitality and courtesy that obtains in this camp. (A friend of mine who'd once been in the field with some of these same men, when I asked her about camp etiquette, told me never to ask for seconds. That would mean the cook wouldn't eat.)

Ngeneo and Onyango are at a game of checkers. Onyango always wins. Exasperated, again, Ngeneo will soon be off to visit with young Turkana women at a nearby settlement, where he will perhaps have better luck. Kamoya and Nzube are off somewhere, probably sleeping under the acacias. Wambua is flat on his back, enjoying his cigarette slowly, the way some men enjoy a drink straight up at the end of a day of work.

It doesn't take me long to pack. I've picked up a number of small stones, one or two each day while we walked. I wrap them in scraps of toilet paper to keep them from scratching each other and bind them together in a handkerchief by cross-tying its corners. These are my surrogate fossil bones.

I have a small deerskin pouch with me from home in which I keep a few stones from my own home ground, several covert feathers from birds there—winter wren, northern flicker, Swainson's thrush—a black bear claw, several small mammal bones, and seeds. Together, they remind me that I come from a real place and that I have responsibilities there. I'm not very aware of this now, as I pack, but will be in a few days when I board a plane in Nairobi to begin the journey home, and three security men find with this pouch an opportunity for sport.

Seating on that Nairobi flight to Harare is all economy class, but some of us have paid a little extra to be closer to the exits. We wait in a separate lounge at the gate where we can see all the other passengers go through a pat-down routine and have their hand luggage opened and searched. The tedium of such routine inspections often, of course, affects authorities at border crossings and security gates, and boredom feeds the desire in some of them for diversion. This situation felt like that to me. The five of us in business class were ushered through the gate without a pat-down, without having our hand luggage inspected. The workday seemed to be over as these men prepared to leave the gate. A few of them remained behind, talking together. One of them called to me to come back. He took my bag and one of the others began patting me down. I realized then that I might have made a mistake while sitting in the lounge by writing in a notebook. When traveling in developing countries, one generally shouldn't take out a notebook and write anything down at a security point.

When I'd asked Richard where I might get detailed topographic maps of the land west of Lake Turkana, he told me it was not smart to travel through Kenya with such maps. In the moment I was glad I had no such maps with me, only an ordinary traveler's map of the country. I hoped the two men now leafing through my notebooks would not confiscate them because of the sketch maps I'd made there. I was relieved when they began repacking the bag, but at the same time I became annoyed with a third man who continued to pat me down in the same places in a perfunctory way. He caught my eye and sneered at my annoyance, letting me know that he was in charge here, not the white man this time, the *mzunga*. He was nearly through with me when he came upon the deerskin pouch. He removed it. What was this? Some things from home, I said. A few stones. Some feathers.

I saw right away that he didn't like this. "This is primitive," he said. I asked him where he was from. Had he come to Nairobi to find work? Did he think about his homeland sometimes? Anger flared in his face. He shook the pouch in my face. "This is primitive!" he repeated. "Where is your Bible?"

Talking my way out of this was not going to be easy. I just hoped he would finish with his denouncement in time for me to board the twice-a-week plane. He began to lecture me about the backwardness of people who still live in the bush, and he instructed me in true Christian living until the outrage in him dissipated.

Now it would become a game.

"Go ahead, sir," said one of the other men with mock politeness, his arm extended in a gesture he maintained. "Please, go ahead. You may board the plane now."

I walked over to the man who had taken my pouch and asked for it. He flipped it to another man. I went to him. He tossed it back to the first man. The person I took to be in charge of the security crew, I could see, had grown tired of this game with a white man and wanted to leave. Eventually they gave the pouch back. I was so rattled it didn't occur to me to check my carry-on bag for my notebooks until after we took off. They were all there, shoved into the bag in a jumble, along with clothing and a few books. If they had kept the notebook with the detailed drawing of Kamoya's camp and the sketch of the route to Nakirai from Lodwar, what would they have made of it? What if they'd read my notes about the Pliocene, Turkana hairstyles, or Abyssinian rollers? As they thumbed through, what might they expect had been left out? To whom would they send the notebook for further scrutiny?

When the flight attendant arrived at my row with the offer of a cool drink, I realized I was still clenching my teeth.

JAMES COOK KEPT NOTES few have ever seen. These journal entries of his are different from the entries published in the official record of his great reconnaissance of the Pacific. The Admiralty vetted Cook's journals before they were conveyed to the publisher, partly to ensure conformity with British social and religious mores, partly to ensure the continued good standing of the Admiralty with the House of Lords. The description of any controversy on a naval voyage needed to be phrased in such a way as not to embarrass anyone in a position of authority. Ideally, the published journals should merely expand upon and ornament what well-informed Englishmen already knew, or suspected, about the world. Officers were required to turn their journals over to the captain, and crewmen were forbidden to publish anything about a voyage.

Against the official published record of a naval voyage of exploration such as Cook's, one had to consider the perceptions of literate crewmen who didn't keep journals for fear of reprisals, or who did but never published them. In the second half of the eighteenth century, during the American and French Revolutions, British seamen were in the thick of the political, economic, and social upheaval of the time. They were witness to the injustice of the British system of naval conscription, to brutal corporal punishment aboard naval ships, and to ethical breaches by naval officers, none of which did the Admiralty want publicly exposed.

One must also consider that few British crewmen, even if they could write, had the education, command of language, or acumen to write with insight about what they were witness to. Like Ranald MacDonald, sailing aboard whalers before the age of oil, and aboard merchant vessels in the South China Sea in the wake of the Opium Wars, these sailors could not articulate what the reasons were for their contempt of those who tried to ensure that their voices were never heard. Only a few who, because of their international experience at sea, might carry doubts about the wisdom of British or American attempts to control other cultures, to take over their geographies and re-rig their economies, left discerning manuscripts behind. And fewer still, like MacDonald, managed a grand gesture of opposition. MacDonald wanted the shogun to understand that neither his warriors nor his traditions were a match for what was coming. The Americans would break into the "double-bolted kingdom" of Japan as efficiently and as successfully as British opium merchants had opened a reluctant Middle Kingdom, the Celestial Empire, to the tea trade.

What I had encountered at the gate of Jomo Kenyatta International Airport, from men Kenyatta had fought to free from colonial influence so that Kenyan people could determine for themselves both the direction and the rate of social and economic change in their country, was the residue of colonial resentment and the fanaticism of converts to the new system that colonialism had engendered. Whatever uncolonized Kenyans—Samburu, Maasai, Rendille, Swahili, Kikuyu—might have offered the imperfect world, it had been plowed under by colonials and converts like this.

LATE IN THE AFTERNOON the camp at Nakirai had the look of a parched dream. No movement, no color, no sound. Birds roosting in the acacias, rollers and weavers mostly, were perched there with mouths agape, panting.

I put my bags in the rear of the Land Rover Kamoya drove and settled my shoulder bag alongside my camp chair in the shade of the trees. Soon the others would rise from their naps. Christopher would bring tea. I would miss participating every day in a determined search with these men, looking for something that had meaning. I had seen here with them the outline of something I was after: an effort to cooperate, a deepened sense of purpose in the search for something nearly impossible to find, an emphasis on courteous regard, and the apparent absence of anyone's insistent allegiance to any citadel of the intellect, to any tribe or nation or religion thought to be infallible.

It would be difficult to say goodbye here, to catch a bus the next morning and never see these men again. They had accommodated the mzunga from someplace far away, the *mgeni*. The outlander.

Once I was traveling with three Japanese men in northern Hokkaido. We had been welcomed into the home of an Ainu elder, a traditional man, a wood-carver. We entered accompanied by a translator who spoke Ainu and Japanese but not English. My companions had many questions for the Ainu elder about brown bears (a remnant population of about 1,100 roamed northern Hokkaido at the time; more do today), the traditional Ainu longbow, and other topics. I, too, had questions, but mine had first to be translated into Japanese by the only one of my companions who spoke a little English, and then posed to the elder by the Ainu translator in the Ainu language. It would be inconsiderate to insist on this, so I asked that people make drawings whenever they could of whatever they were talking about. That way, with my Romanized Japanese/English dictionary in my lap, I could follow some of the conversation.

The Ainu elder sat on a straw mat on the floor with his legs straight out in front of him. It's difficult to guess how old he was, perhaps in his mid-sixties. His hair and his beard were white. He seemed to think carefully about each question he was asked. But I could tell from his occasional bursts of laughter that he was enjoying saying derogatory things about the Japanese, his people's longtime colonial nemesis. My

Japanese companions laughed with him, signaling some kind of agreement with his assessments of Japanese people, letting him know he might not be that far off in an objective critique of the Japanese.

During the hour or so we were in his home, the elder was at work on two carvings. The blanks he used were peeled willow sticks about three-quarters of an inch thick and a foot long. He was planing them with a small knife, creating thick bundles of thin, curled shavings at different points along each stick. They were described to me as offerings (called *inao* in Ainu and *gohei* in Japanese), made to honor gods associated with a residence. Strictly speaking, they themselves are not "house gods." (It was hours before I suddenly awoke to the fact of a remarkable cultural convergence here. Though *inao* and *gohei* carvings derive from two separate and distinct traditions, Shinto and Ainu, they look nearly identical.)

I was glad to have been in the elder's home, but as we prepared to leave, of course, I felt a bit dejected at the thought of all that I'd missed in the conversation.

My Japanese companions and the Ainu translator were walking ahead of me when I stepped off the raised platform we'd all been seated on and reached for my shoes. Just then I felt a light touch on my shoulder and turned to see the Ainu man holding up the two carvings, one in each hand. He bowed slightly and handed them to me. At the same time he indicated we should keep going toward the others, standing now at the entrance to his home. There he spoke to the translator, who then spoke to the one of my companions who spoke a little English. He told me the stick in my right hand honored an Ainu hearth god. When I got home, I should place it next to the woodstove in my living room. The other honored a house guardian. It should be placed at the highest point of the ceiling on the top floor of my home.

We all bowed toward our host and the translator, awkwardly touched hands, and I departed with my friends.

AFTER TEA I said goodbye to Christopher, to Ngeneo and Wambua, and got in the Land Rover. Kamoya drove away slowly, bearing a little to the southwest, aiming to cut the road to Lodwar at some point farther to the west. Kamoya watched the ground to his right, Nzube watched to the left from the front passenger seat. Onyango was watching out the right window behind Kamoya, and I was backing up Nzube on the other side. Taking advantage of every opportunity.

If only to pay them back for all these men had given me, for their generous investment of time in me, I wanted to discover something significant before we reached Lodwar.

THE VEHICLE CARRYING US clambers slowly across the uneven plain, skirting thickets of brush, dropping into wadis and continuing on. In some sections of loose sand we temporarily lose traction and the Land Rover begins to settle on its axles before jumping ahead as Kamoya gears down. A couple of hours later we cross the Turkwel River on the main road and Kamoya weaves along rough dirt lanes in the settlement to a wooden gate in front of the Turkwel Lodge.

It is hot by the river, and humid. I suggest we all have a cold drink before saying goodbye. We sit at a small table on the hotel patio with our iced drinks, talking about certain episodes in the time we'd spent together, but the conversation is full of long silences. The Kamba men finger their sweating glasses. I offer that in a place like this there is as much to be unlearned as there is to be learned. Kamoya raises his head, looks squarely at me, and nods. Nzube says it's strange, the way you expect to find something in certain places, but you find nothing. And then the other way around.

At a nearby table a few men are listening to music on a portable radio, a song about Nelson Mandela, whose name is repeated several times in the chorus. It is the time in Africa just before he is released from the prison at Robben Island. While the song plays, the four of us share furtive glances, wry smiles. This is not a topic that has come up among us before, but by their gestures the others convey the trend of their thinking. One day all the Bothas and racist Voortrekkers will be gone. Then will come the Mugabes, the Idi Amins, the Savimbis, and after that, they hope, the Mandelas. It will take a few generations to work it out.

The men finish their drinks, swirling the last ice cubes, taking each one in to crush it between their teeth. They shake hands with me. We embrace lightly, self-consciously. I thank each man individually, perhaps in too many words. They drive off to fill their water barrels, to pick up some millet and cigarettes and batteries, and then cross the bridge again and turn east on the road to Nakirai.

I watch them pass out of view from an open gate in front of the Turkwel Lodge.

I stare past the gate, up and down the dirt lane, narrow as an alley, indecisive—what next?—and enter the hotel to ask the desk clerk about the matatu, the one leaving the next day for Kitale. Yes, he says, it will come at 5:30 a.m. There are often last-minute changes with things like transport schedules up-country, so I ask the clerk how dependable "5:30 a.m." is. "Oh, very dependable, sir. Right here at the hotel." I inquire about a good place for supper. He gives me the name of the same restaurant Kamoya recommended. When I ask what time the place closes, I realize I'm back in the world of the wristwatch, the white tourist, identity papers, and unfamiliar currency. I don't want to lose the rhythms I've been living by, but they're quickly falling apart.

The room key is superfluous. The latch on the door does not lock. An opening in a side wall serves as the room's only window, though there is no frame or glass. Above the iron bedstead and stained mattress is a hook for a mosquito net. The lone socket in the ceiling of the room has no bulb. After lighting a mosquito coil and placing it in the middle of the mattress on a metal tray, I tuck the mosquito net in smartly all around the mattress and leave. I've put my passport, airplane ticket, travel papers, notebooks, and binoculars into a day pack along with my wallet. I slide my padlocked duffels under the bed and depart the hotel.

As I exit the patio yard, I see, parked parallel to the chest-high wall that surrounds the lodge, an immaculate, pale green overland truck, a Mercedes-Benz cab-over vehicle with a cargo bed. A light tan canvas roof and side walls enclose the cargo bay, which is empty. Not considering the impropriety of it, I begin to inspect the truck more closely. I take in the massive bulk of the front and rear differentials, the drum winch mounted behind a formidable front bumper, and the capstan winch in the rear, both wound with 7/16-inch aircraft cable. Twin forty-gallon saddle tanks for fuel. All-terrain spare tires, one mounted on the roof of the cab, another between the side rails underneath the cargo bay, a bank of driving lamps on a stout push bar in front of the radiator.

I glance around to see whether anyone is watching before stepping on a foot rung that lets me peer into the cabin, to see a pair of shifters for the transmissions, the layout of dials in the dashboard. I circle the truck, searching for any sign of livery that might identify the owner or offer me a phone number. Nothing.

A fully capable machine. I have several questions for the driver, however, about the gearing, the size of the clutch plate, and the wading depth for river crossings. And where is it exactly that he's heading?

I can't locate the driver. The desk clerk has been watching me, and I think finds my inspection of the truck presumptuous. He says he has no idea where the driver is and abruptly turns back to his work. When I return from supper, the truck is gone.

Standing in a large unpartitioned room with sloping concrete floors, I take a shower with a spray of water from a solar-heated roof tank. The inclined floors—the room also serves as an open latrine—slope downward toward a central waste pit.

I find the crowded, overloaded matatu at 6:00 a.m. at a different location in the village than the hotel and we leave from there for Kitale, an hour and forty minutes late. At the bus park there I spot, in among some thirty idling buses, the one bound for Nakuru. At Nakuru, I locate the bus for Nairobi.

WHEN I REACH the New Stanley Hotel, there's a note from Richard. Could I possibly meet with him and Alan Walker the following morning at the museum to look at some australopithecine skulls? Underneath my weariness, I am starting to feel something strange, like a flu. In my room I begin to place a call to Richard's home, but decide it will be better to get a shower and go to bed and call Richard later, or maybe in the morning. In the shower I begin to sweat and then to shiver. My joints tighten. I phone the hotel doctor. Malaria, the doctor says.

Debilitating but not life-threatening. Try to ride it out. He'll check in from time to time.

In my third-floor room I sweat, sleep a little, wrap myself in a blanket against the chills, and wait. When I'm awake, I drink room-service tea and bottled water and eat saltine crackers, which is all I can keep down. I read distractedly in two books, *Gerard Manley Hopkins: Priest and Poet* and *Robinson Jeffers: Selected Poems.* In my early teens I had been drawn to Hopkins (1844–1889), and in college I had discovered the American poet Jeffers (1887–1962), whom I initially took for a misanthrope. On this reading I get something different from Jeffers's lines, something braver and more complicated. Hopkins, a Jesuit priest from a well-to-do family in Essex, is by comparison too much with God, I come to think, and not enough with man. And Jeffers, a recluse sensitive to the metaphysical insistence of the natural world, and critical of human failing, is more a realist than a misanthrope.

The Jeffers poem that will stay with me after this reading, four long stanzas that will follow me for years, is "Apology for Bad Dreams." In it Jeffers describes a scene he observed from a clifftop along the Northern California coast, near Big Sur. He watches while a man and his mother beat a horse the son has snubbed by its head to a tree with a length of barbed wire. He is whipping the horse bloody for its failure to do what his mother wants done. Beyond the couple, the sun is setting in banks of brilliant color over the tranquil Pacific. The horizon is a panorama of stupefying beauty. Later in the poem, Jeffers writes:

> *It is not good to forget over what gulfs the spirit*
> *Of the beauty of humanity, the petal of a lost flower blown seaward*
> *by the night-wind, floats to its quietness*

For two days I crawl to the bathroom from my bed and back to the bed over the tile floor. When I am finally able, I call the museum. Richard has returned to Nariokotome, but Alan Walker would be glad to show me the skulls in the museum's vaults.

I am moving slowly the next morning when I leave the hotel to meet with Walker. The wet sidewalks along Moi Avenue are spangled

with sunlight, and I feel the freshened air cool against my skin following a night rain. I watch flocks of Kenya sparrows scattering through
the limbs of the frangipani and flame trees that flank the avenue, dense
with morning traffic. The waves of chirping are like another kind of
spangling.

At the museum, Walker lines hominin skulls up on a dark velvet
runner he's spread across a trestle table, two robust australopithecines
and a single gracile australopithecine. Walker handles the skulls with
great care, as though they were works of Steuben glass. He handles them
with respect and tenderness. (In camp with Richard, I recall, Walker
was understated, the less inclined of the two men to speak. He seemed
more open to possibilities that he'd not considered before.)

After a long moment of silence, as though we were standing at an
altar bearing relics instead of at a sideboard bearing fossils, Walker puts
the tips of the fingers of one hand on the brow of the single gracile
skull. His contact with the fossil bone is as delicate as the touch of a
camel-hair brush. He has just been talking about the long path of the
hominin, a reeling backward through millions of years to a time before
H. ergaster, the first truly human species to use rudimentary utterances,
some speculate, and coming forward in his commentary to *H. sapiens.*
He is speculating about the emergence of human language, unknown
to the australopithecines sitting before us, the seven thousand spoken
languages of the nineteenth century—the four-toned Mandarin of the
Chinese, 14th arrondissement French, verb-rich, noun-poor Navajo. He
is describing the roots of vocalization in the development of hominin
throats, minor changes taking place in the hyoid bones that support the
tongue, which hominins would have used to shape tonal phrases of particularized meaning, which preceded the development of a full-blown
spoken language—"phonemicized, syntactical, and infinitely open and
productive," as one linguist has written.

In the beginning, before language and complex thinking, hominins
pushed air up the trachea, shaping meaningful and comprehensible
alarm calls and breeding calls. And then came something else, something that conveyed the meaning that melody can carry but which
words can't. These were tones strung together around emotion and
thought, runs of sound like the trills and vibrato in birdsong. Homi-

nins came to know one another as individuals in a new way. The level of cooperation possible between hominins expanded, by an order of magnitude.

With his fingertips on the cranium of an australopithecine skull not much larger than a grapefruit, on the forward part of the vault where one day frontal lobes would rise up in *Homo,* he says, "Barry, I can't prove this, but I believe we sang before we spoke."

Port Arthur to Botany Bay

State of Tasmania

Northern Shore of the Southern Ocean

Southeastern Australia

———

State of New South Wales

Western Shore of the South Pacific

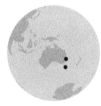

43°09'16" S 147°52'02" E to 34°00'11" S 151°13'32" E

My hands are slotted loosely in the pockets of my trousers. I've turned my head and tilted it so the sun's rays fall full on my face while I gaze, squinting, across the expanse of a former prison grounds. It might be the face of someone hearing but not listening to a eulogy. Inwardly elsewhere. Neither I nor my companion, a poet from Hobart and a recent acquaintance, is paying much attention to the cricket match going on in front of us. It's a pickup affair, being pursued on a greensward where picnickers stroll carrying paper plates of potato salad and fried chicken. The action on both sides—a white ball sailing over the head of a boy racing after it, another ball dismantling a wicket with a clatter—raises desultory cheers from the crowd, some of whom have their backs to the game. They're gnawing breast and thigh meat off the bones, cracking down hard on buttons of resistant cartilage. Those sitting have settled their drinks securely between tufts of green grass to keep them upright. Some have paused with their mouths full to take in, like us, the bleak north wall of a burned-out hulk, a roofless three-story building across the way. They peruse with us its mute facade, perhaps imagining an angry face there at a barred window port, a figure emerging from the interior of one of the cells and glaring down indignantly from behind the twisted steel. Long ago someone took a sledgehammer to most of these windows, a furious, determined effort to dismantle, to render into rubble, what had happened here. Others, the ones who eventually prevailed, saw the commercial potential in preserving what remained of the penitentiary and its surrounding grounds, the parsonage and the cemetery, the watchtower at Scorpion Rock, and the guards' former barracks (now the Frances Langford Tea Room).

My companion is pointing away to the east, toward the remains of the prison church. He offers me some points of its history. While two prisoners were excavating its foundation, in December 1835, one of the convicts, he says, killed the other with a pickax. And it was also a convict, he tells me, one Henry Laing, who designed this building, capable of holding a thousand Christian prisoners for service on a Sunday morning. When construction was completed, however, the Church of England claimed the house of worship for its exclusive use. Catholic convicts objected so violently that extra guards had to be brought in to reestablish order and maintain it until an arrangement could be worked out. The decision the appointed committee made was to accommodate all Christian denominations at the same service.

Roman Catholic prisoners who walked out during the first generic service were sentenced to thirty-six lashes each.

The church was later destroyed in a fire, in February 1884.

I'd asked the poet to bring me here to Port Arthur from Hobart so I could see this storied transportation prison, one of the best preserved of nearly three dozen such penal colonies built by the British across their empire during the eighteenth and nineteenth centuries. These "convict dumping grounds" were meant to house that portion of the British citizenry the Crown had identified as "undesirable" for leading a life in the homeland. By the mid-eighteenth century, the population of these outcasts had grown too large to be warehoused in the carcasses of decommissioned ships—the "hulks," moored in the Thames. Among those sent instead to Port Arthur was the Hobart poet's great-great-grandfather, John Frimley, at the age of fifteen.

The poet and I both consider Port Arthur a monument to the absolute power of an imperial state to purge itself of criminals, the mentally ill, political protesters, paupers—of anyone who posed a threat to the authority of the state or to its right to impose civil order.

ON THIS PARTICULAR WARM AFTERNOON in the fall, the hundreds of visitors spread out across the former prison grounds appear to be enjoying themselves, to be at their ease wandering through the buildings, lolling on the grass. A group of Chinese travelers has just disembarked from a tour bus. Their guide is starting them off with a visit to the formal gardens. Shoppers in the Port Arthur gift shop, housed in a building that was once the prison's insane asylum, are considering souvenir trinkets and postcards. In the flogging yard nearby, two young women stand together motionless before a discolored post with a sheen to it. Finally they step away, looking somewhat relieved.

The history of the criminals once housed here has been slightly jiggered for public consumption by the Port Arthur Historic Site Management Authority, simplified and expurgated to not offend a modern sensibility. Many, perhaps most, of those picnicking on the cricket pitch believe that what happened here at Port Arthur was cruel, unjust, and unenlightened. Some, no doubt, are also inclined to regard most of the former convicts as the unfortunate victims of a misguided experiment in social engineering. Whatever iniquity might have characterized Port Arthur in the years it operated, I believe the consensus among visitors here today—as it might be among visitors to Choeung Ek, the killing field outside Phnom Penh—is that this extreme public violence toward people is no longer openly tolerated. Whatever its faults, civilization has advanced too far to allow it.

The Port Arthur Historic Site has been laid out to edify and reassure the visitor, to isolate its evils in a distant time. It is not set up as a caution to the visitor, nor meant to suggest the existence in the world of inhuman criminal intent or of punitive governments, or to remind the visitor of contemporary governments who regard those who protest their actions as "an affront to the Crown," which was the case with the Irish Ribbonists and members of the Young Ireland movement who were punished here.

No connection is to be made at Port Arthur with Bashar al-Assad's underground prisons in Damascus or of America's off-site destinations for "terrorists." Visitors to Port Arthur are meant to skate easily through and around all this darkness. Most are glad, one assumes, not to be forced to encounter any complexity or paradox here. They are, after all, many of them on holiday.

The poet and I stand about near the pitch, both of us looking slightly out of place. We do not speak to each other, though each has much to say about the situation—about the scourging ground, where the grass growing close around the whipping posts is so lush; about the intimidating presence of the commandant's perfectly restored house, with its spacious veranda, its carefully manicured grounds. Fitful breezes off the water of Mason Cove rustle the dry leaves of massive eucalypts shading the commandant's quarters, and the rising hush of this breeze tempers the shouts of the cricket players and the applause of the crowd.

MY COMPANION, Pete Hay, has been commissioned by the Tasmanian government to compose a long poem about Port Arthur, about the particulars of the experiments conducted here in penal discipline, the incidents of physical assault in the mining pits, the extent of sexual violence, the despair of adolescent victims of serial rape, and the furtive pickpockets, alert for any opportunity to advance their standing in the prison population at the expense of someone else.

The poet is a generous man. He has reams of notes, folders filled with research, the raw material from which he will fashion his poem; but he readily offers me a few of the discoveries he's made. The poet is not an artist obsessed with ownership. He does not hide the knowledge he's acquired. He's eager to share. I liked him right away. When I spoke to him over the phone from my home in Oregon and told him of my interest in the prison, he volunteered to drive me there from Hobart and to guide me through the site.

From the picnic ground we walk off slowly to the southeast, away from the main compound. The poet says he has something to show me, a part of the penal colony closed to tourists but to which he has access because of his research needs, and because it is his intention to look squarely into the factors that complicate the history of this place.

We amble down a service road, enjoying the balmy March weather and exchanging views, like two people walking along the Malecón in Havana or the Marmara Denizi in Istanbul, where the aspect of an adjacent spacious sea encourages latitude in the conversation. I ask the poet who among Australian writers he most admires or enjoys, and he asks the same of me. Neither of us is looking for a critical appraisal. We're feeling our way toward shared ground, so that our acquaintance might develop into a friendship.

We continue on to a locked gate, to which the poet has the key. Beyond it we pass through a copse of gums to emerge on a deserted field. A few buildings once stood here, says the poet, adjacent to a cliff which plunges straight down to the sea, a hundred or so feet below. The view west from this point of land jutting into Carnarvon Bay is toward the town of Port Arthur, and to the penitentiary grounds at Mason Cove. To the east are the waters of the northernmost arm of Maingon Bay, the bay a part of the Tasman Sea. From there, only the open ocean imposes between here and the George V Coast of Antarctica, 1,635 miles away over the southern horizon.

The poet begins his story. This is Point Puer, he says, the tip of a short peninsula that protects the prison compound from storms coming in off the Tasman Sea. And it is here, he says, that boys as young as eight, brought in on the transportation ships, served their time as adolescents inimical to the Crown or to the wealthier classes. The buildings here, set apart as they were, gave the boys some protection from pedophiles housed with the main prison population. The poet describes the general horrific atmosphere of sexual predation and debauchery at the prison. I stare at the ground. It's covered with narrow leaves from the eucalypts, a carpet of fallen leaves, shed bark, and fallen seed capsules. I gaze out across the water at a ridge of mountains on a peninsula miles to the east. I nod stiffly, acknowledging the repugnance of the history, which the poet continues to present.

The poet takes no pleasure in setting out the graphic details. His intention is to establish a context for what he now has to say: it was from these cliffs that some of the boys jumped to their deaths, hand in hand. Pursued through the labyrinth of buildings in the main prison compound during the day, cornered in closets and storage rooms where they were overpowered and raped by sexual psychopaths, hounded and beaten at night by their dormitory guards, who enforced the boys' adherence to a regimen of prayer, penance, and physical labor, some chose death. They slipped away from the dormitory at night and leapt into the darkness masking the water below. Some surely must have jumped during the day, says the poet, perhaps on a March afternoon like this one, when the warm air, soft breeze, and sun-spangled water embodied feelings of relief for them, of salvation. Running for the cliff edge, hand in hand, they ignored their warders' counsel to resist the temptation to end one's life in this hellhole, and sailed into the air.

With a gesture of my hand I stay the poet and walk away by myself to the clifftop, where I stand and watch the water below for a while and listen to the breeze in the eucalypts, the fondly remembered trees of my childhood. On the way back to where the poet waits, making notes, I pick up two dark eucalypt buttons and pocket them.

We quit the peninsula and return to the main prison complex. I do not tell the poet I understand why the boys held hands. If the poet had suggested to me that the suicides were unwarranted, I might have spoken up; but he seems aware of the human capacity to inflict humiliation and pain, and also of the depth of suffering some are forced to endure. Whatever he might have recalled of the forms of harm he's come across in his research at Port Arthur, he offers me only a bare outline of the story, and acts as though he might have compromised himself by saying only that much. His physical gestures speak of his own grief and compassion, so I feel great affection for him and admire his decency.

We walk west along the sea-skirting road, past the commandant's quarters and the site of the first prisoners' barracks, to arrive at the flagellation yard, where Pete explains the procedure at this spot for any prisoner who broke the rules. Prisoners were secured to a triangle frame, arms straight overhead, hands tied at the apex, then the feet spread and tied. The customary lash was a flail, comprising nine lengths of knotted cord. The blows fell across the bare back from neck to waist. The usual sentence of twenty to forty lashes left the victim unconscious and the back a jellied mass of pulverized flesh.

I ASK PETE where we might get a map that shows the layout of the prison, one that might pinpoint where the original buildings once stood, and now that it's a tourist destination, what changes have been made since the prison closed in 1877. He says we can get a map later, on the way out. Just now, however, he thinks we should continue on through the penitentiary complex, examine some of the cellblocks, and look at a few of the curated exhibits, including one featuring portraits of some of the more notorious prisoners. Just before we enter the old cell barracks, Pete points out the Broad Arrow Café on the far side of the cricket pitch. Later, he says, we'll get a map there, as well as a bit of lunch. I can see Pete's red Toyota Corona sitting by the café in the sunshine. Just beyond it is a yellow Volvo 244 with a surfboard strapped to the roof rack.

To get a full sense of what was once in operation here, says Pete, we need to visit the penitentiary itself and then walk through the Model Prison, and then afterward take in the prison museum, in the old hospital. With that sequence now organized in my head, I step into the partially restored ruins of the penitentiary building, three floors of cells in which the most dangerous prisoners were confined. The cells denied prisoners open space and forced them to comply with the prison's daily routine. Stepping in and out of them is like swimming in and out of sleeping compartments on a sunken ship, the rooms emptied now of the bodies of the drowned.

What most transfixes me in the penitentiary building, however, is not this poignant austerity but instead a set of photographic enlargements hanging in one of the corridors, portraits of former inmates: the defiant faces of the insane, the duplicitous gaze of the pedophile, the vapid stare of the murderer.

The images call forth in me neither compassion nor condemnation, only astonishment. My intention had been to ferret out salient details of prison life here, but the day is hardly half over and I am thinking I've already seen enough.

Whatever was decent in men and women in England in 1786, it was not sufficiently strong to condemn this idea when it came before Parliament and to terminate the experiment before it began.

The two of us head for the Model Prison and the museum and then proceed to the Broad Arrow Café for lunch and to look for a map. I could not anticipate how vividly details of this café would remain in my imagination in the weeks and years to come.

———

MY FRIEND MARK TREDINNICK and I are following Annamaria Weldon, a woman from Malta, down the Old Coast Road, a hundred miles south of Perth on the west coast of Australia. We're headed for Lake Clifton, one of the Yalgorup lakes, this one situated in the middle of a national park. We're passing through an undulating landscape of sand dunes, of isolated stands of mallee forest and limestone outcrops, south of the city of Mandurah. The country hereabouts once belonged to Bindjareb Noongar people (or they to it). Some of them are still here, engaging chroniclers, articulate about the history and nature of their home landscape. They speak of it as you would a relative, a close companion or confidant, referencing a sphere of time inclusive of them and indistinguishable from this geography.

Annamaria, up ahead, has listened with the respect and curiosity of an ardent acolyte to the Bindjareb custodians of Noongar culture. She has tried out the templates of her own Mediterranean ideas on this land, and alongside that put in some twenty years of apprenticeship in the place, grappling with an epistemology different from her own, an Aboriginal way of knowing. The fulcrum of her imagining is the complex ways in which people are married to a place, whatever the tradition might be, whatever the place. And also her awareness of the threats to these marriages, from whatever quarter. Once she loved Malta, to the exclusion of all else I think. Now she is a student of Bindjareb love of place. She's graceful, light on her feet, full of deep water, a person dedicated to something outside herself. Or so it appears to me.

I marvel at the way Annamaria and Mark pronounce "Bindjareb Noongar," as easily as they might refer to "Italians"; and how they refer to other local cultural traditions as if they constituted a second subterranean culture, contiguous across Australia, from Anggamudi in the northeast to Wardandi in the Margaret River country, in the continent's southwest, country toward which Mark and I are bound; from Wimambul in Kimberley, in the far north, to Tyerrernotepanner in Tasmania in the far south. An alternative tradition to the national Australian tradition.

A spur off the main highway takes us through a dying tuart forest.

We move into thickets of swamp paperbark and peppermint willow, finally arriving at a spot where we park the cars. Annamaria leads us down a trail—there are a few acacias growing here, acquaintances I recall from Africa—and then there is Lake Clifton. We approach it on an elevated wooden boardwalk above a sedge flat that borders the water.

The summer sun is full upon us. The heat is terrific, but this apparition Annamaria has brought us to is so forceful, the discomfiting heat becomes incidental. The lake has the aura of an exalted being. Its presence is insistent, like that of a wolverine one has suddenly come upon, asleep in a forest clearing in North America. I cannot recall in the moment ever having seen a body of water as ethereal. It's austere but benign, still as a mirror, its surface a color some painters call French gray, the reflection of a haze in the air above it.

In the beating heat and palpable stillness, the overpowering sight of the elongated lake, less than a mile across, undoes me. It's bounded by greenish and gray-brown brush growing on long sand ridges that strike north and south. I identify the faint odor of rotting vegetation. For many minutes no one of the three of us says anything. A line of black swans passes by, bearing north over the water, and I become aware of voices piercing the silence suspended above the stillness—fairywrens. In the trees behind us. Southward along the lakeshore, red-necked avocets emerge from a deckled sheet of dunnish color at the water's edge. Small clutches of hooded dotterel race the shore beyond them. Animated by the color of the sky sifted through a haze, the bowl of our space rounded by flocking birds and birdsong trembling in the heated air, all of this framed by earth-like colors, I grasp in its near fullness the lake Annamaria has brought us here to see.

A billion years or so after debris from a supernova started to coalesce into the rocky planet we now occupy, the possibility of lives like ours was set in motion by an as-yet-unidentified species of cyanobacteria. The cyanobacteria oxygenated an atmosphere otherwise poisonous to the forms of life that were to come. A good guess, according to some, is that these bacteria built stony habitations for themselves in wet environments. Today these now abandoned habitations are called stromatolites. Stromatolites, which accrete in shallow waters in the same way coral reefs do, are related to structures called thrombolites. Together, these two structures set a benchmark in Precambrian time. It was modern-day thrombolites, still being built up by cyanobacteria in the near-shore waters of Lake Clifton, that Annamaria particularly wanted us to see.

This annunciation, Lake Clifton itself, was so overwhelming that for many minutes I didn't notice the thrombolites. These pale hummocks of "living stone" in the lake are believed to be about four thousand years old. They stretch away from us in the near-shore water in both directions, looking like a scatter of white mushrooms at or just below the surface of the water. The circumference of each mound expands by about a millimeter (one twenty-fifth of an inch) per year. The cyanobacteria that create them are photosynthesizers, and it's the residue from that process, a secretion rich in calcium carbonate, that comprises the thrombolites. The surface of each one is marked by whorls and cracks, patterns so distinctive that Annamaria can point out the difference between one structure and another. To my uneducated eye they seem identical. The rise and fall of lake water during wet and dry seasons makes the thrombolite reef more or less visible to a visitor, according to the season. The water is clear, however, so even when they're fully submerged, the thrombolites closest to the visitor are still readily apparent. A line of sunken white pillows at the beach's edge.

This freshwater reef is the largest such reef in the Southern Hemisphere, and the Australian government has designated it as critically endangered. Its future existence is threatened by a complex combination of natural and Anthropocene forces, principally real estate development and global climate disturbance.

Annamaria describes for us what it is like to see the lake and the thrombolites in full moonlight. She describes her encounters here over time with ibis and white-faced herons, with banded stilts and tawny frogmouths (all these creatures are birds). And she says the southern boobook owls and Carnaby's black-cockatoos that once lived here, along with several others, aren't around anymore.

The Bindjareb call the thrombolites *woggaal noorook,* "eggs laid in the Dreamtime by the female creation serpent." A prominent mentor embedded in their mythology, this serpent is difficult to imagine separately from the Bindjareb's home geography. She is an instructor, they say, who illuminates for them the threshold between the inanimate and animate worlds.

Leaving the lake, I thank Annamaria for her "translator's introduction" to the Yalgorup country. How wonderful to imagine making your way across all of Australia with respectful guides like her.

MARK AND I SPEND a couple of days in the Margaret River country, around Yallingup on the Indian Ocean, and then return to Perth to meet up with a few others—a landscape painter, a photographer, another American writer, and two escorts who will lead us on an overland trip north to the Pilbara, a district in the northwest corner of the state of Western Australia. Here we will find the scale and intensity of iron-ore mining staggering; commercial mining's insistence on its justified (and legal) place in the countryside intimidating; and the industry's burgeoning growth regarded as inevitable by local whites.

In the Pilbara, the depauperate, bewildered, and disrupted original residents—Wajarri and Banyjima, remnant Jaburrara from the Burrup Peninsula, and Kariyarra—explain the deracination that has overtaken them by saying, "Natural resource extraction happened to us." The more informed and sympathetic among mining company executives will say that the injustices and lack of charity that affect Aboriginal people in the Pilbara are occasionally deplorable; but the world—especially China—is hungry for more steel. And the fly-in/fly-out (Caucasian) workforce, with their homes and families set up in Bali or Perth, the people who actually extract the ore and haul it away, will simply shrug off the collateral damage to local people. They offer us a rationale they consider irrefutable—their astounding wages. "The money's really good," they say.

On the drive up to Perth, in anticipation of what we would be seeing in the Pilbara, I asked Mark to talk about his idea that geography exerts a powerful and unacknowledged influence on the human psyche. He believes it frames and encourages certain behaviors, human activities, and social arrangements, so much so that one is justified in speaking of a moral geography, a geography that over time inaugurates a social ethic among the people wedded to that place. I know this idea is bound to come up when we reach the Pilbara, between us at least, so I want a little bit of a head start in a discussion about land-based morality and the implacable force of the mining industry.

Annamaria had said to me, when I told her that in the moment I first saw Lake Clifton I had the feeling of stepping into a dream, that an Australian painter named Tom Carment had said once that he painted trees because he was interested "in the emotional content of the light around them." I'm strongly drawn to the intelligence behind this statement, and aware that some industrialists consider Carment's sentiment daft, even socially disruptive.

Mark and I met the others at Perth Airport, the photographer Paul Parin, the painter Larry Mitchell, both from the suburbs of Perth, the American writer Bill Fox, an expert on land art, and our escorts, Mags Webster and Carolyn Karnovsky. Mags and Carolyn are employed by FORM, a small nonprofit investing its energies in the creative work of artists in the Pilbara. FORM sponsors photography, painting, and writing workshops for local people, indigenous and non-indigenous, in the belief that an effective way to reduce tension in the Pilbara, generated largely by the social changes large-scale mining has brought to the area, is to encourage creative expression. FORM operates an art gallery in Roebourne and is heavily involved in collaborations with a number of communities trying to find a way out of the pattern of social destruction that industrialization has wrought here. FORM's earnest partner in this search for another way of living in the Pilbara is BHP Billiton Iron Ore, one of the world's largest mining companies and a principal investor in industrial infrastructure in the Pilbara.

The flight from Perth takes us to the mining town of Paraburdoo, where we rent two four-wheel-drive vehicles and head north to Tom Price, a mining town just west of Karijini National Park. It's late February, coming to the end of summer, but the temperatures here are well over 100° F every day, and the humidity, I learn, will continue to increase the farther north we go. The ore ports at Dampier and Port Hedland on the Indian Ocean are our final destinations.

I had originally planned to drive up to the north coast of Australia by myself. I wanted to visit a square-kilometer array of radio telescopes near Murchison, several hundred miles north of Perth, where astrophysicists have entered into an agreement with local Watjarri people to build a huge phalanx of deep-space probes on traditional land. This technology was intended to locate dark matter and dark energy in the universe, partly to better understand the evolutionary dynamics that create galaxies. I'd spoken with a professor of physics at the University of Western Australia involved in the project and now wanted to speak to the Wajarri Yamatji, to hear their views about the importance and meaning of the array of equipment they'd permitted to be erected on their land.

Unfortunately, construction of the array had been delayed. It is not far enough along at this point for me to follow up on these plans before joining up with the others in Perth. I decide to come back to Murchison the following year, when I can walk with the Aboriginal owners through the radio-antenna farm and listen to what they think of a search for dark matter. What would be their metaphor, their simulacrum to make sense of the scientist's quest? What sort of "white-fella walkabout" did they think was in play here?

Midway between Paraburdoo and Perth was a second place I wanted to see, a remote draw in the Jack Hills that had no name. At that time it sheltered the oldest known fragments of Earth, tiny zircon crystals dated to 4.27 billion years ago, about 250 million years after the formation of the planet. Shortly after the existence of the site was announced in scientific journals in the mid-1980s I made arrangements to travel there, a plan that eventually came to fruition. Now, some twenty years after my first visit, I was curious to see how the geography of the place might have changed, whether people had by now built a road into the area, or whether the manager of the sheep station on which the site was located had put up a locked gate to keep the curious at bay.

Once in Perth, however, I shelved these plans, too, and opted to travel to Paraburdoo with everyone else, a decision that gave me the afternoon at Lake Clifton with Annamaria and Mark and, later, time in the Margaret River country with Mark, an Australian author and poet living in New South Wales, on the opposite coast.

The seven of us were up and gone from Tom Price before sunrise, driving unpaved roads that took us into the heart of Karijini, a landscape that looked as if it were still damp from the creation. Paul had been all through this country before, photographing, and Larry had seen a good bit of it as well. We descended into and climbed out of deep desert canyons, narrow fissures in a rolling plain that rose to imposing heights in the Hamersley Range to the north and west of us, a brilliant magenta shield, dotted with white-barked gum trees and clumps of golden-colored spinifex grass. The crystalline air was shot through with birdcalls, and flocks of white little corellas and pink galahs passed so closely over us we could hear the creak of ligaments in their wings.

We swam in still pools of cool water and in slow-moving rivers in the bottoms of several of the gorges, out of the direct rays of the sun but with the brightness of its incident light ricocheting off walls of two-billion-year-old rock, walls in every shade of purple one might catalog, from damson to heliotrope, from hyacinth to raisin, each purple changing hue as the hours passed and as the sun's reflected rays became direct, creating a harsher light.

In one canyon we hiked into, we stopped for a while where a narrow river bar had formed and several river red gums and weeping paperbarks had taken root. Jeweled geckos ran the vertical walls on either side of the canyon, great slabs of incinerated blacks and bronzed purples. As I lay back on cool, sandy soil, a flock of about twenty zebra finches passed across the narrow strip of blue above the canyon, small, brightly colored passerines with large bills and barred tails, quick as a school of mackerel. They were gone before their wooting cries reached the floor of the canyon where we lay.

Up on the plain, in great reaches of hummocky grassland to either side of the red-orange dirt tracks on which we drove, termite mounds rose higher than the tallest of us stood. Red kangaroos and common wallaroos, suddenly alert as our trucks rolled to a stop, bounded away, drawing the eye to bare escarpment walls rising to the base of the Hamersley Range. The traditional owners here are Banyjima, Yinhawangka, and Kurrama people. Their occupancy goes back more than 30,000 years. To an outlander, the countryside appears primordial; but it has been shaped by these peoples' hunting and gathering strategies and by the fires they've deliberately set, pursuing their "fire-stick" style of farming.

Mags and Carolyn had wanted the five of us to see Karijini to establish a kind of baseline awareness of the country we would be traveling through, by comparison with which Karijini appears completely untouched. Leaving there, we swung north toward Dampier, situated at the far end of the Tom Price railway road, which parallels the tracks that convey the iron-ore trains out of the Hamersley Range toward the deep-water harbors at Dampier and Port Hedland. As we left the park—an open-pit iron-ore mine called Marandoo shares a boundary with Karijini on three of the mine's four sides—the signs of human infrastructure quickly emerged from the otherwise apparently untenanted land. Fence lines, declaring exclusive ownership, and the many modern conduits for water, electricity, fuel, and information: power-line pylons, cellphone towers, and pipelines. Inscribed amid these was a maze of improved and primitive roads, from sealed bitumen two-lane highways to roads too dangerous, our maps warned, to attempt in a single vehicle. (In such spots a convoy is preferable, where mutual aid ensures some success, at least, in dealing with flash-flood washouts, drifting sand, and broken axles.) The maps also warned us away from areas closed to the public, abandoned mining towns like Wittenoom, where asbestos fibers still swirling in the air are too easily inhaled.

Loaded iron-ore trains gain on us and roll past as we drive north, trains too long for us to be able to see both ends in the same moment, each end lying as it does over the curve of the earth. The longest train ever documented passed this way in 1993. It consisted of 682 iron-ore cars, pushed from behind and pulled ahead by eight diesel-electric locomotives. It was nearly four and a half miles long and was carrying 110,000 tons of ore. During each twenty-four-hour cycle in the Pilbara, approximately eleven such trains arrive at Dampier and Port Hedland. Most are composed of more than three hundred cars, each bearing about 160 tons of pulverized ore. The dark umber mounds sitting inert in the cars, leveled off in the shape of neat trapezoids, follow each other like a series of mountain ranges.

At some point on the Tom Price road we cross a broad dry watercourse, its floodplain bounded on either side by a gallery forest of gum trees. On our right at that moment an ore train is passing. On our left a group of eight Aboriginal people are standing still in the riverbed, fixated on the train. Dressed in threadbare clothes, carrying cloth bundles holding their possessions, the group has the look of an extended family. For people whose psychological anchor in stressful times is in part the immediacy, the intimate closeness, of the physical country they were born to, and whose guiding stories are inextricably woven into that land, the passage of the train is traumatic. Its very presence signifies their loss of ownership and denial of access to their ancestors' lands. This is an old story in Australia, in the Americas, on the Tibetan Plateau, and elsewhere. But now, before them in the cars, is the very country itself, being shipped off somewhere. For a Christian, a Muslim, or a Jew, it would be as if Jerusalem and the ground it stood on had been put through a rock crusher, and the gravel of the tombs, the temples, the churches, and the prayer walls had been hauled away by NASA to build dormitories on the moon.

We were enveloped regularly by dust storms as we drove the Tom Price. They forced us to slow down, in order to keep the near left edge of the dirt road in sight. These rolling storms of fine powder—of "bulldust"—arrived with, and then trailed far behind, a kind of vehicle I've not encountered anywhere else but in Australia's parched outback. A road train, a conventional long-haul road tractor pulling a long string of trailers. Most of the trucks that pass us are hauling mining equipment—steel drums of engine lubricants and fuels, dry stores, pipe sections, machinery, and motorized vehicles—utility tractors, small pickup trucks, D4 bulldozers. The mixed load of material is chained down on flatbed trailers. As many as four of them might be lined out behind a tractor, each trailer riding on six axles (three forward, three aft), each axle supported by six tires, three to each side. Sometimes the cloud of dust that shrouds the road is being thrown up by more than a hundred and forty tires. The tractors, most of them sporting twin chromed smokestacks, are festooned with banks of driving lights and protected forward by a rampart of chromed steel bars called 'roo bars, mounted there to ensure that the impact of a large animal struck head-on won't damage the radiator or slow down the tractor. Behind the tinted glass I could never quite catch the cast of the drivers' clean-shaven faces, only the suggestion of the body's determined lean, the glare of mirrored sunglasses, a sleeveless white shirt.

We stopped one afternoon for lunch at an oasis of shade trees and limpid cold flowing water called Millstream, once the headquarters of a homestead and now part of Millstream Chichester National Park. A pipeline originating here carries water sixty miles north to Dampier, that town today an entrepôt for the mining industry. Before that it had been a pearling center (until the pearls were gone), and before that a whaling town (until the whales were gone).

As we pulled into Dampier and drifted slowly down its streets toward our motel, I recognized the familiar trappings of towns I'd passed through in other places where the extraction of natural resources—fish, trees, coal, rare earths, oil, diamonds—fuels economic development. An isolated terminus erected on a pounded landscape. Someone here has planted but not regularly watered a few wilting saplings. In a weed lot, which seemingly belongs to no particular building, millions of dollars' worth of derelict machinery sits rusting. The air is rank with hydrocarbon fumes and is headachy to breathe. Ramshackle houses abut tidy prefabricated warehouses. The parking lot at the motel is littered with cigarette butts, crushed beer cans, fast-food wrappers, glass shards, and bits of clothing. It gleams with spills of food grease and oil from dripping engines.

The air in the bar attached to the hotel is thick with cigarette smoke and hammered by thudding, raging music. Women in tight, skimpy clothes sashay past tables where knots of men fall silent, as if a shark were passing their life raft. The seven of us take our beers outside, onto a shaded patio, out of the air conditioning and into the night heat and humidity. A safer, less crowded, less desperate commons.

I feel a strange affection toward most of the men in the bar, that more than disdain. I can easily imagine anyone here might say, out of hearing of the others, that they feel trapped by the circumstances they find themselves in. Love gone sour at home, mortgages to be paid, college for the kids to save for so the kids don't have to invest in some version of the treadmill work their dads are indentured to. You work every day, then search out an anesthesia that will bury the anger and ennui the work fills you with.

The deeper one pushes into the pall of violence and despair created among too many working people by the extractive industries that employ them—corporations off the leash of government restraints, their policies framed by a relentless quest for strong profit margins, all of it driven hard by men and women on trading floors in Hong Kong, New York, and Frankfurt—the more difficult it is to identify a villain in the fin-de-siècle morality play unfolding here.

The truth, one tends to think, is that all of us, drunk or sober, sedated or not, aggrieved or manic, live consciously or unconsciously within this maelstrom, which no one really wants to risk shutting down. Some of the men I sit with for a while at the bar in Dampier tell me they are making $250,000 a year working in the ore pits, two weeks on, two weeks off. They believe they will outdistance Death, and that those lying dead by the side of the road are just unlucky, no concern of theirs. They're happy with what the job provides, and they are confident that anyone who wouldn't trade places with them is simply dense. When they head for the men's room, they walk with a swagger I'm guessing they learned from watching cowboy movies. They sit at the barstools with that same swagger. They remind me of the two men who compete in a footrace in a story by Jorge Luis Borges, "El Fin." The men, racing furiously side by side, are both beheaded in the same instant. The money down is on which body will run the farthest before collapsing.

THE FIRST MORNING in Dampier, our escorts from FORM have arranged for us to see the Burrup Peninsula. A mammoth desalinization plant has been built here alongside a petrochemical plant, which is the terminus for a natural gas pipeline laid out on the bottom of the Indian Ocean. Plans for a nitrate plant, working in tandem with the petrochemical plant—which uses salt from the desalinization plant, combined with ammonia, to produce chlorine gas—will produce explosives to clear more land on the peninsula for additional plants and industrial infrastructure. The industrial development is considered economically feasible because funds from iron-ore mining are expected to last at least another forty years. (On the day after we arrive in Dampier, the local paper carries a story about a new billion-ton iron-ore deposit located recently just south of the Pilbara.)

The people with FORM have no agenda for us. Their sole interest,

it seems, is in gaining our impressions of the Pilbara as writers and artists, and in the publication of all this—our photographs, words, and
paintings—to advance their primary goal of promoting a better conversation about the fate of the Pilbara and its people. They want something
more enlightening than a fistfight between a drunken Aboriginal and a
drunken truck driver in one of the Dampier bars about what's at stake,
what's being lost and gained. They want a better conversation than a
shouting match with a middle-management executive at BHP Billiton.

It's hard to understand how anyone can look at what's happened to
Aboriginals and their culture on the Burrup Peninsula and simply turn
away, as if this cultural detonation is not happening. To merely bring up
the subject of slavery or genocide, which underlie the varnished history
of so many nations, mine included, is apparently to offer a calumny
of some sort. The Burrup Peninsula, many academics maintain, was
once the geographic center of the greatest array of rock art ever created. Thousands upon thousands of depictions of animals, of humans
interacting, of spiritual drama and historical events once existed here.
It was a Musée d'Orsay of petroglyphs and pictographs.

That morning in Dampier we were taken to see the fraction of
what is left of this extensive outdoor gallery. Mark and I climbed a
slope of boulders beneath a cliff face to get a close look at a four- or
five-thousand-year-old pictograph of a now-extinct marsupial called a
thylacine, and another of a second extinct animal, the flat-tailed kangaroo. All about us were dozens of other ancient glyphs. For diversity
of imagery and density of depictions, the area has, arguably, no equal
in the world.

In order to clear land for the nitrate plant, developers began removing hundreds of pieces of rock art with bulldozers, perhaps with the
exculpating belief that a great deal of Aboriginal art would still remain
intact elsewhere on the peninsula. When the traditional owners of this
site, the Jaburarra, learned they couldn't stop the bulldozers, they asked
that, at the very least, the rock art not be crushed for use as foundation
material. Might the rocks, instead, be moved to a kind of "graveyard,"
to be surrounded by a cyclone fence, so that the artwork might in this
way be cared for? They were told their wishes would be accommodated.

On a peninsula jutting out into the Indian Ocean, a place where the first white settlers repeatedly poisoned the Jaburrara's water holes with arsenic and where, when that didn't kill enough of them, they just started shooting the people, developers broke down 25,000 years' worth of rock art and dumped it like so much construction debris in a single spot, which they surrounded with a cyclone fence. Like a quarantine station. The Jaburrara were left to sort the jumble out any way they could. The flayed walls of the caves at Lascaux and Chauvet, dumped in a barrow ditch.

The Jaburrara sorted it out. Some of the pieces were too large for them to reposition without heavy equipment, which they did not have access to. Images that were not meant to be seen by Jaburrara women, or not meant to be seen by male Jaburrara, were placed facedown. Images of lovemaking, which the construction crews ridiculed and lampooned as "Abo porn," were wrestled over so they could not be seen. Images of beings who should not be forced to stare at the nitrate plant were turned to face elsewhere.

Our party was escorted by two local guides, one Aboriginal, one white. Sparks flew between them periodically, for example when the white man began interpreting the meaning behind some of the glyphs and the Aboriginal man, very disturbed by this, all but shouted at him, "They can't be *interpreted*!" In the tension that persisted between the two throughout the afternoon, we saw the age-old collision of "scholarship," on the one hand, and deference to mystery on the other. Without room for mystery and uncertainty, the Aboriginal man felt, there cannot be any truly intelligent conversation.

Paul wanted to climb a nearby rise to photograph the rock art graveyard from that height, and I accompanied him to gain the same perspective. "Mind the adders," cautioned one of the guides, reminding us the loose rock and heavy grass here was a preferred habitat for *Acanthophis antarcticus,* the common death adder.

The cyclone fence around the catastrophic jumble of rock art sagged in several places where local white children, we were told, had scaled it to get inside the compound, in order to see the "dirty" Aboriginal art. Perhaps these were the same local children who had scaled the rock faces we'd clambered over that morning to paint their names and dates among the glyphs.

———

THE DAY FOLLOWING, we drove east through Roebourne, where we visited with Aboriginal artists at a center run by FORM. Many of them were working in the well-known dot style peculiar to generic Aboriginal art; but others were experimenting, painting in different, mostly modern styles. Some paintings were so accomplished, and so evidently *of* the landscape we'd been traveling through for several days, they were spellbinding. A few appeared to be from the artist's memory of a place that had once existed here, before roadways, industrial infrastructure, and permanent settlements had altered the view.

From Roebourne we proceeded to Port Hedland. In terms of tonnage of material exported, Port Hedland qualifies as the largest port in Australia. As we approached on the North West Coastal Highway, Mark, staring at it, spoke one word quietly: Mordor.

It did look like some version of Tolkien's hellish fortress. In suspension above the town was an enormous cloud of orange dust, rising up from machinery that inverted ore cars to empty them onto one end of a conveyor belt. At the other end of the conveyor belt, the dust rose up again from the bulk loading of ore ships. Off to our right as we entered the town were gargantuan piles of white salt from a desalinization plant. The ground on either side of the road approaching Port Hedland was denuded and scarred by errant vehicles in a headlong rush, each driver bent on an errand of some sort, the lot of them raising a second layer of dust into the air. In my hotel room a placard warned not to drink the tap water, which is contaminated with heavy metals, and for the same reason, to limit myself to a single brief shower once a week.

Mark's image of Mordor—the natural end, as Tolkien saw it, of the road to worldwide industrialization—was a bit of a stretch for me; but Tolkien's evocation of the heathen brutality inherent in industrial development, and of the tyrannical rule of the psychopath Sauron, a figure of puerile greed and the pursuit of power for its own sake, was a characterization of scenes like this that I could not shake. Mordor is one of the most dehumanized landscapes in all of English literature. Once you've slipped behind the curtain in Port Hedland—the well-kept lawns, the fine restaurant, the comfortable motel we stayed in—the town seemed not only to be the future that many dread but also to represent the marriage of ruthless desire and short-term gain that has laid waste to villages all over the world.

The seductive power of this system of exploitation—tearing things out of the earth, sneering at the least objection, as though it were hopelessly unenlightened, characterizing other people as vermin in the struggle for market share, navigating without an ethical compass—traps people in a thousand exploited settlements in denial, in regret, in loneliness. If you empathize with the Jaburrara over their losses, you must sympathize with every person caught up in the undertow of this nightmare, this delusion that a for-profit life is the only reasonable calling for a modern individual.

The last evening we were together on our trip, our FORM escorts arranged for us to tour the Port Hedland harbor with the harbor master. Home construction in Port Hedland, we had already learned, is not able to keep pace with the expanding workforce needed here, where every hour of every day is a work hour. The harbor master is going to show us areas of the port, we're told, that are closed to other visitors. (An underwater fence in the inner harbor protects workers from sharks and poisonous sea snakes; on days when the heat here is stifling, the men jump into the water to cool off.)

Iron ore, the "red gold" of Western Australia, is the state's—and the country's—most lucrative export. The Pilbara iron deposits were discovered in 1952. In 2009, BHP Billiton and Rio Tinto, another giant of worldwide industrial mining, shipped 330 million tons of iron ore from this port, most of it to China. Despite the magnitude of this project, relatively few people are employed in the loading operations. Automated machinery and software programming, which create efficiency, substitute for a labor force. (During our two-hour tour of the harbor, I see only one person, a deckhand smoking a cigarette and watching us from the shadows of his ship's superstructure.)

The ore arrives from the mines on the south side of Port Hedland. The load from each car moves several miles on conveyor belts to the docks. A gantry system there directs the ore into the ships' holds at the rate of 140 tons per minute. Each vessel loads 200,000 tons or more in about twenty-four hours. As we motor through the harbor, six ships are being loaded and five wait at anchor in the outer harbor. The average draft of an empty ore ship is twenty-five feet; of a fully loaded ship, fifty-five feet. In the process of being loaded while we cruise the harbor are the *KWK Exemplar* from Hong Kong, the *Mineral Shikoku,* the *Silver Bell* from South Korea, the *Spar Leo* from Norway, the *Onga,* registered in Panama but, like many of the others, headed for Asia. Emblazoned across the headwall of the *Mariloula's* superstructure, in letters nine feet tall, are the words PROTECT THE ENVIRONMENT.

THE TOUR THROUGH the Pilbara has been tempering for me. Like many others, I hope to stay abreast of the scale of economic and social change in the world, and to be aware of the rate at which things are changing; but the scale and the speed of the changes is frequently beyond anyone's grasp. Too much of what we expect to see appearing on the horizon with sufficient time to take preventive action has already become a part of our lives, entrenched before we notice that anything has happened.

When we returned to our rooms after the tour of the harbor, I sat out on the veranda that my room shared with several other rooms and watched the last of the sun's light lose color on the ocean in the direction of the Burrup Peninsula. My appreciation of rock art aesthetics is not great, but the art vividly represents for me the long effort humans have made to understand the world they were living in at the time. Most of the petroglyphs and pictographs that I've seen evince, for me, both a sense of wonder about the nature of the world and, more subtly, an understanding that human beings do not control their own fate, that in some fundamental way humans are powerless to do so. From this, perhaps, springs both the notion of the existence of influential gods, beings to whom one can appeal, and a contrary notion, that a person is able to control his or her own destiny and, in some cases, the destiny of others.

Many people regard rock art as "primitive," meaning the techniques are not refined and the ideas behind the art are unsophisticated. From there it's a short step to believing that "primitive" man's sense that his powers were limited and his fate beyond his control are fears made obsolete by sophisticated technologies. The vast majority of people in the world, however, find every day, sometimes in harrowing ways, that their fate is not theirs to control. A relative handful of people, primarily in business, engineer the plans for social and economic change that determine the fate of most ordinary people.

In my experience, even the most decent people in positions of power believe that ultimately they know best, that their experience, education, intuition, and instincts have made them authoritative. I am forced to object, with my memories of the slums of Jaipur and São Paulo, of ravaged landscapes in the Texas oil fields around Midland or of carbonized air in Beijing, and the sea ice gone from Arctic seas in late summer, to say that perhaps they do not know best.

Once, years before I traveled to the Pilbara, I had the opportunity to accompany a guide through the Paleolithic art galleries at Altamira, a cave in northern Spain. For half a day I had these galleries to myself, time to linger with Magdalenian Cro-Magnon imaginations. When I emerged from this prolonged encounter with their work, I was so acutely aware of the humanity of these people—their capacity for courage, for love, for innovation, for amazement; their ability to provide for one another here on the Cantabrian coast, 14,000 years ago—so awed I lost my orientation in time. From the edge of a cliff just past the entrance to the cave, I stared down into the gardens and stock pens of domesticated animals that belonged to those living there in two-story stucco houses, a semi-rural neighborhood of Santillana del Mar. Between the lives of the artists working in the cave and these people with their small plots of land—the rows of corn, chicken yards, milch goats, grape arbors, and fruit trees—there seemed to be no time at all. I indulged myself with the thought that both groups wanted the same—a feeling of allegiance with one another, relationships that were just, openness to the numinous character of their worlds, from time to time a quickened heart, and the ability to give as well as receive love.

These thoughts about Altamira, which came to me on the veranda that evening, gazing out toward the ruined galleries of the Burrup Peninsula, led me to think of my stepfamily. Their ancestral home sat on a hill above the town of Cudillero, a hundred and twenty miles up the coast from Altamira, to the west. Their casa del Indio, with its formal chapel, its extensive arbors and gardens secured behind high walls and a massive set of gates, always spoke to me of inordinate wealth and aloofness. I did not feel superior to those people and their families. Their God approved of their murdering los indios in the Americas, and their wealth made them seem good in the eyes of those who believed as they did; but I wanted a different path, one less violent, less indifferent. As the decades passed for me, I began to think that the path many of us now share, a path of self-realization and self-aggrandizement, might eventually leave us stranded, having arrived at the end of exploitation, but with most of us standing there empty-handed. And what is it that we have found through the injustice of exploitation that these Magdalenians at Altamira did not already possess?

WHILE WE WERE VISITING the Aboriginal center at Roebourne, I met a woman named Loreen Samson, a thirty-seven-year-old Aboriginal artist and perhaps the most accomplished painter at the center. She had faced a certain amount of personal hardship earlier in life, I was told, but her focus was now elsewhere—on the goals FORM helped her set for herself. She told me she wanted to use her creative energies to open people's imaginations to better ways of responding to conditions in the Pilbara. She did not want more despair or anger, or to encourage the feeling that the people here were trapped. She taught art as a way of seeing, of being in the world. She was an instructor at both the Roebourne art center and at the Roebourne Regional Prison, where between 90 and 95 percent of the male inmates are Aboriginal.

Loreen had prepared written statements to accompany eight or ten of her landscapes, some of which hung on the walls at the art center. They were not explanations of the work but continuations of the paintings, set out in idiosyncratically punctuated sentences with misspelled words, her syntax the syntax of spoken rather than written English. Bill Fox and I sat in an annex of the studio where Loreen and four or five other Aboriginal women were working, reading half a dozen of her statements, handing them back and forth to each other.

She wrote in one statement:

> Every day I see my Dream Times grounds are drifting away from me. I cry out too stop[,] you are taking away the heart of my people. This lands is what we have[,] the grounds of knowledge for you and me too learn each other about culture. Our children would come and teach their kids of the great lands of knowledge. It hurt me of what I see now[.] We don't have any solid ground to stand on just the tears that come down from my eyes[,] of shame of what our people have done for the prices of dollars. Look at this land[.] Now trains go by night and day with the richness of this wise old country [. . .] Look around[,] that what have mining done to it will hurt people [who] shouldn't [have] aloud mining [to] destroy the land of [the] knowledge that our ancestor put on rock art many years ago.

THE EVIL BEHIND colonial subjugations, from Portugal's Brazil to France's Algeria, has generated a wealth of contemporary anti-colonial writing, criticizing the quests for material wealth and economic control. In the New World, the criticism might be said to have begun with Bartolomé de las Casas and to have come forward to Eduardo Galeano and writers like Jhumpa Lahiri. The centuries of moral and ethical outrage over the colonial politics of race, over ethnic and national exceptionalism, have recently produced eloquent critiques of cultural ignorance and indifference to human life, both of which lie at the core of colonial expansion. These critiques of exploitation and profiteering have continually been deemphasized as international concerns because their logic makes people in power uncomfortable.

As much as anti-colonial literature looks back on colonial history's injustices, it also looks forward today to what is arguably a far more important question: What are we to do? Or put another way: What is going to happen to us as temperatures change, as our numbers climb toward eight billion, as the Pacific becomes more acidic, and as more freshwater goes into making jet fuel from Canada's tar sands?

While "turbocharged" capitalism is routinely singled out as the villainous cause of so much of what is socially and environmentally evil, eliminating hypercapitalism doesn't seem to answer the central question, which is: Why do we harm one another so grievously? Or to phrase this differently, What is the root of our fundamental disagreement, now that the size of our population and the scarcity of essential supplies like uncontaminated freshwater have come into play?

As I read anti-colonial literature from different parts of the world, in English and in translation, what I've come to feel is that a disagreement over which path leads to the more desirable future is a disagreement about the place of empathy in public and corporate life. On the one hand are the ideals, unfortunately, of capitalism: progress, profitability, ownership, control of the marketplace, consumption. On the other hand are the ideals not of a system of economics but a system of social organization best represented in modern times by a group of flawed individuals who nevertheless became the iconic representatives of tolerance, respect for beauty, a preference for reconciliation over warfare, and compassion: the Reverend Martin Luther King Jr., the 14th Dalai Lama, Mahatma Gandhi, archbishop emeritus Desmond Tutu, Nelson Mandela, Mother Teresa, Archbishop Óscar Romero.

The question the latter group consistently addressed was, Why is there so much suffering in the world? Each person in their way tried to do something about suffering created by intolerance, injustice, and ethnic and national exceptionalism. This has always been a hard road to keep. It is not that most people don't support such ideals but that the implementation of the ideals is so extremely difficult that it makes people cynical. Reflecting on the social, economic, and environmental harm that fracking engenders, or on Russia's aggressive efforts to regain its stature as a world power, one must consider whether allowing human misery to develop further in order to gain some sort of short-term economic or political advantage isn't an incurable, systematic problem. Perhaps the actual source of humanity's trouble is genetic. Meanwhile, attempts to address these questions continue to be ridiculed, held in suspicion, or patronizingly dismissed by many people who have the power to make a major difference in the way disenfranchised people live.

Apparently it's regrettable but finally all right to let thousands starve in order to ensure that a few have the yachts they require. Apparently it's all right for thousands to die of lung cancer and for tobacco companies to withhold the evidence that would incriminate them, as long as the companies can show a profit. Apparently it's all right for China to dam a tributary of the Brahmaputra River and endanger the flow of freshwater to Bangladesh if this will help develop a wealthy middle class in China.

WHEN I FIRST began traveling in Australia in the 1980s—not that Australia stood at the nexus of any particular world problem then; it was just that at this time I began to see more clearly the outline of what disturbed me—I was trying to understand a question about human fate. With the horsemen of a coming apocalypse so obviously milling on the horizon, riding high-strung horses, why was there so little effort to bring other ways of knowing—fresh metaphors—to the table? Why was it that, for the most part, it was only the well-groomed, the formally educated, the economically solvent, the white, the well-connected that were invited to sit at the tables of decision, where the fate of so many will be decided? Why, for example, were there no elders invited from indigenous traditions like the Sami, the Mapuche, the Onondaga, the Iñupiat, the Nuer, the Kuku Yalanji? These were people who valued wisdom as much as or more than intelligence, whose traditional concerns lay with the fate of the group, not the self. Were they not worldly enough? Were they too obscure?

I went to Darwin once, a coastal city in Australia's Northern Territory, to listen to an assembly of anthropologists present papers, mostly to one another. These men and women were studying the last few remnant hunting and gathering cultures in the world—Hadza in Tanzania, for example, and Inughuit in Western Greenland. They get together regularly to share what they know. (The more humble say what they *think* they know.) Over a period of three days I listened to about sixty presentations. When I left, I felt informed by many specificities, and warned again of how perilous it can be to try to collapse any single complex idea, like feeding the members of your family, into one sentence that will serve everyone.

Among those I met at the international meetings in Darwin were a few Australian anthropologists with whom I later became friends. Eventually I returned to the Northern Territory to travel with them. They introduced me to Warlpiri, Arrernte, Pintupi, Luritja, and Pitjantjatjara peoples and their not-entirely-congruent ideas about the world and reality. I appreciated and savored the experiences we shared, but I never felt, as I'd never felt among Eskimo people or Native Americans on my home continent, that I wanted to trade places. What I wanted to understand, really, was what they might know that would be of use to my own people, whom I saw traveling very fast on a spavined road. I was looking for anyone who could speak the language of the god of no particular religion.

I returned to Australia a year later to dive on the Great Barrier Reef, to immerse myself in the mostly benign waters there, the blazing colors of tropical fish, the transparency of the water, and the expanse of the coral reefs, in order to remind myself that no matter how steep the spiral of despair might become, beauty without design, without restraint, was all around. You only had to step into these realms—Alexandra Fjord lowland, Roca Redonda, Victoria Falls in the moonlight—to remind yourself of the possibilities, of the things you too rarely thought of. Then you go back to the world of people who are not able to travel to Lake Clifton, to watch unsuspecting caribou through the clear air of a High Arctic summer evening, to behold the panoply of life flickering, like thousands of small colored lights, in an anodyne basin of tropical water.

You feel while you are witnessing such things that you must carry some of this home. That what you've found are not *your* things but *our* things.

WHEN I RECEIVED an invitation to speak at a literary festival in Hobart and was told that my escort for those days would be a Mr. Peter Hay, I began reading Mr. Hay's poetry. I admired the integrity of his lines and was charmed by some of his Tasmanian locutions. When I reached him by phone and asked about going to Port Arthur together, I told him I was trying to understand why those in power so often seek to humiliate and punish those who disagree with them. Or perhaps I only wanted to look at the evidence of an episode in Western history that's too little known in my country. And to consider how class distinctions, the harsh divide between the rich and the poor, for example, lead people to believe that "solutions" like Port Arthur are both just and sane. He said whatever I might make of Port Arthur, grasping the essence of the place would help me greatly in understanding what it means to be Australian.

I could never hope to understand what it meant to be Australian, but appreciated the latitude he seemed to be giving me to think about it. To be freighted off to Port Arthur in the eighteenth century was more than to be banished. It was to be seriously punished—and, it was hoped, to thereby be reformed. Parliament believed it could create at Port Arthur a valued and self-sustaining commercial trading enterprise, and that the penal colony could be profitably operated. It proposed reforming the "dregs" of its own society to run the place, and to achieve that reform through public whippings, long periods of solitary confinement, hard labor, and instruction in Christian living. The notion, of course, was daft, but Parliament thought it could work. Whatever happened, the experiment would cost Parliament very little, if you just ran the numbers with the right attitude and kept the details quiet.

This is an old story. It's unfolded in many places, under many guises.

So one April morning I went with Pete Hay to Port Arthur to take in the aftermath of the experiment. We drove down from Hobart to Eaglehawk Neck, a narrow isthmus that connects the Forestier Peninsula to the Tasman Peninsula. From there it was just a matter of twelve miles to the historic site.

Eaglehawk Neck gives you a good sense and a quick read of the place. In the years the colony was in operation, the authorities maintained a picket line of dogs here to attack and savage any convict attempting to escape to the north. (Fewer than half the convicts at Port Arthur were held in cells. The others, encouraged to ascend the ladder of "good attitude and good works," were permitted to roam freely over the peninsula, logging trees, digging coal, building a railroad, constructing buildings, and competing with one another through their endeavors for privileged positions in the penal hierarchy.) Writing in 1837, the artist Harden Melville said of the picket-line dogs, "The white, the brindle[d], the grey, and the grisley, the rough and the smooth, the crop-eared and the long-eared, the gaunt and the grim [stood chained to wood platforms erected across the isthmus and its salt marshes]. Every four-footed, black fanger individual among them would have taken first prize for ugliness and ferocity at any show."

Mostly ill-tempered to start with, the Eaglehawk dogs were spaced so that two dogs might attack a man together but not get their chains crossed. Their handlers abused them to keep them vicious and edgy, depriving them of food, water, and affection, no doubt making them all the crazier by this treatment. When they wore out, they were shot.

Prisoners confined at Port Arthur were generally held to be "ignorant, stupid, revengeful, hardened, sullen, cunning, thievish, restless, disobedient, and idle." To improve the character of each man, to make him "clever, informed, cheerful, contented, simple, cleanly, obedient, industrious, and faithful," the prison staff implemented a plan for the reform of its incarcerated prisoners based on a system employed at Britain's Pentonville prison, in London. Each inmate in Port Arthur's Model Prison was isolated in solitary confinement. They were not allowed to speak, to make eye contact with the guards or, indeed, to make any audible sound. All prisoners and guards were required to wear slippers to muffle their footsteps. Attendance at religious services was mandatory, with each prisoner quartered in his own small closet within the chapel. One hour of daily exercise was permitted, but prisoners were roped together in the prison yard in such a way that no prisoner could draw closer than five yards to another. All inmates were required to wear visored hats, the visor to be lowered upon leaving the cell, to prevent anyone's seeing anyone else's face. It was believed that anonymity, depersonalization, and silence would provide the right environment for the "divine spark" of enlightenment to ignite in each man the desire to see and accept the road to personal reformation.

Historians believe the decision later to build an asylum on the prison grounds came in response to the number of inmates driven to insanity by residence in the Model Prison. At the end of its years of official usefulness, in 1877, Port Arthur housed mostly the mentally ill, paupers, and the indigent, men who had served their time but who had no family and nowhere else to go.

The question that begs to be addressed at Port Arthur is not about the reasons for the sentence of banishment from England—for the psychopathic child rapist, for the too-assertive Quaker, for the nine-year-old petty thief—but the reasons for so severely *punishing* these people.

In eighteenth-century England, banishment was a tool of large-scale social engineering. The "nether region" to which the Crown first sent its banished was the American colonies; more than 50,000 prisoners were transported there before the American Revolution ended this option. Prisoners were then warehoused in derelict ships moored in the Thames. When the English courts ran out of space there, the transportation system was inaugurated, which initially transferred prisoners to the southeast coast of Australia. The First Fleet, carrying 759 prisoners, sailed from Portsmouth for Port Jackson, just north of Botany Bay, in 1787. (By the 1860s, Great Britain was operating thirty-five such penal colonies, from Corradino in Malta to Glendairy in Barbados to Port Arthur in Tasmania.)

The overarching idea behind promoting the transportation system was that these penal outposts might function as self-sustaining colonies and provide some income to the Crown. While the incorrigible were to be isolated there in actual penitentiary buildings, and while these settlements were expected to accommodate experiments in reformation like the Model Prison, they were also meant to serve a strategic purpose. Prisoners were to be taught trades—blacksmithing, coopering, printing, tailoring, carpentry, and husbandry. They were to sink wells for water, raise their own food, and build a transportation infrastructure on the peninsula. Selected prisoners, both female and male, were to be transported with their families; others were instructed to marry after serving their terms, to have children, and to take up in these regions the duties of the servant and working classes. They were also expected to provide assistance to a separate group of volunteer emigrants, people of "good character" and from the educated classes, who were to be employed as administrators in the penal colonies, developing lumbering and mining operations, constructing shipping facilities, and establishing an export infrastructure. Together, reformed convicts and volunteer emigrants were to extend civilization into the outlands surrounding the penal colony and to cultivate and improve those lands.

While this stratagem struck many as an ingenious and efficient way to establish civilized life in a "terra nullius" like Australia (land that, legally speaking, belonged to no one), it couldn't be expected that all prisoners would grasp the wisdom behind the plan. Enlightening them would in all likelihood require a certain amount of discipline, to ensure conformity with the vision. Britain drew on yet another class of citizens, men with a taste for administering the necessary beatings, scourgings, and humiliations, to take on this task.

The transportation plan was arrogant, immoral, and foolish, and its many injustices sparked nearly constant rebellion in the prison system. One reason some Australians identify today with well-known rebels in the transport prisons, many of whom were in fact criminals, was that these individuals attempted to defy a brutal system of class distinctions. They attacked the presumptions of those who chose to assign every person a state in life, to dictate the shape of each person's future. Although the settings and circumstances were different, the prisoners' yearning to be free of these presumptions was fueled by the same feelings of outrage that precipitated both the American and the French Revolutions.

Among the most remarkable—and for some, heroic—of those in the prison system who refused to be yoked was a "clever, informed, industrious" woman named Mary Bryant. Her defiance and her determination to escape along with her family from the grip of those who sought to make her submit, symbolize, for many modern-day Australians, the will to establish one's own way in an unjust world.

MARY BRYANT (NÉE BROAD) was convicted of assaulting an older woman and taking her purse. Initially sentenced to death by hanging, she was subsequently transported to Port Jackson, where she met William Bryant, a fellow convict and her future husband. The living conditions for everyone at Port Jackson—prisoners, guards, immigrant British citizens, and assorted functionaries and administrators—were miserable. (In making his recommendation for the southeast coast of Australia as a site for a transport prison, Sir Joseph Banks, perhaps waxing nostalgic, gave Parliament the impression that the country was well watered and suited to growing crops, which it was not.) The Bryants and their two young children, Emanuel and Charlotte, experienced the worst of conditions there, with rationed food, insufficient water, and the burden of onerous labor. On the night of September 26, 1790, five convicts unaffiliated with the Bryants managed to steal a boat and equip it for travel. They were caught the same night attempting to put to sea and were savagely punished in front of the other prisoners. Their effort resonated, however, with the Bryants and with a few of the Bryants' friends.

The authorities suspected the Bryants might be planning some sort of escape, but for unknown reasons, no one kept very close watch on the couple. Will Bryant succeeded in stealing a coastal chart (based on Cook's 2,000-mile survey from 1770) and was able to acquire a compass. Mary secretly stockpiled food and water and, with her husband, began inviting a few convicts with crewing skills and navigation experience to join them.

On a moonless evening in March 1791, members of the group stole a cutter belonging to the governor of the colony. It had recently been rigged with a new mast and sails, and fitted with six new oars. The historian C. H. Currey writes, "The cutter breasted the Pacific in the early hours of Tuesday, 29th March 1791," and then, very quickly, they were gone—the Bryants, their two children, and seven other convicts, some serving life sentences. After a harrowing journey up the east coast, facing storms and shortages of food and water, and after clearing the Great Barrier Reef and doubling Cape York, they sailed 1,200 miles across the little-known Arafura Sea to arrive, sixty-nine days later, in Kupang, a relatively large city and Dutch entrepôt at the west end of Timor. According to Currey, their escape and 3,254-mile journey was "a masterpiece of organization and cooperative effort."

The Bryants were inspired to make for Kupang by a group of Englishmen who'd sailed for there from Port Jackson two years previously—and who, they'd heard, had made it: Lieutenant William Bligh and the eighteen crewmen who had remained loyal to him after mutineers forced Bligh to relinquish command of his ship, HMS *Bounty,* in the Tongan Islands. When Bligh, who had a terrific temper, reached England, he had a ship, HMS *Pandora,* under the command of Edward Edwards sent out to track down the mutineers. Sixteen of them had disembarked from the *Bounty* in Tahiti and had made homes there while their leader, Fletcher Christian, had sailed away with eight others. The crew of the *Pandora* captured fourteen of the sixteen on Tahiti but could not find the other two. On the return voyage to England, the *Pandora* had its hull torn open by coral spikes as it attempted to cross the Great Barrier Reef. Most of the mutineers, caged and shackled in the ship's hold, would have perished had it not been for the efforts of the *Pandora*'s crew, who defied their captain's order to let the men drown. (Despite the crew's efforts, four of them did drown.)

After the ship sank, its officers, the survivors among its crew, and the remaining prisoners made their own remarkable journey across the Arafura Sea in four lifeboats. When Edwards arrived in Kupang and encountered the Bryants, he became suspicious. He informed the authorities that the Bryants and their friends were not the shipwrecked survivors that they claimed to be but escaped convicts. Determined to bring them to justice, along with the *Bounty*'s mutineers, Edwards hired a Dutch vessel, the *Rembang*, and on October 5, 1791, sailed for England with them, the mutineers, and those of his crew who'd survived the sinking of the *Pandora*. En route to Jakarta (Batavia at the time), the *Rembang* nearly foundered in a typhoon. Had it not been for the courage and skill of the Australian convicts, Currey contends, the ship would have sunk. The heat, humidity, and pestiferous conditions at Jakarta—malaria, dysentery, typhus—had taken the lives of four of the *Bounty* mutineers and four of the Port Jackson convicts, including Will Bryant and the Bryants' son, Emanuel, not yet two years old.

Edwards, abandoning the damaged *Rembang*, put those who remained aboard four other ships and continued his journey across the Indian Ocean for the Cape of Good Hope, at the southern tip of Africa. One of the four, the *Horssen*, with Mary and her daughter aboard, encountered another typhoon, and another of the convicts was swept into the sea. Arriving in Cape Town on March 18, 1792, Mary, her daughter, Charlotte, not yet six, and the four surviving convicts were transferred to HMS *Gorgon*. According to Watkin Tench, a historian who sailed with Mary Bryant and the others from Batavia, Mary was greatly, but discreetly, admired by passengers aboard the *Gorgon*. She was twenty-seven at the time. "I never looked at [her] without pity and astonishment," wrote Tench. "They had miscarried in a heroic struggle for liberty; after having combated every hardship, and conquered every difficulty." The six of them were regarded, wrote Tench, as people who had manufactured their own dignity. He thought they should be set free once in England, not forced to stand trial as escaped convicts and sent back to Port Jackson.

The punishment for escape from a transport prison, Tench knew, was imprisonment for life. Mary Bryant, whose daughter died shortly after the *Gorgon* left Cape Town, was eventually set free by the court. Numerous people, including the biographer James Boswell, supported her release, citing her courage, the inhumanity of the transportation system, and her personal bearing.

ONE OF THE REASONS Port Arthur stands out so prominently among the images many Australians have of themselves, I believe, and why a rebellious figure like Mary Bryant is appealing to so many is that a significant number of Australians have ambivalent feelings about their country's origins, lying as they do with the penal colonies and, later, with the violent usurpation of Aboriginal lands. Some Australians would just as soon Port Arthur remained obscure in the national memory, and that any mention of one's "convict ancestors" be left to Australia's working class to ponder. Just as many, probably, appreciate knowing the truth about the transportation system and about the lethal violence directed, historically, against Aboriginals. One occasionally hears in Australia comparisons made with Native American genocide in the founding history of the United States. Americans, generally, are loath to acknowledge what invading whites did, and caused to be done, to indigenous people in North America. And are equally as uncomfortable discussing the early decades of their country's slave-based economy, or the treatment of the thousands of indentured servants England shipped to America to assist in the labor of establishing the new colony.

It's harrowing business, trying to sort out the foundations of any nation. Bigotry, genocide, violence, and greed always emerge to play major roles in the narrative, and assigning ethical responsibility for acts that took hundreds or thousands or millions of human lives is always divisive. But without the effort, all nations eventually founder. The decision to stick with civil war, righteousness, denial, and exploitation keeps the door open for tyrants to rule and loyal citizens to become refugees.

The fact that an outsider can hardly detect any regional accents in Australia makes it easy to assume that the country's populace (forgetting, for the moment, Aboriginal people) is in some way seamlessly united. Two significant extremes, however, are apparent in the general population. One prefers to cling to what is essentially English; the other wants to discover a purely Australian destiny, the way Revolutionary era Americans wanted to discover a uniquely American destiny. The former would prefer to avoid the disruption that comes with having to examine the treatment of Aboriginals in the past; the other wants the injustices addressed. A similar division is apparent in the American population, where blacks and Native Americans are concerned. (Interestingly, the majority of Aboriginals in the one instance and of Native Americans and blacks in the other with whom I've spoken would be pleased if the injustices were simply widely acknowledged, and if the persistent barriers to equality were systemically dismantled.)

America revolted successfully against its parent country, declaring its opposition to colonial impositions of any sort and enshrining a "melting pot" folklore that, while it claimed to welcome the oppressed, remained suspicious about and resistant to diversity. And America, the most successful of England's former colonies, went on to become a formidable colonizer itself, imposing its system of political organization and its policies for economic growth on other nations, to the point of authorizing assassinations and supporting juntas and coups that agreed not to interfere with the international operations of American corporations. At the same time, America also ignored institutionalized social injustice around the world, like apartheid, and strong-arm dictators like Suharto and Syngman Rhee, if raising an objection might create significant economic tension or disruption.

Australia, like Canada, has yet to decide how to expend its revolutionary energy. The courage to politically confront what most of the world's peoples consider wicked—the appropriation of other people's lands, the exploitation of human beings for profit, the enforcement of policies that perpetuate economic servitude and promote cultural or racial superiority—is Australia's step to take. It's arrogant of course to suggest this. I mean the observation only as a respectful salute to the citizens of a sibling country who have said similar things to me, for example, after Prime Minister Kevin Rudd's now-famous apology to Aboriginal peoples. When, they asked me, would such a thing happen in the States?

THE DAY I VISITED Port Arthur, Pete parked his car in what was then called the upper car park, a short distance from the lower car park, in which a yellow Volvo 244 sedan stood. A white surfboard was mounted upside down on the left side of a roof rack and the kind of cartoonish stickers one associates with young children adhered to the left rear side window.

After our tour of the grounds we crossed the cricket pitch again and entered the Broad Arrow Café for lunch. (The image of a broad arrowhead on prison equipment and clothing was a sign of Crown ownership.) I found some materials to read later—Marcus Clarke's famous 1874 novel, *For the Term of His Natural Life,* was available—but I couldn't find a decent map, and after we ate, we departed. Pete wanted to use a restroom at the edge of the lower car park. Physically and emotionally drained, I started to lean back against one of the cars there to wait for him. Almost instantly I sprang erect again, as if recoiling from the vehicle. The feeling that I had actually been pushed away from the car was so strong, so strange a sensation, that I turned around to study the vehicle—the surfboard, the small decals in the window, the Tasmanian license plate, CG 2835—as if one of these details might explain what I'd felt. I was still standing there when Pete walked up. I told him what had happened.

He gave me a friendly smile, said, "Sure, mate," and we walked on to his car.

A few weeks later, back home in Oregon, I was walking up the driveway to my house with the day's paper when I stopped to read a story about a massacre that had taken place the day before in Australia. A gunman named Martin Bryant had shot and killed thirty-five people at Port Arthur, twelve of them inside the Broad Arrow Café. He'd wounded nineteen more, many critically, and had been arrested by local police when he came running out of a house he'd set fire to near the entrance to the historic site. As I quickly scanned the following paragraphs my eye fell on a sentence stating that during the melee Bryant had "abandoned his yellow Volvo sedan."

I called Pete immediately. He remembered the moment. A few minutes after I hung up I got a call from the Tasmanian state police. I faxed them a copy of the article in my American newspaper, so they could see precisely what information the story had provided me with, and then told the officer who called, a woman, additional details about the car that I recalled. She said the car Pete and I had seen was undoubtedly Bryant's. She speculated that he might have been there that day to determine the most effective way to kill a lot of people quickly. (The attack was phenomenally lethal. From the moment he fired the first bullet from his AR-15 style semiautomatic rifle inside the small café until he stopped momentarily, seventeen seconds later, Bryant killed twelve people and wounded ten.)

I later learned, during his trial, that Bryant had considered two other locations before settling on Port Arthur. One was the ferry from Lauceston, Tasmania, to Melbourne, Victoria, across the 225-mile-wide Bass Strait. His plan had been to kill everyone aboard the ferry but the pilot, whom he would kill as they were docking. The other venue Bryant considered was the grounds of the annual Salamanca festival for writers and artists in Hobart, an international event that draws a large weekend crowd and to which I had been invited that year.

Pete reminded me, when I talked with him about our having seen Bryant's car there that day, that many Australians had come to believe that the Port Arthur prisoners—especially the ancestors of some of them—hadn't really been bad people. This was a widespread type of denial, he said, of the real nature of Port Arthur. And he cautioned me that Bryant's rampage should not be taken as Bryant's comment on what many Australians refer to as the "hated stain," the nation's convict history. He was just killing people.

This was only one more example, Pete and I agreed, of how some angry or unstable people express their distress. Around the same time as the incident at Port Arthur, an Indonesian soldier killed nineteen people and wounded thirteen at an airport in Timika. And a former scoutmaster, Thomas Hamilton, shot and killed sixteen children, their teacher and then himself at a schoolhouse in Scotland.[1]

SEVEN MONTHS AFTER the massacre, Bryant was sentenced to prison "for the term of his natural life." The Broad Arrow Café was torn down and the prison grounds were reconfigured. The local community would spend the ensuing years sorting out their emotional response to Bryant's descent on Port Arthur. There would be much speculation about why he murdered so methodically and relentlessly, following two small children around the base of a tree, for example, until he'd shot them dead next to their dead mother.

Bryant's name came up for me during dinner once with an Australian friend visiting in Oregon. She was from Tasmania and told me she'd actually taught Bryant in a special class for children with learning disabilities, in Hobart, along with four other boys. All five, she said, were prone to violent behavior. According to her, two later killed themselves, and one of the other three, like Bryant, also committed murder. She characterized Bryant as obtuse and withdrawn, a brooder. He seemed always distracted, she said. And lonely. She believed he'd bought the surfboard—he didn't know how to surf—in order to join a group of surfers who'd rejected him. When he inherited some money, she told me, he used it to travel to California several times so he could visit Disneyland. By himself.

Her recollection of him as slow-witted and someone with limited social skills was later confirmed for me by a woman, a psychologist, who had employed Bryant as a yard worker for a while. She and her husband could finally no longer put up with Bryant's interminable mumbling monologues, she said, and with his wandering errantly all over their lawn with the riding mower. She did not dislike him, she wrote me in a letter, but could do nothing to help him. (Around the time of his trial, to judge from news reports, it was popularly believed that Bryant was suffering from Asperger's syndrome. A court-appointed psychiatrist agreed.)

The policemen who arrested Bryant when, clothes ablaze, he ran out of a bed-and-breakfast he'd set fire to—he'd killed two people inside and left a third behind, tied to a staircase—smothered his flaming clothing and got him airlifted out for treatment to a hospital in Hobart that was also the destination of eighteen of the people Bryant had wounded. Numerous death threats were made against the officers who tried to save Bryant instead of killing him, and many Tasmanians said during the trial that they would kill Bryant in the courtroom if they could get close enough.

The proper response to Bryant and the mayhem he was responsible for is grief. He is apparently unable either to comprehend the immorality of what he did or to understand his culpability. (He giggled in the docket during his sentencing and attempted to engage people in the courtroom to goof around with him.) To feel grief for everyone who was part of the tragedy. And admiration for those who managed to, or are still trying to, find their way out of the trauma. And gratitude to those who did not respond in violent ways but worked toward some semblance of a resurrected moral order in the midst of the hurricane Bryant unleashed.

Australia does not kill people convicted of capital crimes. Bryant, who has spent virtually all of his years since being convicted in solitary confinement, has several times tried to kill himself. If he had been killed for what he did, it is unlikely he would have understood why.

WHENEVER I RECALL Port Arthur, though I'll always carry images of the swift, indifferent violence of that April afternoon, I think most often of how beautiful the landscape there is, and how much of human endeavor, human endurance, is apparent in humanity's effort to rebuild in the wake of catastrophe. In the boreal fall of 2008, fearful that in my own comfortable life in the United States I had lost a sense of real human plight, I traveled to Lebanon to visit refugee camps and then on to Tajikistan, at that time the most impoverished of the old Soviet Republics; then to Afghanistan and finally to northern Sumatra, where people were still trying to put their lives back together after the Boxing Day tsunami of 2004.

The local people I was directed to for interviews in these places by my host, Mercy Corps, the international relief agency, were men and women with great poise, great compassion, great capacity for understanding. Each day they systematically addressed themselves to meeting their neighbors' basic human needs—physical safety, food and clean water, employment, and affection. These were people others deferred to naturally, because they knew them to be the most fully aware, the most trustworthy. And it was they who returned to me a sense of deep faith in the human capacity to overcome nearly every threat to the dignity and possibility of human life, in circumstances that call into question the ability of people to survive severe trauma emotionally and physically intact.

Despite the fact that none of them was, formally, a tribal person, and that some were still in their thirties, these were elders. Not surprisingly, none was officially affiliated with their national government's often-compromised plans to aggressively develop their country's commercial infrastructure. Instead, in developing jobs, they leaned strongly toward arranging small loans for enterprising individuals from not-for-profit sources, declining to link up with large-scale business and government infrastructure.

Standing there in my driveway that morning, reading about an unpredictable moment in the life of the psychopath Martin Bryant, I would have appreciated the counsel of an elder. How does one manage horror like this without cynicism, denial, or indifference? How does one not feel incapacitated by the inevitability of more Martin Bryants, more Stephen Paddocks (fifty-nine killed, more than five hundred wounded at the Route 91 Harvest music festival in Las Vegas), more Omar Mateens (forty-nine killed, more than fifty wounded at the Pulse nightclub in Orlando)?

It's little different from what happens when some militia's rocket tears through a house, burning, maiming, and decapitating. You get up off the floor, tend to the wounded, bury the dead, clear the debris, and start over again. You seek consolation with your neighbors, help them recover from *their* disaster, and discuss with them strategies to mollify the angry, the indignant, the headstrong, the self-important, and the righteous. You nurture the belief that this is not all we are.

At least this is what I heard men say one afternoon in a meeting hall in a northern Afghan village, Dūābī Ghōrband, when I asked about the Taliban. They spoke of their opposition to the Taliban, to militias of any kind, and of farming, and of the importance of caring for their children. What confounded me was that they were seemingly entirely unaware that others, all over the world, victimized by militias, thought as they did.

ONE EVENING I came across a brief article in the British journal *Nature* about the discovery of tiny zircon crystals in the Jack Hills of Western Australia. The crystals were dated to 4.27 billion years before the present. I immediately wrote to the two authors, asking if it might be possible to visit the site. I wanted to see the line and color of the place, find out how it was situated in the surrounding land, and of course learn from them how they found the crystals and which process they'd settled on to date them, and so on. I told them that, as chance had it, I was actually going to be flying to Perth, where they were both teaching at Curtin University, in the weeks ahead, en route from Zimbabwe to the Northern Territory. Could we get together?

I never heard back from them. Several years later, when I returned to Perth with the intention of exploring the Jack Hills, we did catch up. One of the authors told me he hadn't responded to my letter because "this is the sort of lunatic request you might get from an American."

His point made, he then went out of his way to help me. He drew a detailed map of the roadless area in the Jack Hills where he and the other scientists had worked. He showed me samples of the rock formation in which they had found the crystals, so I might recognize the terrain and geology when I got there. He arranged for me to stay with the manager of the sheep station on which the search area in the Jack Hills was located. He also insisted on paying for our lunch that day.

I flew up to Meekatharra from Perth, rented a four-wheel-drive vehicle, and drove west about 120 miles on an unsigned dirt track, arriving at the manager's house in the late afternoon. He and his daughter had a meat pie in the oven, and he wanted to know whether I took milk with my tea. They could not have been more accommodating.

That evening after supper I sat out on the veranda with the manager, taking our final cups of tea. When I asked, he set out for me some of the logistical problems he faces running a sheep station of this size (about 83 square miles) by himself, having to control feral animals grazing on the station as well as predators going after the lambs. He had to make sure in this dry country that the sheep had enough water. We ended up talking about how fortunate we'd each felt in the lives we'd chosen, and agreed that no matter what you did with your life, there was always more to know.

A stone walkway led from the veranda across a small, neatly trimmed lawn to a hip-high hog-wire fence and, beyond that, an open yard. His plane, not a car, was parked there, under the converging crowns of two massive gums.

ON THAT SECOND EFFORT to connect with the geologists at Curtin University, I flew from the States to Sydney and then took the *Indian Pacific* passenger train to Perth. It departed Sydney every few days, crossed the Blue Mountains in the Great Dividing Range, and then went on to Adelaide before starting across the Nullarbor Plain on the longest stretch of straight railway in the world—296 miles. On the far side of the Nullarbor, a treeless, semiarid, hardly inhabited landscape, lay the historic gold-mining town of Kalgoorlie and the hills of the Darling Range, before the train dropped into Perth at the East Perth Terminal. The journey from Sydney, situated in a maritime climate, takes nearly four days and crosses the drainages of two of Australia's three longest rivers, the Murray and the Darling, in pastoral country just west of the Great Dividing Range. The Nullarbor is casually referred to by Australians as "a desert," but Australia's truly formidable deserts—the Simpson, the Great Victoria, the Tanami, the Great Sandy—lie far to the north of the train line.

Before we left Sydney I asked the porter in my sleeping car if he thought I might ride in the train's locomotive for part of the journey. From there, I said to him, I'd have a view forward of the train as well as away to both sides, and I'd be able to talk to the engineers about their work. He said he didn't think so. Besides, he said, if I was riding up there in the locomotive, I might easily miss a meal.

I said I wasn't worried about the meals. He said I should then step back out onto the train platform and make my case directly to the engineers. I did, and they said it would be fine to ride along. They were going to make a brief stop for water a few miles out of Sydney. When they did, I should step out onto the platform like I was going to stretch my legs, then stroll forward and climb up into the cab. Which I did. As my need for sleep (and the occasional meal) allowed, and the spacing of train platforms that necessitated our stopping permitted, I spent most of the journey across Australia with teams of engineers in the cab of the locomotive.

That first night, when I returned to my own cabin at about three a.m., I found my supper sitting on my bedside table, carefully wrapped in aluminum foil to keep it warm. The following day, when I took the seat assigned to me in the dining car for the first time, my table companions greeted me cordially. Two elderly sisters, accompanied by their teenage niece. One of the women asked if I'd gotten on at Adelaide. I said no, that I'd actually gotten on in Sydney.

"Then you must be quite hungry," she said.

I said indeed I was. And the other sister remarked that I had missed quite a lot besides meals, because, in fact, from the dining car you're able to see to *both* sides of the tracks. I agreed. And we all were in agreement, too, starting out as we were now across the Nullarbor, that this journey was providing us with a terrific education.

I thought their niece, driven to the extremes of boredom by our table banter, which mostly concerned Australian history and had nothing remotely to do with things that truly mattered to her, was several times about to bite through her lower lip.

In the locomotive one day, studying a Michelin map of the state of South Australia, I saw that we would be passing just to the south of Maralinga. Between the cold war years of 1956 and 1963, the British conducted a series of nuclear bomb tests here. (I wondered if the sisters would be mentioning this to their niece.) From Woomera, east-southeast of Maralinga, the British had also launched missiles toward the Great Sandy Desert, hundreds of miles to the northwest. They considered the area a wasteland, though it was actually thinly populated by Walmajarri and several other Aboriginal tribes. The Australian authorities sent representatives out to clear the target areas of Aborigines before the missiles were launched; but the country was vast and they were not certain they'd been able to contact everyone living there. (The tests of the warhead's delivery system were deemed too important to delay.) The Australian authorities also attempted to reach Aboriginals who would be affected by fallout from the warheads detonated at Maralinga. After the testing ceased, Aboriginals who had been removed from their traditional lands to allow for the tests to take place were denied further access to them. (In 1994 the Australian government made compensation payments to the tribes involved, for the usurpation of their lands.)

One day on the Nullarbor the train suddenly ran into a wall of water, a rainstorm so fierce the windshield wipers could not, for some minutes, keep the glass clear. When the storm passed to the east and the sun broke through the trailing clouds, we saw a double rainbow off to the south. It seemed to span a dozen miles of the desert. In the same moment we saw more than a hundred kangaroos bounding north and west across the plain, then veering away to the west as they approached the train tracks and the hurtling train. The sight of it was so exhilarating the three of us in the cab nodded an affirmation to one another. Whatever was wild and lyrical in the timeless world, we were in the middle of it now. For some reason we all felt compelled to shake hands.

IN MEEKATHARRA, men at the shop where I rented the four-wheel drive wanted me to understand that the roads I proposed to follow to Nookawarra, the sheep station, were not all that easy to locate. They were, in fact, easy to lose. I said I was familiar with the problem they were referring to and that I would be all right.

The physical evidence that sets a human-occupied landscape apart disappeared almost entirely a mile west of Meekatharra. Soon I had only the road to guide me and a grid of fence lines, some sections of which proved to be miles long, well-tended wire fencing, as taut and as straight as men could make it. I was usually within sight of a fence line to the north (my right), frequently close alongside it; the road— a less insistent domestication of the arid plain—bent to the shape of the land and so was more graceful than the fence line.

The main problem with navigating dirt roads that traverse extensive sections of fenced land becomes apparent when the road passes through a gate. If the gate is open, you leave it open. If closed, you close it behind you. Because there are often only a few gates in very long stretches of fence line, vehicles converge toward a gate from many directions. Once the vehicle is through the gate, the tracks radiate away in as many directions. Without the help of a landmark it might not be possible to pick out the main track again. Similarly, in isolated villages in arid land the world over, it's easier to drive *into* the village on a dirt track than it is to locate the main track on your way *out* of the village, because local residents with distant destinations create such a fan of diverging routes as they depart.

At several places on the dirt track, headed west to Nookawarra, I drove away from the gate in the wrong direction. A quarter-mile or so out, it was obvious that I'd done this. These "errors," however, never raised the fear that I was "lost" in unfamiliar country, or that I was going to be "delayed." This far from clocks, the fear of being delayed lost much of its urgency. When I'd reached him by radiotelephone from Perth, my host reminded me of this, saying the drive from Meekatharra was three and something hours and that if I arrived "sometime after three," he would consider it being punctual.

I had only one difficult moment, when the narrow track I was on entered the equipment yard of a pastoralist and passed, like a private driveway, between a couple of his buildings. The man stared me to a stop and bluntly asked what my business there was. I told him I was on my way to meet a Mr. Richard Brown at Nookawarra. I showed him my hand-drawn maps, given to me by Bob Pidgeon at Curtin University, who I'm sure had passed this way many times. He glanced at them and then waved them away, as if they were gnats annoying him.

"You tell 'Mr. Richard Brown' to let me know when he's got some bloke coming for a visit, understand?"

I said I completely understood his point. His point was that he owned the land I was driving on, and further, I was intruding on his privacy. Like many pastoralists, he felt a challenge to his legal right to the land was likely to be in the air whenever a stranger showed up, because of the peremptory way the land had originally been acquired, by sweeping its first occupants out of the way. He was irascible, I supposed, because he knew some people thought his hold on the land was ethically tenuous.

I volunteered to him that his station seemed very well kept. His machinery appeared well maintained, his sheds in good repair. In the part of America I lived in, I told him, people value these indications of serious purpose and frugality. We looked for it when new neighbors moved in.

He thanked me and I drove away slowly, so as not to raise any dust.

WHEN I ARRIVED, Dick Brown pointed me to a comfortable guest room, and after he and his daughter and I had supper and Dick and I had had a chat on the veranda, we retired. The following morning, just after sunrise, Dick said I might benefit from an overview of the Jack Hills before I drove into that country, and offered to fly me over the land in his Cessna. With the map Bob Pidgeon had drawn for me open in my lap, and aided by the low angle of the sun, I picked out a route to follow through the hills. I was confident I could find my way now to the draw where the zircon crystals had been found. The ground at the foot of that draw, I saw, was boulder-strewn and too steeply pitched to navigate in a vehicle, but I spotted a copse of eucalypts nearby where I could park out of the sun and then walk the rest of the way up the draw.

The drive from the station's headquarters compound took less than an hour. Shortly after I parked I found the rock matrix Bob Pidgeon had directed me to. With the aid of a magnifying glass I found the tiny zircon crystals in it. I sat there in the draw for a while, next to the matrix, trying to place everything I could see from there into a time frame. The sun was well up by now. The surrounding land was an immensity of windless silence. To dig one of the crystals out of the rock I thought would disturb some sort of magic veil I didn't want to disappear, and anyway, I didn't feel compelled. Many places in the world have been so profoundly altered by development projects of one sort or another that they are no longer recognizable to the original inhabitants—or even the residents of a few decades ago. But making this kind of intrusion wasn't really what was holding me back. It was my desire to steer clear, this time, of a bad habit, the desire to take things you don't think will be missed. I didn't need the crystals, any more than I needed many of the other innocuous things I'd pocketed over the years in out-of-the-way places.

The effects of the jarring ride out here in the four-wheel drive from Dick Brown's home took a while to drain away. It took even longer for my list of questions about this place to evaporate to the point where all I was doing was sitting there beside the rocks. They were like exotic animals. On that June morning I watched the sun bear off into the northern sky from the east above a Serengeti-like savannah of scattered trees and open brushland. The rays of its light seemed to tinker with the pale colors of the horizon. I sank into the time pooled in this shallow draw, sank through the Cenozoic into the Mesozoic, the Age of Dinosaurs, and fell further into the Permian, fell down through the Age of Fish in the Devonian, to the time of the first mollusks in the Cambrian, and then into the Proterozoic, the eon of the relatives of the cyanobacteria I'd seen secreting their homes at Lake Clifton. And finally into the eon when there was no life, the basement of Earth time, the Archean. It's from then that these dazzling grains by my side had come. Theirs is the very long view. My time is not even a hair-thin splinter in the great sequoia of the time that is theirs.

Into this reverie a flock of ubiquitous budgerigars suddenly flew, about thirty of them, small green-and-yellow parrots with blue tail feathers and warbling voices. They zoomed past swiftly, actively maintaining their close alignment, like race cars in the tight turns of a road course. I'd spooked them. I watched until they bore off in a straight line, like a single animal. I was aware now of sounds that must have been there all along but which had not registered—the cries of cockatiels, a crested parrot found nearly everywhere in Australia. A plaintive *qweel*.

The birds' voices break and animate the stillness here but do not overtake it. With my binoculars I scour the open hilly country. It has the general look of untrammeled land, but the admixture of plant life and the dry, broken sheets of friable soil signal that sheep and other ruminants not native to the place have been here a while.

I left the bed of the dry wash and climbed a ridge to get a better view of country to the north. I was there only a moment before I saw a red fox. It looked up at me as it emerged from between two boulders and then was gone. I saw it once more in a pile of loose rocks below, and then it was gone for good.

The fox, like the sheep, did not come with this country; it arrived with colonization, the fox to establish the English tradition of fox hunting, the sheep to create the foundation for a pastoral economy. Today Australia is home to large feral populations of a great array of non-native animals. Among them are pigs, camels, hares, cats, dogs, horses, mongooses, several species of deer, donkeys, goats, zebu cattle, and famously, rabbits. Their grazing, rooting, browsing, and predation on native species has so radically altered the nature of Australia's nineteenth-century plant and animal communities that it is now impossible to say, across most of Australia, exactly what these communities once looked like. Imported birds—sparrows, canaries, Indian mynas, quail, pheasants, starlings—have also played a role in altering the landscape. And poorly managed sheep and cattle ranching operations have exposed millions of acres to erosion and desertification. What Cook saw during his coastal survey in 1770, what Matthew Flinders saw during his epic circumnavigation of the continent thirty years later, and what early white explorers of the interior saw—Ernest Giles, John McDouall Stuart, Edward Eyre, and the ill-fated Robert Burke and his partner, William Wills—will never be seen again. Which is as it should be, of course, in the natural order of things (i.e., if *Homo sapiens* is not to be set apart in the natural order). But the changes have been huge. They came on very quickly, and for many, they have been disorienting to the point of despair.

The modern urge to turn a landscape into "what it once was," to make it "better" by eliminating "pests," to rid it of plants and animals that, because they didn't coevolve with the environment, have a special capacity to devastate it, is a complex desire to appease—biologically, ethically, and practically. It is impossible, biologically, truly to "restore" any landscape. The reintroduction of plants and animals to a place suggests that though human engineering of one sort or another has "destroyed" a place, human engineering can bring it back, a bold but wrongheaded notion: humans aren't able to reverse the direction of evolution, to darn a landscape back together like a sweater that has unraveled. Restoration privileges some animals and plants over others, and therefore presents ethical problems identical to those one faces in examining any project of social engineering or any country's policies of racial and ethnic discrimination. Finally, it is not possible to restore the soil chemistry of lands turned nearly lifeless by decades of irrigation, chemical fertilizers, and overgrazing.

The chief value of restoration projects, perhaps, is psychological. At a time when the extent of serious damage to Earth's ecosystems has ceased to be a topic of special interest, restoration projects, like any act of atonement, fill humans with a sense of self-worth and enhanced dignity. This humbling and pioneering work, despite the biological and ethical challenges, seems to me to mark the beginning of a kind of human behavior that will (partially) restore more than landscapes. It will provide living grounds for all life, including human life, until industrial expansion ends and begins to show signs of drawdown.

It is not possible to consider the question of restoration in any complete way, it seems to me, without confronting the discomfiting issue of intolerance that underlies all efforts to restore. And this places one uncomfortably close to the volatile politics of immigration. The chief objection local people have to "invasive species" is that they can so quickly eradicate the familiar, the valued, and the iconic, can so easily turn what was once thought beautiful into what is considered aesthetically offensive. Some people come to feel that the arrangement of life that was formerly in place was intrinsically more valuable than what has replaced it. These judgmental attitudes toward exotic animals and plants overrunning indigenous animals and plants, of course, differs little from the attitudes of an indigenous human culture toward an invasive human culture, or an entrenched human culture toward an influx of representatives from an "exotic" culture.

Evolution, if it is nothing else, is endless modification, change without reason or end. Notions of preserving racial purity in the twenty-first century, or of maintaining biologically static environments, in which all new arrivals are classified as "invasive" or "foreign" and are to be expunged, or are not permitted entry to start with, are untenable. The obvious ethical issues aside, these arguments deny the flow of time. Landscapes are figuratively, not actually, timeless. And ours is an age of unprecedented cultural exchange, of emigration and immigration. Reactionary resentment around issues of race and culture has no future but warfare. And all landscapes are on their way to becoming something else, with incremental slowness and terrifying speed.

The agitation people feel around subtle and radical change in their home landscapes has only partly to do with physical change in the land—Burmese pythons overrunning Florida's Everglades, say, or copses of fast-growing balsa trees appearing in Galápagos. It has equally to do—possibly more to do—with the time available to absorb such change. It is more psychologically disruptive to be confronted with many changes over a short period of time than to encounter only a few changes over the same period of time, which was the universal human experience until a few hundred years ago. Today, with modern aircraft spreading local viruses all over the world, and commercial shipping flooding international harbors with thousands of new species when they flush their ballast tanks, and with communications technology fundamentally altering, in only a few decades, the way people communicate with one another, a continuously changing environment actually seems, for some now, more stable, more comfortable, than a seemingly static environment. Ideas such as racial superiority, once tolerated, now seem outdated. Further, globalization has created an environment in which mixed-race, mixed-culture, dual-citizenship, and immigrant-status populations are increasingly the norm in cities like Los Angeles, London, Sydney, and Rio de Janeiro, bringing the outsider face-to-face everywhere with an evolving and vital international mestizo culture.

At some critical point, accommodation and cooperation replace violence and exploitation, or humanity's fate is delivered into the hands of barbarians.

WATCHING THE FOX disappear in a boulder field below the ridge, I can appreciate what makes this creature inimical to so many Australians.

When chasing foxes here on horseback with packs of hounds went the way of powdered wigs in American courtrooms, the no-longer-harassed foxes spread far and wide. They killed off many of the small native predators they were in competition with for food, and the environments they entered provided few curbs on their behavior and no effective restraints on the growth of their populations. The red fox stands out on the land today as a symbol of colonial incursion. Certainly they snatch the occasional lamb; but actually, they are only foxes, endeavoring to make their way in a world they were transported to, like the camels overland explorers used and then turned loose when they were finished here. And the cats and dogs that ran away from homesteads. And the zebu cattle brought in in 1880 to supply workers building an overland telegraph line with fresh meat. And the mongooses brought in to control the rabbits and to reduce the population of Australia's unusually large number of very poisonous snakes.

Of the many "exotic" animals Australians mounted eradication campaigns against, once the species' wild or feral populations were large enough to be perceived as a threat (campaigns were also mounted to eradicate native animals like the kangaroo and dingo in areas where their existence threatened the profitability of farming and ranching), the one against the European rabbit (*Oryctolagus cuniculus*) was the most epic. Rabbits arrived in Australia with convicts aboard the First Fleet. Later, in the early decades of the nineteenth century, they were introduced at more than thirty different places by immigrants, for whom they were a source of both food and commercial profit. Some rabbits escaped, others were abandoned by families moving on; by the 1860s, wild populations had colonized about two-thirds of the continent. Between 1885 and 1914, more than 200,000 miles of rabbit-proof fencing was erected across Australia to control the extent of damage rabbits were causing farmers by grazing and by the construction of extensive underground warrens. While the fences were going up, farmers turned to increasingly more violent strategies to slow the spread of rabbit populations—explosives, bulldozing, poisonous gas. House cats and mongooses were loosed after the rabbits, and ways were developed to introduce lethal myxomatosis and caliciviruses into wild populations.

Had I been a conservationist at the height of the campaigns against rabbits in the fifties, I would no doubt have cheered the bulldozers on, unaware of the effect 200,000 miles of wire fence was going to have on the land. If I had been a farmer or a pastoralist, I might have spent whatever money I could afford on gelignite explosives and phosphine gas. If I had been one of the handful who still raised rabbits for profit, I might have gone to Brisbane to complain loudly about the manufacture and distribution of viruses lethal to the rabbits of Queensland. If I had been sentimental, uncomfortable taking on the messy business of unenlightened planning, I might have thought that the rabbits were too cute to kill. Who can say what theologians, philosophers, pragmatists, and proto-environmentalists of the time might have said if they'd been offered room at the table along with the pastoralists, the agriculturists, and politicians?

Sitting that day in the Jack Hills, after losing sight of the fox, I felt a twinge of the nostalgia that might come over any of us when we learn that a wild landscape that was emblematic for us in childhood has been turned into a resort community peppered with condominiums. Whatever one finds in front of herself at the moment, however, is what the given situation is. That other thing, the so-called pristine landscape of a former time, is no longer available; and somehow a person must make peace with that. To go in search of what once was is to postpone the difficulty of living with what is.

An observation I heard several times, in different circumstances, about the makeup of Australia's human population—it's a popular assertion, but not one I was able to confirm—is that half of modern-day Australians are either immigrants or first-generation Australians. The increasing heterogeneity of burgeoning populations around the world, of course, raises the ire of reactionary politicians in many countries. They rail against the loss of putative racial and ethnic purity, just as many of the same people rail against evolutionary theory and global climate change. They read these signs of inevitable change in the status quo as threats to plans for human perfection that they feel responsible for engineering.

What is to become of us if we decide that the only relief from a persistent sense of discomfort or irritation in our individual worlds is to go after the newcomer, to denounce reconciliation as cowardly, and to kill the Arab student in Tel Aviv, the black intellectual in Atlanta, the Caucasian relief worker in Somalia? Will we also be burning down the eucalyptus trees in Florence, the bougainvillea in Caracas, the ginkgo trees in Manhattan? Will we be inventing a rationale to legitimize the use of arsenic against the Jaburrara or the eradication of rabbits? Will we drag our gods in, and our economists, to preside over the division of wealth that comes with each of our victories over those whose only mistake was to have other ideas? Or will we grant the imperfections in our behavior which have for so long been apparent, and examine instead the forms of reconciliation, those already known and those yet to be invented?

Crossing back over the countryside to the copse of eucalypts in the draw where I'd parked, I felt suspended still in that deep well of time at the bottom of which molecules of zirconium silicate crystallized and became tiny brown grains with four billion years of history ahead of them, before some traveler with a hand lens bent down to peer at them. I pictured the emergence of hominins at the distant end of this arc of time, their lone survivor, *Homo sapiens,* standing with its paintings, its music, telling mythic stories, and becoming acquainted with its problematic appetite for triumph, for vengeance, cruelty, war, and acquisition.

An arduous life in a world of gargantuan human mistakes, of realpolitik decisions, and of personal failure might have prepared any one of us to grasp unflinchingly our own capacity to become immoral, to become the terrorist, the seeker after power and extensive privilege, the anointed enforcer of whatever we construe to be right. And enabled us to understand, considering shortages of good water, metallic ores, and arable land worldwide, what many human populations are likely to face long before the century is out. And compelled us to object to the efforts of elected governments to ferret out and review the thoughts of each of their citizens, and to object to the argument that for-profit businesses be accorded the same rights individual citizens enjoy, to oppose the ceaseless manufacture and distribution of lethal weapons, and to consider that our own progeny will have to face harrowing decisions in the future merely to survive.

Are we not bound, now, to learn how to speak with each other?

―――――

RICHARD AND I had tea on his veranda again that evening. He told me stories about seeing flocks of galahs and other birds flying alongside his plane, about the upwelling of affection he felt for them, their guileless effort at life, and how very different this dry country seemed to him after a pelting rain, how fresh. That morning, before I left for the Jack Hills, Richard had brought me a .308 Enfield. He asked if I might take the rifle along and shoot any wild goats I saw. They compete with the sheep for food, he said. I declined, and he nodded his understanding.

"It's not for everybody," he offered, and returned the rifle to its place in the rack on the living room wall with the other rifles and guns.

Richard was such an agreeable companion, thinking so hard about trying to find his way in the world, I was sad to see our last evening end. A year later, in the wake of an airplane crash he had limped away from, he came to visit me in Oregon. I was so very pleased for the opportunity to show him around.

AT A MEETING of hunter-gatherer experts in Darwin in 1988, I met a young woman named Petronella Vaarzon-Morel, an anthropologist who lived in Alice Springs and who had gotten to know Bruce Chatwin when he came through the country with Salman Rushdie to research his book *The Songlines*. She invited me to come to Alice, as local people call it, to learn about the land claims movement in the Northern Territory, an effort to establish a legal basis for returning certain stretches of Crown land to their original owners.

It was some months before I was able to get back to Alice and meet Petra's colleagues, including her then-husband, Jim Wafer, and the writer Robyn Davidson. Robyn, who had a background in anthropology, was teaching with several other Australian women in Aboriginal settlements. Like Petra and Robyn, these women were also providing professional help to the land claims movement. Robyn offered me her home to stay in while she was working in the settlements.

I returned to Alice with the intent, first, of traveling out bush with a group of wildlife biologists and field technicians from the Conservation Commission of the Northern Territory's (CCNT's) office in Alice Springs. They were collaborating on a project with local Warlpiri people, hoping to reestablish a population of rufous hare-wallaby on Aboriginal land in the southeastern Tanami Desert. (The word for this particular wallaby in Warlpiri, and about twenty other Aboriginal languages, is *mala*. A small marsupial, about the size of a hare—thus its English name—it's also known locally as the Western hare-wallaby or pejoratively, as the spinifex rat. To scientists it's *Lagorchestes hirsutus*.)

At the time, mala were endangered throughout their range. They were in direct competition with feral rabbits for habitat and they were preyed upon by feral foxes and cats, which had hunted them to extinction in the Tanami Desert. The spot the Warlpiri and the scientists had chosen for the reintroduction experiment, about 220 miles north of Alice Springs and about forty miles north of Willowra, a Warlpiri settlement on the Lander River, was in semiarid desert country, a savannah dominated by a scatter of melaleuca trees and tussocks of spinifex grass.

The biologists placed several groups of mala, raised in captivity in Alice, in a 240-acre enclosure adjacent to the (usually dry) Lander River and the sprawling Tanami Desert Wildlife Sanctuary. Electrified fencing kept rabbits and predators out. Once they had acclimated to the area, the idea was to release the mala into the countryside.

We set our camp up out of sight of the enclosure, next to a billabong. The purpose of this particular trip to the mala enclosure was to observe the animals from a distance, using the nearby brush for cover and approaching the enclosure only to check the drip tanks along the fence line, to be sure they were providing the mala with enough water. When we walked back into camp from our observation posts that afternoon, a few people headed for the billabong for a swim, a relief from the heat. I assumed my hosts had made certain it was all right to swim there. Places this obvious, no matter how remote the country you find them in might seem to be, always have the filaments of the Tjukurrpa attached to them, the Dreamtime narratives. It would be tragically easy to "pollute" such a place without realizing it. Just before I spoke up, the leader of our small group called the swimmers back.

———

IN THE YEARS before I made this trip, I'd had the opportunity to travel, mostly in the Arctic, with a number of field biologists who'd developed close working relationships with indigenous people. They'd apprenticed themselves to native hunters in order to learn more about the animals they were studying, mostly, in my experience, wolves, bears, caribou, and marine mammals. They came to believe that even though Western field biology had traditionally ignored, or simply disparaged, indigenous field observations, this enormous body of native knowledge was as precise and rigorous as what Western science had built up. It often, in fact, went deeper and was more nuanced. In the minds of many of the field biologists I accompanied, taking both kinds of research into account provided a more complete understanding of the animal. (It surprised no one that given periods of close observation exponentially larger than Western scientists had been able to manage, native knowledge sometimes corrected erroneous conclusions scientists had reached or challenged assumptions they'd made.)

The biologists I joined for the field trip to the Lander River mala enclosure had taken this kind of mutually respectful cooperation one step further. Knowing that mala were threatened throughout their range by feral animals, biologists at the CCNT approached Warlpiri elders about helping the CCNT restore the Tanami Desert population. The Warlpiri thought this was a good idea and helped the scientists locate a place on Warlpiri land that would provide excellent mala habitat. The scientists told the Warlpiri, however, that they needed a special kind of help to restore the mala to this part of its traditional range. They said they could get the *biological* part of the restoration project right (i.e., captive breeding and the selection of suitable habitat in which to build the enclosure, in order for the mala to make a successful transition from the breeding shed in Alice to wild land); but they felt their efforts would eventually fail because they had no knowledge of the spiritual nature of the mala, of its place in the Tjukurrpa. They asked the Warlpiri elders to assist in the reintroduction by "singing the wallaby up," by ritually calling mala back into the country. After some hesitation, the Warlpiri said they would do it. The older men—no children, no women, no "white fellas"—would go out to a certain place in the Tanami, they said, "paint up," and sing the mala back into the country.

The degree of awareness the biologists showed here was, in my experience at the time, unprecedented, and the approach they took spoke in a profound way to Warlpiri elders. The principal reasons for the collapse of mala populations in the Northern Territory were predation by feral mammals and rabbits that took over mala denning complexes and competed with them for spinifex seeds and other favored foods. The root cause of the collapse, however, was more complicated than this. Aboriginal people had practiced for millennia a sophisticated land-management technique called fire-stick farming on lands where mala lived. They used controlled burns—slow-moving grass fires—to remove dry spinifex brush and encourage new growth, thereby creating a mosaic of old and new spinifex vegetation which served their needs as hunter-gatherers. The practice also served mala well. They denned in the older patches of spinifex and fed in the new sections.

When Aboriginal people began coming in off the desert during a prolonged period of drought in the Northern Territory in the 1950s, taking up residence in settlements and mission stations, fire-stick farming no longer affected the ecology of the desert as much. This had a deleterious effect on mala populations, and because of this foxes, feral cats, and rabbits had a greater impact on remnant mala populations than they otherwise might have.

In the Dreamtime narratives of the Warlpiri, Luritja, Arrernte, Pitjantjatjara, Pintupi, and other desert traditions, Mala plays an important role in bringing Aboriginal lands to life. Like other Creation Beings, Mala was a traveler, and a Mala songline, marking part of his travels, runs roughly north and south from the Lander River country to an isolated monolith called Uluru (Ayers Rock). The collapse of mala populations along this songline threatened the spiritual foundations of many Aboriginal traditions, and as the mala population headed toward extinction, Mala ceremonies began to atrophy. The sensitivity of the CCNT biologists to this relationship between the spiritual and material world, and their telling the Warlpiri that they had neither the authority nor the ability to act in this realm, but that they understood that without the help of the Warlpiri here the reintroduction effort would collapse, was mind-boggling for the Warlpiri.

I later had a conversation with one of the men who'd traveled out into the Tanami to sing up the mala, prior to their release from the enclosure. He told me that the idea of an animal being "locally extinct," as the biologists said, was a difficult concept for him to understand. It's possible, he told me, that the *body* of an animal might not be visible to someone traveling through a certain country, but the animal was still there. In its corporeal form it might be "finished" in a particular place, but it wasn't "gone," the way white people use that word. If you couldn't see it, I asked, couldn't find its tracks or scat or signs of its feeding, wasn't it "locally extinct"? No, he said. He waved his extended left hand quickly in a sweeping arc. "It's all out there, everywhere." After he and the other men sang the mala up, he said, the spirits of local mala who were present entered the bodies of the mala in the enclosure.

Someone entirely wedded to a Western way of knowing might find this story fatuous, but in interviews with Western field biologists over the years, I've found that the issue of local extinction is, for many of them, not entirely clear. There are too many cases of animals being declared locally extinct only to have them turn up again. "Singing" an animal back into existence is a metaphorical expression for some as-yet-unplumbed biological process of restoration, quaint only in the minds of those who believe they already know, or can discover, precisely how the world is hinged.

ON THEIR RETURN to Alice Springs, the CCNT field party dropped me in Willowra. Petra had long been conducting research in this settlement, and with her intercession I was able to stay there for a few days. Some weeks later she arranged for me to visit with Pitjantjatjara people in Mutitjulu, the community at Uluru. My experience in these places was little more than incidental, and of course I missed a great deal of what was right in front of me, not being familiar with the traditions or the "ways of seeing" of Warlpiri or Pitjantjatjara people, and not knowing the physical geography of either culture. Reading the work of anthropologists who were studying these two desert groups, however, and later interviewing a few of the anthropologists, as well as indigenous people in both those settlements, I came to appreciate their detailed knowledge of the physical world of which they were a part.

I was able to get away from Mutitjulu on several occasions and into the surrounding country with a small group of Pitjantjatjara men. They were bilingual and patient with my questions, and never appeared bored or offended by my effort to understand how very different this place was from other places I'd visited. One day when a couple of us were north of Uluru a mile or so, I asked my companions if they were permitted to tell me about the songlines that converge here, about the Creation Beings of the Dreamtime who came here. What direction did they come from, and where were they going? I wanted to know specifically about Mala. Had Mala been here?

Oh yes, one of my guides answered. Mala was here. With a tilt of his forehead he indicated a spot at the base of the north side of Uluru, an indentation like a grotto, marked at that hour of the day by a long vertical shadow. Mala had slept there, they told me. They told stories about Mala for a long while. The four of us were sitting in very strong sunlight on a sandy rise. They spoke in English and I resisted the desire to impose questions to clarify what they were saying. When they were finished, I was reluctant to break the silence.

I cannot recall all they told me. To have stepped away at the conclusion of one of the narratives and gotten out my notebook to write down what they said would have been rude, and, I thought, it might have given me the appearance of being a thief. And to have interrupted the experience itself with questions might easily have disrupted or truncated the stories, broken them off in such a way that the storytellers would have been reluctant or unable to submerge themselves again in the particulars of their emotional and intellectual history.

We continued to sit in silence on the sandy rise. I had the sense when they were speaking that the three men were really talking to one another, that I was not there, that they were reminding each other of the great breadth of Mala in their lives, in the life of their community.

We continued to sit facing the north side of Uluru. Then one of the men began pointing out other features of Uluru and explaining their places in the Dreamtime narratives, how each was related to the activities of other of the Dreamtime Beings. Somewhere in this explication I realized that they were describing features I couldn't make out, because they were features on the *other* side of the rock, a part of Uluru

I couldn't see but which they apparently could easily imagine. Some time later they returned in their recollections to the place they'd first spoken about, the place where Mala slept, and I understood then that they had circumnavigated the rock. They had taken me completely around Uluru without referring to any shift in perspective that might be required for me to understand this. What was seamless for them was broken for me into two separate parts, what I could see with my own eyes and what I could not.

From childhood on, these three men had heard the Tjukurrpa stories that included Uluru. What they could see in any particular moment, and what they could remember having seen, which only memory, in fact, could give them, constituted a piece of whole cloth. In this way, not only did memory function as one of the senses for them but the way they described Uluru made it clear that they, far more than I, lived in three spatial dimensions. Their view of the physical world had no correct or privileged point of view. Sitting together on the north side of Uluru as they spoke with me about Uluru and the Dreamtime was actually incidental to their story, to accurately describing the way Uluru fit in their world. For them there was no "front" or "back" side, no "right" or "left" to the phenomenon. They were not hampered, as I was in my perception of the rock, by a lifetime of learning from flat surfaces, reading about the world mostly left to right and, as often as not, top to bottom—books, maps, drawings, and computer screens.

Traveling with these three men around Uluru—we also drove off one day about twenty miles to the west, to visit Kata Tjuta (the Olgas), a rock formation with a female identity for my companions—our conversations ranged across many topics: pop music, the virtues of the Toyota Land Cruiser in comparison with the Nissan Patrol, tourists swarming Ayers Rock (whom they referred to uncharitably as *minga*—ants), petrol sniffing, and the fortunes of the various rugby league and football clubs these Pitjantjatjara men followed closely. It took time for me to recognize that it was only when we were riding in the vehicle, or back in someone's home in Mutitjulu, that our conversations became this topical and animated. When we were walking across the land together, no one said much at all.

Wherever we walked, our steps always seemed to fit the landscape. The pace never felt rushed or uncertain. Our movement was like water's—measured, responsive to the topology of the ground. If someone began telling a story while we were walking, the story would be about a place we were then moving through. The story would start just as a prominent feature of the place came into view. The duration of its telling matched our pace through the region, with the story tending to end in the same moment as the prominence that was at the center of the story passed out of view. A rhythm (the pace of the story) within a rhythm (the pace of our walking) within a rhythm (diurnal time passing).

My intuition, that for my Pitjantjatjara escorts being fully present in a place meant not only a high degree of sensory awareness but being acutely aware of one's *memories* of the place (or of what one had been told about it by a trusted voice), led me to another intuition, or at least to a fuller explanation of the meaning of our exchanges. Even though we all spoke English—the English spoken by two of the three men I was with was excellent—I couldn't help but feel that something was not coming across. Something elusive in the conversation made me think I was missing important points my companions were making. What I came to believe was that the Pitjantjatjara were so cognizant of the third dimension of the landscape around us that for them, the land we were passing through was never a *projection*. They were never outside a place looking in, they were incorporated *within* whatever we were seeing. To them, some of my questions about the places we were in were too strange to answer easily. For example, my questions about "aspect," seeing something from a particular point of view, often seemed to present them with difficulty. These questions of mine grew out of my habit of flattening the third dimension (depth) in order to create a bounded scene, something a painting or photograph provides. When I lay awake in my bed in Mutitjulu one night, I tried to imagine the way in which our conversations on some occasions took place in the dimension of 2.5. And it became my opinion that they were not as eager to find and hold this two-dimensional view, the one I was most often comfortable with, as I was to learn how to stay with them in the realm of three-dimensional perception, once I located it.

The possibility of being able to see a country more fully in this way was clear.

ONE AFTERNOON IN Willowra an opportunity I had not anticipated presented itself. Sometime in the late 1920s—I was asked not to present all the details of this story—a small group of Warlpiri men and women were murdered by territorial police at a water hole in the Tanami Desert. The murders spiritually contaminated the water hole and Warlpiri people stopped going there. Previously, it had been an important stopover, because surface water in that region is scarce.

Elders in Willowra had decided the time had come to return to this place and to "clean it up," to spiritually cleanse the water hole and the land around it through ceremony, and to physically remove any natural debris that might have accumulated in the water hole.

It would take four or five days to travel there, then there would be ceremony, then four or five days to travel back. Did I want to travel with them? Yes, I said, I very much wanted the experience of being on the land in their company, and of watching, and also helping if that would be all right, while they cleaned up the place. I was certain they had extended this invitation because of the affection they had for Petra, more this, I think, than whatever impression they had formed of me during my short stay in Willowra. (When Petra and I traveled with her Warlpiri friends, it was imperative, she explained to me, that we conduct ourselves as if we were brother and sister. It wasn't a matter of what we actually were to each other—good friends—but a matter of fitting into Warlpiri society in a way that showed respect for Warlpiri mores. In order for us to place our sleeping areas next to each other on the ground, it was necessary that we be brother and sister. In the same vein, before Petra left Willowra for Alice Springs, she pointed out some places outside the settlement I should not approach or inquire about, Dreamtime places. They were off-limits to someone like me. I followed her instructions precisely.)

Despite what I felt was the honor, as well as the gift, of what the Warlpiri offered me, I finally decided not to accept the invitation to travel with them to the water hole. An acquaintance in the settlement helped me frame my explanation so that it wouldn't be read as either a rejection or an insult, and the party left without me. To this day I do not know if I made the right decision. The argument for going was that I would be able to report an extraordinary story of Warlpiri vitality, of human passions and historical perspective, of racism and perseverance. The argument against going was essentially my discomfort with being a witness. The Warlpiri, I decided, did not fully appreciate what I did as a writer. For them, the invitation they extended was an invitation to be with them socially during extremely significant days. It was not an invitation for me to describe for strangers what I had witnessed—or at least it was my impression that this was the case. If I wrote about it, I argued with myself, I would only be putting myself in the position of interpreting something spiritually important about which I had only the most superficial understanding.

If I had to make the decision all over again, I think I would have gratefully accepted the invitation, let it inform my general thinking about traditional people, and never have tried to interpret publicly what I had witnessed. I asked the people who went to describe their experience to me when they came back. I wanted to leave the decision about what was said about the event to them. It didn't seem to me that I was going to miss something important by not going. It seemed I would miss something important by not waiting for them in Willowra, and letting them, for once, be the sole reporters, the only interpreters, of their culture.

IN THE YEARS I was growing up, paleoanthropologists like Louis and Mary Leakey, with their pioneering work in Tanzania and Kenya, provided people with a rough sketch of the physical evolution of *Homo sapiens*. The view since then has been greatly refined; in recent years, however, the attention of scientists interested in the history of mankind and what it means to be human, has moved more toward cognitive research. Their focus is not on the outward appearance of our human ancestors but on the evolution of the mind within. Their work offers us new ways to understand human beings by moving us away from long-simmering questions about racial and cultural superiority and toward more pressing issues, such as the development of empathy and the human capacity for cooperation.

Research into the development of the human mind is a pelagic and not infrequently contentious field of inquiry. It's easy to become completely lost in its neurological pathways and psychological implications—not to mention having to deal with the utility (or inutility) of altruism, and a related assertion, that compassionate governance and altruistic behavior constitute "socialism" for some on the political right. To successfully address major international problems like freshwater availability, however, for all human cultures in the decades ahead, will require empathetic understanding. But, one must ask, How is empathy on this scale to operate in cultures that remain suspicious about the predicaments of strangers? Or in cultures that are already on the verge of falling apart because of war, environmental stress, and the abuse of dictators? Or, indeed, in cultures that are indifferent to the fate of those living beyond their borders, believing their final disintegration is of no real consequence?

Empathy and compassion would seem to be requisite components in the development of any new politics that aimed to place human welfare, for example, above material profit in a restructuring of national priorities, or in the redesign of domestic economies.

I return to this topic of the capacities of the human mind, originally presented in the chapter set at the Jackal camp, because research into the psychological development of personality and into the phylogenetic development of the human mind suggests that certain people within the same social group—psychologically fit and emotionally mature Australian pastoralists, say, or psychologically fit and emotionally mature Pitjantjatjara Aboriginals—have a greater capacity to empathize with others in that group across a broad spectrum of ideas, such as the utility or appropriateness of certain ethical positions. In any given group, then, some people will be more capable than others of understanding what another individual is trying to say. They are able to help make that person's position or reasoning clearer to others in the group. Again, this ability to listen closely and empathetically, to ameliorate social tension and increase understanding in a group, is not necessarily associated primarily with a listener's relative level of intelligence or his or her ability to perceive and then explain complex patterns. Success here depends as much or more on something harder to define: the ability to see the world from someone else's point of view without fearing the loss of one's own position.

For me, the ability to listen carefully to another person's perspective, rather than summarily deciding what that person means, is in keeping with the behavior one expects of an elder. And the ability to understand what someone else is thinking is the foundation of stable social order.

I'm often fearful, listening to discussions about human fate arranged around the agendas of government and international business, that "the best minds" are infrequently present when critical decisions are being made. If Theory of Mind psychology is correct in saying that minds operating at the higher levels of intentionality have the greatest capacity to be discerning and empathetic, and if it is wise to take seriously the idea that global climate disturbance, ocean acidification, and other planetary environmental problems cannot be successfully addressed without the highest level of international cooperation, what are we to do in our time about ultranationalists and xenophobes in positions of power and authority? Or more important, if the best minds are not at the table—because of prejudices about race, ethnicity, gender, formal education, urbanity, and material wealth—what is the process that will place them there?

Theory of Mind speculation supports—almost inevitably, it seems to me—the observation that in traditional societies the selection of elders, the people who are widely supported when it comes to making decisions (with other elders) for the group, has relatively little to do with how old a person is or with how intelligent they might be. More important is an ability to empathize, an ability to respect other views. (Another common attribute among elders is that they have historical imaginations. They draw on the details through memory of what has worked and what has not worked in the past when people were faced with challenges.)

In order to imagine a successful conversation among people who deeply understand one another and who can bring into play the metaphors and patterns of thought upon which their enduring cultures are founded, it would be necessary to set aside the Western commitments to progress and improvement. (When Darwin argued that the arc of biological evolution for any particular species was not about improvement but instead about successful adaptation to a new or changing environment, he was making a point about the evolution of *Homo sapiens* fundamentally at odds with much of Western thinking.) Furthermore, for such a group to function productively, it could not begin by embracing any one culture's views, or by differentiating, for example, between "advanced" and "primitive" cultures, or by favoring any one religion's sense of human destiny. Confronted with the task of discovering a path to reconciliation and cooperation in a time of unprecedented threat to human existence, elders focus on the idea that the primary organizing principle for human achievement is stability, not progress, meaning that balance, symmetry, and regularity are more to be valued than change, growth, deviation, and ambition.

The idea that people without an overriding allegiance to any particular form of governance, economic organization, or religious conviction, and with no great investment in personal advancement or cultural superiority, can come together and achieve what neither government nor business nor armed combatants are able to do today, to put human physical and mental health first, not their own welfare, is of course anathema to most governments, corporations, and armies. Until we can do this, we remain stuck with the often venal aspiration that drives many first-world nations—to triumph. To win.

I'VE HEARD SOMETIMES that unless you're truly interested in the physical landscape you find yourself moving through, it's tedious to travel with traditional people, that when you make camp in the evening "there is no intelligent conversation to be had." That is not my experience, and the complaint ignores several important considerations. Are things that are human more important to talk about than the things that are not? Is it right, in such circumstances, to pursue a conversation that doesn't make room for everyone? Would it be the case that people from your own culture are more likely to be engaging conversationalists than those from another culture? It's generally true that traditional people are mostly quiet while traveling, because the syntax and vocabulary of spoken language too often collapse the details of a place into meaning, precluding other interpretations. The conversation around indigenous campfires, however, is often metaphorical, or even allegorical. So it engages more than one type of mind. It provides for more than one level of intellection.

It is also true that in every culture there are people who choose not to say anything, though they could say a lot that was worth remembering. In my experience, it is possible to soar in conversation with someone not of your own culture, if you can find a way around the language barrier and if you and the person you're speaking with are focused on a world outside the self; and if both of you are able to empathize with views not your own and to incorporate them into the great reality of human experience. For this to work, both sides of the conversation must be informed by curiosity, respect, and an understanding that the world we find around us is too changeable, too multifaceted, too ramulose, for anyone to fully comprehend. It is not meant to be understood.

I've implied much here about the ability of elders in traditional societies to guide their people down the perilous roads all societies must travel (and left it to the reader to imagine that some elders, engaging their own egos to too great an extent, or seduced or corrupted by the secular world, fail at the work). I should emphasize, then, that all elders know they're fallible, that they know there are "no guarantees in life," and that some dangerous situations simply cannot be circumvented. But the thing with them is that they also know that once they are chosen, they must never quit out of despair or fear. To do so would be an act of betrayal. And they are chosen because people agree, *every day,* that this person is the best mind they have. It is not possible to make oneself available to serve as an elder, and I've been told that no one really seeks the position anyway, because the responsibility is so great.

With the disintegration of traditional societies the world over, the model they represent of wisdom passed on through a series of elders whose decisions were not questioned is in danger of being lost. The democratic model of governance in the West is based on the idea that everyone's voice must be heard. Those individual voices are often drowned out and subsumed in the West, however, by charismatics who say, "Follow me! I know the way!" In traditional societies people come to understand who it is who can really hear another person's voice, so they are comfortable with that person coming up with a plan in conversation with other elders in an emergency, and they feel no loss of autonomy in doing what the elder asks of them. They know the elder is not a "follow me" personality. His guiding thought is no one left behind.

OVER A PERIOD of two decades, partly by accident, partly by design, I visited many of the Pacific landfalls of James Cook and Charles Darwin. I went ashore at Valparaíso on the coast of Chile to walk the streets of that town where Darwin began his journey on foot across the southern Andes. I rounded Cape Horn, as both Cook and Darwin had done, and walked the shore at Kealakekua Bay on the island of Hawai'i where Cook was killed. I picked up Cook's trail in Tahiti, in the waters of Antarctica, and along the coast of northern Alaska. I imagined Darwin present on the streets of Buenos Aires with me, ashore in the Falkland Islands, having a meal in Cape Town, and bushwhacking his way to the top of Isla Santa María in the Galápagos.

Any reasonable traveler or reader, I believe, given the time, can draw his or her own conclusions from all that's been written about Cook and about Darwin, and can enhance their views by visiting places like Kealakekua Bay. I admire Cook for the reasons I've previously set forth. He lost his way, figuratively, on his final voyage, and he was a never-around kind of husband and father; but he encouraged in us all a sense of endeavor and of great imagining. I'd give nearly anything to have had dinner with him, ask him how he imagined the duties of a navigator changed when it was not a journey of exploration but a journey in service to commerce. What, for example, was to be considered a detour?

It is his counterpoint, however, whom I think about more often today, the poorly recollected and uncelebrated Ranald MacDonald, a man born into two cultures, in neither one of which did he ever feel truly comfortable. He arrived in Australia in the early 1850s, setting up in the Victoria goldfields around Ballarat, hoping to make a fortune after his months in Japan and after sailoring for a few years in southeast

Asian seas. We lose track of his whereabouts after that, but not of his life quest. Ranald MacDonald longed to be someone, in part because for most of his life he was regarded as no one. He made his mark in Japan, but history moved him aside to create room for Commodore Matthew Perry. MacDonald didn't have the pedigree, the friends, or the money to establish and promote his claim.

What would we have thought of MacDonald if he'd found the gold he was so eager to acquire in Ballarat, or later in the Cariboo region of southern British Columbia, and if he'd been able to use that wealth in a well-managed campaign to amass accolades and achieve social standing? Would this mestizo roustabout ever even have taken that route of self-promotion? He wasn't any kind of erudite James Cook, a cultured person to sit down to dinner with. It would have been a plate of beans and a cup of sugared tea in a ramshackle mining camp for him; but his conversation, his opinions, would have provided perspective on what we thought we knew about the world when we sat down together to eat. He would have tinted the view we have of ourselves, standing before our gods with the lists of our accomplishments. And we might have felt some tenderness toward him, as his long-lived dreams and the trouble he weathered emerged from beneath the dramatic accounts of his adventuring.

Elders would have understood equally the unfiltered lifelong inquiries of either man, the bicultural mestizo and the Enlightenment hero.

ONE CLEAR, LATE SUMMER afternoon in early March, in Sydney, I was crossing a greensward in a city park, en route to the Art Gallery of New South Wales. I'd been invited to give a talk there about a collection of short stories of mine that had just been published. Fair-weather cumulus clouds were nearly stationary above me as I walked, and the air immediately around me was calm as well. The most fragmentary of breezes occasionally unsettled the leaves of gum trees in the park. The delicacy of the weather reinforced a feeling I had in that moment of unfocused exuberance, a faith that no matter what people had to face in the world that is coming for us, they would fare well. Whatever the nightmare, some group of us would see a way through, for ourselves and others. I recollected bits of conversations I'd had with people well placed in international business who seemed, to my mind, to have no substantial belief, really, in what they told me they believed in—a world of successful commercial strategies and conventional success. Instead they appeared to believe in something quite different, in affection for certain parts of the broken world, and in the possibility of changing the corporate enterprises they ran, so that they would no longer be contributing to the social and environmental wreckage around them.

I recalled some lines of the Peruvian poet César Vallejo, published when he was only twenty-four, in a poem called "El pan nuestro":

> *And in the cold hours, when the earth*
> *smells of human dust*
> *and is so sad,*
> *I want to knock on every door*
> *and beg forgiveness of whoever's there,*
> *and bake bits of fresh bread for him,*
> *here, in the oven of my heart . . .*
>   *(my translation, with Luis Verano)*

An affection for humanity had swelled in me that morning, a hope that we would be all right, find the grace to accommodate each other more completely, invest more deeply in the philosopher's cardinal virtues, the ones that transcend all religions: courage, justice, reverence, compassion.

THE READING TOOK PLACE in a sunlit space that had a view down toward Sydney Harbor. People had a few questions after I spoke and read and then most went on their way while I shook hands with a few people and signed some books. Among the last of those who introduced themselves was a man named Luke Davies, a poet finishing his degree at a local university and teaching in Sydney. He had come over to the museum on his lunch hour with a signed copy of his second book of poems, *Absolute Event Horizon*. He said he'd dedicated one of the poems to me. He didn't appear to be looking for congratulations for his work or for a way to insinuate himself. He seemed at peace with his life, a guileless person. We spoke for a few moments and shook hands good-bye. As he turned to leave, I asked if there was a way I might contact him. He gave me a phone number and I told him I was going to be in the city for a few days and might call.

The following morning I took a long walk in Sydney's botanical garden with an Australian landscape painter, John Wolseley, whom I'd once traveled with through Watarrka National Park in the Northern Territory. Later we went to the Art Gallery of New South Wales together where a number of his paintings were up, a retrospective show. He walked me through the galleries, making humorous, self-deprecating remarks but being serious, too, about the importance of art in a highly industrialized and commodified world. I admired the way he could infuse the static space of a canvas with time, creating images that were not animated by time but where both the presence and passage of time were clear.

I felt a peculiar sense of camaraderie as John and I went along, passing, at one point, the gallery where I'd spoken the day before. Our aesthetics were different, but we were enthusiastic about many of the same questions, like how to render and comprehend the way time gives space another dimension. John's life was so unselfconsciously about art. He was someone who had become his own idea.

That evening I called Luke. I told him I wanted to visit Botany Bay, where Cook had made his first landfall in Australia, on April 28, 1770. Did he want to come along? Luke lived at Bondi Beach, next to the water in east Sydney. He said it would be easy for him to come by the hotel and pick me up.

We drove past the airport, crossed the Georges River on the Captain Cook Bridge, then turned east onto a road that went out to Inscription Point, on Kurnell Peninsula, where we parked the car. The fair weather of the past few days was holding—salubrious might have been the word—behemoth cumulus clouds with flat bottoms and rounded shoulders in a cerulean sky, holding faint shadows in their thick folds.

Cook's landfall here marked the beginning of a major shift in European thinking about possibilities for trade in the South Pacific. In 1606, Willem Janszoon, a Dutch sea captain, made the European discovery of the Gulf of Carpentaria, on the northeast Australian coast, west of Cape York. His landing on the shore there, the exploration of the west coast of the continent in 1619 by Frederik de Houtman, and of the southwestern coast in 1627 by François Thijssen and Pieter Nuyts, established a Dutch claim to a "new Holland," the east and most of the south coasts of which had not yet been seen by Europeans.

In 1642 Abel Janszoon Tasman, in the employ of the Dutch East India Company, set out from Batavia in the *Heemskerck* with a consort, the *Zeehaen,* intending to expand the sphere of Dutch influence to the south. He rounded the northwest and southwest capes of Australia, discovered and doubled the Tasman Peninsula (he assumed it was the continent's southeastern cape, not knowing it was the south coast of an island, Tasmania), and continued on to "new Zealand," leaving the east coast of Australia unexplored. He returned to Batavia in 1643 via Torres Strait, which separates Cape York Peninsula from Papua New Guinea, having discovered neither Bass Strait, which separates Tasmania from Australia, nor the Great Barrier Reef.

In an informal sense, Tasman was the first European to circumnavigate Australia, to prove that it was not the northernmost part of a fabled southern continent. The question of whether New Holland included an inland sea was still unsettled, however, and would remain so until 1798–1803, when Matthew Flinders and George Bass circumnavigated Van Diemen's Land (Tasmania) and then Flinders and Nicolas Baudin charted all the rest of Australia's coastline that remained undemarcated. Flinders's surveys, made in several ships, confirmed that there was no entrance to an inland sea north of the Southern Ocean's Great Australian Bight. Australia was one landmass, not two.

Early in his naval career, Cook established himself as a cartographer by creating impressive charts of the coasts of Newfoundland. His charts of Australia's east coast—Flinders, a great admirer of Cook, was full of praise for their accuracy—became the basis for England's claim to eastern Australia, called New South Wales by the British to distinguish these lands from those of New Holland.

When Cook looked beyond the entrance to Botany Bay at dawn on April 28, he thought the embayment, beyond high headlands on either side of the entrance, comprised a large, sheltered harbor. He entered the bay later that afternoon and anchored off its southern shore. Perhaps it was Eora people he saw camped there on the beach. Whoever they were, they all but ignored the ship, continuing to go about their business. When a landing party put ashore, the Eora walked away, leaving two men armed with spears standing to meet the sailors. When people from the boat tossed iron nails and colored beads to them, they ignored the trinkets; when someone shot off a musket, the Eora barely flinched. They had never seen, probably, nor ever heard about sailing ships like the *Endeavour*—or people like these Europeans.

As the party stepped ashore—the first to do so was eighteen-year-old Isaac Smith, Cook's wife's cousin—the two Eora men retreated from the beach and joined the others who'd withdrawn into the fringing forest of gum trees. As members of the landing party approached the Eora's bark huts, the owners stepped back farther into the trees. The sailors found a few children hiding behind a warrior's shield in one of the huts and gave them strings of beads. When Joseph Banks picked up some fishing spears to examine—he had them taken back to the *Endeavour*—he suspected they might have poisoned tips. He cautioned Cook to stand well away from the Eora.

Cook's officers and crew, shouting at the Eora, denouncing them as cowards, went off in search of freshwater and wood, which they continued to load aboard the *Endeavour* over the following week. Banks and others with an interest in natural history explored the perimeter of the bay and the lower reach of the Georges River. The field parties made a large collection of plants, work that later inclined Cook, long after he'd left the area, to change the name he'd originally given the place, "Sting ray's harbour," to Botany Bay.

When the onshore winds that had kept Cook penned in Botany Bay longer than he wished finally abated, he sailed out between the headlands and northward up the coast. Over the following four months he would chart nearly the entirety of the shoreline—and his expedition would almost end tragically when the *Endeavour* ran aground on the Great Barrier Reef.

Cook wrote that Botany Bay was "Capacious safe and commodious." The field parties' notes describe waters rich with oysters, clams, and mussels, trees of imposing size (likely casuarinas), and impressively large flocks of cockatoos, parrots, and waterfowl. The naturalists judged the soil's potential to support farming poor and said they were puzzled by the Eora's lack of interest in the gifts they were offered (the Eora having drifted back onto the beach to occupy their huts, opposite the anchored *Endeavour*). "All they seem to want," Cook wrote in his journal, "was for us to be gone."

An Ordinary Seaman from the Orkneys, thirty-year-old Forby Sutherland, died of tuberculosis while the *Endeavour* lay at anchor in the harbor. Cook named the inner point of the south headland for him and he was later buried at sea.

The general impression of Botany Bay that Cook, his officers, and the supernumeraries aboard took away was favorable, even enthusiastic. Nine years later Banks recommended the area to Parliament as a destination for prisoners. Eight years after that, the First Fleet sailed for Botany Bay. On arriving the captain decided instead to go ashore at a smaller harbor just to the north, Port Jackson, a place that would one day become the city of Sydney.

LUKE AND I WALKED OUT to Sutherland Point and read the inscription there on a plinth, erected close to a spot where Cook had ordered his date of departure, May 6, 1770, and the name of his ship carved in the bole of a tree. The public park where the plinth stood—the tree is long gone—seemed now part of a thoroughly humbled place at the tip of the Kurnell Peninsula, in contrast to the wild land Cook had found there more than 230 years before.

Luke and I stretched out on the grass and talked about the books we'd each been reading. I asked him for a list of writers whose feel for things Australian impressed him. I told him I'd been reading David Malouf, Helen Garner, Tim Winton, and a few others, and thought that Malouf had a great gift. Luke agreed. I asked him more about his own work. He'd just finished a novel, *Candy,* which would later become the basis for a movie, for which he would cowrite the script. The novel was a fictionalization of his battle with heroin addiction.

I listened to his description of the broken life out of which the book had come, a life of thievery and scams, of manipulation and self-hatred, of suicidal despair. But here he was, just out of university after the long delay occasioned by his addiction. I'd read his *Absolute Event Horizon* and thought the poems were very good, the work of a singular imagination. In the middle of our conversation about literature—given a kind of wrenching twist by his telling me about the background for his novel—he said the turning point in his heroin addiction—he'd been clean for three years at this point—came when he read a book of mine called *Arctic Dreams*. It changed his perspective, he said.

I understood then why he'd come to the reading.

I told him about a conversation I'd had a couple of years before this with a few writers at the festival of arts in Adelaide. The festival committee had put some of us up at a resort about thirty miles outside the city in the days before the festival got started, so we could get to know one another a little. One morning after breakfast five or six of us began talking about what we thought we were up to as writers. The group included Canadian novelist Susan Swan, a young writer from India named Vikram Chandra, John Coetzee from South Africa, the American writer Annie Proulx, and David Malouf. Someone asked whether, despite the differences in our cultural backgrounds, despite the difference in gender, in literary taste, in the genres we liked to work in, and in our politics—despite all these differences, was there some subject we were all writing about, one way or another? Everyone immediately said the same word. Community. Why does it fall apart? Can you put it back together? What makes that smallest of communities, marriage, cohere? How do we go on with life when we've chosen to remain cut off from our traditional communities, or chosen to remain with others who have no interest in knowing who we are?

Luke said he could understand that, could see how all of us, including himself, were writing about the functional and dysfunctional dynamics of different sorts of communities, the integrity of which, or the possibilities for reconciliation within which, provided us with a promising, or at least a believable, future.

The idea seemed so big, so close to self-celebration, we dropped it. But we believed in it.

We fell silent, basking on our backs in sunlight on the great lawn there, the rain-softened ground. A Greek chorus of lorikeets, of turquoise and king parrots, of cockatoos and galahs, sailed back and forth above us, beautiful, dazzling, streaming colors, the birds babbling and calling sharply, as if they had not yet gotten the word that we were all civilized now.

Occasionally a 747 or one of the smaller Boeing jets, a 727 or 737, decked out in Qantas or Singapore or KLM or Air Canada livery, lumbered overhead on final approach, pushing against a wind out of the north that we could barely feel on the ground, and floated down to touch the runway at Sydney Airport, which jutted a mile out into Botany Bay like a quay.

I showed Luke a bit of technology I took out of my backpack and was now about to use, an early version of a handheld GPS device. I said I liked the air of authority it seemed to be equipped with.

"It'll tell you right where you are, you know," I said.

"Really? It's that accurate?"

"Well, it's very precise, but I don't know how accurate it is. It says we're at thirty-four degrees, zero minutes, eleven seconds south latitude and one hundred fifty-one degrees, thirteen minutes, and thirty-two seconds east longitude."

Cook made this spot 34° 16' South and 151° 21' East. But it's the same place, where we lay on the south shore of the bay that afternoon, many years later, watching clouds, the birds, the planes, each of us glad of the other's company.

The device has no power to determine any further where we are, by noting that cumulus clouds, with their involuted heads of cauliflower florets, were passing through. Or how the spaciousness of the sky here changed when flocks of birds flew over us in a rush. Or how all this might look if it happened to be raining. The numbers marked a portal, like the address on a house.

Graves Nunataks
to Port Famine Road

Queen Maud Mountains

Central Transantarctic Mountains

Northern Edge of the Polar Plateau

Antarctica

———

Brunswick Peninsula

Shore of the Strait of Magellan

Southern Chile

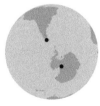

86°43'39" S 142°07'39" W to 53°25'43" S 70°59'22" W

To get oriented here is difficult. The light is flat because the sky is overcast. The sun's weak rays create only a few anemic shadows by which to judge scale and distance. Far-off objects like mountain peaks have crisp edges because the atmosphere itself is as transparent as first-water diamonds; the mountains, though, are not nearly as close as they seem. It's about –12° F, and the wind is relatively calm, moving over the snow distractedly, like an animal scampering.

Four of the six people living here are in their tents now, next to their cookstoves, two by two, warming up and preparing their suppers. I'm the fifth of the group, almost motionless at the moment, a hundred yards south of the tent cluster, kneeling on a patch of bluish ice in the midst of a great expanse of white. I'm trying to discern a small object entombed there a few inches below the surface. Against the porcelain whites of this gently sloping landscape, I must appear starkly apparent in my cobalt blue parka and wind pants. I shift slowly right and left, lean slightly forward, then settle back, trying to get the fluxless sunlight to reveal more of the shape and texture of the object.

The sixth member of our group, wearing a turquoise windbreaker and yellow wind pants with red knee patches, is working at a fuel cache some ways to the west. He's rolled a snow machine over on its side and is adjusting one of the bogie-wheel trucks, mechanisms that tension the vehicle's drive belt. He is gauging the tension by tugging on the belt with his bare hands. When the light breeze that's blowing falls off a bit, the ratcheting clicks of his socket wrench carry several hundred yards to where I am, but I can barely hear them through the fabric of my balaclava and the hood of my parka.

The three yellow pyramid-shaped tents the six of us are bivouacked in form the points of an isosceles triangle with a long base. They all face north, their backs set against a prevailing katabatic (gravity-driven) wind from the south. The generous space between them is insurance that a fire in one is less likely to spread to another, and the arrangement ensures there won't be a constricted area where the wind might eddy in a blizzard and pile snow against a tent entrance.

The arrangement of the camp is simple, tight, and to my eyes elegant. Food caches and equipment stored in the open are flagged so they can easily be located again after a storm. We have an emergency shelter—one tent and its full complement of supplies—buried in a snow pit fifty yards to the west should, for example, a cookstove's flames somehow ignite a tent wall and high winds whip the fire into a conflagration. The place is designed for safety, convenience, and economy of movement

Seventy-five feet to the north we've dug a latrine.

This arrangement comprises field quarters for a National Science Foundation deep-remote cold camp, established at 7,460 feet in the Transantarctic Mountains, 220 miles from the South Pole. We've been living on the rim of Antarctica's Polar Plateau, part of an ice cap four times the size of Greenland, which forms the continent's vast interior. The U.S. Geological Survey (USGS) quadrangle map of the region immediately to the south of us, between us and the Pole, depicts the two southernmost outcrops of bedrock on the planet, Mount Howe and D'Angelo Bluff. The lower half of the map is empty white space. An irregular line, marked "Limit of Compilation," separates the two outcrops and a few associated crevasse fields from all that lies to the south. Attempting to define this blank space would be like trying to sketch contour lines on a map of the ocean.

We have no source of heat but our cookstoves, and the four men and two women have been here for nearly two weeks.

On this summer day in mid-January no one other than the six of us is to be found in the surrounding region. Scientists and support staff living at Amundsen-Scott South Pole Station would be the closest. It's eleven-thirty at night according to my watch, but this far south the hour of the solar day is of no help in trying to understand the situation we're in, or our sleep rhythms. In a few minutes the never-setting sun will break through all but the last layer of cloud cover. It will hang there, burning in the sky like a molten coin, nineteen degrees above the horizon. Its light will strengthen the triangular shadows of the tents and those of the two of us still working outside.

I'm convinced now that what I've found buried in the ice in front of me is a meteorite, a small dark rock, an object not of this Earth.

WE'VE JUST RETURNED from seven hours of searching at Graves Nunataks, two miles distant. (*Nunatak* is an Inuit word for an isolated mountain peak standing proud of a large ice sheet.) It is mostly the wind that determines the length and shape of our days here. When it quietens, dropping below about ten knots, no longer thickening the atmosphere with loose snowflakes and creating an obscuring ground fog, it's safe for us to leave camp and navigate with confidence between our camp and ice fields around the nunataks, where we're looking for meteorites. We're not concerned about suddenly losing sight of camp behind the contours of the surrounding hillocks of snow and ice; what we're watchdog alert for is any sustained change in the intensity of the wind.

If it should pick up and hold, our workday is done.

With twenty-four hours of sunshine available almost every day during the height of the austral summer, from late November until mid-January, it's not the advent of "night" and "day" that tells us when it's time to take up again the search for meteorites. And it's not just the wind that can hamper us. It's the temperature as well. The colder it gets, generally, the shorter our workday will be.

We've all set our watches to the same minute, an additional precaution in a situation like this, where coordinating with one another is critical. We keep to New Zealand Daylight Time (NZDT), the time at Christchurch, 2,900 miles to the north, the city most of us rendezvoused in more than a month ago, and from which we flew to

McMurdo Station. McMurdo, 719 miles north of our camp, the major American scientific base in Antarctica, also runs on "Kiwi time." At 08:30 NZDT every morning we establish radio contact with McMurdo, conveying information about our weather and assuring them that all is well. If we're not up on the radio at that time, someone in McMurdo will be opening a manual of emergency procedures to determine what the next step will be.

Three years before we arrived, four scientists, the first people to visit this part of Antarctica, landed nearby in a Twin Otter. They off-loaded two snow machines and together searched several square miles of bare-ice fields at the foot of the nunataks. They wanted to determine whether there were enough meteorites sitting on these stranding surfaces to warrant putting in a full team to collect them and to conduct a more thorough reconnaissance. An expensive and major operation, that. The four men flew back to McMurdo and later decided the area was rich enough in extraterrestrial material to warrant putting in a camp for a full forty-day field season.

We are that projected field party.

We spend our days at this moderately high altitude, in the coldest and most remote of Earth's deserts, looking for bits of debris from a great shower of stones that daily peppers the planet. It's the allure of this simple empirical task together with the ur-remoteness of the region that have brought me here. Also my friendship with the man in the turquoise windbreaker, John Schutt. For years we've been hoping to hatch a plan that would put us together in the field in Antarctica for a few weeks.

John, a geologist and alpine guide, is the director of our field party.

I get to my feet—at this point in my life the cold saps my strength a little more quickly than I'm comfortable admitting—and return to the westernmost of the three tents. I'm certain I've located the expedition's 156th meteorite just a short ways from my and John's tent during my evening stroll. I'll describe it to John. After supper we'll chip it free of the ice and collect it, observing all the necessary formalities.

I wait by the tent's entrance—a collapsible canvas tunnel—until John finally looks my way. I mime putting food in my mouth and point sharply with my mitt to my chest. "I'll get supper going," I mean. John signals with a wave and a salute, and goes back to work on the snow machine.

In the middle tent at this same moment, the one at the apex of the triangle, the two women in our group, Nancy and Diane, have finished their supper and are preparing for bed. In the easternmost tent, Paul and Scott are playing pinochle. A little more than eight hours from now the four of them will drift over to our tent to hear the weather reports from twenty-two American field parties, all but three of them living in semipermanent heated camps set up across many thousands of square miles of ice, mostly on the western side of the continent. We will be especially attentive to the weather reports from camps within a thousand miles of us but we will identify most closely with the reports coming from the only other deep-remote cold camps. Like us, these two small field parties are living closer to the weather than those in the other camps.

Using our limited ability to predict the weather, based on changes in local barometric pressure, together with reports from the other camps and information McMurdo relays to us from their weather station, we've been reasonably successful in anticipating the sort of storms that will make a day of work impossible for us. In the morning, if the wind is up and a ground blizzard is blowing, the others will still make their way to our tent for "weather ops," using numerous red, blue, and green cloth flags tied to bamboo poles anchored in patches of wind-hardened snow to guide them. (Nancy, Diane, and Paul have never camped in conditions like this and are especially anxious about having to use the latrine in a storm. It's sheltered on three sides by snow walls and its perimeter is demarcated by flagged poles. During heavy weather, though, all you might be able to see is the next flag on your route.)

SOME DAYS I wonder where the rest of us would be without John. He's been out there for more than half an hour fixing one of the snow machines, going at it bare-fisted in the tight places that won't admit a hand wearing a glove. At every turn—four-cycle engines, electronics, crevasse rescue—his knowledge far outstrips my own. Ever since we first met, eleven years ago at McMurdo Station, we've enjoyed our experience together "on the ice." Whenever I've arrived here to accompany scientific field parties he's always been staging gear for one of the meteorite teams.

An appetite for physical engagement with the world of snow, ice, and rock beyond our tent, and having an opportunity to work together, almost always in silence, are desires John and I share. We're comfortable being confined in the limited space of a Scott tent, we split the cooking chores easily, and we observe the same unwritten rules that ensure each person a bit of privacy. I like the rhythm of our daily problem solving and the hours of stories and reminiscence we share in the tent on storm-bound days, the physical and technical challenge of the work the six of us all do, and the deep sleep that comes with exhaustion. Humans, I think, were built for this. We can do it superbly.

WE'VE HAD A ROUGH TIME with weather. Five of us flew into McMurdo from Christchurch on December 4 to meet John. Since then, local weather, either at McMurdo or at our put-in site on upper Klein Glacier in the Queen Maud Mountains, has been too stormy for us to fly. Or the logistics to get clear of McMurdo have been too challenging, with four or five other scientific parties all trying to get out into the field at the same time.

We spent nineteen days in McMurdo before we were able to fly south, most of it working to stay busy. We didn't need nineteen days to stage the expedition. Six or seven would have done it. We had to tune up and test-drive the snow machines. We had to get three of our party through two days of safety-and-rescue training. And we needed to prepare a dozen pallets for the cargo planes, strapping down our food boxes, personal gear, collection equipment, camp gear, and the snow machines and sleds. Two ski-equipped LC-130 Hercules aircraft would get us up to Klein Glacier. Once there, we'd off-load the pallets, set up a temporary camp, then start loading supplies onto our Nansen sleds, which we'd then tow behind the snow machines on a thirty-five-mile traverse to Graves Nunataks.

While those new to Antarctic camping learned how to refuel cookstoves, set up Scott tents, use portable radios, and navigate crevasse fields, John and I requisitioned food and inspected every piece of equipment we were taking into the field with us, looking for excessive wear or any flaws.

It frequently goes like this for scientific parties seeking to deploy from McMurdo. Waiting. All flight plans are tenuous because of the continent's notoriously undependable weather. Some field parties, having secured the financial and logistical support of the National Science Foundation (NSF), and having worked out, painstakingly, a research plan and flown in to McMurdo, find their field season finally canceled. In the end, they were never able to get out of McMurdo—except for the return flight to Christchurch.

I used our long delay as conscientiously as I could. I continued my reading about chondrites and achondrites. I studied inorganic chemistry texts to better understand what my five companions were talking about. Ralph Harvey, the team's principal investigator, offered me several blackboard sessions, which took me to the margins of my ability to comprehend the sometimes esoteric chemistry involved. (Because of the long delay in McMurdo, Ralph would not make it to Graves Nunataks with us. Scott, scheduled to replace Ralph midway through the expedition, did make it, but this arrangement limited Ralph to only three days in the field before he had to depart.)

Most every morning during the delay John and I walked over to the weather center at McMurdo to learn where we stood. With the elevated view the weather operations tower afforded us, we could quickly see whether our pallets had been loaded that morning or were still sitting in the same spot out there on the ice runway, so we had the beginnings of an answer to our question before we asked. (About a dozen anemometers that were torn apart by winds in excess of a hundred knots are mounted around the room on wooden plaques near the ceiling, like a frieze.) Are we still scheduled to go today? we ask. No, canceled. Maybe tomorrow? The answer to that is usually just "We'll see" or "Maybe." The meteorologists and flight planners don't commiserate with people in a delayed field party. They don't offer explanations or encourage hope. There are rumors enough around town, of course, without this. John and I leave Mac weather ops (McMurdo Station Weather Operations), nodding politely to members of other field parties in the room, all hoping for better news than we've just gotten.

One morning at the weather center a flight planner I'd worked with on earlier trips to Antarctica said she could offer me a seat on a helicopter ferrying supplies to a field camp at Cape Crozier, forty-five miles away on the east end of Ross Island. We weren't going south that day, I knew, so I was immediately interested.

During the winter of 1911 three members of Robert Falcon Scott's ill-fated 1910–1913 polar expedition had made the round trip from Scott's winter quarters at Cape Evans, on the west end of Ross Island, just down the coast from McMurdo, to Cape Crozier. They would endure weeks of very cold temperatures—as low as –77° F—and the scientific research they went there to conduct at an emperor penguin colony at Cape Crozier would, regrettably, never be put to use.

I'd long hoped to be able to visit the ruins of a rock shelter that this small party had built at the cape to shield themselves during a storm, and had in fact made an unsuccessful attempt several years before this to locate it. As it happened, the flight planner who spoke to me that morning remembered this disappointment of mine and offered me a second opportunity, which I took.

Sometimes serendipity cancels out bad luck at McMurdo.

ON THE MORNING of December 23, our bad luck ran out and we were airborne. The Hercs landed us and our gear at the upper (southern) end of Klein Glacier. We set up a temporary camp and prepared for the traverse to Graves. The next morning a fresh storm blew in shortly after we left. We were each towing two or three fully loaded sleds behind our snow machines as we climbed moderate slopes in thirty-knot winds, which became more and more challenging as the day went on. At one point a gusting wind punched through at the edge of my snow goggles. The sudden rush of violent, cold air caused my eyes to water. Tears splashed across the inside of the plastic lenses in my goggles and froze there, completely obscuring my view. I had to stop on a steep incline I was traversing to clear the lenses. While I did, the wind, blowing broadside to my sleds, began pushing the two of them down the slope, swinging them below me. In the swirling snow I could barely make out the others ahead. No one was following behind me, I knew, and the heavily loaded sleds were beginning to pull me down the incline.

Moments of mounting tension like this tend to stand out in retrospect, but in fact such moments are little more than predictable interruptions in a plan you can never assume will work smoothly anyway, especially in situations like this. To keep from panicking, you focus on what you have to address first, then move on to the next thing. I cleared my goggles, anchored them firmly against my face, inched ahead on the slope to get out in front of the sleds hanging below me, and followed the tracks of the sled ahead of me. Soon I had everyone in sight again.

They'd pulled up and were waiting.

We traveled sixteen miles that day, southeast from the head of Klein Glacier, gaining 1,600 feet in altitude before setting up a second temporary camp, hoping for better weather the following day. We positioned our tents near a small rock outcrop in the icecap called Inuksuq and remained there all the next day, pinned down by high winds. When the wind dropped, a hard, depthless blue sky replaced the layer of overcast above. From the height of Inuksuq, John picked out a route that would circumvent a crevasse field and take us to Graves Nunataks, which we could now see clearly, about fourteen crow-fly miles away (nineteen actual miles by snow machine, in order to bypass the crevasse field). We arrived there late that afternoon.

The clear weather and dintless skies held for another day and a half, during which we located and collected several dozen meteorites. Then a new storm confined us to our tents again, this time for six days. Our original plan had been to have our camp set up at Graves by December 12. By this point, then, we had lost nearly three weeks of time in the field. As it turned out, we would get in only eight more full days of searching for meteorites before our scheduled pullout date. There was nothing for it but to let the frustration turn to bemusement and to utilize any time we had, whenever the wind dropped, to pursue our search.

———

OUR CAMP AT GRAVES is isolated geographically, but we're also cut off electronically from the outer world. We have no satellite phone and no means of tuning in to an international news program, like the BBC's. Our solar-powered radio communications with McMurdo are rudimentary, and McMurdo's policy with deep-remote camps is not to pass on any personal news except of a death in the family.

I enjoy the sort of mental space this kind of isolation affords. There are no intrusions here, no unexpected inquiries or announcements. One can unfurl a thought without fear of interruption, unfurl it until one decides he's finished with it. No phone rings. No doorbell, pager, or intercom sounds. No one knocks.

The isolation encourages you to think in a different way about what it means to be human, and to consider the long stretch of humanity's epoch. And the strangeness of this place. Nearly everything on Earth can be referenced by using chemistry, physics, and biology. But that's not the shape of reality here. The interior of Antarctica is about chemistry and physics, not biology: the rock exposed above the ice has a chemical composition; gravity, in the province of physics, causes the ice to flow downhill to the ocean. And it's the pressure of accumulating snowfall that turns snow to ice in the firn zone below us. More physics. This is Earth without life. No birds fly across the sky. No plants grow. The spoor and tracks of animals do not appear. The wind scours. There's no gurgle of flowing water. The polar night, like the polar day, is months long.

We, the six of us, and the scientists and workers at the South Pole, are all there is of biology here, for tens of thousands of square miles.

We're camped in an abiotic ocean of seemingly stayed time and nearly undifferentiated space, beneath a fall of Archean light. Our presence seems as inconsequential as the death of a mayfly. Yet I am as comfortable in this place as one hand would be resting in the cradle of the other.

It feels so oddly safe here.

———

METEORITES LAND IN a random pattern across the entire surface of Earth, falling day after day no more in one place than another. Most are lost to view immediately in the world's oceans, lakes, and rivers. Many don't stand out sufficiently amid Earth's ordinary rocky debris to draw any notice. And of these, most weather and erode into fragments more or less quickly. In Antarctica, however, the unusual dynamics of the ice environment not only preserve an inordinate number of meteorites but actually concentrate many of them in clusters on top of the ice, in areas called stranding surfaces. (The blazing object we commonly see in the night sky, the "shooting" or "falling" star, is a *meteor*. A *meteorite* is what someone picks up on Earth, the metallic or stony remains of a *meteoroid*, which is any random bit of solar system debris with the potential to enter Earth's atmosphere.) It's in Earth's atmosphere that the meteoroid becomes a meteor, because of its friction with the air. Hundreds of millions of meteoroids, many of them the size of sand grains, enter the Earth's atmosphere every day. Most burn up entirely during their descent.

The majority of the meteorites that reach Antarctica make a comparatively soft landing on a cover of snow. Over time, as more snow falls on them (and the bottom of the moving ice sheet below them slowly melts, because of its friction with Antarctica's bedrock and the ice sheet's encounters with geothermal hot spots), these meteorites move down deeper in the layer of snow they landed on, until that layer reaches a transition zone. Pressure from the burden of snow above becomes sufficient at this point to reconfigure the snow crystals, turning them into crystals of ice. The meteorites thereafter lie embedded in a mass of moving ice, like raisins in a cake. As the mammoth ice sheets flow downhill toward the sea, they encounter bedrock obstructions, the most formidable and prominent of which is the continent's spine, the Transantarctic Mountains. To get around these obstructions and continue on to the sea, the ice sheets flow slowly toward areas of least resistance—mountain passes. Where Antarctic bedrock stands proud of the flowing ice sheet—the case at numerous places in the Transantarctic Mountains, such as Graves Nunataks—the deep horizontal layers of the ice sheet are forced to bend upward and flow vertically. Eventually these layers reach the surface of the ice cap with their loads of meteorites, and the wind bares them to the sky.

The dominant winds of the Antarctic interior are katabatic—driven by gravity—not cyclonic. (Cyclonic winds are generated by changes in air pressure.) One way to think of katabatic winds is to picture them as gargantuan cataracts of air moving over the ice sheets. Because the pull of gravity is a constant, the *direction* of the flow of air hardly varies for a katabatic wind—it flows downhill as a river would on a slope. The *force* behind this river of air, however, does change, with the volume of falling compressed air, the changing contours of the ice surface, and the gusting of the wind.

Wherever the deep layers of an ice sheet are forced to flow vertically, they encounter the scouring effects of a katabatic wind as they reach the surface. These winds do two things that make Antarctica a mecca for meteoriticists. They shatter the crystals of any snowflakes that fall on the bare surface of emerging ice, scattering the debris and keeping the ice surface clean; and in a process called sublimation, the winds vaporize the ice as it emerges, turning it from a solid to a gas. There is no intermediate liquid stage. Over time, as the ice sheet continues to flow vertically and as the wind continues to erode the ice, embedded meteorites are left stranded one by one on the surface. Over millennia, the concentration of meteorites on these stranding surfaces can become very large, some concentrations running into the thousands.

Once scientists came to understand how this concentrating mechanism worked, they began systematically searching old aerial photographs of the Transantarctic Mountains for blue-ice fields, as they came to be called. Additional ground reconnaissance determined which blue-ice fields had the highest concentrations of meteorites. Available funding from the NSF, and the logistical complexities associated with putting a small scientific party in the field for five or six weeks, determined which sites might be most readily exploited. In keeping with Antarctic Treaty protocols, each meteorite found by members of a field party belongs collectively to every country that is a signatory to the Antarctic Treaty. The meteorites, including the ones we were finding, are shipped to NASA's Johnson Space Center in Houston, Texas, where they are made available to any qualified scientist. The name of the individual who finds a particular meteorite is not entered into the collection record, out of respect for the spirit of equality and common cause that the treaty embodies.

THREE YEARS AFTER the reconnaissance at Graves Nunataks, the six of us found ourselves at this dependably windy site, prepared finally to begin work midway through what was then the stormiest and shortest field season in the twenty-three-year history of the NSF-sponsored Antarctic Search for Meteorites (ANSMET).

Viewed on a grand scale, our camp at Graves sat on the periphery of what amounted to a back eddy in the flow of the ice sheet moving off the polar plateau. The back eddy was created by the nunataks, outliers of the main mass of the Transantarctic Mountains. Far to the east of us, the flow of ice broke into two separate streams around the La Gorce Mountains, a part of the Transantarctics. The northern stream became Robison Glacier, the southern stream Klein Glacier. Thirty miles farther north and east, these two glaciers flowed into, and became part of, Scott Glacier, which descends through the Queen Maud Mountains and becomes a part of the Ross Ice Shelf. *This* ice, carrying a load of meteorites that never happened to emerge at the surface, ultimately calves into the Ross Sea, an embayment of the Southern Ocean. The calved ice might be released in the form of a tabular iceberg more than a hundred square miles in extent, or set loose as a small ice floe that soon melts somewhere in the vicinity of Antarctica, dropping its meteorites to the bottom of the Southern Ocean. From there, these fragments from the asteroid belt, from Mars, and from Earth's moon, find their way eventually into the planet's upper mantle.

ONCE TAKEN IN HAND and placed under a microscope, each meteorite is revelatory. The overwhelming majority of them come from the asteroid belt, between Mars and Jupiter, and are so distinctive, one from the other, that scientists have been able to create a kind of geography of the asteroid belt, a geologic map that allows them to push deeper into our still-hazy understanding of how the solar system evolved. In short,

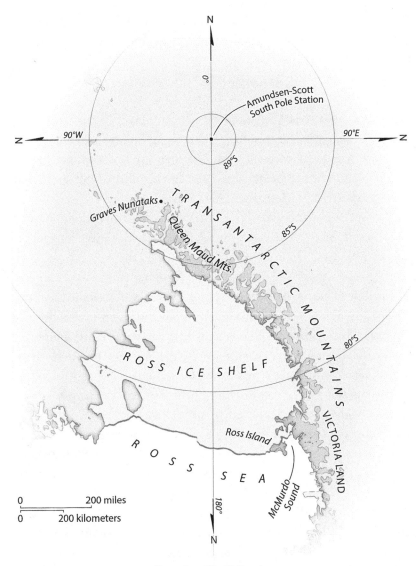

Ross Ice Shelf Region

every meteorite represents an important contribution to the unraveling of the mystery of Earth's origin. Therefore, though the six of us will find only 186 meteorites—the fewest ever by an ANSMET field party—our weather-compromised effort will still be viewed as successful.

There are several nunataks at Graves Nunataks, all peaks of the same mountain. Each one is slowly shedding its weather-exposed face, which falls to the stranding surfaces below as rocky debris. To search efficiently for meteorites also stranded on the ice, each of us must learn to visually separate this terrestrial debris from extraterrestrial material. On our first day here, then, we climb up to the exposed ridge of one of the nunataks to examine and memorize the color and grain patterns of the rock there. Later, when we're lined out on the stranding surfaces and walking six abreast, we'll be able to visually sort the rocks resting on the ice, mentally discarding everything but a meteorite. A shout from one of us will bring the others to a stop. We'll each mark the place we're standing at that moment and then gather around the find.

We fix the location of the meteorite using a GPS device and record its general characteristics in a field notebook—the species of meteorite, its size, color, shape, and whatever else about it that seems noteworthy. One of us then picks the meteorite up with a pair of sterile tongs and deposits it in a sterile transparent collection bag, which is then sealed. Back in camp, certain meteorites might be examined once more before being packed in reinforced cases for transshipment, first back to McMurdo aboard an LC-130, then to the Astro Materials Acquisition and Curation Office at the Johnson Space Center, and eventually to the Smithsonian, in Washington, D.C.

John has been doing this for so long he's able to make an informed guess about the pedigree of nearly every meteorite we find. And he'll often notice that three or four meteorites we've just collected are all part of a single meteoroid that shattered on impact. It's as if the images of meteorites he's seen over all those years drift through his memory like the faces of people he remembers.

JOHN CRAWLS IN through the snow tunnel and begins stripping off his weather gear. With both cookstoves going, the tent is relatively warm, about 40° F just above floor level and over 50° F where the tent walls meet at the tent's apex. (We hang damp socks, boot and glove liners, handkerchiefs, and scarves up there to dry.) It'll stay that way until we shut the stoves off when we go to sleep.

Now that John's fixed the track assembly on the snow machine, he's fiddling with our radio, which has been acting up. At this point, in the middle of his nineteenth season, John is the best-known scientist associated with the ANSMET project, after Bill Cassidy, the visionary who started these searches in 1976. (It was during his second field season that John and Ian Whillans found the golf-ball-size meteorite near the Allan Hills that turned out to be a lunar breccia, the first piece of the moon identified on Earth.)

I tell John about the meteorite I found on my walk around the perimeter of the camp that evening. While I cook and he rewires the antenna lead to the radio, we talk through a plan for the days remaining to us. He tells me we have had so many delays to start with, and so many tent-bound days since we arrived, that ANSMET will have to come back the following year, or maybe the year after that, to finish searching the stranding surfaces here. With just a few days left, John's leaning toward more reconnoitering, with less time spent actually collecting. Tomorrow, he says, we should search around the southern flank of the nunataks and probe bays in the steep east face of the mountain for any concentrations of meteorites. We might collect some of the larger ones but should concentrate on flagging as many meteorites as we can and sketching them in on a map we'll draw for each bay.

While I finish up the dishes, John visits with the others in their tents. He tells them that if the wind doesn't pick up tonight and we still have sufficient contrast under these cloudy skies to read the surface of the snow for crevasses, we'll leave again in a few hours. He wants us to climb a steep ice slope on the west flank of the nunataks and then search to the east, along the nunataks' south flank, for any meteorite concentrations. He wants at least this one quick look at the only side of the nunataks we've not visited yet. Three of our group are uneasy about climbing the slope with snow machines, but John exudes a kind of understated confidence in the three of them, which they trust.

While he's out, John chips a bucketful of ice from the glacier our tent sits on to replenish our water supply. (When we're weathered in, we still must leave the tents to get ice to melt for water. We also always refuel our cookstoves out there, to be absolutely certain we're not close to anything flammable in case of an accident.)

When he returns, John has Scott in tow. He asks if the two of us would mind collecting the meteorite I've just found while he concentrates on transferring some of the GPS coordinates he's been writing down onto sketch maps of the bays we've been working in.

Whenever I'm the one tasked with taking notes while we collect a meteorite, I'm aware of the precision and finitude of the numbers I'm writing down, of how infinitesimal these particular data points are in the overall effort scientists make to understand what happened 4.5 billion years ago, during the early stages of the geological development of a planet on which the conditions to support biological life would develop, 93 million miles from a nuclear furnace that circles us each day along the horizon, the track of a halo tilted slightly to the south.

The following evening, after we return from our reconnaissance on the south side of the nunataks, John and I fall into conversation about the ultimate significance of our work. We're at a point in the expedition where a question like this often arises, during the closing days, when what still needs to be done comes under intense scrutiny. Like most good scientists, John is not entirely convinced of the ultimate authority of the rational mind, and he recognizes the potential for peril in strict cause-and-effect reasoning. He doesn't like the way much of science, particularly laboratory science, discounts awe and mystery, as though the capacity to respond to reality in this way was something to outgrow. I tell John that in the years I've been coming to Antarctica and working with different field parties, I've watched the scientific respect for "data sets" supplant scientific respect for firsthand field experience, and have wondered where this trend will lead. I've worried about the impatience with which the inevitable loose ends and inconclusiveness of fieldwork is often met, and the modern preference for theory, and the recruitment of numerical data to support one or another theory.

Back in McMurdo we've both witnessed changes as the hallways of the old science building, perennially crowded with camping gear, have given way to the antiseptically tidy and brightly lit hallways of the Crary Science and Engineering Center. The corridors of the new building buzz with the ceaseless clicking of keyboards, a kind of white noise accompanied by the electronic beeps that signal a task has been completed or information is now awaiting retrieval. The numerical results of a theoretical approach, of someone's plumbing the nimbus of numbers surrounding a little-understood event, are both esoteric and arcane; and the speed with which they're produced, and the sheer volume of them, is intimidating. The process suggests that knowledge has been obtained, but in fact there is not much more here than staggering precision and a quantity of numbers significant enough to support statistical probability. Massive data sets represent irrefutable truth for some, or insights that transcend previously established boundaries, but the data might be no more than intensely self-referential. Impressive but unconvincing.

The belief that one can reach a state of certainty about anything acts as a goad for those who regard the anomalies that inevitably turn up in their data not as a caution but as an inconvenience.

"I had a theology professor once," I said to John, "who told us that religion was not about being certain but about living with uncertainty. It was about being comfortable with doubt, and maintaining the continuity of one's reverence for a profound mystery."

I wasn't sure John heard me. He was reclined on his sleeping bag with only his lower legs visible to me past a pile of gear. Perhaps he'd fallen asleep. It'd been a long day.

"We gain deeper knowledge," he responded. "But no guarantee that we're any closer to wisdom."

A FEW HOURS after we fell asleep that night, the wind woke me, punching the tent wall behind my head, a creature moved to a state of rage. Its caterwauling, its screaming wail, the pitch of it rising and falling, the decibels of it, would all suddenly collapse nearly to silence, then mount again. The sound of it shimmered in my ear, like light striking the eye from a sheet of shaken foil. The tent shuddered on its stout poles and the tent fabric strained at its triple-sewn seams, seething and popping. The inconstant tympanic thrumming of the fabric was an intonation underlying one shrieking run after another of banshee notes, some of them single tones within the squalling wind that sustained themselves for several seconds before dropping an octave. It might be hours of this before stillness returned. Or days. Or it might all fall apart and cease in a few minutes.

The day that we climbed to the summit of one of the nunataks at

Graves, to orient ourselves and to examine shattered slabs of sedimentary and metamorphic rock on the ridge, our footsteps generated the sounds of broken crockery. I turned one rock after another over in my gloved hands, to get its measure, to take it in more completely. In the absence of any other kind of life, these rocks seemed alive to me, living at a pace of unimaginable slowness, but revealing by their striations and cleavage, by their color, inclusions, and crystalline gleam, evidence of the path each had followed from primordial birth to this moment of human acquaintance. Each rock I examined, all of them ostensibly remnants of the same dark slabs, was nevertheless distinguished from the others by some rosette of color, some angularity that made it stand apart. As I sat there, reluctant to put down a single one of these "undistinguished" rocks, contemplating the history of each one in the gigantic sweep of time that was for them a "lifetime," they suddenly seemed wilder than any form of life I'd ever known. Like the wind, they opened up the landscape.

SINCE THERE WAS no telling when a storm might make travel back to Klein Glacier difficult, or indeed impossible, and with the cargo planes scheduled to land there on January 20, John has decided to take advantage of a period of clear weather on the eighteenth to break camp. We're so conscious of how many days we've been tent-bound, feeling so aggrieved about the attenuated search for meteorites, it's hard for us to capitulate, but John's right. We're done. We pack the sleds and leave for the landing zone under cloudless azure skies, sailing through great intermontane basins of calm air. A trip that had taken us more than fourteen hours, stopping and going, several weeks before, we now make in just four and a half hours. We erect a temporary camp on Klein Glacier. John gets the radio going and tells McMurdo we're safely there. McMurdo says only one Herc will be coming for us on the twentieth. The second will come in on the twenty-first, so we'll go out in two groups.

It takes most of the afternoon to unload the sleds and restage our gear on the pallets we left here, to prepare everything for the planes. When we've finished, we depart on our snow machines, headed for a valley at the foot of the La Gorce Mountains, several hundred square miles of unexplored, steeply pitched heights and deep valleys, first sketched out on a map of Antarctica in 1934, during an aerial reconnaissance.

To enter what is actually a cirque on the southwest side of one of the range's most prominent ridges, we must descend a steep slope of glacial ice. Moving laterally from one patch of snow to another—the snow machines have better traction here than on bare ice—keeps us from losing control on the descent; climbing out, we'll use the same snow patches, like stepping-stones, skittering from one to another. The mouth of the cirque is about four miles wide, an amphitheater about as deep as it is wide. The floor of the valley is a felsenmeer, a sea of shattered rock that has fallen over many millennia from heights a thousand feet above us.

The dark granite rubble of the felsenmeer, warmed by the sun, radiates an impressive amount of heat. John and I each find narrow slots between boulders in which to lie supine. We're protected from a light breeze that's blowing and bathed in sunlight. The air temperature is about 5° F, but it feels twenty degrees warmer in these "solar ovens." They offer us a kind of threshold, a road to another country.

All around us is the silence of deep space.

Nancy calls out, "John, what do you call this place?"

"Heaven."

"No, no. That's what *you* call it. What do *people* call it?"

He doesn't answer.

The cirque has no name, nor do the peaks above or the spurs radiating from the main ridge, the arêtes. Descriptive, eponymous, fanciful, memorializing, and valorizing nomenclature has not made it this far. The place seems so indifferent to our presence that for one of the few times in my life I find myself gazing into an enormous space with no sense of the time in which it exists. Most of the interior of Antarctica actually seems like this to me, not just uninspected or unnamed but unknown. It has not yet been snared in a catalog of designations and coordinates, of metes and bounds. The state of relief I feel, resting on my back in a slot in this boulder field, out of the wind and staring up at the surrounding ramparts of the cirque, causes me to hear John's response—*heaven*—not as a synonym for conventional feelings of ecstasy but as a word characterizing the absence of disintegration.

We rested there for an hour before climbing up out of the cirque and entering the other world.

AT THE FOOT of the Wilson Piedmont Glacier in northern Victoria Land, on the west coast of McMurdo Sound, some eighty-five miles by air north-northwest of McMurdo Station, is an indentation called the Bay of Sails—more of a bight, actually, than a bay. It is commonly assumed that the name derives from a recurrent natural phenomenon, the grounding here of large icebergs. Seen from a distance on a bright summer day, the bay appears to be crowded with sailboats racing before the wind. In fact, it was named after a technique members of a sledge party used, in the austral spring of 1911, to assist them in crossing the sea ice. They rigged masts and sails to help propel their man-hauled sledges.

Eight years before I made the trip to Graves Nunataks, I joined a team of benthic ecologists working under the sea ice in McMurdo Sound. Our field season eventually took us to the Bay of Sails and to a series of dives we made around the bases of grounded icebergs there. (Benthic ecologists study communities of organisms living on the bottom, or benthos, of bodies of water.) The National Science Foundation was funding the project in order to determine, in part, how polluted Winter Quarters Bay was, McMurdo Station's small natural harbor, which is ice-free for only a few weeks in late summer. (The harbor basin, about the size, shape, and depth of a large European soccer stadium, was named by members of a 1901–1904 British expedition, the first serious effort to reach the South Pole, who overwintered on that shore.) Early in the history of McMurdo Station, when it was being operated by the U.S. Navy, the military dumped construction debris, raw garbage, and barrels of toxic waste on the sea ice of Winter Quarters Bay. When the ice melted during the short summer each year, all this waste sank to the bottom of the bay. Eventually the bay gained the distinction of being the most contaminated harbor in the world, its bottom covered with decommissioned shipping containers, transformers leaking PCBs, rusting barrels of corrosive fluids, broken machinery, and discarded furniture and mattresses.

We spent the early days of our field season collecting samples of organic life from the bottom of the harbor, which we then compared in the lab at McMurdo with bottom samples taken from unpolluted waters in close proximity to Winter Quarters Bay. During the latter days of the project we began studying places in McMurdo Sound where the benthos had also been radically disturbed but by natural or nonanthropogenic events, such as iceberg groundings.

On the day I'm remembering, we crossed McMurdo Sound by helicopter and landed alongside a massive iceberg in the Bay of Sails. Its sheer, brilliantly lit walls, abutting the frozen sea at a 90-degree angle, were impossible to take in without sunglasses. We drilled a four-inch-wide hole through about seven feet of sea ice next to the berg, then placed explosives there to create an opening large enough for us to use as a dive hole. Underwater here, we discovered the same extraordinary clarity we'd seen everywhere else in McMurdo Sound, except in the self-contained basin of Winter Quarters Bay. This early in the austral summer, before the annual plankton bloom, we were able to measure an astonishing 900 feet of lateral visibility. This meant that if the letters were large enough, we could read a billboard at this distance, or looking right and left underwater here, we could take in 1,800 feet of the side of the iceberg we were working alongside.

Above the surface of the sea ice, the regal presence of the icebergs, like austere blocks of alabaster, had an altogether different look than they did underwater. Below the surface they seemed malign, vaguely threatening, the nuptial whites of the carapace above being a deceptive guise. No one was comfortable swimming close to them, especially in the shadows where they had plowed into the bottom, like ships run aground.

We surveyed the communities of mostly motile sea life here—starfish, scallops, urchins, nemertean worms—at the depth we routinely worked at (65 feet), and then began photographing a series of small scientific platforms. These square metal tiles, mounted on the tips of steel posts embedded in the bottom, supported communities of sponges and other small sedentary life-forms. Most of the platforms, erected twenty years before, had been destroyed by icebergs; those that remained gave some hint of what type of sessile, or sedentary, communities come and go in this periodically disturbed environment.

The water is extremely cold, 28.6° F (3.4° F below the freezing point of freshwater and about 0.2° F above the freezing point of seawater). After an hour down here, most of us are ready to surface and warm up. On this day, however, with the day's work done, I wanted to linger. Stretched out horizontally on my back in the currentless water, forty feet below the dive hole, I started to roll over slowly, like an idling dolphin. My eyes traveled first across the underside of the sea ice cover above me, with its dark communities of epontic (ice-associated) algae, until they reached the distant edge of the ice cover, meeting the dark water there and then the field of black cobbles below me. The cobble plain sloped away beyond into the depths of an unlit ocean. My eyes roved the potholes and undulations of the cobble plain, where massive red starfish, pale green urchins, and a kind of large, long-legged sea spider called a pycnogonid moved with nearly imperceptible slowness. Winding their way among these creatures were white nemertean worms, three and four feet long and as thick as my forearms. Between me and these benthic creatures, pods of slow-swimming scallops pass by, like flocks of birds in slow motion. At the end of this rolling, I see the walls of the iceberg rising, and then I am looking straight up through the roughly round, six-foot-wide lens of still water in the dive hole. Through it I can see a cobalt blue sky and the dark-clad bodies of the other divers in their dry suits, their faces glowing in sunlight reflecting off the sea ice.

They're milling about, basking in that torrent of sunshine. No one is in a hurry. I begin the slow roll of 360 degrees again. Tomorrow we'll go somewhere else in the Bay of Sails and examine a similar situation. That night I'll go over in my mind what I saw, and a desire to study it again more closely will overcome me.

Traveling and working in Antarctica, I'm residing in the only place on Earth without an aboriginal history, a place with only the very fewest threads of modern human history. An old piece of Gondwana. Inspiring, not empty. Almost everything one sees here is new. The familiar but very misleading separation between *human* history and *natural* history has no footing here.

WORK UNDER THE SEA ice brings all of us into a richer world than any of the early explorers suspected was here. They didn't have the technologies to enter this space. Despite a cover of solid ice two meters thick, the bottom, down to about twenty meters, is surprisingly well lit because the ice is translucent and the water column is largely devoid of life before plankton blooms in the austral spring. In areas undisturbed by icebergs, the organic life is astonishingly rich—we estimated that each square meter contained between 100,000 and 150,000 organisms, a density of life that compares with the floor of a temperate-zone rain forest.

One day, at the tip of an island in McMurdo Sound called Big Razorback, I slid into a tidal crack with two companions to see what the benthic community looked like in an area scoured regularly by sea ice as the tide ebbs and flows. In an environment like this one, divers check on each other three or four times a minute, more often than usual. The tidal nature of these waters means large slabs of sea ice might suddenly collapse or shift. Or a piece of equipment, already working near its limits in these frigid temperatures, might malfunction. We're miles from the edge of the sea ice, where the ice cover of the sound meets the pack ice of the Ross Sea, and where orcas and predatory leopard seals prowl for penguins and seals; but it's not good to forget that these sea mammals, who need to surface regularly to breathe, could somehow also turn up here, deep in the semipermanent ice cover.

A more subtle danger divers face in Antarctica is that because the water is so transparent, it's easy to misjudge distances, and therefore easy to swim off too far from the dive hole. Based on your experience with clear tropical water (not very helpful), you could easily swim over to a place you believe is about 100 feet away. It turns out to be 300 feet away, and suddenly you're much lower on air than you planned and faced with a 300-foot swim back to the dive hole, the only access you have to the surface.

When divers are swimming together in areas less dangerous than those around tidal cracks, they tend to check on each other less frequently. To get the others' attention, a diver might tap on her tanks with a dive tool, alerting companions to something she's found. On the dive that day at Big Razorback, three of us were swimming nearly shoulder to shoulder when we all heard that familiar sound. Who had made it?

Not me, we signaled each other. Strangely indifferent to the sound—we had to be more than a thousand miles from any other diver—we swam on! Then we heard the sound again. This time we all turned around and found ourselves face-to-face with an adult Weddell seal, all ten feet and 600 pounds of him or her. The sound was coming from the seal—we could see its throat muscles moving in synchrony with the clanks. We knew Weddell seals were around, but we didn't expect them to approach us this closely, to be this curious. (We had discovered a few caverns on the underside of the sea ice that day where seals had surfaced to breathe. When we located these domed cavities we made a surmise and surfaced in them ourselves. Removing our regulators, we inhaled the stagnant, vaguely fishy air of multiple seal exhalations.)

When I look back, what I most often recall about those weeks of diving is the contrast between the apprehension I brought to this sort of technically challenging work and the euphoria that so quickly replaced it once I saw the density, the bright colors, and the variety of organisms in the enormous carpet of life spread out below me. To be suspended above it was trance-inducing, what I imagine it must be like to drift over the Serengeti in a hot air balloon, several hundred feet above herds of unalarmed wildebeest and impala, the lions, giraffes, and stalking hyenas.

It's in the nature of places like Antarctica, regions of Earth where relatively few people are to be found, to present themselves in unexpected and unique ways, even to casual observers, to slightly contradict conventional wisdom. On all of our dives, for example, I saw small patches of bottom life flattened against the underside of the sea ice. How did they get up there? Could they survive there? It was later explained to me that in some spots a weak bottom current might eddy around a cluster of benthic creatures, rocks, and bottom sediments and come to a complete halt in some crevice or notch. Here, a few molecules of seawater might freeze. Over time this initially small platelet of frozen freshwater might expand (as seawater crystallizes into platelets, it squeezes out the sea salts that keep seawater from freezing at 32° F), creating a growing matrix of freshwater ice crystals. (The specific gravity of freshwater allows crystals of it to float in seawater.) At some point the expanding mass of freshwater ice becomes large enough to exert an upward force sufficient to uproot a section of the bottom. This scrap of the benthic community continues to float upward until it lodges on the underside of the sea ice cover.

One day I almost swam straight into a dark basalt cobble floating in the water column in front of me. I assumed it was encased in freshwater ice, but I could find no angle of observation that made this apparent. Had I not learned what can happen in these very cold waters, I would have had to conclude that here in Antarctica, dense rocks float.

WANDERING THROUGH THE LABS at McMurdo Station one night, talking to people at random about their work, I came upon a desiccated jumble of benthic life sitting on an examination table. When I asked the researcher tugging at it with a large pair of tweezers what it was, she explained that parts of the benthos not only occasionally rise up to press against the sea ice but can actually migrate through it. (The process is somewhat comparable to what happens during a meteorite's journey from its fall on a snowfield to its emergence on a stranding surface.) In the water, as new ice forms on the bottom of the sheet of sea ice, parts of the benthos that have risen to its underside are entombed. Surface winds, meanwhile, are exposing the uppermost layer of sea ice to sublimation. Whatever is embedded in the sea ice, therefore, moves slowly closer to the surface. This researcher had collected the mat of sea life she was examining weeks before on the surface of the Ross Ice Shelf. In some places the ice shelf is over 2,000 feet thick. It could have taken centuries for these desiccated communities to emerge.

She hopes to learn what the benthic community looked like down there when, all those years ago, these organisms were alive on the bottom.

When benthic life emerges on the surface of the Ross Ice Shelf, it is fragmented and scattered by the wind, so how these creatures and plants once fit together is hard to determine. This is also, of course, the case with attempting to determine the shape of certain Paleoeskimo dwellings in the High Arctic. If a group of Thule people arrives at the site of an old Dorset camp and uses those stones to build dwellings of their own, how can one be sure what the parent Dorset dwelling looked like?

When someone comes upon a couple of bones from the rear leg of a mala, searches for and is able to find most of the rest of its bones, and then studies the imprint of tracks in the soil nearby, along with tooth marks on the bones and other clues, it's possible to figure out what killed the mala and what happened afterward, as other animals became involved in the event. To put back together what nature or man has taken apart, to make something whole and integrated again out of its remnant parts and puzzle out how it came to be dismantled, constitutes the area of science called taphonomic research.

One can argue, I think, that such a formal inquiry into disintegration and the attempt to achieve reintegration is an emerging force in the arts today, just as it is in archeology and field biology. It's part of a much larger determination, far and wide across human cultures today, I believe, to understand a familiar type of menacing levigation, the fission of integrated communities.

MCMURDO STATION IS LOCATED on one side of Ross Island's Hut Point Peninsula. A New Zealand station, Scott Base, is situated on the other side, at a place where the McMurdo Ice Shelf, a small spur of the Ross Ice Shelf, meets the enormous expanse of sea ice covering McMurdo Sound. Our group hoped to dive in this spot where the ice shelf, nearly one hundred feet thick at this point, abuts the sea ice, only six or seven feet thick. The relatively thin sea ice offered us light; the ice shelf, on the other hand, presented us with unusual under-ice structures. We hoped to pick out subtle shifts in the composition of the benthic communities here as we ventured farther into the side-lit darkness underneath the ice shelf.

We entered the water through a tidal crack in front of Scott Base and soon were swimming through massive convolutions in the ice shelf's seaward wall. It was like swimming through the interior of a drowned cathedral, gliding above the aisles and the nave, peering into the grottoes of side chapels, floating past the choir stalls, and rising into the domes of the ceiling bays. Looking up eighty feet or so into the irregular geometry of the ice shelf front, bathed as it was in late evening sunlight

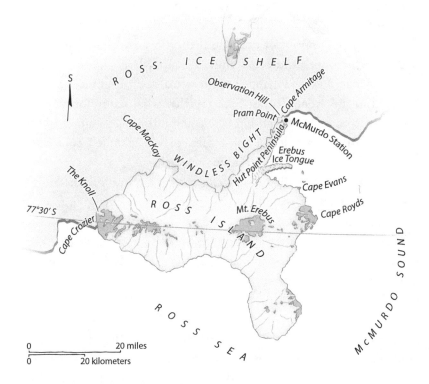

Ross Island

from the northwest, I felt like I was standing in the apse at Chartres, gazing up into groined vaults between the capitals of the columns, complexly curved surfaces lit by the cathedral's clerestory windows.

In Antarctica there was no end to the wonder.

SOME EVENINGS, after finishing up work at the benthic lab in McMurdo, identifying and counting organisms, I'd look up John Schutt. I'd usually find him around the Berg Field Center, staging gear for that year's ANSMET expedition and awaiting the arrival of other members of the team. One responsibility he had was testing the snow machines people were going to use, so some evenings he and I went out together, running them at full throttle across the Ross Ice Shelf (more testing, perhaps, than they might have really needed). Often this

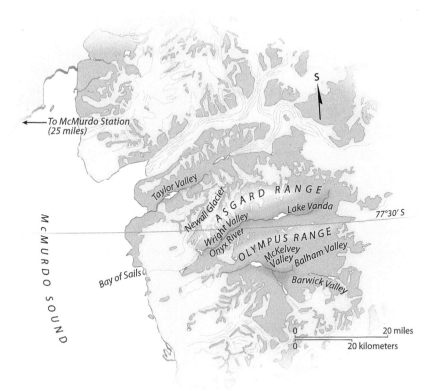

The Dry Valleys

was after midnight. The sun would by then have come around to the northeast, front-lighting the peaks of the Royal Society Range on the far side of the sound. The textures of these mountains would sometimes come out so forcefully in the crystalline air that we'd stop the machines, lay back on the seats as though they were couches, and just watch. The rays of the sun brought out more strongly a third dimension in the peaks and intensified their colors. I remember on one of those nights talking to John about the dives, about how easily one might come to think of Antarctica as geographically desolate, not knowing of this vivid underworld of color and movement. Emperor and Adélie penguins rocketing past as you worked. Levitating rocks. He told me about a meteorite crater he'd been working at on Devon Island in the Canadian High Arctic, during the boreal summer preceding the austral summer we were now in. The meteoroid had hit with such force it melted the ground around it, making it look like foundry slag. The heat turned quartz intrusions in the bedrock into chunks of colored glass.

We went on like this for hours, as though there was no being done with the astonishment landscapes might offer us, or to the potential for any seemingly inconsequential thing out there to startle and inform, or for the most ordinary event to kindle one's hunger to experience again what was still undisturbed and beautiful. One brilliantly clear afternoon when we saw the sun crossing behind the Royal Society Range we quit whatever it was we were doing at the Berg Field Center and sat down to watch. At the head of the Koettlitz Glacier, where the ice of the polar ice cap crowns slightly as it flows over a shoulder of bedrock, the low light of the sun was passing through the glacier's interior. For a few moments it seemed that this part of the glacier was lit from within. The sun burned there like a lightbulb shining through a parchment shade.

IF THERE WAS a siren landscape for me in my forties and fifties, it was Antarctica. It was difficult and expensive to get to without major help; weather still played a challenging role in whatever you wished to do; most of its particular geographical corners were still unexplored (though each year field parties reached a few more of them). It was relatively easy, unless you were part of the workforce employed at McMurdo, to get off by yourself or get out in the field with a few others doing things you never imagined. Importantly, Antarctica is the only continent that has not been nationalized and which might still be characterized as virtually unpopulated. Anything might turn up here, like the most sane, equitable international treaty humans ever negotiated and signed, the Antarctic Treaty. This treaty places the gathering and sharing of knowledge ahead of subjugating the land (a ridiculous thought, here, to begin with); forbids military maneuvers; and encourages ethical comity.

All that's missing, for me, is a wider understanding of how little human history this place has. When I looked at the mountain ranges to the northwest of our camp at Graves Nunataks, I saw not only mountain ranges no one had ever been in but a land in which no animals familiar to me had ever made a home. When we left the southern end of Klein Glacier, headed for Graves Nunataks, and had to make an emergency stop at Inuksuq, I was reminded of the Thule, pressing on eastward from Skraeling Island in summer. When I closed up an unknown, unnamed gelatinous ctenophore underwater in a two-quart specimen jar, to be photographed in the lab that night and then released back into the sea, I felt like William Bartram, exploring the St. Johns River in Florida with his father in 1765.

ON MY FIRST JOURNEY to Antarctica I accompanied two other journalists on a press tour. Our host, a gentleman from the Office of Polar Programs at the National Science Foundation, escorted us on three or four brief trips, showing us in just a few days much of what is iconic about the continent. The time was adequate to acquire a general sense of Antarctic history and of the scientific research efforts the United States was funding there. We traveled by helicopter to Cape Royds, where Ernest Shackleton's polar expedition overwintered in 1908, and then to Cape Evans, where Robert Falcon Scott's party overwintered in 1911 and 1912. Both capes, located on the west coast of Ross Island, front McMurdo Sound and are relatively close to McMurdo Station. The diligence of several New Zealand historical societies, a paucity of visitors, and the extremely cold, dry weather have combined to preserve the buildings at both places in a nearly pristine state. (Scott's pocket watch was still hanging on a hook by his bunk the first time I visited.) The immediacy of the past at sites like this is eerie. If you felt free to do so at Cape Evans, you could have stretched out on Captain Titus Oates's upper bunk, bunched up his pillow, and, removing his bookmark, begun reading at the page where he left off in a book he never saw again. The hardy sun-bleached buildings with their heroic gloom still evince the dutiful lives of resolute men, a hundred years after the abrupt departure of those who survived the tragedy of Scott's final expedition.

We were also flown out to McMurdo Sound's ice edge, where inquisitive emperor penguins approached us cautiously, waddle-walking across the ice; and pods of orcas passed close by in the open water and paused a dozen yards away to spy-hop, curious about our presence. We were flown 840 miles south to visit Amundsen-Scott South Pole Station and also had time at a cacophonous Adélie penguin rookery on Ross Island. On several occasions we were helicoptered out from McMurdo to inspect scientific camps in Victoria Land. Most of the geologists, glaciologists, limnologists, chemists, and other scientists working at these sites lived in heated semipermanent tent shelters with plywood floors, called Jamesways. They were equipped with cots, tables, and chairs, and occasionally garrisoned with a small support staff, which would include a cook if the camp was large enough.

Our weeklong exposure to Antarctic history, to the diversity and importance of the scientific work there, and to the scale and breadth of the land itself, left the three of us astonished and admiring. We returned to the United States to write—insightfully, we each wanted to think—about what we'd seen for *The New York Times, The Atlanta Journal-Constitution,* and in my case, *The Washington Post.* Amid all that I had found exciting and fascinating during my introduction to Antarctica, however, I also observed several dark threads, the kind of undercurrents that move beneath any enterprise of this scale once it becomes institutionalized. By the 1980s waste management had become a major problem at McMurdo Station, and the NSF had been pressured by a contingent of Greenpeace representatives to attend to it. (Greenpeace had by then set up their own base near McMurdo Station, to monitor NSF's compliance with the treaty.) The military infrastructure at McMurdo, primarily responsible for airborne logistics, a tradition held over from Operation Deep Freeze in the 1950s, had become a disruptive and frustrating presence for many scientists by the 1980s. The military had worked tirelessly from the start of scientific research in Antarctica to prevent female scientists from working there, arguing that the conditions were too challenging for women. By the end of the 1980s, military protocols, such as no flying on Sundays, were creating too great an impediment to scientific work, and eventually much of the military presence at McMurdo was phased out. More efficient private contractors came in to replace them. During the years the military maintained its presence there, however, only two ways of imagining the world were encouraged at McMurdo: scientific inquiry, with its focus on rational analysis and data; and military order, with its focus on chain of command and strictly enforced regulations.

As military operations in Antarctica were reduced, the NSF started up an Antarctic Artists and Writers Program, an effort to support the work of writers, painters, and photographers, primarily, who wanted to interpret the continent. The NSF's goal with the program was to have artists and writers convey to the lay public a more complete image of Antarctica than science alone was able to provide. After my initial visit as a journalist, under the auspices of the NSF, I applied and was accepted several more times as a participant in the Artists and Writers Program. My first opportunity, as it happened, followed immediately on my first stint there as a reporter. The NSF invited me to stay on an additional week at McMurdo and provided me with logistical support that got me out to a couple of additional field camps.

OF THE HALF DOZEN or so outstations I'd visited during my first week in Antarctica, the one that had worked most on my imagination was a New Zealand base called Vanda. I was able to make arrangements to visit it once more during my second week there.

The tight cluster of five or six pale green one-story flat-roofed huts at Vanda stood out starkly on a dark, snowless plain, a tiny green checkmark in a vast basin of space, a dot in a long, barren, narrow valley slung between the ramparts of two mountain ranges, the Olympus Range to the north, the Asgard Range to the south. The buildings were chained to the ground at all four corners to hold them against ferocious katabatic winds that fell off the polar plateau and sometimes swept down the entire eighteen-mile length of the valley at sixty knots or more.

Though it's almost completely covered by ice and snow, Antarctica is in fact a desert. Snowfall is scant. And in several places along its coast, where valley glaciers have retreated, the land surface has been scraped bare by the wind. (These particular glacial retreats occurred long before the modern-day glacial retreats caused by global climate change.) These dozen or so expanses of exposed ground are called dry valleys. They're found at two very infrequently visited spots in East Antarctica, the Bunger and Vestfold Hills, and also, famously, in Victoria Land, where they form a series of parallel valleys running perpendicular to the western shore of McMurdo Sound. Vanda station is situated in the middle of one of them, the Wright Valley, next to a permanently frozen body of water, Lake Vanda. (The eponymous Vanda was a lead dog for a sledging party, the first to arrive here, in 1958.) To the north of the Wright are several more dry valleys, three of which—Barwick, Balham, and McKelvey—are closed to human entry in order to preserve them in a completely undisturbed state for as long as possible.

It hasn't rained in the Wright Valley for two million years, and what little snow falls here is immediately shattered and sublimated by the wind. A geologist's eye would note right away that rocks lying loose on the valley floor have never been sorted by the force of moving water—small rocks sit beside big rocks, and grains of rock are not clustered together. Their distribution is completely random. Generally speaking, biological life in Antarctica is restricted to the coast; life in the interior is represented only by lichen, algae in the snow, and microscopic creatures living in protected environments. Deep within the microfractures of some rocks in the dry valleys, for example, where they're shielded from desiccating winds and where concentrated solar energy can provide moisture from a melted snowflake, microorganisms called cryptoendoliths are able to survive where logic once dictated life was not possible.

I arrived at Vanda station by helicopter, shook hands all around with people I'd met the week before, and was shown my bunk. The interior of the main building is tidy, spartan, and snug. Solar panels and wind generators produce electricity quietly, complementing the soft-spoken atmosphere researchers maintain here. I'd already gotten a good look at the station on the press tour, and had interviewed the scientists working there. I was actually most eager to go for a long walk.

What also drew me to Vanda was the administrative simplicity of this outpost for the curious, and the starkness of the place—the buildings looked like a handful of fishing dories gammed up on the North Atlantic without a mother ship. The classical lines of the Wright itself, too, called out to me, so fundamentally different from the rococo lines of a tropical jungle. The austerity of the valley was made even bolder by the transparency of the dustless air, through which reflected light from the farthest mountains reached us with no loss of detail. And it was here too, in the Wright Valley, that Antarctica's longest river, the Onyx, flowed.

I set off northeast along the course of the Onyx with two companions, men bent on addressing some scientific responsibilities they had farther up the valley than I intended to go. While they attended to that, I planned to climb up to the saddle of Bull Pass in the Olympus Range for a look into the sequestered McKelvey Valley. If time permitted, I'd hike west a short ways along the spine of the mountains. I was in radio contact with the other two men, but the landscape was so naked and the air so clear, we were rarely out of binocular sight of each other.

The milk-blue sky was clear of obvious clouds, the winds were light, and the temperature pleasantly warm, around 20° F. Near the top of the pass I came upon a smooth, shallow slope. Its sandy surface was pocked by rocky fist-size debris shed from the mountain walls above. Most of these rocks had come from a thin vein of dolerite, a fine-grained black igneous rock similar to basalt. Those that had not rolled farther down the slope were settled here on a layer of compacted sand. Over time, katabatic winds, bearing grains of sand and slivers of ice, had abraded some of the rocks and eroded the ground around them, leaving each one mounted on a small pedestal of sand, like a faceted gem sitting on a jeweler's dop. They resemble small deliberately shaped modernist sculptures, the convergence of their flat, dark planes suggesting the form of a pyramid. They're called ventifacts. Things made by the wind.

At the top of the pass I radioed my companions below to tell them I was going to walk a mile or two to the west, along the crest of the range. The fact that Bull Pass had a name was no indication that anyone had ever been here, and in effect, Bull Pass was a dead end—the McKelvey Valley just to the north was unexplored and permanently closed. I had not seen anything like a human footprint since I left the banks of the Onyx, and in the moment, I had the thought that I might be the first person ever to hike west along the crest of the Olympus Range, toward Mount Jason. As I set off, I allowed myself to hold on to this thought, partly because the history of human exploration in the interior of Antarctica is so exceedingly thin. I was intoxicated with a sense of the great space surrounding me, the deep Earth history on display, a scene easy to come by in the Transantarctics. Some time into this reverie I came upon a camera case lying on the ground.

I picked it up. I looked it over thoroughly. The upper half of a black Nikon 35mm SLR case. Chastised, I put the case in my backpack with my survival gear and went on. The urge to make an exclusive claim runs deep in a culture like mine, where individuals fear more and more a loss of identity, the onset of anonymity. I was privately mortified by the fact that I had entertained the illusion that I might be the first to walk this ridgeline. To what degree had this adolescent daydream of mine taken me away from what was actually here?

I doubled back to the pass some time later. From that height I could see the morning sun glinting on the river below, a line of silver neatly bisecting the lower Wright Valley. Across the way I could see the instructive mountain walls of the Asgard Range, a deck of sedimentary layers, mainly sandstones, called the Beacon Supergroup. In one of these layers, formed about 200 million years ago, scientists have found the fossil remains of a reptile, *Lystrosaurus,* and a fern, *Dicroidium.* The identical organisms have been found in rocks of similar age in South Africa, Australia, India, and South America, evidence that supports the modern idea that these continents once formed a single continent before drifting apart in the early Mesozoic.

I descended from Bull Pass to the valley below without having gotten a view into the McKelvey Valley. It appeared I'd have to hike a ways to the north before I could see into it, and I didn't want to take the chance of crossing an invisible line and ending up in the protected area.

TECHNICALLY, THE ONYX is a meltwater stream. When the glacier that once filled the Wright Valley retreated west toward the polar plateau (its remnant is called the Wright Upper Glacier), it left an ice mass at the lower end of the valley (the Wright Lower Glacier) which is the major source of the Onyx River's water. In January, at the height of the austral summer, the Onyx flows unobstructed in gently sinusoidal sweeps southwestward to Lake Vanda. In its lower reach, the section I'd hiked beside that morning, the river is about thirty feet wide and less than a foot deep. (Years later, at Graves Nunataks, I thought of the wind as the only animal living in the interior of Antarctica; on this day the Onyx had made a similar claim on my imagination, a thing living where no other thing lived, but no more in the realm of biology than the wind.)

I emptied my canteen of McMurdo Station water and filled it from the river.

On their return, the two scientists met me with chunks of ice chipped from the nose of a glacier that descends into the valley from the Asgard Range, and we headed back to the base together. Glacial ice fizzes and pops as it melts, releasing Earth's atmosphere of many thousands of years ago in bursts. The small tote of glacial ice the scientists were bringing back would be greatly appreciated by the other scientists. At the time, glacial ice, whenever it was available, was used in a ritual at Vanda station, a ceremony meant to bring a visiting guest into the exclusive company of those in the so-called Royal Drambuie Society. This ritual was little more than a friendly drink together, but it had an edge to it. The New Zealanders disliked the overfriendly presumption and swagger of the young American military pilots who flew scientists and visitors to Vanda. The pilots couldn't join the Royal Drambuie Society because they couldn't drink while on duty, so this was a way for this otherwise friendly group of people to send them a message.

Every visitor, on the other hand, including the pilots, was offered freshly baked scones with their tea upon arrival, and an invitation to join the Royal Vanda Swim Club. Candidates were escorted out onto the ice covering Lake Vanda, where they were asked to remove all their clothes except the heavy wool socks most everyone wore, which provided traction on the ice. Those who were still game descended a spiral staircase carved into the walls of an ice shaft that ended twelve feet below in the extremely salty waters of Lake Vanda. Full immersion—in over your head—was required for membership.

Not everyone opted to join. The ambient air was always colder than the water, and if the wind was blowing, the chill factor could easily be well below 0° F. Back in the hut, successful candidates were given a jacket patch and an official wallet card attesting to their daring.

One year, a few of the military pilots, for whom joining the club was a rite of passage at this duty station, began surreptitiously filming women while they were undressing for the Vanda plunge. Admonished by the New Zealanders to behave better, the pilots responded with jeers. Thus was born the impulse to create the Royal Drambuie Society. When some of the videotapes, including images of female staff members of the National Science Foundation, began circulating at McMurdo Station, the commanding officer of the flight squadron ordered an end to the practice. The military pilots, once they were officially reprimanded, became the most vigilant enforcers of the new "no photography" protocol at the Royal Vanda Swim Club's outdoor facilities.

The morning I reluctantly left Vanda—I felt like I'd had three glorious days at an exclusive spa, despite the necessarily spartan accommodations and plain meals—a female geologist, whose patience I had apparently tried, took me aside to inform me that I was inexcusably confused about the difference between a stone and a rock. The terms are not interchangeable, she said. A stone was a rock that had been put to some utilitarian or cultural use by a human being. Thus a headstone, a paving stone, a cornerstone, and Stonehenge. A rock was something that had *not* been handled by a human being. Thus a rock-ribbed coast and highway warning signs about falling rock. I didn't inquire about rock gardens but said I was grateful for the clarification. In the years following, I myself was able to annoy a number of people by requesting that the distinction be observed.

DURING THE WEEK following my visit to Vanda station I spent a few days in a camp in the Taylor Valley, just to the south of the Wright. One reason I'd asked to visit this camp was that the scientific party here was headed by a female principal investigator, one of the few at that time who'd been successful in getting funding from the NSF for remote fieldwork in Antarctica. Diane McKnight's team was composed mostly of geochemists and biogeochemists studying stream flow and stream chemistry at Lake Fryxell, Lake Hoare, and Lake Bonney, all frozen lakes in the Taylor Valley. Melt streams in the Taylor are little more than rills, but the chemistry being done here, crucial to understanding the larger overall impact of climate change, doesn't depend on analyzing large volumes of water.

In my usual role as a low-level field technician, I helped with the work going on in Diane's camp but often used my free time in the evening to explore the valley. One day, hiking alone across a slope above Lake Fryxell, I found a mummified seal, a young crabeater that had died there. Its carcass had been dried out and preserved by the wind, about seven miles inland from McMurdo Sound. At that time more than forty mummified seals had been located in the dry valleys, some much farther inland than this one. No one knows why some seals meet this fate. Of the several hypotheses offered, the one that makes the most sense to me is the one that would occur to an Inuit hunter—the presence occasionally of a "water sky" over the valley.

If a dozing seal, especially an inexperienced one, awoke on the sea ice to find that the crack or hole it had hauled out from had closed or frozen over, it might attempt to reach open water by searching for a "water sky." In an ice-covered landscape, a dark patch appearing in an overcast sky reveals a place below where light is reflecting poorly off open water, compared to the reflection of daylight off a surface of snow or ice. Absent any other opening in the ice cover it might find nearby, the seal would continue to move toward that dark patch. The sky above McMurdo Sound, however, contains a *second* set of naturally occurring dark patches, those above the snow-free dry valleys. If a seal headed off in the direction of one of the dry valleys' water skies, and never turned back, it would find itself stranded far from the open sea, where it would eventually die of exposure, hunger, and dehydration.

The taut skin of these freeze-dried animals feels smooth and hard to the touch, like the surface of a river cobble. No scavenger but the rare south polar skua, a seabird, ever tears into these mummified seals. Some bodies, unevenly dried out by the wind, arch upward from the ground in the shape of a semicircle, the rear flippers and the head both pointed toward the sky. Others lie prone, facing the wind, eye sockets bored out, mouths agape. With the retreat of its lips, a crabeater's molars and premolars, an ornate frieze of tiny, closely fitting cusps, stand out boldly in highly evolved but now useless efficiency.

Whenever I encountered these animals, I found it difficult to leave them. And when I finally left, often as not I stopped and looked back. They were inconsolable in their error. Most all of them had died alone. Some lay there with the clouded eyes of the blind.

———

EARLY ON A SUNDAY MORNING, the quietest time of the week in McMurdo, I met up with five companions at the snow machine shed. We lashed our survival gear, packed in bivouac sacks, to six Ski-Doos and proceeded in tandem slowly down a muddy, half-frozen road to the sea ice, turning south there at the shoreline. We rounded Cape Armitage at the tip of Hut Point Peninsula and climbed the low wall of the McMurdo Ice Shelf. From there we headed east-northeast to Cape MacKay, the halfway point of a fifty-mile journey that would take us to Cape Crozier at the east end of Ross Island, the site of one of the largest emperor penguin colonies in Antarctica.

By coastal Antarctic standards, this was a perfect summer morning, with temperatures in the mid-twenties and clear skies. The forecast was for continuing good weather.

The urge to make this journey came from reading Apsley Cherry-Garrard's *The Worst Journey in the World* (1922), a classic of Antarctic literature, written toward the end of the so-called Heroic Age of (mostly British) Antarctic exploration. On June 27, 1911, Edward Wilson, the father figure of Scott's 1910–13 expedition, Cherry-Garrard, a wealthy patron (and full-fledged member) of the expedition, and a feisty, diminutive British lieutenant named Henry "Birdie" Bowers left Cape Evans, bound for Cape Crozier and the penguin colony. (Bowers and Wilson would both perish nine months later on Scott's attempt to reach the South Pole.) The cape was only seventy-five miles away, over flat terrain; but it was midwinter and their clothing and equipment were barely adequate. Pulling two man-hauled sledges, initially laden with 757 pounds of food and equipment, they endured gale-force winds and brutal low temperatures on the monthlong trip. On July 15 they reached their goal, a prominence known as The Knoll, high above the penguin colony on the sea ice below.

To better shelter themselves at the cape, the men built a stone enclosure roofed over with sail canvas. Wilson's obsession—the rationale for the journey—was to collect penguin eggs. On July 20, with a break in the consistently stormy weather, they were able to descend a precipitous cliff face and collect six eggs, three of which broke on the climb back up the cliff. For the next five days the men remained in their hut while a storm tore the canvas roof apart and scattered their belongings, some of which, in the darkness, they never recovered. On July 25th they began their return journey. They met with fewer stormy days than they had on the trip out and arrived back at Cape Evans on August 1.

This story is widely known and most often retold in Antarctica with a mixture of awe, disbelief, and mild contempt. By modern standards the trip was a mindless bit of Edwardian bravado and scientifically pointless. (The eggs ended up at Edinburgh University, where they sat unexamined for decades.) However ill advised the trip was, though, Wilson, Bowers, and Cherry-Garrard had made an arduous effort to achieve something they believed in. The expedition was further driven by Wilson's genuine conviction that they were conducting important scientific research about embryonic development and what it revealed about phylogeny and evolutionary biology. Whatever judgment one might finally make of the advisability of the journey, it was epic. And Wilson's straining after new information is the archetype today for the process that produces Antarctica's massive and only export—knowledge.

Whatever my feelings about what had happened that winter at Cape Crozier, I had a strong urge simply to see the huge penguin rookery there, spread out on the sea ice below the gargantuan seaward wall of the Ross Ice Shelf, the crèche that, in winter darkness, these three men had never been able fully to take in. I had a similar desire as well to pay my respects to the three explorers at what remained of the stone-walled redoubt where they'd struggled to stay alive. One day at McMurdo it occurred to me that a small, experienced group of people, using snow machines, could retrace the route of "the worst journey in the world" in high summer, and do it in less than twenty-four hours, not thirty-five days.

To bolster my chances of getting approval from the lead NSF representative at McMurdo for such a trip, knowing some were sure to declare it too risky, I selected a group of highly qualified friends and acquaintances with backgrounds in search and rescue, and then added in McMurdo's head snow-machine mechanic. I presented my case to the NSF supervisor in his office on a Saturday afternoon, a few minutes before his office closed (and, I knew, just before he and his staff would be gathering at one of McMurdo's bars to socialize). I handed him a list of names with each person's credentials, along with all the required paperwork describing several tasks we intended to perform and signed forms from each person's supervisor, releasing them for the trip. Everyone in the group, I told the NSF representative, had read the guidelines governing entry to Specially Protected Area Number 6 (Cape Crozier) and had signed a form I'd prepared saying that they had. As for the route there and back, it had been flagged by a New Zealand group some weeks earlier and the flags were still in place. I'd checked.

The supervisor nodded thoughtfully at all the paperwork and finally said okay. Early the next morning, long before any supervisor might reconsider and ask for further discussion, we were gone.

After we doubled Cape Armitage on the sea ice, climbed the front of the McMurdo Ice Shelf, and doubled Pram Point on Hut Point Peninsula, we were flying. The twenty-nine miles between Pram Point and Cape MacKay took us across the Ross Ice Shelf's Windless Bight, an embayment on the south side of Ross Island where, mysteriously, Antarctic winds rarely stir. We were driving six abreast, ten yards apart, raising rooster tails of powder snow behind us. Past Cape MacKay the shelf ice begins to buckle, creating swales and some crevasse fields. We proceeded carefully here, coaxing the machines through the worst of the fractured ice. Eighteen miles past Cape MacKay we swung up to our left, climbed a snow-covered slope, and parked the snow machines at a previously agreed-upon spot outside the boundary of the Specially Protected Area. We doffed some of our heavy-weather gear, shouldered survival packs, and began hiking north, toward an overlook that lay beyond The Knoll.

What we saw when we got there had the same two effects, it seemed, on each of us. No one in the group was talking as we approached, but in that moment we all came to a standstill and remained in utter silence, motionless, for many minutes. Each person finally sat down on the snow apart from the others. In a kind of vast amphitheater on the sea ice below us was the sort of wildlife spectacle one fantasizes about seeing one day, and then gazes at in disbelief, as though confronted by an illusion, a scene that would resolve itself into ordinary reality when the spell broke.

The spell never broke.

Our view is east across the frozen headwall of the Ross Ice Shelf, about eighty feet high here. Icebergs that have recently calved from the ice shelf stand frozen in the nearby sea ice, like stranded buildings. To our left we can see the shoulder of Post Office Hill; to the right, the north face of The Knoll. Below us all is ice, radiant under a uniform fall of direct sunlight. The frozen sea is all gray and white—the grays of fog, of smoke; the whites of gypsum. The "alleluia plain" of sunlit sea ice here carries heterogeneous patches and narrow lines of dark charcoal, with dabs of light brown within the dark patches. With my binoculars, the dollops of brown resolve into fuzzy emperor penguin young, the charcoal patches and lines into adults. The orange blotches at the back of the adults' heads and the yellowish glow of their upper chests sharpen in the glass of the binoculars.

Useless, really, trying to count them. Hundreds stand on the sea ice amid the icebergs. The silence that had overcome us was only our bated breath; the air here virtually hums with the clatter and blare of the penguins' voices, their nasal cries. Perhaps these are cries of alarm set off by our arrival. For more than an hour we do not move. Eventually the penguins quieten.

The hour we spend with them is intimacy without narration, an experience without increments of measured time. The unvoiced emotions we felt, which we mention to one another later, include inexplicable tenderness, moments of soaring elation. In Antarctica, where death seems to lurk more than it does in other places, each of us is drawn strongly to anything as clearly alive as these birds. Feelings of affinity with these free animals, a sense of shared fate with them, seemed to go deeper and to come on more quickly here than elsewhere.

From my vantage point I can see tints of green and turquoise in the barrier face of the ice shelf, phantom colors, some of them. The changing angle of the sun lifts these pastels out of the ice and then releases them.

I've gotten cold, sitting here the hour. Finally I stand up. I work my binoculars slowly across the penguin colony, isolating individual birds and following their socializing among the others for a while. When you watch wild animals, it's impossible to know what they're really doing, or when they started doing whatever it is you see them doing, or when they've started doing something else. The minutes or hours a human might intently watch them don't create a valid framework for the animals' lives.

No one has said anything since we arrived, but some kind of end point arrives for the few still sitting and for the ones standing, and we begin to leave. Leaving here feels like walking out on a piece of music before it's finished, music that is so beautiful it fills you with something unbearable.

BACK AT THE SNOW MACHINES we dig out thermoses of hot soup and coffee and have our first meal since breakfast, then set out to find the stone shelter. We search for two hours, south and west of The Knoll, but can't come up with it. We repeatedly check the detailed map of Cape Crozier we'd brought along but are defeated. Either the map is wrong or we're inept. We give up.

It'll be colder on the way back. We work the machines down off the snow-covered slope and through the maze of ridges and fissures in the pressure ice. This part, we now know, will take longer to negotiate than we'd originally planned, but in a few hours we're back at Cape MacKay at the edge of the Windless Bight. By eight that night, fourteen hours after we left, we're back at McMurdo.

Years later, waiting in McMurdo for the weather to clear at Klein Glacier, I was offered a ride aboard a helicopter bound for an ornithologists' camp at Cape Crozier. After dropping supplies there, the pilot set me down about a hundred feet from Wilson's stone shelter. It seemed we had come within about thirty yards of it ten years before. I still can't imagine how we missed it.

The pilot told me to take whatever time I needed, that he had no further duties that night. In the eighty-seven years since Wilson, Bowers, and Cherry-Garrard had built it, all but the lower few courses of stone had fallen, these last still crudely mortared with scraps of vegetation. Bits of clothing—a sock, a piece of knitted wool—lay about. Shreds of sail canvas. The paper drawer from a small box of matches.

I circled the structure slowly twice. It was like circling the body of someone left on a battlefield.

In his much-praised study, *The Last Place on Earth,* comparing Roald Amundsen's and Robert Falcon Scott's separate quests for the South Pole during the same few months in 1911–12, Roland Huntford mercilessly eviscerates Scott for his mistakes and heaps praise on Amundsen, casting the one as a criminally negligent amateur and the other as a peerless professional. A British historian told me once that to really understand Huntford's book I would have to understand more fully what class envy and class hatred had to do with Huntford's vitriolic attack on Scott; that aside, I find myself agreeing with some of Huntford's analysis of why one man succeeded and the other died, along with four other men in his party, in an effort to reach the pole. Essentially, Amundsen crossed the Ross Ice Shelf, the Transantarctic Mountains, and the polar plateau as a party of Inuit would, using dog teams and outfitted in fur clothing and traditional Inuit foot gear. He had no real interest in science, only in fame. On the return leg from the pole, he fed his dogs to the other dogs. Scott undertook the trip with an attitude of cultural superiority, eschewing sled dogs for Manchurian ponies, all of whom would die miserably along the way, and championing man-hauling, which fatally exhausted him and his companions. Like many of his countrymen then, he considered the Inuit an inferior race, people with nothing much to teach an Englishman. He, too, was interested in fame, but he took the time en route to do pioneering science. And his expedition, compared to Amundsen's, was tragically thwarted by unusually harsh weather.

A large wooden latinate cross, a memorial to Scott and his polar party, stands atop an elevated prominence at McMurdo Station called Observation Hill. The view from there, out beyond White Island, is over the enormous Ross Ice Shelf. The last members of the ill-fated polar party—Bowers, Wilson, and Scott—died 119 miles south of here. The final line of Tennyson's "Ulysses" is carved in the horizontal beam of the cross: "To strive, to seek, to find, and not to yield."

By the time I finally arrived at the stone hut at Cape Crozier, I'd experienced a little of what Scott and his men had had to face on their polar journey, including man-hauling sledges at 9,000 feet on the polar plateau, in –30° F weather. In truth, I'd barely brushed up against the full prolonged experience of what they'd not been able to endure; but I had grounds enough to be absolutely stunned, given the planning errors in their attempt to reach the pole and the inadequacy of their equipment, that they had nearly made it back.

I've never stood atop Observation Hill without taking off my hat. They were so intensely and unmistakably human, these furiously determined men.

MY COLLEAGUES Paul Mayewski and Cameron Wake were working at the bottom of a snow pit, about twelve miles from Amundsen-Scott South Pole Station one December morning. I was perched eight feet above them on a snow ledge. The fourth man in our party, Mike Morrison, was looking down from the edge of the pit, eight feet above me. Our tents were pitched about three hundred yards downwind (where the fumes from our cookstoves wouldn't contaminate the snow samples we were gathering). It was the start of an overcast day, light winds, –31° F. It was colder at the bottom of the pit, near –40° probably. (The average temperature at South Pole in those years was –57° F, which was also the temperature year-round of the surrounding snow cover at a depth of about fifty feet.) In my experience it becomes noticeably harder for people to work efficiently at temperatures below about –30° F, a point I was about to prove.

Paul, the principal investigator in the field party, was extracting a series of snow samples from the wall of the pit we'd dug, working his way down ten millimeters at a time. The sterile samples, reaching back many years in the history of Earth's atmosphere, would give him a chemical record he could compare with other data he was compiling from other sites in Antarctica and Greenland, much of it ice-core data. I'd spent the previous month with Paul, Cameron, and four others on upper Newall Glacier in Victoria Land's Asgard Range, at about 5,000 feet. That work had culminated with the successful extraction of a 177-meter (581-foot) ice core, now packed for shipment and sitting in a refrigeration unit at McMurdo.

Having finished that work at Newall Glacier, we were now here, on the fourth and final day of Paul's attempt to secure a painstakingly assembled string of uncontaminated snow samples. Paul and Cameron were both wearing sterile suits, gloves, and masks over their cold-weather gear. They deposited each sample in a sterile numbered bottle. The sample in bottle 481 was taken .39 inches (ten millimeters) below sample 480 and .39 inches above sample 482, and so on. Mike lowered a crate of fifty numbered sample bottles to me and I lowered them to Cameron. At some point it entered my mind that I could shorten Cameron's time in the extreme cold at the bottom of the pit by lowering the bottles to him in the plastic bags in which they were triple sealed. I'd just get rid of the crate.

The problem with this idea was that if I removed the bags from the rigid crate, the sequential arrangement of the numbered bottles would be destroyed. Reaching in, Cameron was as likely to pull out bottle 451 as bottle 473. Before this realization hit me, unfortunately, I was leaning down from the ledge, handing Cameron the bag. He and Paul stared at me silently from the frigid depths in which they would now have to stand idle, waiting for me to correct my mistake.

I climbed out of the pit, donned sterile garments, placed one of the three sterile bags in the crate and began arranging the bottles inside it in numerical order. Fifteen very long minutes later I handed the crate down to Cameron.

The day following this fiasco with the bottles, two Sno-Cats came out from Pole (as it's called by most here) to pick us up, along with our gear and the 505 samples. Back at Pole we learned we had a choice of accommodations. We could return to McMurdo that evening and spend three or four days there before catching a flight to Christchurch, or we could stay at Pole and fly to McMurdo later. (At the time, ski-equipped LC-130s flew into Pole almost as often as weather permitted, mostly ferrying fuel for the generators that keep the station running year-round, but also transporting basic supplies, construction materials, machinery, spare parts, scientists, and visitors. The window for them to do this is usually no more than ten weeks.)

We opted to stay at Pole. At that time the station supported only a small resident community during the austral summer, and the opportunity to live without noise and disruption was far greater here. McMurdo was loud, sprawling, crowded, awash in regulations, and entirely too social for all of us. In addition, for me, Pole was a center for research in areas I knew little about—geodesy, solar plasmas, dark matter. The extreme isolation of this circumscribed station, its relatively small human community, and the nature of much of the scientific inquiry being carried out here gave Pole the feel of a research platform traveling onward in deep space.

I OFTEN BROODED, wherever I happened to sleep in Antarctica, about things that always seemed to be on my mind. Global climate disruption; refugee camps in the Horn of Africa, with their listless, crumpled lives and bewildered, wandering children; the avarice behind corporate exploitation of what some describe as the world's "commons," places like the open ocean that belong to all people; the mendacity and self-ishness of national governments, including my own; the plague of underreported femicides in Juárez tied to drug cartel activity, which ends for hundreds of them at the cartel's crude mass graves, *narcofosas;* the acre of crudely caged animals and the baskets of their body parts I'd encountered once in an all-night market in Yueyang. The depression this brought on sometimes left me with feelings of guilt, however unwarranted that might have been, and feelings of anger, which occasionally polluted the memories of what I had been doing that day.

It did not escape me that I returned as often as I did to Antarctica because it offered a kind of relief I could find nowhere else. In Jaipur, my hotel sat isolated behind a distant wall that separated me, I knew, from the lives of the truly destitute. In untenanted reaches of the Mojave in Arizona, I rarely passed a vehicle on the road without wondering whether it carried beleaguered "illegals," or whether la Migra would knock on my motel door one evening and ask for identity papers because of my surname. Even living in a rural part of the Cascade Mountains in Oregon, I knew I had to come this far to drink water from a stream without a second thought about giardia.

I can say that nearly every day in Antarctica I was astonished by something—picking up part of an asteroid from the ice cap; being escorted through the blue-light tunnels at Pole that housed part of the

AMANDA project; the huge penguin colony at Cape Crozier; placing my hand on the forehead of a mummified seal. Against the horrors I'd seen elsewhere or knew about, these things were a balm. I wanted to respect and absorb the experience, and I wanted to give it away to whoever might need it.

ON MY FIRST VISIT to Pole, the one I'd made with the two journalists, I brought a tubular map case with me. I told our NSF escort, who was inquiring more and more pointedly about the contents of the tube, that it contained a fly rod. I told him some of my friends were fly-fishermen and I was going to assemble my fly rod at Pole and make a few casts, just to have a good story. I knew he didn't believe me and sensed he'd lost patience with my trying to humor him. As we disembarked our plane at Pole, he confronted me. What was in the case? I told him it was a kite. I wanted to fly a kite at the South Pole. He said kite flying was not permitted at the South Pole, that flying a kite here might interfere with aircraft operations. I said the only aircraft within hundreds of miles of us was the one we flew in on, and which was now sitting on the ground here. I assured him, however, that I would walk off a good ways before sending the kite aloft. He said that if I attempted to do this he would contact his superior in McMurdo by radio and I would be reprimanded. I said that was okay. He left for the radio room inside the station's geodesic dome and I walked off a couple of hundred feet from the plane and flew the kite. No one, then or later, ever reprimanded me.

I wanted to fly the kite at Pole because, for me, one of the most impressive aspects of the Antarctic Treaty, a document meant to guide all human activity in Antarctica, is its insistence on equality. Sign the treaty and you may share in the bounty of what people here are learning—about the ice, the planet, our solar system, the Milky Way galaxy, and the universe beyond. Flying a kite over the semicircle of national flags in front of the station (representing the treaty's twelve original signatories) was a whimsical and private gesture of disagreement. If Antarctica belongs to no one, then no national flags should fly here. And if the continent is to be held in common, then the flag of the United States should not be flying by itself, as it did then, next to a brass-capped rod in the snow that marks, precisely, the location of Earth's south geographic pole.

It would have been disrespectful of me to have asked our escort to listen while I explained my reasons for bringing a kite to Pole. Not the place, not the time. As it was, most thought flying the kite was a stunt, a wry comment on the NSF's humorless emphasis on safety issues. During lunch at Pole that day my host, by way of apology, said he was concerned only about everyone's safety—and that flying the kite, even at −26° F that day, actually seemed like fun.

After I finished my interviews with the staff and a few of the scientists at Pole that afternoon, I realized I still had a few minutes before the other two journalists and I were scheduled to leave. I went in search of a staircase that led to an observatory I had heard about called the Skylab. At the top of these stairs was a small square room with a couple of battered club chairs in it, and triple-paned windows looking out over the polar plateau in three directions. The windowsills were set low, so as not to cut off the view for someone sitting in one of the chairs, and tinted shades hung in front of the windows to cut the sun's glare and reduce the intensity of its radiation, softening the ambient light in the room.

I stood transfixed in the doorway, seeing that the view of the polar plateau ran unimpeded to the horizon. From this height, as if from a ship's bridge, the prospect was of another Pacific, though there was no indication that life had ever been here, nor was there any suggestion that it might come. The view was of a void so utter it seemed empty even of space. The geography for an anchorite. As far as you could look, it was unstoried, free of history on a human scale. A land not yet saturated with laws.

There were two chairs in the small room. My escort, Jack, was sitting in one. After he gave me a nod, I took the other chair. Here sat a man staring at what he loved. I looked out at the sky with him for a while, out across the crusted plain of wind-wrinkled snow. Where the waist of the sky met the white plain, the air was lit by a belt the color of lapis lazuli. High above it, past succeeding shades of light blue, over the shoulders of the sky, hung a few wisps of mare's tail cirrus, parallel to one another.

"I'm so sorry about the kite," I said.

"And I'm sorry about the confrontation," Jack said.

THERE IS NO LONGITUDE at the South Pole. Its lone coordinate is 90° South. From here, every direction is north. East and west come into play, technically, as soon as one steps away from the actual geographic pole, but such coordinates are without meaning here, they're too difficult to imagine. People orient each other outside by referring to the wind, as in "upwind of that research hut," or to the movement of the polar ice, as in "just downstream of that front loader." The ice on which Pole station sits is moving seaward at the rate of about 33 feet per year. Annually, on January 1, a representative of the U.S. Geological Survey locates the precise point in Earth's bedrock, 9,300 feet below, that marks the southern end of Earth's axis of rotation, and then he or she drives a metal rod into the snow above. For the next 365 days the rod is assumed to mark the South Geographic Pole.

A few hundred feet from the actual pole (where all of Earth's twenty-four time zones converge) is a ceremonial pole, a short barber pole of red, white, and blue stripes supporting a chrome sphere the size of a basketball (it's flanked by the half circle of twelve flagpoles), a setting often used for group photographs. Off in another direction a sort of folk version of the South Pole has been established, with yard ornaments like pink flamingos, FOR SALE signs, a placard warning that NO LIFEGUARD IS ON DUTY, bouquets of plastic flowers, a municipal bus stop sign from suburban Boston, and a tall post bearing a plywood cutout of a salmon on which the number 9,512 is printed, the distance in miles to Salmon, Idaho.

On the first day the temperature at Pole drops below –100° F, which it does regularly in winter, any one of the winter-over staff can elect to join the most exclusive of Antarctica's social clubs, the 300 Club. Candidates enter the station's sauna, which has been heated to "200 degrees" (more like 130 degrees), then sprint for the marker at the true pole wearing only their boots. One turn around the pole marker and then back inside the station. Most candidates develop frost nip in a few spots—the first stage of frostbite—typically in gender-specific areas. The 300 Club initiation and gag T-shirts (*South Pole: One Inch of Powder, Two Miles of Base*) take some of the edge off winter life at the station, a place that does not fit easily into any one of the geography boxes one is handed as a child.

It does not actually snow, for example, at South Pole. What lies on the polar ice cap around Pole is "diamond dust," ice crystals drifting down on perennially light winds. And a visitor is able to see farther over the surface of the planet here than he or she can anywhere in the middle latitudes, almost twice as far, because Earth flattens out at the poles, making the planet an oblate spheroid. The atmosphere above the pole also flattens out, making the atmospheric layer thinner and creating an effective pressure altitude at Pole of around 11,500 feet, not 9,300 feet. Some visitors arriving by plane from sea level at McMurdo suffer from altitude sickness for a few days before they adjust. (Others never adjust and must return to McMurdo.) The stars, moon, and sun do not rise and set here each day but rise and set instead on a 365-day schedule, a primary reason why so much celestial research is supported here at a place that is difficult to reach and very expensive to maintain. Telescopes of several sorts trained on celestial objects overhead can track them for months without ever losing sight of them; and the relative dryness and thinness of the flattened atmosphere makes the optical images they record clearer than images made at lower latitudes.

Arrayed around Pole, rather like satellites, are a dozen or more data-gathering stations with long-running programs. Situated anywhere else on the planet, they would be far less useful—or of no use at all. For example, because Antarctica is seismically the quietest of Earth's continents, seismographs located in snow pits at Pole are able to pinpoint earthquakes all around the world that are too faint to register at other seismic research sites. And the enormous reservoir of snow and ice beneath Pole has proven ideal, as I've said, for trapping neutrinos, near-massless particles that are a key component in competing theories of dark matter. Finally, the transparency and electromagnetic stillness of the atmosphere at Pole—it has the lowest water vapor content and the least "sky noise" of any place on Earth—makes it an ideal place for locating the 13.8-billion-year-old edge of the expanding universe, and for researching the chemical composition and behavior of Earth's upper and lower atmosphere.

The deployment of instruments here—helioseismographs, optical and gamma-ray telescopes, ozone depletion recorders—represents some of the more technically complex work being carried out by Edward Wilson's progeny. If he were to visit here today, however, Wilson might initially feel at a loss. He'd find himself far from the questions in biology that were dear to him; but I think he'd quickly gain a sense of the relevance of research on the internal structure of the sun, and the importance today of studying atmospheric chemistry. And sitting around the mess-hall tables with other scientists, he might even understand why some of their research, like that into global climate disruption, irritates the members of certain religions and an international class of profiteers who've supplanted, in some ways, the oligarchs of his own era as social tyrants.

And he would be puzzled by the tension between science and popular culture.

Wilson, a person in the mold of Darwin, might not fully appreciate the versatility of computer software or be surprised by the ability of a chromatograph to determine the chemical nature of objects in the asteroid belt, but a difference in tools is not the striking difference between the science of his time and ours. It's not the scientific topics—how Niels Bohr used quantum theory to reorganize the interior of the atom, for example—that would baffle him. Wilson was intensely curious about the material world. He would understand the drive and the urgency—though he himself was not an urgent man—behind all of this. What he would have trouble absorbing is the aggressiveness with which modern governments and for-profit businesses selectively promote debate about scientific research, and how avidly they pursue the development of technologies that this research makes possible. He would have been appalled by the lack of an ethical framework for the development of the atomic bomb, or by the dissemination of genetically engineered food, or by the dumping of chemical waste in urban water supplies. He would have been made anxious about the emergence of identity theft and the eclipse of personal privacy that these new technologies had made possible.

The differences between his world and ours that would have most kept Wilson awake, I tend to think, were not the advances in science that would have required him to reeducate himself, but changes in human behavior and human aspiration, primarily the development and promotion—and enthusiastic acceptance—of problematic technologies without regard to their long-term consequences. He would have had difficulty matching his sense of what was moral—one reason every member of his 1910–13 British Antarctic Expedition, including Scott, regularly sought his counsel—with the rapaciousness that generally characterizes twenty-first-century quests for material gain.

Wilson was a calm man, comfortable in his own skin. Were he able to sit around Pole and listen to people's conversations at the start of the twenty-first century, and to listen in the evening to BBC broadcasts of world news, as we did, I think he would not have felt so much scientifically uninformed—he could catch up on that—as morally quaint.

A SECOND REASON the four of us elected to stay on at Pole instead of returning to McMurdo that day was that it was easier to think here. Inside the heated buildings, which were situated inside a sheltering dome at that time, life was protected from the weather. A chef made meals, someone else melted snow for water. When you opened your eyes in the morning, you had no agenda to address. You could sit undisturbed at a table with a map or a book in front of you for hours, or go upstairs and sink into a club chair in the Skylab.

Well fed and freed of my responsibility to assist, I felt served by fortune during those few days at Pole.

Once or twice a day I would leave the shelter of the dome to take a walk. I'd long since adjusted to cold temperatures, and wind chill was infrequently a factor because the air at Pole is usually calm. This is not to make light of low temperatures in the interior of Antarctica, only to say that you become accustomed to them, to the numbness in your hands and the patches of frost nip on your face.

The first night I spent on Newall Glacier with Paul, Cameron, and the others, I headed to my one-man tent to sleep. On my knees, I began to take off my outer clothing, stripping down to expedition-weight long underwear and my socks. As I maneuvered in that tight space, twisting out of my clothes, I began to doubt that I had either the tolerance or the stamina for twenty-seven days of this. It was –20° F, I was all gooseflesh, involuntarily hunching over in a ball to conserve heat as I crawled shivering into my sleeping bag.

The warning story for me about living with cold temperatures—and all this was taking place in the Antarctic *summer*—came in the week following those moments of frigid introduction. Cameron and I and the others were studying the uneven flow of Newall Glacier. (This was part of a two-year effort to determine where the less-stressed sections of ice were in the glacier, so we could locate a promising spot at which to drill, one that would produce the fewest stress fractures in the ice core.)

Cameron and I were perched on the flank of a sheet of exposed bedrock high above the glacier and employing a laser theodolite to pinpoint the exact location of a two-mile-long series of bamboo poles that had been set out the year before in a straight line at 90 degrees to the flow of the glacier. Our colleagues below were traveling from pole to pole by snow machine and holding up a reflector at each spot for the laser to find. Between sightings, the two of us retreated to the interior of a small tent we'd set up to shield us from thirty-knot winds. We "spooned" there, hugging each other for warmth. The temperature was about –25° F and the high winds were creating wind chills of close to –60° F. At one point while we were outside surveying, Cameron lost, for a split second, his grip on one of his mittens. A few moments later, lying together in the tent, he said, "In Antarctica, you make the big mistake only once."

ON THE DAYS I left Pole station, I'd walk out a ways across the plateau, enjoying the simplicity of the view, whichever way I looked. Some arresting arrangement of solar light would often be apparent in the sky, refractions of various sorts, which gave rise, for example, to bright spots of pale pink and lime on either side of the sun—sun dogs—or to a ghostly pillar of vaporous light between the sun and the horizon, a column the color of the moon's gray soil.

The forever setting sun, the squeak of my boots on the snow as I walked, the sound of my breathing over the extensive silence of the plateau, suggested somehow that the buildings around me were only insubstantial projections of mine. They might wink out at any second.

Some days I watched as pearly opalescence bathed an entire cloud. The interior of an abalone shell, mounted in the sky.

IN AN EFFORT to understand what might have driven Robert Falcon Scott, a person few knew well, to be the first to reach the pole, I visited the National Archives in Washington, D.C., one spring morning to examine the only existing facsimile copy of the diary he was keeping when he died. Passages from the diary are often quoted in Antarctic histories but, even when words from a handwritten document are quoted accurately, they do not convey the information that the point of a pencil or pen can leave behind.

I wanted to look especially at two sentences in the diary, the place where he expresses his anguish after learning that Amundsen had beaten him to the pole by thirty-four days; and the last sentence he wrote, composed in the tent where he was later found dead with Wilson and Bowers (Evans and Oates, the other two members of the polar party, having perished on the trail behind them). The three of them were only eleven miles from food at the One Ton Depot and 147 miles from Cape Evans, but Scott knew this was the end. He wrote: "For Gods sake look after our people."

At the pole, Scott had written, memorably, "Great God! this is an awful place [. . .]." It was clear from the way his pencil moved across and down the page leading up to that sentence that he was simply dutifully recording what had occurred, writing as if detached. As he begins the first *G* in the words *Great God,* the pencil bears down hard. Taking in the appearance of the entire page, looking at what precedes and what follows this sentence, I felt that this abrupt burst of intense feeling (from someone who usually kept his emotions tightly in check) represented the sudden full realization that he had irrevocably "lost the prize."

As I read biographies of Scott, I came to believe that he'd long envisioned the achievement of the pole as the looming apex of his life. To arrive there first, before anyone else, would secure him a knighthood, retirement from the Royal Navy, international renown, and some significant amount of money. To lose it, he thought, to come in second, would burden him with a reputation for mediocrity for the rest of his life. Scott pictured his conquest of the pole, I believe, as an assault on a great mountain, a vertical quest, with the summit of the mount his prize. When Birdie Bowers spots Amundsen's first black flag, tied to a standard jammed in the snow about ten miles from the pole, Scott knows he's been beaten. In that moment, the quest is over. (The flag marked one corner of a survey box Amundsen employed to ensure there would be no doubt about his having stood at 90° S, not a mile or so away.) But it's not until the following day, when Scott encounters the tent Amundsen left behind (with a note in it for him) that it sinks in. It's then that he writes: "Great God! this is an awful place and terrible enough for us to have labored to it without the reward of priority." The long, perilous ascent of his figurative Everest having been accomplished, Scott now sees the place for what it actually is, an undistinguished, unbounded, flat, and anonymous expanse of snow. He is standing somewhere—it could be anywhere—on the polar plateau. The mountain has collapsed. It's −22° F and he is more than eight hundred miles from the shelter of his winter quarters at Cape Evans, which he will not live to see.

The second entry I wanted to examine might have been made on March 29, 1912—this is the last date Scott wrote out, but his final entry might have been made as much as two days later. There is no period after the sentence "For Gods sake look after our people"; exactly whom Scott meant by "our people" has long been debated. The least generous interpretation is that he was referring only to members of the expedition. Another interpretation—a line at the front of his diary requests that the notebook be given to his wife, Kathleen—is that he was asking Kathleen to look after their class of people, including his infant son, and not incidentally, Scott's reputation. What I understand his words to mean is less narrow than this, and perhaps overly generous. Anyone who seriously engages with the landscape of Antarctica learns that the geography here is indifferent toward humanity. It is not "antagonistic," seeking to thwart human effort. It made no distinction between Scott and Amundsen, the latter an obsessed, ruthlessly efficient, emotionally cold individual with as many character faults as Scott. Antarctica personified didn't care who arrived where or what happened. It was *people* who built these constructs to sort "winners" from "losers," and it was people who might have just as easily not built them. What Scott meant, I think, was for Kathleen to attend to those whose lives had not ended in Antarctica, and to support those whose questing might one day carry them to Antarctica's shores.

The bitterness of his defeat, I came to think, had burned out in Scott by the time he lay dying. We assume sometimes that whatever the dying say at the end, or last write down, represents a conscious final thought, but I don't believe this is very often true. What is really going on at the end mostly goes unspoken and unwritten, and what actually happens at the end remains unknown to the living. Final written thoughts are not likely to be profound summaries of everything that has come before. What was on Scott's mind at the last, I believe, was what many men of my generation came to understand in Vietnam, in circumstances as harrowing in their way as Scott's. The strategies for victory—and the handing out of medals—are as nothing compared to what people learn about looking out for one another under duress. No calling was higher in Vietnam than covering for another soldier. And the adolescent daydream, that the world is there for the taking—Scott wrote repeatedly on the trek back to Cape Evans about how his dreams had been dashed—or that the world acquiesces before strong men, is delusional.

The world outside the self is indifferent to the fate of the self.

IN THE BOREAL SPRING of 1992 I boarded a 308-foot ice-breaking research vessel, the *Nathaniel B. Palmer* (named for an American captain who codiscovered the South Orkney Islands in 1821, and was among the first to sight Antarctica, in 1820), and accompanied it on its first voyage to the Southern Ocean. We sailed from the coast of Louisiana in March, passed through the Panama Canal and a little more than a week later put in at Punta Arenas, Chile, on the Strait of Magellan.

After refueling and taking on freight and a small party of scientists, we crossed the Drake Passage and entered the Weddell Sea, the first ship to do so in the austral fall since Shackleton's *Endurance* was crushed there in 1915. Our objective was a joint U.S./USSR (at that time) scientific camp operating on an ice floe deep inside the Weddell.

A Soviet icebreaker had put the camp in in the austral summer of 1991–92. Ours was the first resupply mission. The *Palmer* had been built in the Louisiana bayous the year before and outfitted for nearly every type of scientific research liable to be conducted in the Southern Ocean, from drilling for sediment cores on the bottom to observing the lives of sea mammals. Its extensive lab space supported studies in water chemistry, sea-floor mapping, ocean ecology, geodesy, and phytoplankton and krill distribution. This initial voyage was a shakedown cruise; there were only a few scientists and support staff aboard, in addition to the normal complement of officers and crew, and a few supernumeraries like me. The scientists were all bound for Ice Station Weddell (ISW). The ones they were replacing there would return with us to Punta Arenas.

On the way to ISW, sailing in loose pack at the ice front between the Southern Ocean and the Weddell Sea proper, we encountered dozens of whales, mostly minkes, orcas, and a few southern rights. Hundreds upon hundreds of crabeater seals—the most numerous large mammal in the world—were hauled out together on ice floes around us, along with smaller clusters of two other Antarctic seals, the gregarious Rosses and the Weddells. Leopard seals, hunters of penguins, we nearly always saw by themselves.

One evening I stepped out on the *Palmer*'s aft deck to take in the night before retiring. The deck is open and spacious, and built close to the water's surface to make the lowering and retrieving of instrument platforms easier and safer. Work lights illuminated both the surface of the deck and the *Palmer*'s prop wash, the standing wave aft of the ship created by the ship's propellers turning just enough during the dark night to keep the ship snug in a temporary slip it had made in the edge of a large ice floe. After a few moments I noticed four emperor penguins standing on the edge of the ice floe, staring at the ship's standing wave. Suddenly one dove into the water. The other three followed immediately. I assumed that they were swimming off, mistrustful of the ship, until I saw a head emerge from the standing wave. Then three others. They were surfing the prop wash of what was probably the first ship they had ever seen.

As winter came on and we continued to force our way through the pack ice farther and farther south, we had fewer and fewer hours of daylight with which to work. Without light from the sky to help us spot open leads and read fault lines in the ice that the breaker could wedge open, we were compelled to lay up at night along the edges of large ice floes in order to conserve fuel. On some of those long nights I took advantage of the permission the captain gave me to step overboard and go for a walk. He only asked that I take a couple of companions with me, that I carry a radio, and that we all wear life jackets.

The air was often cold—20 or 25 below zero Fahrenheit—and we were far from any well-known place on Earth, nine hundred or so miles southeast of Cape Horn, in a sea roughly the size of the Mediterranean, one in which no ship but ours was to be found. Commercial shipping lanes, which run east and west in the Southern Ocean, were far to the north of where we were. No air traffic passed overhead (nor, at that time, did any satellites). Walking away from the ship across the frozen sea provided me with an unfamiliar perspective. On nights that were clear, light from the stars, radiating through the immaculate air, offered us all the illumination we needed to make our way safely across the ice. Within the hemisphere of remote space we occupied on those nights, there was only the sky, the ship, and the ice.

For our safety (and for our reassurance, no doubt), the captain ordered every outside light on the ship turned on when we stepped overboard—deck lights, navigation lights, work lights, searchlights. Seen like this, silhouetted against the blue-black vault of a sky pulsing with stars, with the murmur in our ears of the ship's diesel engines idling, the *Palmer* could have been an intergalactic vessel of some sort, settled here for the moment on an inhospitable moon. It bristled with antenna wands and satellite communication globes. Its deck cranes gave the ship the appearance of a deep-space freighter. Steam rising from its ventilation grills and the plume of pale smoke wafting from its smokestack gave it a look of behemoth vitality. Behind its walls, past thermal-pane windows glowing in its superstructure, there was, I knew, daunting electrical, mechanical, and electronic wizardry.

On these walks I often raised my binoculars to draw closer to galaxies just beyond the Milky Way. Imagining my way out to them mentally, from the pindot of a planet I stood on between two spiral arms of my own galaxy, named Sagittarius and Orion, out past the halo of dark matter and dark energy surrounding our galaxy, out to the galaxies in our Local Group, brought the entire scene—the ice, the ship, the dark water underneath—into a continuum of time and into an unbroken expanse of space that overrode any information I might have gotten from my wristwatch or handheld GPS at that moment.

The only thing my watch could offer me was the time I told the officer on deck we would be back. For the moment, however, I was not in Antarctica. I was afoot on some planet's moon.

IN THE END, the captain decided to halt the *Palmer* about twenty-two miles shy of our destination. The supplies we were bringing to ISW were all small enough to fit inside a helicopter, and the fuel required for a few helicopter runs would be very much less than the fuel required to get the 308-foot ship through pack ice to the edge of the ice floe the camp was built on. We were far enough south, and it was late enough in the solar year, that our "days" were now dominated by darkness. This hampered our search for navigable water, something else the captain now had to take into consideration.

The resupply of the ice station that April night had, again, the feel of an interstellar docking. The Bell 206 Jet Ranger appeared from and then disappeared into the utter dark, ferrying cargo and personnel and taking with it, each time, the turbines' terrific roar. The helicopter's turnaround times were so brief, the pilot never shut the engines down. The whirring rotor blades and the whining of the turbines added an element of urgency to the scene, and the pilot intensified the urgency by hand-signaling his constant impatience. Beneath a high canopy of silent stars, with the ship's halogen work lights holding the night at bay, crewmen, hunched over and anonymous in their heavy parkas, scuttled back and forth under the prop wash. Men shouted. Cold steel banged in the numbing air. The five ice-station-bound scientists, wearing too much clothing and waiting anxiously in the ship's interior, were perhaps wondering what they had gotten themselves into.

The helicopter pilot, clearly irritated by something not apparent to us, began shouting at his passengers as soon as they boarded about safety, as though needing to emphasize his authority over them in this chaotic situation, or as if he believed, because these men were academics, that they needed confrontational instruction in routine matters, or that, for him, these were people whose ideas about harsh weather and lunar-like isolation had come from watching television. His strange outburst spoke eloquently to tensions that frequently characterize such situations, ones in which working-class and middle-class people are forced to share the same confined spaces when they hold slightly different ideas about the nature of the mission. I'd felt this tension aboard the *Palmer* for some weeks, disgruntled crewmen resenting the condescending treatment they had to endure from a few of the scientists, people who strolled the ship with the air of owners. This theme, the tension between two social classes, is the infrequently reported story, in my experience, of numerous fractious scientific and adventuring expeditions.

The Soviet Union collapsed shortly after ISW was established. The Soviet scientists at ISW were very suddenly back to being simply Russians. When I landed in the camp and began interviewing people, I gained the feeling that most of the Russian scientists were both confounded and depressed about the collapse. What were their futures to be now? Many of them were still flying the Soviet flag from the tops of their wall tents.

The scientists here—chemists, oceanographers, sea-ice experts, meteorologists, scuba divers—had been funded to come to the Weddell because, with global climate change, it was suddenly extremely important to learn more about this sea. The Weddell is sometimes referred to informally by researchers as Earth's "main engine." To oversimplify, cold water from its depths flows into the depths of the South Atlantic and is primarily responsible for driving that ocean's circulation, the temperature gradients of which directly affect worldwide weather. The more the chemistry of Earth's atmosphere changes, with ongoing infusions of carbon dioxide from the combustion of fossil fuels, and with the buildup of manufactured chemicals like chlorofluorocarbons, the more important it has become to understand what is actually going on in the Weddell. (Prior to this, no scientific party had ever overwintered here.)

As I took tea with some of the Russians and chatted with Americans, all living a spartan life here in insulated shelters, I felt a growing respectful affection for them. Their work was physically demanding and was being conducted with a level of cooperation and courtesy between people from—back then—two superpowers, which was encouraging. With the demise of the Soviet Union, their project now had an extra-national dimension: they knew they were addressing together a challenge that no single nation alone could effectively manage.

ISW was the most isolated scientific camp I'd ever been in, and its occupants seemed to me to be even more cut off from the outside world than is usually the case. Global climate disturbance, at that time, was a contested issue. The work of scientists like these was being criticized and dismissed, even ridiculed, by religious leaders, industrialists, and politicians. The scientists knew it would take years for poorly informed people to accept and appreciate the nature of the emergency that had brought them here. At the time, and ever after, I thought of the researchers at ISW as heroic. Perhaps what made the helicopter pilot so irritable, so given to exasperation, was that he himself had doubts about the value of the research, but could find no one in the camp with whom to share his contempt. Or it could have been that as a (mere) helicopter pilot, he was not accorded the full respect he felt was due him at ISW.

The ongoing refusal of some governments and many politicians and business leaders to take global climate disruption seriously is part of a movement in some first-world countries to denounce any form of "politically inconvenient" science. The ongoing resilience of this obdurate denial, of course, is an indication of the deteriorating state of public education in these countries.

A MORE ENGAGING, less autocratic, and less secretive British Antarctic explorer than Robert Falcon Scott was Ernest Shackleton, a person more working-class Brits identified with at the time than identified with the aristocratic Edwardian. When an opportunity arose to follow the track of Shackleton's famous 830-mile open-boat journey from Elephant Island in Antarctica's South Shetland Islands to the coast of South Georgia, I was delighted to accept an invitation from an ecotourism company to do so. They requested that I deliver several lectures aboard their chartered vessel, the *Hanseatic*. I asked my oldest stepdaughter, Amanda, twenty-two, to go with me.

In the austral summer of 1914, Ernest Shackleton sailed the *Endurance* into the Weddell Sea with the intention of landing an expedition party on what is now the Luitpold Coast. These men hoped to be the first to cross the continent from Coats Land to the pole and then from there to Ross Island. The *Endurance* was nipped in the ice short of the Luitpold Coast, in January 1915, and was subsequently carried farther and farther north until, so badly crushed it was no longer seaworthy, it was abandoned, in late October of that year. Twenty-seven days later, on November 21, the *Endurance* sank. Shackleton and his men had managed to transfer enough food and supplies onto the pack ice to be able to survive for five months, enough time, they hoped, for them to reach open water with the ship's cutters. From there they could row for Elephant Island, which they did, putting ashore on April 14, 1916.

On April 24, Shackleton and five of his men launched one of the 22-foot cutters, refitted for ocean voyaging and christened the *James Caird*. They left twenty-two men behind, camped in a penguin colony on a small patch of rocky shoreline. The six of them sailed and rowed more than 800 miles north and east before reaching King Haakon Bay, on the southwest coast of South Georgia. The cutter was all but done in at that point, and two of the men were sick. Shackleton knew there were several whaling stations on the opposite, or northeast, side of the island. Leaving one man behind to tend to the sick, he and the other two set out to climb up and over the island's crest, which was more than 9,000 feet high in several places. This singular feat of mountaineering brought them to the Norwegian whaling station at Stromness Harbour on May 20, 1916, ten days after they'd stepped ashore at King Haakon Bay.

Frank Worsley, the former captain of the *Endurance* and the navigator aboard the *James Caird,* boarded a whaler at Stromness to guide it to the rescue of the three men they'd left on the other side of the island. Shackleton then arranged for all six to sail for Elephant Island aboard a Norwegian whaler. Pack ice—it was now the middle of the southern winter—forced the whaler to retreat instead to the Falkland Islands, where Shackleton was offered the use of a fishing trawler. Once again the pack ice forced him to turn back, this time for Punta Arenas, the Chilean port on the Strait of Magellan. With the help of local residents who subscribed to a fund, Shackleton was able to hire a motor schooner, the *Emma,* and to set off for a third time for Elephant Island. When the *Emma* broke down, about a hundred miles from its goal, the Chilean government lent Shackleton a steamer, the *Yelcho,* with which he finally reached his crew's encampment, on August 30, 1916, 129 days after he set sail from the same beach, hoping to reach South Georgia.

Famously, Shackleton—and importantly, Frank Wild, who had led the party Shackleton had had to leave behind on the beach—did not lose a single person on the expedition. All members of the party who had entered the Weddell's pack ice aboard the *Endurance* twenty months before were now sailing for home aboard the *Yelcho.*

THE PARTY ABOARD the *Hanseatic* intended to trace the reverse of Shackleton's open-boat journey, starting at Stromness Harbour and ending at Elephant Island. On the first leg of the trip, from Port Stanley in the Falkland Islands to the northeast coast of South Georgia, the 403-foot ship ran into a Beaufort force 11 storm, one step shy of a hurricane on the Beaufort scale.

For more than an hour during the height of the storm, I stood on deck with Will Steger, the great polar explorer of my generation. Together we watched the gargantuan tumult of slow-motion seas before us, as the ship yawed sideways, its stern periodically coming clear of the water and sweeping across fifteen degrees of gray sky while the ship buried its bows again in a forty-foot wall of water and the stern fell. The ship plunged on. The surface of the ocean had no single point of stillness, no transparency. Veils of storm-ripped water ballooned in the air around us, and the high-pitched mewling of albatrosses, teetering impossibly forty feet away from us on the wind, cut through the sound of the storm rising and collapsing around the ship's superstructure. Fifty-knot winds ripped the crests off waves in the cross seas and wailed without letup through passageways in the upper deck.

Shackleton, I knew, had faced weather like this crossing to South Georgia. I now had a more informed respect for what he and the others had accomplished.

High winds were still with us when we came around on the northeast coast of South Georgia, but the seas had dropped and the following day the skies cleared. We sailed out of Grytviken, the abandoned whaling station on the shore of Cumberland Bay where we visited Shackleton's grave and spent the morning exploring, cruising to the west and anchoring a few hours later in Stromness Harbour.

Once ashore, a group of us set off on foot, following the meltwater stream down which Shackleton and his two companions had come on that day in 1916. Our goal was the final obstacle on their trip, a thirty-foot-high waterfall. Shackleton's party could see the whaling station off to their right from the top of the falls, but to reach it they would need to descend to the valley floor. The rock face around the waterfall was heavily encrusted with ice. The only way down was to lower themselves on ropes through the full force of the falling water.

Where my daughter and I stood at the base of the falls, we could see wildflowers blooming for hundreds of feet around us. Lush sea grasses were rolling like horses' manes under the press of the wind. It must have been 50° F. We shed our windbreakers and sweaters. Some passengers, still woozy from crossing the Drake, stretched out on the sea grass to gather more of the sunshine and feel the brush of soft air across their faces.

I suggested to Amanda that we climb the rock face to the left of the waterfall, to get a view of the country inland. From up there we could also see the protected harbor where our ship was anchored and the now-collapsed and rusting whaling station. After we reached the top of the falls we continued on a ways, across a cushion of mosses and wild grasses, until we reached a level area. From here we could see clearly the intimidating ice wall farther inland that Shackleton and the others had had to inch their way down. The softness of the ground underfoot, the intensity of the sunshine, and the balmy air, however, overwhelmed us. We lay back on the carpet of moss. Amanda handed me a long stem of sea grass. I didn't comprehend the gesture until I saw her rolling a similar stem from one side of her mouth to the other, imitating the casual air of a bucolic. It made sense, having passed through that storm in the Drake and to be sitting here in this idyllic setting in, historically, a harsh place. I did the same, flat on my back like her, and we said not much of anything until there was a shout from below.

It was time to head back to the ship.

We were in no mood to leave. As I reached down for my parka, folded up on the ground for a pillow, I caught sight of someone half a mile farther inland. He seemed to be scouting the icefall for a route up, some way to go even farther.

Will Steger. Nearly gone into the hinterland. Neither Amanda nor I wanted to be the one to shout to him that it was time to go.

ONE AFTERNOON I accompanied some of the other passengers on a cruise through an ice-choked embayment in the Antarctic Peninsula. My mind was wandering in the moment, thinking of what I'd seen on previous trips to the continent. The landscape around us, I knew, was the great teacher here. You just had to step into it, with an open mind and an eager heart. Steger was off in one of the other boats and the photographer Galen Rowell, with his wife, Barbara, were in yet another boat, motoring through the loose pack ice. Will, Galen, and I had spoken together privately on several occasions about how strange this trip was feeling to us. We'd all brushed up against death here, had known bivouacs in Antarctic storms we hoped never to have to repeat. We also knew people who had died here after making a single simple mistake in a moment of inattention. And yet here we were, aboard a luxury ship, with comfortable beds, heat, five-star meals, and no reason to be anxious about a coming storm, even one as violent as the one we sailed into in the Drake. We had had the privilege of knowing, more intimately than most, what Shackleton and the others had managed to work through.

It was my sense of gratitude for the safety I'd experienced in Antarctica, and my appreciation for what I'd already been given, and not wanting to ask for more, I suppose, that kept me from saying anything about the meteorite when I spotted it. It was embedded in a piece of floating ice, like a stone mounted proud in a setting. This scrap of ice had surely been calved from the seaward face of a glacier or an ice shelf and had been in the water awhile. Eroded by waves, melting, the remnant looked like a miniature iceberg, with its underwater protuberances and asymmetric spires.

The dark cast of the meteorite, as large as a soccer ball, along with something else ineffable, suggested to me that it might be a chondrite (a type of stony, as opposed to iron or stony-iron, meteorite). I hesitated to point it out to anyone, however. I feared there might be an accident if five or six Zodiacs were suddenly to converge there, with people holding up cameras and jockeying for position. Better, I thought, to find another way to bring it to people's attention. The unique shape of the piece of ice it was in, its nearness to the ship, and the fact that the sea was calm and the weather fine, meant I could return with a few of the ship's staff and collect it, possibly.

Frankly, I really didn't know what I wanted to do. No one could say from where in Antarctica the ice had come, so we wouldn't have that information. Technically, we'd be in violation of Antarctic treaty protocols if we collected it, which could get the tour company I was working with in trouble. And keeping the meteorite sterile would present a problem. The question of ownership would arise. Around and around I went. Ultimately, I told no one except Steger, Galen, Barbara, and my daughter, after finally deciding just to leave it be.

Maybe we should have quietly taken this heaven stone back to South Georgia and set it by Shackleton's grave. Someone someday would probably walk away with it, but the theft wouldn't compromise our gesture of respect.

One evening on the deck of the *Hanseatic,* bundled in our parkas and nursing our coffee, I told Amanda about the trip I'd made, years before, aboard the *Palmer*. Antarctica, I told her that night, was like a great island, separated in so many ways from the world of our everyday lives. I began to describe the anticipation I'd felt that day when the *Palmer* entered the west end of the Strait of Magellan and began to approach the town of Punta Arenas. This was to be our last stop before

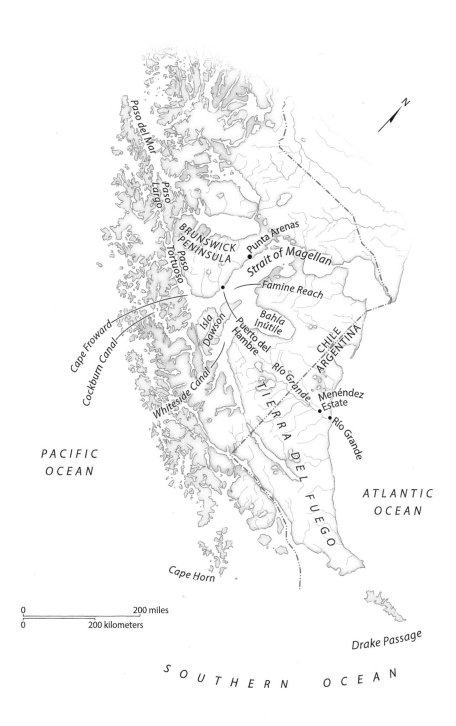

N

Paso del Mar

Paso Largo

Paso Tortuoso

BRUNSWICK
PENINSULA

Punta Arenas

Strait of Magellan

Famine Reach

Cape Froward

Cockburn Canal

Isla
Dawson

Puerto del
Hambre

Bahía
Inútile

Whiteside Canal

Río Grande

CHILE
ARGENTINA

Menéndez
Estate

Río Grande

T I E R R A   D E L   F U E G O

PACIFIC
OCEAN

ATLANTIC
OCEAN

Cape Horn

0   200 miles
0   200 kilometers

Drake Passage

S O U T H E R N   O C E A N

Brunswick Peninsula and the Strait of Magellan

crossing the Drake Passage and pushing our way into the fringe of loose ice at the edge of the little-known Weddell. That particular crossing of the Scotia Sea, I told her, was very much calmer—just five-foot swells—than our experience in the same Drake Passage ten days before. Sometimes you get the weather you want.

I decided not to tell her then about the encounter I'd had with a man on the road between Port Famine and Punta Arenas, the day before we set sail in the *Palmer* for the Weddell.

IN HIS NOVEL *Who Will Remember the People . . . ,* the French writer Jean Raspail dramatizes the plight of Kaweskar people in southern Chile in the middle of the nineteenth century. The lament in his novel is over the loss of yet another way of comprehending the unknowable world, as colonizers forced their way, literally and figuratively, into the homes and homelands of some of the last undisturbed inhabitants of what Europeans called the "new" world. The tone of the book is elegiac but not sentimental, aggrieved but not angry.

At the southern end of the system of Chilean channels, or "canals," that offers ships a protected passage south to the western entrance to the Strait of Magellan from Puerto Montt, the *Palmer* had entered traditional Yámana land at Paso del Mar. Paso del Mar gives way to Paso Largo, Paso Largo gives way to Paso Tortuoso. At that point in the southern Andes, the mountainous spine of South America disappears beneath the sea, taking with it, among other things, ghosts from the mining tunnels of Potosí.

Raspail's novel was one of twenty-five or so books I'd brought aboard the *Palmer,* for what would be a sixty-eight-day voyage, by the time we arrived back in Punta Arenas. I often read as I go books set in the places I'm traveling through, and for this trip had brought along Edgar Allan Poe's *The Narrative of Arthur Gordon Pym of Nantucket,* Rockwell Kent's *Voyaging: Southward from the Strait of Magellan,* E. Lucas Bridges's *Uttermost Part of the Earth,* and Thomas Bridges's *Yamana-English: A Dictionary of the Speech of Tierra del Fuego.*

One of the advantages of reading about places like the Strait of Magellan during the time one is visiting them—Antonio Pigafetta's firsthand account of Magellan's famous voyage, say—is that what is not said or only implied or left unnuanced in the book can suddenly become important or more meaningful when one is looking at the actual place.

As we sailed through the eastern end of the strait—we were just northwest of Cape Horn here—I was able to comprehend, in just moments, how very easily a sixteenth-century ship of exploration, approaching from the east in heavy weather, could entirely miss the entrance to Paso Tortuoso. The mountains come down to the water so steeply here, they effectively close off the view to the northwest, forcing one to sail south into the dead end of Whiteside Canal, or into the constricted waters of Cockburn Canal.

Raspail calls the western end of the strait a "blind alley of the Stone Age." Other authors have described the landscape as a "rain-sodden mass" and "a ruthless desolation of tundra." Darwin wrote: "Death, instead of Life, seem[s] the predominant spirit" of the place. Nevertheless, this country worked well for the Kaweskar and the Yámana, and its maze of waterways and strong contrary winds make it a modern mecca for serious sailors. The day I saw the western entrance for the first time, from the bridge, I went out on the main deck of the *Palmer* to appreciate it more fully. The authority, the boldness of the land, draped in all its subtle dark hues and sylvan textures, was so strong, trying to take it in through windows on the bridge was like trying to see Paris from the windows of a taxicab.

The *Palmer* doubled Cape Froward, rounding the southern tip of South America, and bore off northeast up Famine Reach to the roadstead at Punta Arenas. Captain Bouziga brought the ship up gently against the west side of the city's concrete pier, as softly as a parent might put a hand to a child's face. (When we departed Punta Arenas for the Weddell Sea five days later, the ice pilot for that leg of the journey, watching Bouziga ease away from the pier using only the *Palmer's* bow and stern thrusters, then align the nimble ship in the strait before he brought the main engines on line, said, "My God, you could waltz with her!")

There are few places better than this area of South America in which to dramatize succinctly the slow disintegration of indigenous societies in the Americas, the dismantling of their spiritual and economic cultures, including the ones living here in the nineteenth century, people widely regarded then as the most primitive in the world. Collectively referred to as Fuegians, they compromise the Kaweskar (or Alacaluf), the Yámana (or Yaghan), and the Selk'nam (or Ona). Each of those traditions, each one's singular idea of what being human meant, fell apart because no one of these cultures could cope with the colonizer's culture (a culture shaped by someone *else's* idea of what being human meant), which was imposed on them "for their own good." No pitched battles were fought. There is no history of guerrilla warfare to chronicle, and few anthropological efforts were made to understand how any one of these three tribes conceived of itself in the world. Each eventually became another torn prayer flag, snapping in the wind over burned ground.

Raspail gave this familiar story of unraveling a cosmopolitan context in *Who Will Remember the People . . .* , eschewing the usual indictment of Western civilization and concentrating instead on a more relevant point—the way local geographies are heedlessly torn up in order to make room for more efficient, more economically viable cultures. To his credit, Charles Darwin, who sailed these waters aboard the *Beagle* in the austral summer of 1832–33, speculated that only the thinnest veneer separated some gentleman walking briskly down Lombard Street in London one mid-nineteenth-century morning from some Yámana, Selk'nam, or Kaweskar hunter making his way through the bogs and dense beech forests here. In the egalitarian gaze of Charles Darwin, the two were brothers, not, as some then held, different species.

During the days we were docked in Punta Arenas, often when I was out walking the streets and alleys of the city, I could not clear the images of these native peoples from my mind. What perished with their cultures were their unique ideas of what it meant to be courteous, reverent, courageous, and just. What disappeared with them were their thoughts about what could be expected to be going on in the places into which we cannot see. As our own cultures continue to unfold around the riptides of aggressive commerce and heedless development, it seems these thoughts might have been good things to have made note of.

———

AS SOON AS we came alongside the quay and cleared Chilean customs and immigration, the cooks went ashore to secure fresh vegetables and the ship began taking on fuel. At the airport we dropped off members of the *Palmer*'s shakedown crew, their tasks aboard the ship now completed, and picked up air freight for the *Palmer,* including "absolute essentials" (as the manifest had it) that had accidentally been left behind in Louisiana. The ice pilot, a German, flew in from Frankfurt, accompanied by Peter Wilkniss, the head of Polar Programs at the NSF at the time. Walter Sullivan, a seasoned reporter from *The New York Times,* accompanied them. The haste and intensity with which we were gearing up to sail for ISW had an imperative about it and an otherworldliness that appeared to mark us all as we went about our chores in port. Others, the residents of Punta Arenas, were all the while painting their houses, driving city buses, changing diapers, making love, or reading magazines at trestle tables in the library. But *we,* in just a few days, were headed for Antarctica!

In our own eyes, at least, weren't we more like Ferdinand Magellan and the crew of his *Concepción* than the stevedores on the quay below us, staring up at the bridge of the *Palmer* with their looks of mild incomprehension? Or was it bright-eyed envy?

I was nearly beside myself with expectations about what we would witness in the Weddell Sea, and with feelings of privilege about being given a berth aboard the *Palmer.*

AHEAD OF ME when we came into Punta Arenas was an awkward task I'd volunteered for at the start of the voyage. About a hundred people, with return addresses all over the world, had sent the *Nathanial B. Palmer*'s owners letters of request, asking that the ship's seal be stamped on an enclosed 3×5-inch card and mailed to them from Punta Arenas in an enclosed self-addressed envelope. I thought I would learn something uplifting from the accompanying letters, but most of them, as I interpreted them, were handwritten pleas expressing a desire to be relieved of chronic loneliness. I understood now what the captain had meant when he told me in Port Fourchon, Louisiana, as the mailbag arrived, that he wanted it thrown overboard. It was presumptuous, of course, for me to judge the correspondents in this way, but dozens of letters into the process, I began to see these hopeful writers as people who'd lost a sense of purpose in their lives, to see them as correspondents who desired some tangible proof of their presence in the world.

I spread the letters out on a table in one of the wardrooms, organizing them so I wouldn't misplace anything, and began stamping each card carefully with the ship's official seal. I'd earlier set aside the American dollars and the international postal coupons that had accompanied the letters of request, to cover the cost of Chilean stamps. After I'd sealed all the envelopes and put the correct postage on them, I asked the captain what I should do with the money left over, about US $70.

"Throw it overboard," he said.

Ashore, on my way to the post office, I decided to deposit the money in a donation box at the cathedral. I was passing through an open-air farmers' market on my way to the church when I saw a Chilean couple and their child coming toward me. They looked robust and handsome together. The woman was holding the hand of a girl about three or four; the man had a large bundle strapped across his back, and a small cardboard suitcase in his right hand. The woman also carried a large bundle on a shoulder strap, held close to her side. Their jet-black hair, dark brown skin, and simple cotton clothes suggested they were full-blooded Indians, perhaps Selk'nam, arriving in Punta Arenas from some outland settlement much smaller than this city.

It was the look of innocence in their faces, the wonderful air of expectation about them and their cautious movement, the *couplehood* of them, that stopped me in the street.

I would like to say that in that moment I had the presence of mind to take the $70 out of my pocket and hand it to them. To say to the three of them, formally and respectfully, *"Para ustedes, de la Madre de Dios,"* like a messenger, and then to walk away. But I didn't. I gawked at them as they passed me by. And then it was too late.

———

I POSTED THE MAIL, put the money in the poor box at the cathedral, an embarrassing act of indifferent generosity, and went off to look at the neighborhoods of Punta Arenas I'd not yet seen. A short distance from the center of town I came upon a large house, open to the public. It seemed strangely familiar to me, a house not fully incorporated into its surroundings. A patrician residence. Once the home, I speculated, of wealthy Europeans, people afraid of losing touch with Europe, its cultural history and ambience. It reminded me of the *casa del Indio* in my stepfather's family.

Inside, I learned the home had once been the residence of the Braun-Menéndez family, scions of one Don José Menéndez. In the late nineteenth century, Señor Menéndez was operating two sheep ranches in Tierra del Fuego, one on either side of the Río Grande, a river that flows into the Atlantic. These two huge estancias were about 100 straight-line miles southeast of Punta Arenas, on the Argentine side of Tierra del Fuego. I was curious whether this Don José might be a member of the branch of the Menéndez family that was linked by marriage—as the Brauns were linked by marriage to the Menéndezes here—to my stepfather's family in Asturias. My stepfather had often spoken to me with great respect of a "José Menendez" he had known as a boy. But I could not find any information in pamphlets at the house or learn anything from the docents.

The furniture, the art, the appointments and embellishments—all were beautiful. Everything here was meticulously ordered.

I walked from the Braun-Menéndez house to Hotel Cabo de Hornos, the town's major hotel, for lunch. Many of Punta Arenas's well-dressed businessmen and merchants appeared to take their midday meal here, men who had adjusted well to being neither European nor Selk'nam but something else. I liked Punta Arenas. As we'd approached the town at sunrise that morning, with the Southern Cross fading away directly overhead, I'd found the sight here of hundreds of bright pastel homes rising in tiers up the hillsides from the waterfront very appealing. And the lobby at the Cabo de Hornos, with its between-the-wars architecture and heavy furnishings, felt very accommodating. I sat in one of the chairs reading for a couple of hours after lunch. Some people offered a polite nod as they passed.

I liked the setting so much I almost inquired about a room, and later daydreamed of settling in here at some time in the future to write a novel like Raspail's.

I would write the story of the biblical Samaritan again. He's someone who feels we are living in the last days of human restraint, and in the last days, too, of the northern bald ibis of the Ethiopian highlands and the spoon-billed sandpiper of the Chukotski Peninsula. He sees thievery, indifference, and suffering in every direction, but in his own small world he decides to help, not to turn his back. His life will become more painful, more compromised than he had imagined, and he will disappear into anonymity. In writing him as a character, I'd use as a guide the remark Camus made after he wrote *The Plague,* that Earth is all we have, that there is no other sanctuary.

Over lunch I consider the traps his story presents, with his efforts and naïveté. Maybe I should get a room at the Cabo de Hornos and learn to sail instead, ply the waters of Tierra del Fuego, hope for mostly good weather, and ignore the trouble to the north. But this image of disengagement does not have the pull I still feel toward Zagajewski's "mutilated world." I am still alive, sitting here in the restaurant, and want to be an emissary for *all* of its life, even if that idea has now become obsolete.

ONE AFTERNOON THE CAPTAIN asks me to put on a pair of slacks and a dress shirt and join him, the first mate, and the chief engineer. They've been invited to have tea with the captain of the *Irbenskiy Proliv,* moored opposite us on the quay. The *Irbenskiy Proliv* is a Russian fish-processing ship, out of Murmansk. It has been in port for two days, refueling and provisioning, perhaps taking advantage of the hospital. The rust, the patched hull, the badly damaged deck crane, the paint peeling off the superstructure, all point in one direction: We who sent you do not care about your fate. Just bring back the fish.

Before we sit down for tea, the captain of the Russian ship gives us a short tour, mostly of the engine room, perhaps because our chief engineer is along. Its narrow passages and grime-encrusted bays are lit by a few bare incandescent bulbs suspended at the ends of wires descending from gloom in the deck braces above. The room has more the feeling of a machine shop, with its metal lathes, drill presses, and welding table, than an engine room; and where one might expect to see a compact display of LED lights and computer screens bright with panels of digital information and schematics, there is only the proletariat bulk of aging machinery, the handles and edges of which, here and there, had been polished to gleaming by the swipe and grip of human hands. At several spots below the waterline the sea is spraying in between hull plates. The hammer of diesel generators makes the heated air seem even closer around us, and the metal shavings and welding slag, the tools lying loose on the oil-slickened floor, speak of constant labor, of endless maintenance.

We've brought some gifts. Medical supplies, some storm gear, rubber boots. Articles of clothing. A chocolate cake one of the *Palmer*'s cooks has made. The tea is good. We drink from glasses in silvered tin holders, each designed with scenes from Russian folklore in filigree. When we finish, one of the Russian officers motions for us to take the glasses and the ornamental holders with us. A gift. They give us each a brand-new pair of white cotton work gloves.

While we are still in the wheelhouse, the captain ushers us outside, onto the starboard wing bridge, to point out that the railings there are made of wood, and that the floor is also wood. Very reassuring, he says, to step out here during a storm, to feel the wood.

We say a cordial goodbye, everyone shaking hands. On the gangway the captain says to me over his shoulder, "These guys have nothing. Nothing."

On the pier he turns to address the three of us, saying, "None of us has the balls to sail with that ship." He speaks as if we should feel ashamed, not lucky.

I don't want to go back aboard the *Palmer*. Instead, I walk into town. I sit in a restaurant with my coffee and write in my notebook about the *Irbenskiy Proliv*, what I saw, the particulars of it, so I won't forget. Our captain, a veteran of Vietnam, had invited me, as I saw it, to accompany him on a personal errand, a ceremony aboard the *Irbenskiy*

Proliv about mutual respect and generosity, which I was not entirely sure I understood. I want to thank him again, and ask his permission to write about it.

In a bookstore on a side street in Punta Arenas the next day I find what I am looking for, a used copy of Robert Cushman Murphy's *Oceanic Birds of South America: A Study of Species of the Related Coasts and Seas, Including the American Quadrant of Antarctica, Based Upon the Brewster-Sanford Collection in the American Museum of Natural History.* It is a two-volume work, "illustrated from paintings by Francis L. Jaques, [and with] photographs, maps and other drawings." Published in 1936. Twelve hundred and forty-five pages. I buy the volumes and also a more manageable popular guide to the birds of the Southern Ocean.

I'll offer the captain the popular guide, to place among the books he keeps handy on the bridge, and give him Murphy's *Oceanic Birds of South America* for the ship's library, which he's asked me to help him develop. He'll like Murphy's prose. It is conversational, the tone informed but not recondite. The book provides a thorough answer to anyone with a serious question about oceanic birds. When we were entering the western end of the Strait of Magellan, Captain Bouziga had asked me about kelp gulls. There were flocks of them wheeling around us at the time. He said he'd heard that kelp gulls will drop clams on the sea ice from fifty feet in the air to crack them open. Might we see that on our trip? I answered him as best I could, the short answer being yes, but who knows whether we'll see them farther south, after we enter the ice.

Now I can hand him Murphy, who has fifteen pages on the kelp gull.

When I get back to the ship, Russell is taking a nap. I put the books on the bridge with a note for him and then ask the second mate, who has the watch, what the weather is looking like for us in the Drake Passage. We go over the weather faxes together. At that moment, he says, there were only five-foot swells, and it looks like the pressure cells south of us are stable. But he shrugs off any certainty. He hopes for good weather, but the waters of the Drake are notoriously rough. Squalls and worse come in from the west with little warning. He's seen sixty-foot seas there.

We are scheduled to sail in two days.

———

THE CHIEF ENGINEER and I take a cab into town one evening with a few others from the engine room to have supper. Afterward, he invites me to accompany them to a nightclub members of the crew have been frequenting. I decline, sheepishly. The invitation is a sign of the crew's genuine acceptance of my presence on the ship and I am reluctant to say no at this point in the voyage. I know an invitation to accompany them to a club like this is not commonly offered to an outsider. I say it's late, thank them all, and wish them a good evening. As I turn for the ship, the first mate comes up to me and deliberately shakes my hand. I don't know what for, perhaps to release me from my embarrassment. Or to emphasize that the degree of their acceptance is unwavering.

The *Palmer* is docked at the far end of a long pier. A crewman is on watch at the top of the gangway, to make sure no unauthorized person comes aboard, and that everyone who's gone ashore is back before the watch changes at four a.m. He checks off my name. I say good night and am turning away when out of the corner of my eye I catch sight of a tall, well-dressed man in a dark wool overcoat coming up the gangway hesitantly. He is hatless and the unbuttoned overcoat flares to reveal a white shirt open at the throat and a pair of charcoal trousers.

"¿Señor López?"

I gesture toward my chest. "¿Sí? Soy Señor López, pero mi español es pobre. Pocos palabras solamente. Digame, do you speak English?"

"Yes, I do. And your English has improved. The accent is better."

"I see. Yes, thank you. Uh, do we know each other? I'm sorry, perhaps I should recognize you. The light here is not very good."

"Do we know each other? Well, yes. A few years back we used to have lunch together every week or so. At the university."

"I'm sorry, but this is actually my first time in Punta Arenas. I've never been to the university."

The man looks about as if trying to get his bearings, and now seems a little more irritated than confused.

["You're making me look like a fool, you know, in front of this American crewman. I don't understand the reason for your rudeness."]

I don't completely understand his Spanish but grasp enough of it—"fool," "rudeness," "this American crewman"—to guess that he feels offended and also perhaps embarrassed in front of the crewman.

"I'm very sorry about the confusion. I've never visited here before, and I apologize that I don't recognize you. Have we met before, somewhere else possibly?"

"You are the writer Barry López?"

"Yes, I am a writer. Soy un escritor, sí."

"Yes, you were a guest here at the university several years ago. The students read your novels. You came to my home for supper several times."

"No, I don't think so. I think you have me confused with someone. I've never written a novel."

"But I *recognize* you, your face, your hair. How could you not be who you are? Why are you doing this?"

"I'm not doing anything. I'm trying to get to the bottom of this. I told you, I am in Punta Arenas for the first time. I arrived here three days ago, aboard this ship—"

He waves me off. A liar, rude, a disappointment to him now. He turns abruptly and descends the aluminum gangway with long, loud strides. At the bottom he turns to regard me, glaring, the way you might glare at a willful child, and then he walks off.

Standing there at the ship's railing, I don't know what to think. His ideas made no sense. Was he just a deluded person who had perhaps seen my face on a book jacket and was now operating here in a reality of his own? If I came back to Punta Arenas in a few years to write a novel in rented rooms in the Cabo de Hornos, as I intended to do, would we meet again? What would I say?

Years later I mention this incident to a friend in Bogatá, a Cervantes scholar. "Once again," he says, "the two Americas collide, North and South. The logos and the mythos."

IN MARCH 1584, the Spanish explorer Pedro Sarmiento de Gamboa founded a settlement on Chile's Brunswick Peninsula, about halfway between Cape Froward and what would later become the town of Punta Arenas. He called it Ciudad del Rey Don Felipe. His purpose was to establish a military presence in the Strait of Magellan, to prevent the British from entering the Pacific, as Drake had in 1578, an ocean the Spanish regarded as their own.

Gamboa landed about three hundred soldiers and settlers and departed as winter was coming on. The settlement faltered. When the English navigator Thomas Cavendish visited the site three years later, he found everyone had perished from starvation and exposure. He renamed the place Port Famine. Today local people call it Puerto del Hambre (Port Famine, Port of the Starving, Port Hunger).

Early in the nineteenth century, Puerto del Hambre began to serve as a base for members of the Royal Navy surveying the coasts of Patagonia. In 1828, while the survey ship HMS *Beagle* was anchored there, its commanding officer, Pringle Stokes, killed himself. Command of the *Beagle* passed, after some delay, to a flag lieutenant, Robert FitzRoy. FitzRoy sailed for England in 1830 with an impressive collection of nautical charts and four Kaweskar people: Yokcushlu, Orundellico, and El'lelaru were the names FitzRoy wrote in his notes before rechristening them, respectively, Fuegia Basket, Jemmy Button, and York Minster. The fourth Kaweskar, a twenty-year-old man christened Boat Memory, died in England of smallpox. His body, pickled in a barrel, was presented to the Royal College of Surgeons in London.

In 1831 FitzRoy was given command of the *Beagle* once more and ordered to continue the British survey of the coasts of South America. On December 27 he sailed from Plymouth with a young naturalist, Charles Darwin, aboard.

After surveying the southeastern coastal waters of South America, the *Beagle* anchored again at Port Famine in midwinter, on June 1, 1834. Darwin wrote of the place that he "never saw a more cheerless prospect; the dusky woods, piebald with snow, could be only seen indistinctly, through a drizzling hazy atmosphere." After a visit ashore, during which he attempted to force his way through dense thickets of heavy brush, he described the surrounding area as a "death-like scene of desolation exceed[ing] all description."

I'd always thought of Port Famine as a profoundly desolate place, a negative monument to human efforts to seize and possess on a monumental scale. The site of the old Spanish camp was only about forty miles down the coast from Punta Arenas. Darwin's having been there was also something of a draw for me, a second prompt to see the place, as were some historical descriptions I'd come across. What had come to interest me most about the site, though, was a small chapel there, built, I believe, in the 1950s. Individuals disabled by despair regularly make pilgrimages to this chapel. Its walls, I'd heard, were thickly crowded with *milagros,* the corsage-like assemblages of fresh flowers, religious medals, holy cards, ribbons, and handwritten notes imploring the saints, especially the Blessed Mother, to intercede for them in heaven. (*Milagro* is the Spanish word for miracle.)

Small folk chapels like this one, their walls and sometimes even their ceilings crowded with milagros, can be found all over South America, their interiors lit by hundreds of votive candles. For me these chapels transcend religion. They speak to a fundamental human need, the need to be reassured. Whatever we may say to each other about living well, about enjoying the fruits of our labors and the closeness of our families and friends, these chapels insist that the experience of human suffering known to us all, the universal suffering that takes more lives than anyone would have the stamina to hear about, not be ignored.

The chapels are as eloquent about deep-seated human fears as they are about deep-seated faith.

I find it impossible to visit such places and not feel compassion. To regard the milagros there as evidence of superstition, or to describe these out-of-the-way chapels as backward, seems to me to dismiss what it means to be human, which is to live in fear in a world in which one's destiny is never entirely of one's own choosing.

Puerto del Hambre is now a part of a Chilean national monument. The setting is humble, just a few cleared acres, the chapel and several other buildings clustered together near a parking lot. A few sections of post-and-split-rail fencing keep drivers from wandering off into a field of unmarked graves. The dominant element in the clearing is a monument that iterates Chile's claim to a pie-wedge-shaped slice of Antarctica. (Like Argentine, British, Australian, and three other national claims to Antarctic territory, the Chilean claims are not internationally recognized. The Antarctic Treaty placed all such national claims in abeyance in 1959.) Puerto del Hambre is about 2,450 miles south of Chile's border with Peru and about the same distance north of the South Pole, the apex of Chile's pie-wedge-shaped claim. The monument was erected here to mark, defiantly, the country's geographic center.

I rented a car in Punta Arenas one morning and found my way out of town on the coast road leading to Puerto del Hambre. The road is not paved, but it is good and has been routed high enough on a slope that rises up from the Magellanic shore to provide a traveler with a grand view of the strait and of the island of Tierra del Fuego to the southeast. Green hills stand on the mainland interior to the northwest, many of them cleared for pasture, the paddocks separated from one another by copses of beech trees. It was cloudy when I left the ship, but as I drove along the clouds began to clear, and I stopped several times to get out and look through my binoculars into the vastness of the strait. Directly across from me was Bahía Inútile (Useless Bay), an indentation in the northwest coast of Tierra del Fuego. To the south of there I could make out the tip of Isla Dawson.

Beyond those far shores was the immensity of the southern horizon. The land of Tierra del Fuego and the waters beyond, where I could see them, folded over the edge of Earth here like a waterfall. What lay off still farther, on the other side of the last water, was a place I had visited three times before, but which, nevertheless, still remained just out of reach for me. The journey ahead, into the Weddell aboard the *Palmer,* promised further illumination, but, I thought, there could be no end to the illumination of what existed out there beyond that line.

I was eager to see it.

THE RAYS OF THE SUN were intense now, coming over my left shoulder as I drove south toward Puerto del Hambre. Under the press of its light, the wind-buffed waters of the strait gleamed like a jostled tray of black ink. I was keeping track of the birds I saw, a few shearwaters and petrels mostly, and was thrilled to see a crested caracara at one point, sitting on a fence post ahead of me, a falcon with a streaked black-and-white chest and long legs. Despite its being technically a falcon, the caracara has the appearance of a long-legged hawk and is mostly a carrion feeder. It's red-faced, with a black crest, and its dark wings are tipped with white.

I saw a caracara once, many years before this, on Matagorda Island off the Gulf Coast of Texas. The bird is more common in South America, more apt to be seen. A thread might be said to connect the two places, however. I was pleased to find a caracara here on the Brunswick Peninsula, the southernmost extent of its range, the northernmost point of its range being on the Texas coast, 5,700 miles away. A gossamer thread, I thought, holding an abstract idea together.

Sighting the caracara was really a kind of trivial thing. What made it more memorable that morning was that I had barely passed the caracara sitting on the fence post by the road when I saw a second caracara, sitting on another post in the same fence line, about a quarter mile ahead. And then, farther on, was a third. In all I slowly passed seven or eight of them at these regular intervals, all sitting on fence posts alongside the road. Each was facing in the same southerly direction the others were. I wondered why, of course, but most of the time you can't figure such things out. Neither logic nor a grasp of falcon behavior and ecology far better than my own, I thought, could prise anything loose here. Perhaps the Kaweskar would know something.

Brief rain showers passed through as I drove along, the heavy drops beading up on the dusty road. I hadn't seen another car past the outskirts of Punta Arenas, nor any person walking in the pastures or standing in the yards of homesteads or at work within corrals visible from the road. The sky along the horizon above Tierra del Fuego bulged gray and black, with streaks of amethyst, henna, and puce, the colors of a black eye. Occasionally, as cumulus clouds unblocked the sun, the dirt road ahead would brighten up, the tan and umber shades there turning chalky as it did.

And then I saw someone, a man walking toward me on the left side of the road. I was driving slowly and he was walking the same, so it took a while to reach him. There was no farmstead about, and then suddenly he was under a rainbow. The colors formed for just a few moments over the road, a short, low span of vapor. I was so amazed I took my foot off the accelerator.

He came on steadily. His nondescript shoes were worn out. He wore black pants and a dark shirt and was hatless. Perhaps he was sixty or seventy—or, considering the place we were in, maybe younger. He came to a stop some feet in front of the car but paid me no attention. He was grimacing at the waters of the strait, as though the strait was animate, willful. Defying him. Or maybe it was the unsettled weather he was reacting to. He passed a few feet from the driver's side window, which was rolled down—should I speak? offer him a ride?—but he never glanced over.

I watched him grow smaller in the rearview mirror, a determined walker in a sunlit and otherwise deserted landscape. I wondered if he was someone who was mentally disturbed.

WHENEVER I RECALL this man, I imagine him as a small figure under the Magellanic sky and picture him wearing a white shirt, though it is there in my notebook that it was a dark shirt. I see him in that plein air panorama with all its bold colors, the high sky of cumulonimbus and the distant darkscape of Tierra del Fuego, the crêpe de chine surface of the water. I see the shock of white hair and the improbable rainbow and know that, for me, this was a portal, one that I did not enter. For now, it remains an inscrutable memory. I hold it against the day when something will cause the scene to suddenly open.

I think of him on the road to Puerto del Hambre, mad though he might have been, as no different from most of us, doing what we all do when the scaffolding of the certainties we carry with us, and by which we navigate, collapses, when indisputable truth suddenly reassembles itself in front of us, like the images in a kaleidoscope. We go on professing confidently what we know, armed with a secular faith in all that is reasonable, even though we sense that mystery is the real condition in which we live, not certainty. We forge ahead, stating what we know, watching for, hoping for, those who believe as we do, and trying to keep peace with those who see it differently. And even as the tension mounts, above all this the blue sky towers, masking, during our waking hours, the dark voids of space beyond, as we are accustomed to think of them. The sky, with its anomalous waist and that horizontal line, where what we take to be real—the ocean, the land, the ice—encounters what we regard as only speculation.

Cook, at sea aboard HMS *Endeavour,* writing up his thoughts on the

Maori; the mestizo traveler and historical footnote Ranald MacDonald, carefully pronouncing his English words in the court of the shogun; the young Darwin, picking his way through *Cordia lutea* thickets on Isla Isabela, searching for a finch. The pioneering of those few who have altered the way we see is known. The pioneering of others remains unknown to us, or barely noted. What we say we know for sure changes every day, but no one can miss now the alarm in the air. Our question is, What is it out there, just beyond the end of the road, out beyond language and fervent belief, beyond whatever gods we've chosen to give our allegiance to? Are we waiting for travelers to return, to tell us what they saw beyond that line? Or are we now to turn our heads, in order to hear better the call coming to us from that other country? It arrives as a cantus, tying the faraway place to the thing living deep inside us, a canticle that releases us from the painstaking assembly of our milagros, year after year, and from a faith only in miracles.

Notes

Introduction

1. If I'm certain about the identity of a particular plant or creature for which I offer a common name in the text, as here with "white oak," I've listed its genus and species alongside its common name in the Appendices. When the term I've used in the text refers to one of several animals, all in the same genus, but where I'd be guessing about the species, I've used the scientific name for the genus followed by "spp." In cases where I've used a common term, like sea lion, but can't be certain of the genus—a "sea lion" might be either a California sea lion (*Zalophus californianus*) or a Steller sea lion (*Eumetopias jubatus*), whose ranges overlap—I've not included a scientific binomial. An animal already identified in the text by a binomial is not listed again in the Appendix. For domestic or feral animals, I've not provided scientific binomials.

2. A memoir describing a period of traumatic sexual abuse in my childhood, "Sliver of Sky," appeared in the January 2013 issue of *Harper's*.

3. Though the state of Alaska could not be considered an international destination for an American writer, that vast landscape seemed a uniquely untrammeled part of the larger world when I first visited it, in March 1976. I think of having gone there then as a greater leave-taking of my country than earlier trips to Europe, in 1962 and 1966. Over the following seven years, I continued to travel widely in both Alaska and the Canadian High Arctic, working on a book and on magazine articles and essays. Outside those earlier trips to Europe, and boyhood jaunts to the California borderlands in Mexico, I didn't feel I had any real international experience (not in the usual sense of that term) until I decided to travel to Japan, in 1984. That exposure to rural and urban Asian culture began a period of heavy international travel, a way of working that hardly slowed until 2016, when I had to adjust the way I travel for health reasons.

When a marriage of twenty-nine years ended for me in 1996, I went on living in the house I had shared with my first wife since 1970. It's situated on a white-water river in temperate rain forest on the west side of Oregon's Cascade Mountains. My first wife and I didn't have children; in the years following our divorce, I found myself away from home even more often than before—back in Alaska or Antarctica or traveling without a definite purpose through Indonesia, the Middle East, and Central Asia. During those years I developed a relationship with Debra Gwartney, a writer who would later become my second wife, a single mother with four young daughters.

Because I had had no children during my first marriage, and because I had mostly set my own schedule for decades as a freelance writer, I was able to travel the world as few others my age ever could. When Debra and I became a couple, and I got to know my four stepdaughters as a stepfather, my sense of how most of the world actually lives—in families, with all their inherent complications and responsibilities, and with the joy and illuminations and expressions of love that family life brings—and my perspective on human life began to change. I took my youngest daughter to Cuba with me. I took my oldest to Antarctica, and the six of us traveled together to Belize. Traveling with my family gradually changed the way I understood the complexity of social forces at work in the modern world. Debra traveled with me to Greenland and to the Canadian High Arctic. The two of us traveled to Mexico, to South America, and to Europe together. With this experience I began to see even the remoter parts of the world through which I had traveled earlier (without them) through the eyes of these people I loved.

Whatever it is in me that requires—demands, Debra might say—the kind of travel experience one can find only by traveling alone—journeying to physically demanding places, where following the story is everything, or choosing arduous situations, where the schedule for eating and sleeping is haphazard—I need to thank both my family and my first wife. I benefited enormously from their understanding and support.

4. My mother had no children with her first husband, Sidney van Sheck. She had two boys, myself and my younger brother, Dennis, with her second husband, Jack Brennan, a New York advertising executive. When Jack and my mother (née Mary Frances Holstun) were married in Atlanta—in 1942, I think—Mary Holstun van Sheck became Mary Holstun Brennan. Jack was married at the time to another woman, whom he never subsequently divorced and to whom he returned after he walked out on my mother, my brother, and me in California in 1950. In 1956, my mother married Adrian Lopez, a New York publisher, and took his surname, as did both my brother and I. Jack and his first wife, Anne, had had one child, John Brennan, born in 1938. He and I were unaware of each other's existence until 1998, fifty-three years after I was born. My mother died in 1976. Jack, whom I never saw again after his divorce from my mother was finalized, died in 1984. Adrian Lopez died in 2004. My younger brother took his own life in 2017.

5. After his experience in the First World War and his work on military aircraft for Bechtel-McCone, Sidney became a pacifist and humanitarian. His mural is entitled *Youth's Strife in the Approach to Life's Problems,* and it bears the following inscription: "Gloried be they who forsaking unjust riches strive in fulfillment of humble tasks for peace, culture, and equity of all mankind."

When I saw the mural for the first time, in 2011, it was being restored and conserved by a Birmingham firm, a project for which Woodlawn High School alumni had raised $281,000. The work was finished in 2013.

6. It is slightly disingenuous, of course, for an American writer to denounce foreign dictators like the Shah of Iran or Pol Pot without pointing out the ways in which his own country has been complicit in the mayhem some of these dictators were responsible for. Since the United States overthrew the Hawaiian Monarchy in 1893, deposing Queen Liliuokalani, it has acted decisively to remove the legitimate governments of Luis Muñoz Rivera in Puerto Rico, José Santos Zelaya in Nicaragua, Salvador Allende in Chile, Ngo Dinh Diem in South Vietnam, and Mohammad Mossadegh in Iran. Though these interventions were routinely presented to the public as efforts to dethrone dictators or spread

democracy, they were also attempts, especially in Central and South America, to protect American business interests. In some of these cases, the United States failed to categorically denounce the brutal regimes of some of the very dictators it installed, like Mohammed Reza Shah and Augusto Pinochet.

There is also the question of America's "look the other way" support for repression in countries considered important economic partners or allies, like South Africa and Saudi Arabia.

My condemnation of inhumane behavior, then, must include an indictment of my own country for its legitimization of slavery and genocide—about which it remains embarrassed but not formally repentant; for its promotion of a robust international trade in arms; and for its history of self-aggrandizing economic intervention in the affairs of other countries, which is in fact a continuation of nineteenth-century colonialism.

Puerto Ayora

1. When Darwin visited the Galápagos in the nineteenth century, individual islands in the archipelago were commonly identified by their English names. Over time, Spanish names have largely supplanted these English names, though one still hears, for example, "Tower" for Genovesa, "Hood" for Española, and "James" for San Salvador. The list below privileges the Spanish names, and would be useful to anyone trying to sort out the occasional confusions here.

SPANISH NAME	ENGLISH NAME
Baltra	South Seymour
Bartolomé	Bartholomew
Española	Hood
Fernandina	Narborough
Floreana (also Santa Maria)	Charles
Genovesa	Tower
Isabela	Albemarle
Marchena	Bindloe
Plaza Norte	North Plaza
Plaza Sur	Plaza
Rábida	Jervis
San Cristóbal	Chatham
San Salvador (also Santiago)	James
Santa Cruz	Indefatigable
Sin Nombre	Nameless
Tortuga	Brattle

2. The seven-ton vessel has a beam of 21 feet and a draft of 3 feet and 1 inch. Its mainmast (*kia hope*) is 34 feet and 6 inches tall and set in tandem with a foremast (*kia mua*) of the same height. A crew of eleven to thirteen steers the vessel by adjusting the cotton sails, using a sweep (*hoe uli*) aft, and port and starboard steering blades aft of the main mast (respectively, *hoe ama* and *hoe'akea*). The deck platform (*pola*) covers about 300 square feet.

Hōkūle'a is the Hawaiian word for Arcturus, a bright star that passes directly over the island of Hawai'i. The word means "star of gladness."

3. The fact that the number of visitors to Galápagos continues to climb—215,000 arrived in 2014—doesn't necessarily mean the overall experience of the islands has been

seriously diluted. This is especially the case on islands other than Isla Santa Cruz, which receives and accommodates the majority of visitors who arrive in Galápagos and have signed up for land-based tours. The national park is more strictly managed now and the number of approved visitor sites has been increased. Each tour vessel is given an itinerary by the Park Service—which sites to visit, when, and for how long—which vessels must strictly adhere to. One group of tourists, therefore, rarely encounters another at the same site. According to one of the most responsible tour operators in Galápagos, a person whose experience goes back four decades, the impact of greater numbers of visitors on the wildlife in recent years has been "relatively light."

There is no longer a limit on the number of people who may visit the islands in any given year.

4. There are three domains of life: Eukaryota (which includes all plants and animals), Bacteria, and Archaea. Darwin's theory applies almost exclusively to eukaryotes. It applies poorly to the evolution of bacteria and archaea, which dominated the first two billion years of life on Earth.

Jackal Camp

1. The coordinates for Jackal Camp, 3°06'08" N 35°53'18" E, are approximate. I've obscured the camp's actual location to protect the search area.

2. The somewhat confusing terms *hominin, hominid,* and *hominoid* refer, respectively, to ever larger groups of human and human-like creatures. You and I are hominins, along with Neanderthals and australopithecines. Gorillas, chimpanzees, and extinct forms such as *Proconsul* and *Dryopithecus* are hominids. Gibbons are hominoids. The term Hominoidea refers to a *superfamily* that includes hominins, hominids, and hominoids. The less inclusive term Hominidae, a *family,* refers to all great apes, including humans, chimpanzees, and gorillas. The *tribe* Homini, a subdivision of the *family,* includes all the genera of bipedal primates that are more closely related to humans than they are to chimpanzees, i.e., humans, extinct species of *Homo,* australopithecines, and species in the genera *Paranthropus* and *Ardipithecus.*

Making sense of this terminology is made more difficult by the fact that until the 1990s most paleoanthropologists had used the term "hominids" to refer to species they now call "hominins."

An easy way to keep track of all this is to remember that "hominins" are us and our near relatives; that "hominids" includes hominins and our *more distant* relatives; and that "hominoids" includes both these categories as well as our *even more distant* relatives.

3. The tree-of-life metaphor is an even more precarious construct than I've suggested here. Recent work in molecular phylogenetics—the study of evolution at the molecular level—and the discovery of horizontal gene transfer—that genetic material moves not only vertically through successive generations but *horizontally* (sometimes with the aid of viruses)—is on the verge of rendering the tree-of-life metaphor too misleading now to perpetuate. Archaea, once thought to be subgroups within the domain of bacteria, are now understood to constitute an entirely separate domain of life, one coequal to the Bacteria and the Eukaryota. (The Eukaryota encompass the kingdom of plants, the kingdom of animals, the kingdom of fungi, and certain other creatures with nucleated cells.)

Antibiotic-resistant bacteria like methicillin-resistant *Staphylococcus aureus* (MRSA), sometimes referred to by health professionals as "a modern scourge," are now believed to appear, often with inexplicable speed, as a direct result of horizontal gene transfer, sometimes referred to as "infective heredity." A book that explores these phenomena and brings them together skillfully into a new understanding of evolution is David Quammen's *The Tangled Tree: A Radical New History of Life.*

4. In describing the emergence of "behaviorally modern" man, I've chosen to follow a particular line of scientific thought about how and exactly where this might have occurred. Other views concerning the timing of this event, the relative size of the human population initially distinguished in this way, and the nature of any supposed encephalic change all offer valuable insights. For example, it could be that the capacity to become "behaviorally modern" was present in *H. sapiens* for a long time but that it went unused; or that the capacity was used in a way that left no trail of obvious archeological evidence. The change in behavior could also have been a function of population density, such as the genetic presence of altruism in most members of a particular band, which might have ushered it across an evolutionary threshold and then been widely imitated.

Port Arthur to Botany Bay

1. During the time I was writing this book, 1,624 mass shootings took place in the United States, over a period of 1,870 days, according to a February 15, 2018, article in *The Guardian*. (The British paper defined a mass shooting as one in which four or more people, not including the gunman, were shot.) Since Adam Lanza killed twenty first graders and six adults at Sandy Hook Elementary School in Connecticut, in 2012, more than four hundred people have been shot in more than two hundred school shootings. (The same *Guardian* article estimated that America now has more guns than people.)

The incident at Port Arthur traumatized the Australian people, and the country moved quickly to ban the sale of the gun Martin Bryant used, an AR-15 style semiautomatic rifle. (The AR-15 has been used frequently in mass shootings in America, including those at the Pulse nightclub in Orlando; the Route 91 Harvest music festival in Las Vegas; Sandy Hook Elementary School in Newtown, Connecticut; Marjory Stoneman Douglas High School in Parkland, Florida; and the First Baptist Church in Sutherland Springs, Texas.) There are so many mass shootings in the United States every year that many Americans are now inured to this sort of violence. They have difficulty identifying with the state of genuine shock most Australians felt in the wake of the Port Arthur shootings.

The overwhelming majority of mass shootings in the world occur in the United States. Shootings like those committed by Martin Bryant in Australia, the solider in the airport in Timika, in Irian Jaya in Indonesia, and by Anders Breiviks in Oslo and Utøya, Norway (seventy-seven killed and more than two hundred wounded), are statistical outliers. Repeated efforts by members of Congress to strengthen gun laws routinely fail, largely because of political pressure brought by the National Rifle Association, a powerful lobbying group that contributes heavily to the political campaigns of pro-gun members of Congress. According to repeated national polls, U.S. citizens overwhelmingly support stricter gun laws.

Selected Bibliography

General

Beaglehole, J. C. *The Death of Captain Cook.* Wellington, New Zealand: Alexander Turnbull Library, 1979.

———. *The Life of Captain James Cook.* Stanford: Stanford University Press, 1974.

Camus, Albert. *Lyrical and Critical Essays,* trans. Ellen Conroy Kennedy. New York: Knopf, 1968.

Cook, James. *A Voyage to the Pacific Ocean in the Years 1776, 1777, 1778, 1779, and 1780 . . . Vol. I and II written by Captain J. Cook, vol. III by Captain J. King,* ed. John Douglas. London: John Murray, 1784.

———. *A Voyage Towards the South Pole and Round the World . . . In the Years 1772, 1773, 1774, and 1775,* two volumes. London: John Murray, 1777.

Earhart, Amelia. *Last Flight.* New York: G. P. Putnam's Sons, 1937.

Flannery, Tim F. *The Future Eaters: An Ecological History of the Australasian Lands and People.* Sydney: Reed Books, 1994.

———. *The Weather Makers: How Man Is Changing the Climate and What It Means for Life on Earth.* New York: Grove Press, 2005.

Hough, Richard. *Captain James Cook: A Biography.* New York: W. W. Norton, 1995.

Kolbert, Elizabeth. *The Sixth Extinction: An Unnatural History.* New York: Henry Holt, 2015.

Lewis, William S., and Naojiro Murakami, eds. *Ranald MacDonald: The Narrative of His Life 1824–1894.* Portland: Oregon Historical Society Press, 1990.

Pickles, Rosie, and Tim Cooke. *Map: Exploring the World.* London: Phaidon Press, 2015.

Roe, Jo Ann. *Ranald MacDonald: Pacific Rim Adventurer.* Pullman: Washington State University Press, 1997.

Schodt, Frederik L. *Native American in the Land of the Shogun: Ranald MacDonald and the Opening of Japan.* Berkeley, CA: Stone Bridge Press, 2003.

Thomas, Nicholas. *Cook: The Extraordinary Voyages of Captain James Cook,* ed. Nicholas Thomas. New York: Walker, 2003.

Withey, Lynne. *Voyages of Discovery: Captain Cook and the Exploration of the Pacific.* Berkeley: University of California Press, 1987.

Epigraph

Saint-Exupéry, Antoine de. *Southern Mail.* London: Heinemann, 1971.

Prologue

Parini, Jay. *John Steinbeck: A Biography.* New York: Henry Holt, 1995.

Introduction

Budde-Jones, Captain Kathryn. *Coins of the Lost Galleons.* Key West, FL: self-published, 1989.

Decter, Jacqueline. *Nicholas Roerich: The Life & Art of a Russian Master.* South Paris, ME: Park Street Press, 1989.

Hochschild, Adam. *King Leopold's Ghost: A Story of Greed, Terror, and Heroism in Colonial Africa.* Boston: Houghton Mifflin, 1999.

Huashi, Shilan, and Peng Huashi. *Terra-Cotta Warriors and Horses at the Tomb of Qin Shi Huang: The First Emperor of China.* Beijing: Cultural Relics Publishing House, 1983.

Krupnick, Jon E. *Pan American's Pacific Pioneers: A Pictorial History of Pan Am's Pacific First Flights 1935–1946.* Missoula, MT: Pictorial Histories Publishing Company, 1997.

Markham, Beryl. *West with the Night.* Boston: Houghton Mifflin, 1942.

Merton, Thomas. *The Wisdom of the Desert: Sayings from the Desert Fathers of the Fourth Century.* New York: Laughlin, 1960.

Saint-Exupéry, Antoine de. *Night Flight.* New York: New American Library, 1942.

Cape Foulweather

Akçam, Taner. *A Shameful Act: The Armenian Genocide and the Question of Turkish Responsibility.* New York: Metropolitan Books, 2006.

Banks, Sir Joseph. *The Endeavour Journal of Joseph Banks 1768–1771,* two vols., ed. J. C. Beaglehole, Sydney: Angus & Robertson, 1962.

Bettis, Stan. "Voyage to World's End." *Seattle Times Magazine,* August 25, 1971, 8–11.

Blainey, Geoffrey. *The Tyranny of Distance: How Distance Shaped Australia's History.* Melbourne: Macmillan, 1968.

Bruckner, Pascal. *The Tears of the White Man: Compassion as Contempt.* New York: Free Press, 1986.

Chapin, Mac, and Bill Threlkeld. *Indigenous Landscapes: A Study in Ethnocartography.* Arlington, VA: Center for the Support of Native Lands, 2001.

Crosby, Alfred W. *Ecological Imperialism: The Biological Expansion of Europe, 900–1900.* Cambridge: Cambridge University Press, 1986.

Davis, Mike. *Late Victorian Holocausts: El Niño Famines and the Making of the Third World.* New York: Verso, 2001.

Devorkin, David H., and Robert W. Smith. *Hubble: Imaging Space and Time.* Washington, DC: National Geographic Society, 2008.

Doczi, György. *The Power of Limits: Proportional Harmonies in Nature, Art, and Architecture.* Boulder, CO: Shambhala, 1981.

Ebony, David. *Botero Abu Ghraib*. New York: Prestel, 2006.

Elliott, T. C., ed. *Captain Cook's Approach to Oregon*. Portland: Oregon Historical Society, 1974.

Faris, Robert E. L., et al. "The Galapagos Expedition: Failure in the Pursuit of a Contemporary Secular Utopia." *The Pacific Sociological Review* 7, no. 1 (Spring 1964): 48–54.

Fay, Peter Ward. *The Opium War 1840–1842: Barbarians in the Celestial Empire in the Early Part of the 19th Century, and the War by Which They Forced Her Gates Ajar*. Chapel Hill: University of North Carolina Press, 1997.

Fitzpatrick, Kathy Bridges, ed. *Beaches and Dunes Handbook for the Oregon Coast*. Newport: The Oregon Coastal Zone Management Association, 1979.

Gifford, Don. *The Farther Shore: A Natural History of Perception, 1798–1984*. New York: Vintage Books, 1991.

Gourevitch, Philip. *We Wish to Inform You That Tomorrow We Will Be Killed with Our Families: Stories from Rwanda*. New York: Farrar, Straus and Giroux, 1998.

Harms, Robert. *The Diligent: A Voyage to the Worlds of the Slave Trade*. New York: Basic Books, 2002.

Harrison, K. David. *The Last Speakers: The Quest to Save the World's Most Endangered Languages*. Washington, DC: National Geographic, 2010.

Hayes, Derek. *Historical Atlas of the North Pacific Ocean: Maps of Discovery and Scientific Exploration 1500–2000*. Seattle: Sasquatch Books, 2001.

Hocking, Charles. *Dictionary of Disasters at Sea During the Age of Steam: Including Sailing Ships and Ships of War Lost in Action, 1824–1962*, 2 vols. London: Lloyd's Register, 1969.

Kamehameha Schools Hawaiian Studies Institute. *Life in Early Hawai'i: The Ahupua'a*, 3rd ed. Honolulu: Kamehameha Press, 1994.

Lees, James. *The Masting and Rigging of English Ships of War 1625–1860*, 2nd rev. ed. London: Conway Maritime Press, 1984.

Leland, Charles G. *Fusang or the Discovery of America by Chinese Buddhist Priests in the Fifth Century*. Albuquerque, NM: Sun Publishing Company, 1981.

Longridge, C. Nepean. *The Anatomy of Nelson's Ships*. Hemel Hempstead, Hertfordshire, UK: Model and Allied Publications Ltd., 1974.

Martin, Paul, and H. E. Wright Jr. *Pleistocene Extinctions: The Search for a Cause*. New Haven: Yale University Press, 1967.

Materials for the Study of Social Symbolism in Ancient and Tribal Art: A Record of Tradition and Continuity. Edited by Edmund Carpenter, assisted by Lorraine Spiess, based on the researches and writings of Carl Schuster. Twelve books arranged in three volumes. New York: Rock Foundation, 1986.

Morris, Roger. *The Devil's Butcher Shop: The New Mexico Prison Uprising*. New York: Franklin Watts, 1983.

Obeyesekere, Gananath. *The Apotheosis of Captain Cook: European Mythmaking in the Pacific*. Princeton, NJ: Princeton University Press, 1992.

O'Brian, Patrick. *Joseph Banks: A Life*. Boston: David R. Godine, 1993.

Parkin, Ray. *H.M. Bark Endeavour: Her Place in Australian History: with an Account of Her Construction, Crew and Equipment and a Narrative of Her Voyage on the East Coast of New Holland in the Year 1770*. Carlton South, Australia: Miegunyah Press, 1997.

Parmenter, Tish, and Robert Bailey. *The Oregon Ocean Book: An Introduction to the Pacific Ocean off Oregon Including Its Physical Setting and Living Marine Resources*. Salem: Oregon Department of Land Conservation and Development, 1985.

Piccard, Jacques. "Man's Deepest Dive." *National Geographic* 118, no. 2 (August 1960): 224–39.

Piccard, Jacques, and Robert S. Dietz. *Seven Miles Down: The Story of the Bathyscaph* Trieste. New York: G. P. Putnam's Sons, 1961.

Plummer, Katherine. *The Shogun's Reluctant Ambassadors: Japanese Sea Drifters in the North Pacific,* 3rd ed. rev. Portland: The Oregon Historical Society, 1991.

Rediker, Marcus. *The Slave Ship: A Human History.* New York: Penguin Books, 2008.

Rehbock, Philip, ed. *At Sea with the Scientifics: The* Challenger *Letters of Joseph Matkin.* Honolulu: University of Hawaii Press: 1992.

Rodger, N. A. M. *The Wooden World: An Anatomy of the Georgian Navy.* New York: W. W. Norton, 1996.

Romoli, Kathleen. *Balboa of Darién: Discoverer of the Pacific.* New York: Doubleday, 1953.

Salgado, Sebastião. *An Uncertain Grace.* With essays by Eduardo Galeano and Fred Ritchin. New York: Aperture Foundation, 1990.

Stannard, David E. *Before the Horror: The Population of Hawai'i on the Eve of Western Contact.* Honolulu: Social Science Research Institute, University of Hawai'i, 1989.

Suttles, Wayne, ed. *Handbook of North American Indians: Northwest Coast.* Washington, DC: Smithsonian Institution, 1990.

Tarnas, Richard. *The Passion of the Western Mind: Understanding the Ideas That Have Shaped Our World View.* New York: Ballantine Books, 1991.

Vairo, Carlos Pedro. *The Prison of Ushuaia.* Buenos Aires: Zagier & Urruty, 1997.

Walsh, Don, Lt. "Our 7-Mile Dive to Bottom." *Life,* February 15, 1960, 112–120.

Webber, Burt. *Silent Siege: Japanese Attacks Against North America in World War II.* Fairfield, WA: Ye Galleon Press, 1984.

Yenne, Bill. *Seaplanes of the World: A Timeless Collection from Aviation's Golden Age.* Cobb, CA: O.G. Publishing, 1997.

Skraeling Island

Bliss, L. C., ed. *Truelove Lowland, Devon Island, Canada: A High Arctic Ecosystem.* Edmonton: University of Alberta Press, 1977.

Blodgett, Jean, ed. *The Coming and Going of the Shaman: Eskimo Shamanism and Art.* Winnipeg: Winnipeg Art Gallery, 1978.

Boeke, Kees. *Cosmic View: The Universe in 40 Jumps.* London: Faber & Faber, 1957.

Bornstein, Eli, ed. *The Structurist: Transparency & Reflection,* no. 27/28 (1987/88). Saskatoon, Saskatchewan: The University of Saskatchewan, 1988.

Figes, Eva. *Light.* New York: Pantheon Books, 1983.

Gobodo-Madikizela, Pumla. *A Human Being Died That Night: A South African Story of Forgiveness.* Boston: Houghton Mifflin, 2003.

Greely, Adolphus. *Three Years of Arctic Service: An Account of Three Years of the Lady Franklin Bay Expedition, and the Attainment of the Farthest North.* New York: Scribner, 1886.

Grønnow, Bjarne, and Jens Fog Jensen. "The Northernmost Ruins of the World: Eigil Knuth's Archeological Investigations in Pearyland and Adjacent Areas of High Arctic Greenland." *Man and Society,* vol. 29. Copenhagen: Danish Polar Center, 2003.

Guttridge, Leonard. *Ghosts of Cape Sabine: The Harrowing True Story of the Greely Expedition.* New York: G. P. Putnam's Sons, 2000.

Herbert, Wally. *The Noose of Laurels: Robert E. Peary and the Race to the North Pole.* New York: Atheneum, 1989.

MacDonald, John. *The Arctic Sky: Inuit Astronomy, Star Lore and Legend.* Toronto: Royal Ontario Museum, 1998.

Mary-Rousselière, Guy. *Qitdlarssuaq: The Story of a Polar Migration.* Winnipeg: Wuerz Publishing Ltd., 1991.

McGhee, Robert. *Ancient People of the Arctic.* Vancouver: University of British Columbia Press, 1996.

———. *Canadian Arctic Prehistory.* Toronto: Van Nostrand Reinhold, 1978.

———. *The Last Imaginary Place: A Human History of the Arctic World.* New York: Oxford University Press, 2005.

Mitchell, Frank. *Navajo Blessingway Singer: The Autobiography of Frank Mitchell, 1881–1967.* Tucson: University of Arizona Press, 1978.

Nadolny, Sten. *The Discovery of Slowness.* New York: Viking, 1987.

Nettleship, David N., and Pauline A. Smith. *Ecological Sites in Northern Canada.* Ottawa: Canadian Committee for the International Biological Programme, Conservation Terrestrial Panel 9, 1975.

Schledermann, Peter. *Crossroads to Greenland: 3,000 Years of Prehistory in the Eastern High Arctic.* Calgary: Arctic Institute of North America, 1990.

———. *Voices in Stone: A Personal Journey into the Arctic Past.* Calgary: Arctic Institute of North America, 1996.

Svoboda, Joseph, and Bill Freedman, ed. *Ecology of a Polar Oasis: Alexandra Fjord, Ellesmere Island, Canada.* Toronto: Captus University Publications, 1994.

Weigelt, Johannes. *Recent Vertebrate Carcasses and Their Paleobiological Implications,* trans. Judith Schaefer. Chicago: University of Chicago Press, 1989.

White, Randall. *Dark Caves, Bright Visions: Life in Ice Age Europe.* New York: W. W. Norton, 1986.

Puerto Ayora

Adams, John Luther. *The Place Where You Go to Listen: In Search of an Ecology of Music.* Middletown, CT: Wesleyan University Press, 2009.

Allen, Jennifer. *Mālama Honua: Hōkūle'a—A Voyage of Hope.* Ventura, CA: Patagonia Books, 2017.

Barnett, Bruce D. "Eradication and Control of Feral and Free-Ranging Dogs in the Galápagos Islands." Proceedings of the Twelfth Vertebrate Pest Conference, University of Nebraska at Lincoln, 1986, http://digitalcommons.unl.edu/vpc12/8.

———. "Phenotypic Variation in Feral Dogs (*Canis familiaris*) of the Galápagos Islands." Manuscript, 1985.

Beebe, William. *Galápagos: World's End.* New York: G. P. Putnam's Sons, 1924.

Browne, Janet. *Charles Darwin: The Power of Place, Volume II of a Biography.* New York: Knopf, 2002.

———. *Charles Darwin: Voyaging, Volume I of a Biography.* New York: Knopf, 1995.

Collis, Maurice. *Cortes and Montezuma.* New York: Harcourt, Brace, 1955.

D'Orso, Michael. *Plundering Paradise: The Hand of Man on the Galápagos Islands.* New York: Perennial, 2003.

Darwin, Charles. *From So Simple a Beginning: The Four Great Books of Charles Darwin,* ed. Edward O. Wilson. New York: W. W. Norton, 2006.

———. *The Voyage of the Beagle,* annotated and with an introduction by Leonard Engel. Garden City, NY: Doubleday, 1962.

Dennett, D. C. *Darwin's Dangerous Idea: Evolution and the Meanings of Life.* New York: Simon & Schuster, 1995.

Díaz del Castillo, Bernal. *The Discovery and Conquest of Mexico 1517–1521.* New York: Farrar, Straus and Giroux, 1956.

Feinberg, Harris. "Adaptive Radiation of Feral Dogs on the Island of Isabela in the Galápagos." Manuscript, 2003.

Finney, Ben. *Voyage of Rediscovery: A Cultural Odyssey through Polynesia.* Berkeley: University of California Press, 1994.

Fraenkel, Gottfried, and Donald Gunn. *The Orientation of Animals: Kineses, Taxes and Compass Reactions.* Oxford: Oxford University Press, 1940.

Grove, Jack Stein, and Robert J. Lavenberg. *The Fishes of the Galápagos Islands.* Stanford, CA: Stanford University Press, 1997.

Harris, Michael. *A Field Guide to the Birds of Galápagos, Revised.* London: Collins, 1982.

Jackson, M. H. *Galápagos: A Natural History Guide.* Calgary: University of Calgary Press, 1985.

Levenson, Jay A. *Circa 1492: Art in the Age of Exploration.* New Haven: Yale University Press, 1991.

Lewis, David. *We, the Navigators: The Ancient Art of Land-Finding in the Pacific,* 2nd ed. Honolulu: University of Hawai'i Press, 1994.

Marx, Richard Lee. *Three Men of the* Beagle. New York: Knopf, 1991.

Melville, Herman. *The Shorter Novels of Herman Melville.* New York: Liveright, 1928.

Merlen, Godfrey. *A Field Guide to the Fishes of Galápagos.* London: Wilmot, 1986.

Nash, June. *We Eat the Mines, and the Mines Eat Us: Dependency and Exploitation in Bolivian Tin Mines,* rev. ed. New York: Columbia University Press, 1993.

Osorio, Jon Kamakawiwo'ole. *Dismembering Lāhui: A History of the Hawaiian Nation to 1887.* Honolulu: University of Hawai'i Press, 2002.

Perry, Roger, ed. *Galapagos: Key Environments.* Oxford: Pergamon, 1984.

Puleston, Dennis. *Blue Water Vagabond: Six Years' Adventure at Sea.* New York: Doubleday, Doran & Co., 1939.

Robinson, William Albert. *Voyage to Galapagos.* New York: Harcourt, Brace, 1936.

Sale, Kirkpatrick. *The Conquest of Paradise: Christopher Columbus and the Columbian Legacy.* New York: Knopf, 1990.

Silva, Noenoe. *Aloha Betrayed: Native Hawaiian Resistance to American Colonialism.* Durham, NC: Duke University Press, 2004.

Sinoto, Yosihiko. *Curve of the Hook: Yosihiko Sinoto, an Archeologist in Polynesia, with Hiroshi Aramata,* ed. and trans. Frank Stewart and Madoka Nagadō. Honolulu: University of Hawai'i Press, 2016.

Slavin, Joseph Richard. *The Galápagos Islands: A History of Their Exploration,* Occasional Papers of the California Academy of Sciences No. XXV. San Francisco: California Academy of Sciences, 1959.

Strauch, Dore, and W. Brockmann. *Satan Came to Eden.* New York: Harper & Bros., 1936.

Todorov, Tzvetan. *The Conquest of America: The Question of the Other,* trans. Richard Howard. New York: Harper & Row, 1984.

Weiner, Jonathan. *The Beak of the Finch: A Story of Evolution in Our Time.* New York: Knopf, 1994.

Wittmer, Margret. *Floreana.* London: Michael Joseph Ltd., 1961.

Jackal Camp

Baron-Cohen, Simon. *Mindblindness: An Essay on Autism and Theory of Mind.* Cambridge, MA: MIT Press, 1995.

Campbell, Joseph. *The Masks of God,* four volumes. New York: Viking Press, 1959–68.

Chauvet, Jean-Marie, et al. *Dawn of Art: The Chauvet Cave, the Oldest Known Paintings in the World.* New York: Harry N. Abrams, 1996.

Clark, Geoffrey A. *The Asturian of Cantabria: Early Holocene Hunter-Gatherers in Northern Spain.* Tucson: University of Arizona Press, 1983.

Doherty, Martin. *Theory of Mind: How Children Understand Others' Thoughts and Feelings.* New York: Psychology Press, 2008.

Eldredge, Niles. *Unfinished Synthesis: Biological Hierarchies and Modern Evolutionary Thought.* New York: Oxford University Press, 1985.

Eldredge, Niles, and Ian Tattersall. *The Myths of Human Evolution.* New York: Columbia University Press, 1982.

Elkins, Caroline. *Imperial Reckoning: The Untold Story of Britain's Gulag in Kenya.* New York: Henry Holt, 2005.

Golding, William. *The Inheritors.* New York: Harcourt, Brace & World, 1955.

Guinea, Miguel Angel García. *Altamira and Other Cantabrian Caves.* Madrid: Silex, 1979.

Hillaby, John. *Journey to the Jade Sea.* London: Constable & Company, 1964.

Huxley, Elspeth. *The Flame Trees of Thika: Memories of an African Childhood.* London: Chatto & Windus, 1959.

———. *Out in the Midday Sun: My Kenya.* London: Chatto & Windus, 1985.

Jeffers, Robinson. *Robinson Jeffers: Selected Poems.* New York: Vintage Books, 1965.

Johanson, Donald Carl, et al. *Ancestors: In Search of Human Origins.* New York: Villard Books, 1994.

Kapuściński, Ryszard. *The Shadow of the Sun,* trans. Klara Glowczewska. New York: Knopf, 2001.

Klein, Richard G. *The Human Career: Human Biological and Cultural Origins,* 2nd ed. Chicago: University of Chicago Press, 1999.

Kurtén, Björn. *Dance of the Tiger: A Novel of the Ice Age.* New York: Pantheon Books, 1980.

Lamb, David. *The Africans.* New York: Random House, 1983.

Landau, Misia. "Human Evolution as Narrative," *American Scientist* 72, no. 3 (May 1984): 262–68.

———. *Narratives of Human Evolution,* rev. ed. New Haven, CT: Yale University Press, 1993.

Leakey, Mary. *Africa's Vanishing Art: The Rock Paintings of Tanzania.* Garden City, New York: Doubleday, 1983.

Leakey, Richard E. *The Origin of Humankind.* New York: Basic Books, 1994.

Lelyveld, Joseph. *Move Your Shadow: South Africa, Black and White.* New York: Times Books, 1985.

Lovell, Mary S. *Straight on Till Morning: The Biography of Beryl Markham.* New York: W. W. Norton, 2011.

Maclean, Gordon Lindsay. *Roberts' Birds of Southern Africa,* 5th ed. Cape Town: John Voelcker Bird Fund, 1985.

Malan, Rian. *My Traitor's Heart: A South African Exile Returns to Face His Country, His Tribe, and His Conscience.* New York: Atlantic Monthly Press, 1990.

Merton, Thomas. *Ishi Means Man.* Greensboro, NC: Unicorn Press, 1976.

Miller, Charles. *The Lunatic Express: An Entertainment in Imperialism.* New York: Macmillan, 1971.

Mithen, Steven. *After the Ice: A Global Human History 20,000–5000 BC.* Cambridge, MA: Harvard University Press, 2004.

Monbiot, George. *No Man's Land: An Investigative Journey Through Kenya and Tanzania.* London: Macmillan, 1994.

Morell, Virginia. *Ancestral Passions: The Leakey Family and the Quest for Humankind's Beginnings.* New York: Touchstone, 1996.

Moss, Rose. *Shouting at the Crocodile: Popo Molefe, Patrick Lekota, and the Freeing of South Africa.* Boston: Beacon Press, 1990.

Pakenham, Thomas. *The Scramble for Africa: White Man's Conquest of the Dark Continent 1876–1912.* New York: Random House, 1991.

Pick, John. *Gerard Manley Hopkins: Priest and Poet,* 2nd ed. New York: Oxford University Press, 1966.

Quammen, David. *The Tangled Tree: A Radical New History of Life.* New York: Simon & Schuster, 2018.

Sarmiento, Esteban E., G. J. Sawyer, Richard Milner, Viktor Deak, and Ian Tattersall. *The Last Human: A Guide to Twenty-Two Species of Extinct Humans.* New Haven: Yale University Press, 2007.

Schlesier, Karl H. *The Leopard Springs of Ussat.* Charleston, SC: Karl H. Schlesier, 2010.

Singer, Sam, and Henry R. Hilgard. *The Biology of People.* San Francisco: W. H. Freeman, 1978.

Spawls, Stephen. *A Field Guide to the Reptiles of East Africa: Kenya, Tanzania, Uganda, Rwanda and Burundi.* San Diego: Academic Press, 2002.

Waal, F. B. M. de. *Bonobo, The Forgotten Ape.* Berkeley: University of California Press, 1997.

———. *Peacemaking Among Primates.* Cambridge, MA: Harvard University Press, 1989.

Walker, Alan, and Pat Shipman. *The Wisdom of the Bones: In Search of Human Origins.* New York: Knopf, 1996.

Williams, John, and Norman Arlott. *The Collins Field Guide to the Birds of East Africa.* Lexington, MA: Stephen Greene Press, 1980.

Willis, Delta. *The Hominid Gang: Behind the Scenes in the Search for Human Origins.* New York: Penguin Books, 1991.

Zunshine, Lisa. *Why We Read Fiction: Theory of Mind and the Novel.* Columbus: Ohio State University Press, 2006.

Port Arthur to Botany Bay

Alcock, John, with illustrations by Marilyn Hoff Stewart. *The Kookaburras' Song: Exploring Animal Behavior in Australia.* Tucson: University of Arizona Press, 1988.

Blackburn, Julia. *Daisy Bates in the Desert: A Woman's Life Among the Aborigines.* New York: Pantheon, 1994.

Blainey, Geoffrey. *Triumph of the Nomads: A History of Ancient Australia,* rev. South Melbourne: Macmillan Company of Australia, 1983.

Brand, Ian. *Penal Peninsula: Port Arthur and Its Outstations 1827–1898.* Launceston, Tasmania: Regal Publications, no date.

Brock, Peggy, ed. *Women Rites and Sites.* Sydney: Allen & Unwin, 1989.

Buttler, Elisha, ed. *The Pilbara Project: Field Notes and Photographs Collected Over 2010.* Perth, Australia: FORM, 2010.

Carter, Paul. *The Road to Botany Bay: An Exploration of Landscape and History.* New York: Knopf, 1988.

Chatwin, Bruce. *The Songlines.* London: Penguin Books, 1980.

Clark, Betty, and Fred Myers. *Report on the First Contact Group of Pintupi at Kiwirrkura.* Manuscript, 1985.

Clarke, Marcus. *For the Term of His Natural Life.* Rosny Park, Tasmania: Book Agencies of Tasmania, no date.

Coman, Brian. *Tooth and Nail: The Story of the Rabbit in Australia.* Melbourne: Text Publishing, 1999.

Compston, W. and R. T. Pidgeon. "Jack Hills, Evidence of Very Old Detrital Zircons in Western Australia." *Nature* 321 (June 19, 1986): 766–69.

Currey, C. H. *The Transportation, Escape and Pardoning of Mary Bryant (née Broad).* Sydney: Angus & Robertson, 1963.

Davies, Luke. *Absolute Event Horizon.* Sydney: Angus & Robertson, 1994.

———. *Candy.* Crows Nest, Australia: Allen & Unwin, 1996.

Estensen, Miriam. *The Life of Matthew Flinders.* Crows Nest, Australia: Allen & Unwin, 2002.

Frost, Allan. *Botany Bay Mirages: Illusions of Australia's Convict Beginnings.* Melbourne: Melbourne University Press, 1994.

Gee, Dennis, J. L. Baxter, Simon Wilde, and I. R. Williams. "Crustal Development in the Archaean Yilgarn Block, Western Australia." *Special Publication, Geological Society of Australia* 7 (1981): 43–56.

Gibson, D. F. *A Biological Survey of the Tanami Desert in the Northern Territory,* Technical Report No. 30. Alice Springs, Northern Territory, Australia: Conservation Commission of the Northern Territory, 1986.

Grishin, Sasha. *John Wolseley: Land Marks III.* [The third volume by Grishin on Wolseley's work. The earlier volumes are *Land Marks* (1999) and *Land Marks II* (2006).] Port Melbourne: Thames & Hudson Australia, 2015.

Groom, Arthur. *I Saw a Strange Land: Journeys in Central Australia.* Sydney: Angus & Robertson, 1950.

Hawley, Janet. *Encounters with Australian Artists.* St. Lucia, Australia: University of Queensland Press, 1993.

Hay, Pete. "Port Arthur: Where Meanings Collide." *Vandiemonian Essays.* North Hobart, Tasmania: Walleah Press, 2002.

Hughes, Robert. *The Fatal Shore: A History of the Transportation of Convicts to Australia 1787–1868.* London: Collins Harvill, 1987.

Kent, Rockwell. *Voyaging: Southward from the Strait of Magellan.* New York: Halcyon Press, 1924.

Latz, Peter. *Bushfires & Bushtucker.* Alice Springs, Australia: IAD Press, 1995.

Layton, Robert. *Uluru: An Aboriginal History of Ayers Rock.* Canberra: The Australian Institute of Aboriginal Studies, 1986.

Lennox, Geoff. *A Visitor's Guide to Port Arthur and the Convict Systems.* Rosetta, Tasmania: Dormasland Publications, 1994.

Lines, William J. *A Long Walk in the Australian Bush.* Sydney: UNSW Press, 1998.

Malouf, David. *The Conversations at Curlow Creek.* Melbourne: Bolinda, 1996.

————. *Remembering Babylon*. New York: Vintage Books, 1994.

Meggitt, Mervyn J. *Desert People: A Study of the Walbiri Aborigines of Central Australia*. Sydney: Angus & Robertson, 1962.

Menneken, Martina, et al. "Hadean Diamonds in Zircon from Jack Hills, Western Australia." *Nature* 448 (August 23, 2007): 917–20.

Moorehead, Alan. *The Fatal Impact: An Account of the Invasion of the South Pacific 1767–1840*. New York: Harper & Row, 1966.

Moyal, Ann. *A Bright and Savage Land: Scientists in Colonial Australia*. Sydney: Collins, 1986.

Myers, Fred R. "The Politics of Representation: Anthropological Discourse and Australian Aborigines." *American Ethnologist* 13, no. 1 (February 1986): 138–53.

Nugent, R. *Aboriginal Attitudes to Feral Animals and Land Degradation*. Alice Springs, Australia: Central Land Council, 1988.

Pyne, Stephen J. *Burning Bush: A Fire History of Australia*. New York: Henry Holt, 1991.

Read, Peter and Jay. *Long Time, Olden Time: Aboriginal Accounts of Northern Territory History*. Alice Springs, Australia: Institute for Aboriginal Development, 1991.

Reynolds, Henry. *Fate of a Free People*. Ringwood, Australia: Penguin Books, 1995.

————. *The Other Side of the Frontier: Aboriginal Resistance to the European Invasion of Australia*. Townsville, Australia: James Cook University of North Queensland, 1981.

Rolls, Eric C. *They All Ran Wild: The Story of Pests on the Land in Australia*. Sydney: Angus & Robertson, 1969.

San Roque, Craig. "On 'Tjukurrpa,' Painting Up, and Building Thought." *Social Analysis: The International Journal of Social and Cultural Practice* 50, no. 2 (Summer 2006): 148–72.

Scot, Margaret. *Port Arthur: A Story of Strength and Courage*. Sydney: Random House Australia, 1997.

Slater, Peter, Pat Slater, and Raoul Slater. *The Slater Field Guide to Australian Birds*. Willoughby, Australia: Lansdowne-Rigby Publishers, 1986.

Stanner, W. E. H. *White Man Got No Dreaming: Essays 1938–1973*. Canberra: Australian National University Press, 1979.

Stokes, Edward. *Across the Centre: John MacDouall Stuart's Expeditions 1860–62*. St. Leonards, Australia: Allen & Unwin, 1996.

Strehlow, T. G. H. *Journey to Horseshoe Bend*. Sydney: Angus & Robertson, 1969.

Sutton, Peter, ed. *Dreamings: The Art of Aboriginal Australia*. New York: George Braziller, 1988.

Toyne, Phillip, and Daniel Vachon. *Growing Up the Country: The Pitjantjatjara Struggle for Their Land*. Fitzroy, Australia: McPhee, Gribble Publishers, 1984.

Vaarzon-Morel, Petronella, ed. *Warlpiri Women's Voices: Our Lives Our History*. Alice Springs, Australia: IAD Press, 1995.

Vallejo, César. *Los Heraldos Negros*. Buenos Aires: Losada, 1918.

Weldon, Annamaria. *The Lake's Apprentice*. Crawley, Australia: UWA Publishing, 2014.

Graves Nunataks to Port Famine Road

Alberts, Fred, ed. *Geographic Names of the Antarctic*, 2nd ed. Arlington: National Science Foundation, 1995.

Bainbridge, Beryl. *The Birthday Boys*. New York: Carroll & Graf, 1991.

Bridges, Lucas E. *The Uttermost Part of the Earth*. New York: E. P. Dutton, 1949.

Bridges, Thomas. *Yamana-English: A Dictionary of the Speech of Tierra del Fuego.* Buenos Aires: Zagier y Urruty Publicaciones, 1987.

Byrd, Richard E. *Alone.* New York: G. P. Putnam's Sons, 1938.

Campbell, David G. *The Crystal Desert: Summers in Antarctica.* Boston: Houghton Mifflin, 1992.

Cassidy, William A. *Meteorites, Ice, and Antarctica.* Cambridge: Cambridge University Press, 2003.

Cherry-Garrard, Apsley. *The Worst Journey in the World.* London: Constable and Co. Ltd., 1922.

Cordes, Fauno. *Winter Survival in the Antarctic as Described by James Fenimore Cooper.* Thesis, San Francisco State University, 1991.

Gurney, Alan. *Below the Convergence: Voyages Toward Antarctica, 1699–1839.* New York: W. W. Norton, 1997.

Huntford, Roland. *The Last Place on Earth.* New York: Atheneum, 1986.

———. *Shackleton.* New York: Fawcett Columbine, 1985.

Huxley, Elspeth. *Scott of the Antarctic.* New York: Atheneum, 1977.

Lansing, Alfred. *Endurance: Shackleton's Incredible Voyage.* New York: McGraw-Hill, 1959.

Mason, Theodore K. *The South Pole Ponies.* New York: Dodd, Mead, 1979.

McSween, Harry Y., Jr. *Meteorites and Their Parent Planets.* Cambridge: Cambridge University Press, 1987.

Murphy, Robert Cushman. *Oceanic Birds of South America: A Study of Species of the Related Coasts and Seas, Including the American Quadrant of Antarctica Based Upon the Brewster-Sanford Collection in the American Museum of Natural History, Vol. I, Vol. II.* New York: Macmillan Company, 1936.

Poe, Edgar Allan. *The Narrative of Arthur Gordon Pym of Nantucket.* New York: Harper & Brothers, 1838.

Preston, Diana. *A First-Rate Tragedy: Robert Falcon Scott and the Race to the South Pole.* Boston: Houghton Mifflin, 1998.

Pyne, Stephen. *The Ice: A Journey to Antarctica.* Iowa City: University of Iowa Press, 1986.

Raspail, Jean. *Who Will Remember the People . . .* San Francisco: Mercury House, 1988.

Smith, Michael. *I Am Just Going Outside: Captain Oates—Antarctic Tragedy.* Staplehurst, UK: Spellmount, 2002.

Solomon, Susan. *The Coldest March: Scott's Fatal Antarctic Expedition.* New Haven: Yale University Press, 2001.

Spufford, Francis. *I May Be Some Time: Ice and the English Imagination.* Boston: Faber & Faber, 1996.

Wilson, Edward. *Diary of the Terra Nova Expedition to the Antarctic, 1910–1912.* London: Blandford Press, 1972.

Worsley, Frank. *Shackleton's Boat Journey.* London: Philip Allan, 1933.

Scientific Binomials

Animals

aardvark	*Orycteropus afer*
Arctic blue butterfly	*Agriades glandon*
Arctic fox	*Alopex lagopus*
Arctic hare	*Lepus arcticus*
Arctic wolf spider	*Pardosa glacialis*
bearded seal	*Erignathus barbatus*
beaver	*Castor canadensis*
beluga whale	*Delphinapterus leucas*
black bear	*Ursus americanus*
black mamba	*Dendroaspis polylepis*
black rat	*Rattus rattus*
black rhino	*Diceros bicornis*
blacktail deer	*Odocoileus hemionus*
black widow spider	*Latrodectus* spp.
blue whale	*Balaenoptera musculus*
bowhead whale	*Balaena mysticetus*
brown bear	*Ursus arctos*
Burchell's zebra	*Equus quagga burchellii*
Burmese python	*Python bivattatus*
camel	*Camelus* spp.
chacma baboon	*Papio griseipes*
cheetah	*Acinonyx jubatus*
collared lemming	*Dicrostonyx* spp.
common bottlenose dolphin	*Tursiops truncatus*
common wallaroo	*Macropus robustus*

crabeater seal	*Lobodon carcinophaga*
crocodile	*Crocodylus* spp.
ermine	*Mustela ermine*
fisher	*Martes pennanti*
ghost crab	*Ocypode gaudichaudii*
giant panda	*Ailuropoda melanoleuca*
giraffe	*Giraffa camelopardalis*
golden jackal	*Canis aureus*
gray fox	*Urocyon cinereoargenteus*
gray whale	*Eschrichtius robustus*
Grévy's zebra	*Equus grevyi*
grizzly bear	*Ursus arctos*
guenon	*Cercopithecus* spp.
harbor seal	*Phoca vitulina*
impala	*Aepyceros melampus*
jeweled gecko	*Naultinus gemmeus*
Komoda dragon	*Varanus komodoensis*
leopard	*Panthera pardus*
leopard seal	*Hydrurga leptonyx*
lion	*Panthera leo*
lynx	*Lynx canadensis*
marine iguana	*Amblyrhynchus cristatus*
marten	*Martes americana*
mink	*Neovison vison*
minke	*Balaenoptera acutorostrata*
mountain gorilla	*Gorilla beringei beringei*
mountain lion	*Puma concolor*
muskox	*Ovibos moschatus*
narwhal	*Monodon monoceros*
northeast African carpet viper	*Echis pyramidum*
northern red-backed vole	*Clethrionomys rutilis*
Norway rat	*Rattus norvegicus*
ocean strider	*Halobates* spp.
orangutan	*Pongo* spp.
orca	*Orcinus orca*
oryx	*Oryx gazella*
Pacific green turtle	*Chelonia mydas*
Peary caribou	*Rangifer tarandus pearyi*
polar bear	*Ursus maritimus*
polar bumblebee	*Bombus polaris*
Portuguese man-of-war	*Physalia physalis*
puff adder	*Bitis arietans*

razor clam	*Siliqua* spp.
red fox	*Vulpes vulpes*
red kangaroo	*Macropus rufus*
red spitting cobra	*Naja pallida*
reticulated giraffe	*Giraffa camelopardalis reticulata*
ringed seal	*Pusa hispida*
river otter	*Lontra canadensis*
Roosevelt elk	*Cervus elaphus roosevelti*
Ross seal	*Ommatophoca rossii*
sally lightfoot crab	*Grapsus grapsus*
sei whale	*Balaenoptera borealis*
southern right whale	*Eubalaena australis*
southern white rhino	*Ceratotherium simum simum*
sperm whale	*Physeter macrocephalus*
spotted hyena	*Crocuta crocuta*
springbok	*Antidorcas marsupialis*
star spider	*Gasteracantha cancriformis*
striped polecat	*Ictonyx striatus*
thylacine	*Thylacinus cynocephalus*
topi	*Damaliscus lunatus*
tube worm	*Riftia pachyptila*
walrus	*Odobenus rosmarus*
warthog	*Phacochoerus africanus*
Weddell seal	*Leptonychotes weddellii*
wildebeest	*Connochaetes* spp.
wolverine	*Gulo gulo*
Yangtze river dolphin	*Lipotes vexillifer*
zebu cattle	*Bos taurus indicus*

Fish

blue-eyed damselfish/yellowtail damselfish	*Stegastes arcifrons*
boxfishes	*Lactoria diaphana* and *Ostracion meleagris*
canary rockfish	*Sebastes pinniger*
clown razorfish	*Novaculichthys taeniourus*
coelacanth	*Latimeria chalumnae*
dusky sergeant major	*Abudefduf troschelii*
Galápagos shark	*Carcharhinus galapagensis*
grouper	*Mycteroperca olfax*
guineafowl puffer	*Arothron meleagris*
Moorish idol	*Zanclus cornutus*

orange-eyed mullet — *Xenomugil thoburni*
Pacific beakfish — *Oplegnathus insignis*
rainbow scorpionfish — *Scorpaenodes xyris*
scalloped hammerhead shark — *Sphyrna lewini*
Sheepshead mickey — *Microspathodon bairdi*
spotted eagle ray — *Aetobatus narinari*
thresher shark — *Alopias vulpinus*
tinsel squirrelfish/sun squirrelfish — *Sargocentron suborbitalis*
whitetip reef shark — *Triaenodon obesus*
yellowtail surgeonfish — *Prionurus laticlavius*

Birds

Abyssinian roller — *Coracias abyssinica*
Adélie penguin — *Pygoscelis adeliae*
Baird's sandpiper — *Calidris bairdii*
banded stilt — *Cladorhynchus leucocephalus*
black guillemot — *Cepphus grylle*
black-necked stilt — *Himantopus mexicanus*
black noddy — *Anous minutus*
black swan — *Cygnus atratus*
blue-footed booby — *Sula nebouxii*
Brandt's cormorant — *Phalacrocorax penicillatus*
brown pelican — *Pelecanus occidentalis*
budgerigar — *Melopsittacus undulatus*
Cape rook/Cape crow — *Corvus capensis*
Cape vulture — *Gyps coprotheres*
caracara — *Caracara plancus*
Carnaby's black-cockatoo — *Calyptorhynchus latirostris*
Cassin's auklet — *Ptychoramphus aleuticus*
cockatiel — *Nymphicus hollandicus*
common eider — *Somateria mollissima*
common murre — *Uria aalge*
common raven — *Corvus corax*
compact weaverbird — *Ploceus superciliosus*
dark chanting goshawk — *Melierax metabates*
double-crested cormorant — *Phalacrocorax auritus*
Egyptian vulture — *Neophron percnopterus*
emperor penguin — *Aptenodytes forsteri*
fairy tern — *Sternula nereis*
fairy wren — *Malarus* spp.
galah — *Eolophus roseicapillus*

Galápagos dove	*Zenaida galapagoensis*
Galápagos hawk	*Buteo galapagoensis*
glaucous gull	*Larus hyperboreus*
glaucous-winged gull	*Larus glaucescens*
gray-headed social weaverbird	*Pseudonigrita arnaudi*
great black-backed gull	*Larus marinus*
great frigatebird	*Fregata minor*
greater flamingo	*Phoenicopterus ruber*
gyrfalcon	*Falco rusticolus*
Hawaiian petrel	*Pterodroma phaeopygia*
hoary redpoll	*Carduelis hornemanni*
hooded dotterel	*Thinornis cucullatus*
Indian myna	*Acridotheres tristis*
ivory gull	*Pagophila eburnea*
kelp gull	*Larus dominicanus*
Kenya sparrow	*Passer rufocinctus*
king eider	*Somateria spectabilis*
king parrot	*Alisterus scapularis*
kori bustard	*Ardeotis kori*
Lapland longspur	*Calcarius lapponicus*
lappet-faced vulture	*Torgos tracheliotos*
little corella	*Cacatua pastinator*
long-tailed duck	*Clangula hyemalis*
long-tailed jaeger	*Stercorarius longicaudus*
magnificent frigatebird	*Fregata magnificens*
masked booby	*Sula dactylatra*
mew gull	*Larus canus*
mockingird	*Mimus* spp.
mourning dove	*Streptopelia decipiens*
mute swan	*Cygnus olor*
northern bald ibis	*Geronticus eremita*
northern flicker	*Colaptes auratus*
northern masked weaver	*Ploceus taeniopterus*
northern wheatear	*Oenanthe oenanthe*
ostrich	*Struthio camelus*
oystercatcher	*Haematopus* spp.
pale chanting goshawk	*Melierax canorus*
parasitic jaeger	*Stercorarius parasiticus*
pelagic cormorant	*Phalacrocorax pelagicus*
purple sandpiper	*Calidris maritima*
red knot	*Calidris canutus*
red-billed hornbill	*Tockus erythrorhynchus*

red-billed tropicbird	*Phaethon aethereus*
red-footed booby	*Sula sula*
red-necked avocet	*Recurvirostra novaehollandiae*
rhinoceros auklet	*Cerorhinca monocerata*
ruddy turnstone	*Arenaria interpres*
sanderling	*Calidris alba*
sandgrouse	*Pterocles* spp.
shearwater	*Puffinus* spp.
short-eared owl	*Asio flammeus*
snow bunting	*Plectrophenax nivalis*
Somali sparrow	*Passer castanopterus*
Somali yellow-backed weaverbird	*Ploceus dichrocephalus*
south polar skua	*Catharacta maccormicki*
southern boobook owl	*Ninox novaeseelandiae*
spoon-billed sandpiper	*Eurynorhynchus pygmeus*
strange weaverbird	*Ploceus alienus*
superb starling	*Lamprotornis superbus*
surf scoter	*Melanitta perspicillata*
Swainson's thrush	*Catharus ustulatus*
swallow-tailed gull	*Creagrus furcatus*
tawny frogmouth	*Podargus strigoides*
Thayer's gull	*Larus thayeri*
tumbler pigeon	*Columba* spp.
turquoise parrot	*Neophema pulchella*
vermilion flycatcher	*Pyrocephalus rubinus*
vesper sparrow	*Pooecetes gramineus*
waved albatross	*Phoebastria irrorata*
wedge-rumped storm petrel	*Oceanodroma tethys*
white-backed vulture	*Gyps africanus*
white-cheeked pintail	*Anas bahamensis*
white-faced heron	*Egretta novaehollandiae*
white-headed buffalo weaver	*Dinemellia dinemelli*
white-headed vulture	*Trigonoceps occipitalis*
white-winged scoter	*Melanitta fusca*
winter wren	*Troglodytes hiemalis*
yellow-crowned night-heron	*Nyctanassa violacea*
zebra finch	*Taeniopygia guttata*

Plants

acacia	*Acacia* spp.
Arctic willow	*Salix arctica*
balsa tree	*Ochroma pyramidale*
beech	*Fagus* spp.
bigleaf maple	*Acer macrophyllum*
black cottonwood	*Populus trichocarpa*
blackberry	*Rubus* spp.
blue gum	*Eucalyptus saligna*
Borassus palm	*Borassus aethiopum*
bougainvillea	*Bougainvillea* spp.
California sycamore	*Platanus racemosa*
casuarina	*Casuarina* spp.
creosote bush	*Larrea tridentate*
domestic holly/English holly	*Ilex aquifolium*
Douglas-fir	*Pseudotsuga menziesii*
dwarf willow	*Salix herbacea*
European beachgrass	*Ammophila arenaria*
evergreen huckleberry	*Vaccinium ovatum*
fireweed	*Chamerion angustifolium*
flame tree	*Brachychiton acerifolius*
frangipani	*Plumeria rubra*
ginkgo	*Ginkgo biloba*
golden chinquapin	*Castanopsis chrysophylla*
great hedge nettle	*Stachys chamisonis* var. *cooleyae*
highland saxifrage	*Saxifraga rivularis*
Himalayan blackberry	*Rubus armeniacus*
jacaranda	*Jacaranda* spp.
lantana shrub	*Lantana* spp.
London planetree	*Platanus x acerifolia*
manzanillo tree	*Hippomane mancinella*
matazarno tree	*Piscidia carthagenesis*
melaleuca/tea-tree	*Melaleuca glomerata*
mountain avens	*Dryas* spp.
muyuyo	*Cordia lutea*
native dandelion	*Taraxacum ceratophorum*
oleander bush	*Nerium oleander*
orchid [Galápagos]	*Epidendrum spicatum*
Oregon grape	*Mahonia aquifolium*
Pacific madrone	*Arbutus menziesii*
Pacific silver fir	*Abies amabilis*

Pacific yew	*Taxus brevifolia*
paulownia	*Paulownia* spp.
pearly everlasting	*Anaphalis margaritacea*
pepper tree	*Schinus molle*
peppermint willow	*Agonis flexuosa*
purple saxifrage	*Saxifraga oppositifolia*
red alder	*Alnus rubra*
red elderberry	*Sambucus racemosa*
red-flowering currant	*Ribes sanguineum*
river red gum	*Eucalyptus camaldulensis*
Russian thistle	*Salsola kali*
salal	*Gaultheria shallon*
salmonberry	*Rubus spectabilis*
scalesia	*Scalesia* spp.
Scotch broom	*Cytisus scoparius*
sea grape	*Coccoloba uvifera*
shore pine (lodgepole pine)	*Pinus contorta*
Sitka spruce	*Picea sitchensis*
smooth yellow violet	*Viola glabella*
spinifex	*Triodia pungens*
swamp paperbark	*Melaleuca* spp.
toothbrush tree	*Salvadora persica*
tuart	*Eucalyptus gomphocephala*
weeping paperbark	*Melaleuca leucadendra*
western red cedar	*Thuja plicata*
white oak	*Quercus* spp.
wild strawberry	*Fragaria virginiana*
woolly mullein	*Verbascum thapsus*
yarrow	*Achillea millefolium*

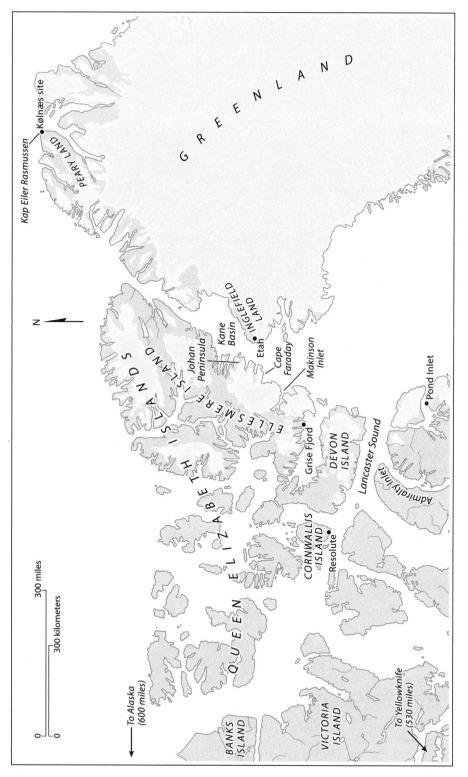

Queen Elizabeth Islands Overview

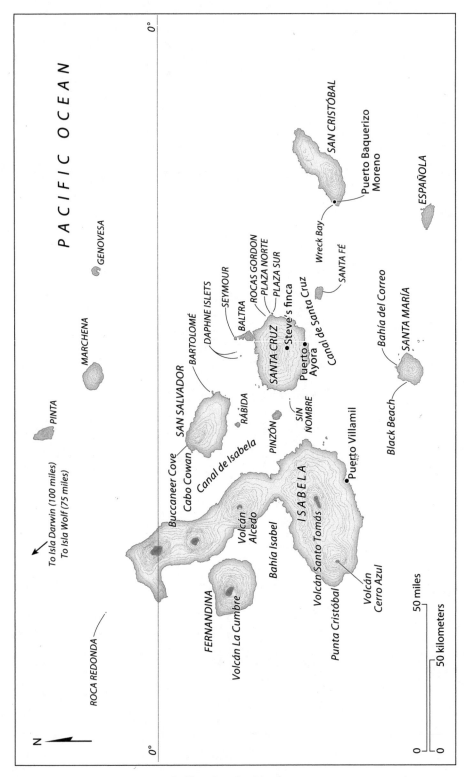

Galápagos Archipelago

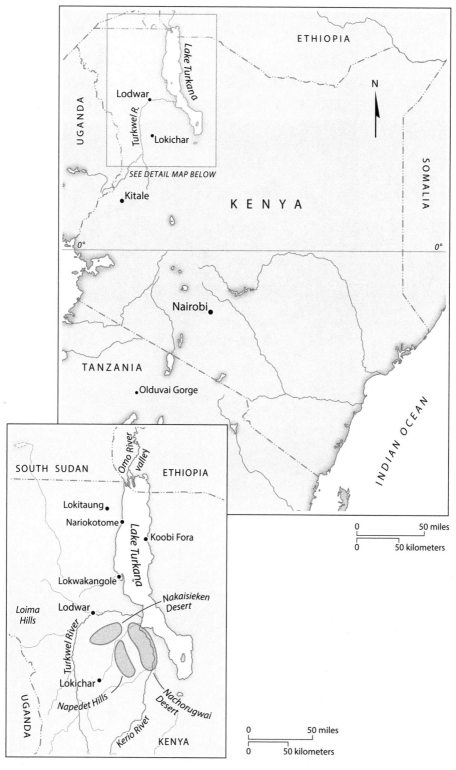

ETHIOPIA

N

Lake Turkana

UGANDA

Lodwar

Turkwel R.

Lokichar

SEE DETAIL MAP BELOW

SOMALIA

Kitale

K E N Y A

0°   0°

Nairobi

TANZANIA

Olduvai Gorge

INDIAN OCEAN

0   50 miles
0   50 kilometers

Omo River valley

SOUTH SUDAN

ETHIOPIA

Lokitaung

Nariokotome

Lake Turkana

Koobi Fora

Lokwakangole

Loima Hills

Lodwar

Nakaisieken Desert

Turkwel River

Lokichar

Napedet Hills

Nachorugwai Desert

Kerio River

UGANDA

KENYA

0   50 miles
0   50 kilometers

Kenya

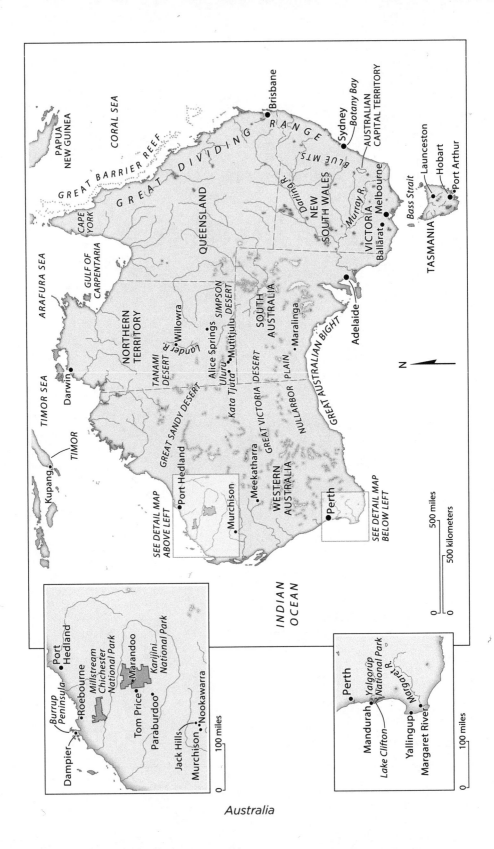

Australia

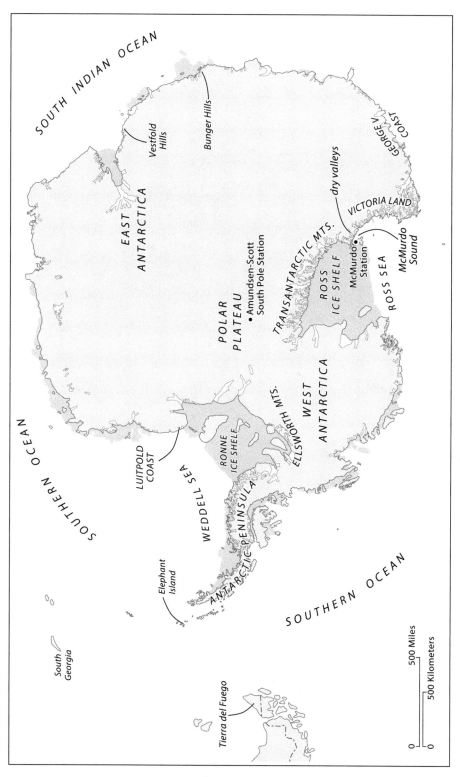

Antarctica

Acknowledgments

After I finished a book called *Arctic Dreams,* in 1986, I started to see more clearly the outline of a loosely related nonfiction project, a work I knew would take a good while to complete because, at the time, I lacked sufficient experience in the field to write it.

The initial research for this book was funded by the Guggenheim Foundation in 1987, under the title "The Shape of Time in Remote Regions," and on five occasions by the National Science Foundation's Antarctic Artists and Writers Program. I'm grateful to both institutions for their sponsorship. I also want to thank Richard Bangs of Mountain Travel Sobek for his early support in Africa; Polar Continental Shelf Program, Department of Energy, Mines and Resources Canada for their sponsorship on Skraeling Island; Bill Roberson at Inca Floats for his generosity in Galápagos; Ray Rodney at Wilderness Travel for support in Antarctica; Neil Keny-Guyer at Mercy Corps for their underwriting of travel in the Middle East and Central Asia; Matthew Swan at Adventure Canada for support in Greenland and the Queen Elizabeth Islands; Kasumasa Hirai for his hospitality and support in Japan; Hilary MacGillivray and Ivy O'Neal of Travel Dynamics for international travel; and Peter Shaindlin for his hospitality in Honolulu. I also want to thank Bobbie Bristol and Cheryl Young for offering me the Bernardine Kielty Scherman Residency Fellowship at the Macdowell Colony in Peterborough, New Hampshire; Deb Ford for a residency fellowship at the Playa writers and artists colony at Summer Lake in central Oregon; and Michael Adams at the University of Texas at Austin, who awarded me the Dobie-Paisano International Residency Prize at a critical time in the final stages of work on *Horizon*. All three provided crucial space and time to write.

From the beginning I benefited from the unstinting support of Sonny Mehta, my publisher at Alfred A. Knopf. I started work on the book at Knopf with Elizabeth Sifton and, after her departure, continued to develop my ideas with Bobbie Bristol, with whom I published a collection of short stories, *Field Notes: The Grace Note of the Canyon Wren*, in 1994. When Ms. Bristol moved on from Knopf in 1997, I began work with Robin Desser, with whom I published a collection of essays, *About This Life: Journeys on the Threshold of Memory*, in 1998, and two short story collections, *Light Action in the Caribbean*, in 2000, and *Resistance*, in 2004. Robin's patience with me during the years it took to research the book, to develop the perspective I believed it needed, and finally to actually write the book, was extraordinary. The intelligence and editorial acumen Robin brought to discussions of early drafts of the manuscript are only a part of the reason why she is regarded as legendary in American publishing. Working side by side with her was more than a pleasure. It defined for me what a collaborative effort with an editor should look like for a writer.

It's unusual to be able to work with the same editor and publisher for so long, and I'm very grateful for the opportunity to have done so with Sonny Mehta and Robin Desser. I've also had the great fortune to work with Peter Matson, my agent of nearly forty years. His understanding of what I was trying to do as a writer always straightened out the road ahead for me, and his representation over the years has been impeccable.

This book is dedicated to Robin and Peter for their deep friendship and professional counsel over that time, but first to my wife, the writer Debra Gwartney. On many occasions, Debra set aside her own work to help me keep on schedule, particularly after my health began to fail in the final stages of writing.

The assertion that without Debra's, Peter's, and Robin's support and advice this book would likely have gone no further than the stage of extensive note-taking is a sentiment easy to express but difficult to adequately underscore.

In addition to Polar Shelf, the Guggenheim Foundation, Mercy Corps, and the National Science Foundation, I want to thank Susan O'Connor for her financial support and her enduring friendship. Mags Webster at FORM in Western Australia, Richard Leakey in Africa, and Fatima Galani and her family in Afghanistan all helped greatly with logistics. My gratitude as well, on Ellesmere Island, to Peter Schledermann, Karen McCullough, Eric Damkjar, Eli Bornstein, and Hans Dommasch. Also to Robert McGhee. In Galápagos, gratitude to Steve Divine, Tui De Roy, Bill Roberson, Orlando Falco, Eugénio Moreno, the late Christine Gallardo, Jack Nelson, Bruce Barnett, and the late Karl Angermeyer. In northern Kenya, my thanks to Richard Leakey again, to Alan Walker, Kamoya Kimeu, Nzube Mutiwa, Onyango Abuje, Bernard Ngeneo, and Wambua Mangao. Also in Nairobi to the late Mary Leakey. In Australia, I'm grateful to Mark Tredennick, Petronella Morel, Pete Hay, Luke Davies, Mags Webster again, John Wolseley, Richard Brown, Bob Pidgeon, Peter Latz, Robyn Davidson, Annamaria Welden, Fred Myers, Loreen Samson, and to my Pitjantjatjara companions at Mutitjulu and my Warlpiri companions at Willowra for their accommodation. Gratitude as well to my other colleagues in the Pilbara, Paul Parin, Larry Mitchell, Bill Fox, and Carolyn Karnovsky. I also want to thank the organizers of the international literary festivals at Perth, Adelaide, Hobart, and Melbourne for providing transportation to Australia.

In Antarctica, I wish to express my indebtedness to Guy Guthridge, the late Jack Renirie, John Schutt, the late Peter Wilkniss, Paul Mayewski, Berry Lyons, Cameron Wake, Mark Twickler, Mike Morrison, Bruce Koci, Ted Clark, Diane McKnight, and Elle Tracy, and, at Graves Nunataks, to Ralph Harvey, Diane DiMassa, Nancy Chabot, Paul Benoit, and Scott Sanford. I also want to thank Captain Russell Bouziga for his friendship and instruction aboard the *Nathaniel B. Palmer*, members of his crew for their accommodation, and Skip Kennedy of the National Science Foundation, who introduced me to executives at Edison Chouest, the company that built the *Palmer*.

Also in Antarctica, gratitude to my companions aboard the *Hanseatic,* the late Galen and Barbara Rowell, Will Steger, and my stepdaughter Amanda. And finally, thank you to my dive companions in Antarctica, Rikk Kvitek, Cathy Conlan, Diane Carney, Hunter Lenihan, Kim Keist, Brenda Konar, and John Oliver, all from Moss Landing Marine Labs in California. And Jeff Bozanic, the divemaster at McMurdo Base.

I want to thank John Beusterien for help in understanding the role of the perros de presa during the Spanish incursion in the New World, Dr. Noa Emmett Aluli in Hawai'i for his guidance on Hawaiian history, Gregory Retallack for help identifying geological specimens, Desirée Fitzgibbon and Christine Wilson for their help with the Martin Bryant material, Stan Bettis for his help with the history of the *Western Trader* foray, and Cort Conley for alerting me to the existence of the Kølnæs site. My thanks also to Dennis Corrigan, the late Wally Herbert, and Dave Fross for their help at various points, and to my colleagues at Texas Tech University.

And gratitude to David Lindroth, for his wonderful maps, and to my colleagues in production at Knopf, Cassandra Pappas, Carol Carson, Rita Madrigal, and Andy Hughes. And to Annie Bishai.

My stepchildren, Amanda Woodruff, Stephanie Woodruff, Mary Woodruff, and Mollie Harger, have been a source of love, enthusiasm, and support throughout the process of researching and writing this book. I'm forever in debt to them.

In a project that unfolds across as many years as this one has, it is difficult to recall each moment in which the work was significantly inflected or illuminated by a conversation with someone. I would particularly like to thank, however, Neal Keny-Guyer, Don Walsh, Bill Roberson, Alan Walker, John Schutt, Jack Renirie, and Neill Archer Roan, the former director of the Bach Festival in Eugene, Oregon, who introduced me to Arvo Pärt. To the other people who provided interviews and support and should expect to see their names here, I apologize for the imprecision of my memory.

I would like to acknowledge the friends with whom I have tried over the years to work through my ideas about landscape and culture, and such things as the difference between autobiography and memoir. These would include, first, my wife, Debra, who, in addition to writing, also teaches memoir; the writers David Quammen, Pattiann Rogers, John Keeble, the late Conger "Tony" Beasley, Rebecca Solnit, Jane Hirshfield, W. S. Merwin, John Freeman, Colum McCann, the late Brian Doyle, Julia Martin in South Africa, and Mark Tredinnick in Australia; Marion Gilliam, Chip Blake, and others at The Orion Society; and a long list of artists whose working lives and artistic endeavors I have found inspiring. These would include the photographers Stuart Klipper, Susan Middleton, David Liittschwager, Lukas Feltzmann, Mary Peck, Ben Huff, Frans Lanting, and Linda Connor; the painters Alan Magee, Tom Pohrt, and the late Rick Bartow; the ceramic artist Richard Rowland; the sculptor Tom Joyce; the book artist Charles Hobson; the filmmaker Toby McLeod; the biographer Jim Warren; the curator Emily Neff; and the composer John Luther Adams. In addition, my brother John Brennan; my good friends Frank Stewart in Hawai'i and Richard Nelson in Alaska for sterling conversation and excellent insights; translators Gary Witherspoon, Anton Fraga, Joe Moll, the late Luis Verano, and B. Mokaya Bosire; Bill Wade, formerly the president of Arco, Will Rogers at the Trust for Public Land, Richard Harvey, M.D., my colleague on a long, around-the-world plane trip, and, for his many years of guidance, the Onandaga elder Oren Lyons. Also Pualani Kanaka'ole Kanahele for the example of her life.

My former assistant Zoë Livelybrooks provided extraordinary help in all phases of preparation of the manuscript. I'm greatly in her debt, and to Candice Landau as well, who helped in the final stages. My former assistants Emma Hardesty, Julie Polhemus, and Nancy Novitski all supported the work in its early stages, and I thank them for their help. Julie, Nancy, Zoë, and I fact-checked the entire manuscript. Anything we might have missed is my responsibility. Special thanks to Isabel Stirling for help with research.

In addition to these, my deep bow of respect to Davide Sapienza in Italy, Alberto Manguel in Argentina, Hans Jurgen Balmes in Germany, and Anne Collins in Canada. And profound gratitude to Julie Graff, M.D.; John Stacey, Ph.D.; Pam Schmid, R.N.; the Comanche/Chiricahua Apache healer Harry Mithlo; and my brother John Brennan, a traditional healer, for their ministrations and counsel.

Finally, the late Robley Wilson at *The North American Review,* Lewis

Lapham at *Harper's,* Chip Blake at *Orion,* Stephen Covey at *The Georgia Review,* Sigrid Rausing at *Granta,* and Joël Garreau at *The Washington Post,* who published my early thoughts about Galápagos and Antarctica, and about traveling with indigenous people.

Steve Frost, Mark Tredinnick, and Guy Guthridge read critically, respectively, the chapters on Africa, Australia, and Antarctica. David Quammen also provided help with the chapters on Galápagos and Africa. I'm grateful to them all for the corrections and improvements they offered. Whatever mistakes or inaccuracies remain in the book are my responsibility.

Index

Page numbers in *italics* refer to maps.

Abbotsbury Swannery, 89
Abbot Ice Shelf, 213
Aboriginal people, 23, 95, 128, 274, 330,
    354, 357, 359, 361–62, 365, 366, 367,
    371–72, 382, 383, 384, 392, 402, 403,
    404, 405–6
    see also specific Aboriginal groups
Abrojos Reef, 40
Absolute Event Horizon (Davies), 418–19,
    423
Abu Ghraib prison, 75
Abu Hureyra people, 181
Abuje, Onyango, 269, 270, 271, 275, 286,
    289, 329, 332, 338, 339
Academy Bay, Santa Cruz Island, 217, 221,
    249
Aceh province, Sumatra, 53
Achebe, Chinua, 289
Acklins Island, 43
Adams, John Luther, 108, 184, 240
Adam's Lament (Pärt), 195
Adelaide, Australia, 390, 391, 423
Aden, Gulf of, 299
Admiralty, British, 96, 264, 334, 335
Admiralty Inlet, 200–201
Aeneas, 324
Aertsen, Pieter, 32
Afar Depression, 316
Afar region, Ethiopia, 301, 302, 316
Afghanistan, 23, 33–34, 80, 104, 274, 387,
    389

Africa, 27, 153, 173, 276, 277–80, 291, 296,
    297, 300, 301, 305, 313, 314–17, 324,
    330, 338, 355, 381
    see also specific countries, regions,
    settlements, and cities
Africa, Horn of, 126, 299, 474
African Rift Valley, 314–17
A German Requiem (Brahms), 107
Aguilar, Martín de, 79
Ainu people, 80, 100, 336–37
Aivik archeological site (Skraeling Island),
    145
Alabama, 18
Alameda, California, 86
Alaska, 19, 30–31, 61, 62, 108, 159, 240,
    330, 416
Alaska, Gulf of, 51
Alaska, University of, Museum of the
    North at, 240
Albireo (double-star), 72
Alcedo volcano, 241–42, 243–44, 264
Aleutian Islands, 51, 213
Alexander the Great, 82, 324
Alexandra Creek, 141
Alexandra Fjord, 133–34, 137, 148, 183
    see also Skraeling Island
Alexandra Fjord lowland, 135, 138–48, *140*,
    149–51, 152–53, 156, 159, 163–64, 174,
    214, 375
Alfonso XIII, king of Spain, 41
Algeria, 330, 372

Alice Springs, Australia, 402–3, 404, 406, 410

Allan Hills Middle Western Icefield, 74, 442

Aloha Betrayed: Native Hawaiian Resistance in American Colonialism (Silva), 251

Alsean Indians, 79, 104, 105, 106, 128

Alsea people, 56

Alsea River, 79

Al-Shouf Cedar Nature Reserve, 66

Altamira cave, Spain, 325, 370, 371

Aluli, Noa Emmett, 254

AMANDA, *see* Antarctic Muon and Neutrino Detector Array

Amanda (author's stepdaughter), 490, 492–93, 494

Amazon rain forest, 306, 330

American Airlines, 12

American Indians, 13, 93, 98, 99, 117, 128, 214, 246, 311
see also indigenous peoples; *specific American Indian groups*

American Museum of Natural History (New York), 280

American Revolution, 246, 335, 378, 379, 383

American Scientist, 322

American South, 19

American Southwest, 287

American West, 23, 44, 114

Amin, Idi, 338

Amsterdam, 215

Amundsen, Roald, 189, 471, 482, 483–84

Amundsen-Scott South Pole Station, 161, 429, 457, 472

Anabase (Léger), 46

Anaktuvuk Pass, Alaska, 30–31

Anatolia, 57, 314

ancestors
animal, 196, 212, 231, 283
Homo, 23, 193, 196–97, 256, 272, 279–81, 283–85, 288, 292–302, 305–6, 316–19, 323, 326–27, 340, 342–43, 401, 411
human, 43, 79, 119, 150, 164, 171, 193, 200, 202, 213, 240, 252, 272, 273, 274, 307, 362, 372, 382, 385
see also elders

Ancestral Puebloan sites, 149, 287

Andes, 38, 496

Andromeda, 292

Angermeyer, Karl, 221

Angola, 43

Animalia kingdom, 293

ANSMET, *see* Antarctic Search for Meteorites

Antarctica, 23, 24, 27, 38, 39, 56–57, 58, 74, 96, 213, 253, 254, 287, 314, 330, 351, 416, 427, 430, 432, 434, 436, 437–39, 443, 445–84, 455, 485–89, 490, 491–96, 499, 508–9
British 1910–1913 expedition in, 434, 457, 466–67, 470, 471–72, 480, 482–84
Graves Nunataks in, 427–34, 435–36, 438, 439, 440–43, 444–45, 447, 456–57, 462–63
McMurdo Station in, *see* McMurdo Station
Shackleton's expedition in, 454, 457, 485, 489–91, 492, 493
South Pole in, 428, 447, 466, 471, 472, 473–80, 481–82, 509
Vanda station in, 459–61, 463–64
see also specific geographic regions, features, and landmarks

Antarctic Muon and Neutrino Detector Array (AMANDA), 161–62, 164, 475

Antarctic Peninsula, 102, 493

Antarctic Search for Meteorites (ANSMET), 439, 440, 442, 454

Antarctic Treaty, 438–39, 456, 475–76, 508

Anthropocene period, 256, 283, 356

Antietam, Battle of, 3, 114–15

Antietam Creek, 114–15

Aparajitas, 47

"Apology for Bad Dreams" (Jeffers), 245, 341

Apotheosis of Captain Cook: European Mythmaking in the Pacific, The (Obeyesekere), 57

Apuriná people, 330

Aqaba, Gulf of, 316

Arabian Peninsula, 314

Arabs, 332

Arafura Sea, 299, 380, 381

Arapahoe Indians, 252

Archean eon, 395

Archeozoic period, 37

Archipiélago de Colón, 210–11

"Archipiélago de Colón" navigational chart, 264

Arctic (journal), 155

Arctic, 24, 32–33, 156, 158, 164, 170, 188, 306, 370, 404
  see also High Arctic
Arctic Dreams (Lopez), 423
Arctic Ocean, 174
Arctic Small Tool tradition (ASTt), 145–46, 152, 153, 157, 164, 170, 196, 202
Ardèche Valley, France, 146
Argentina, 38, 508
Arizona, 474
Arizona, USS, 3, 5
Arlington National Cemetery, 190
Arno River, 261
Arrernte people, 375, 405
Art Gallery of New South Wales (Sydney), 417, 418, 419
Asgard Range, 459, 462, 463, 472
Asia, 296, 297, 299, 369
Assad, Bashar al-, 215, 349
Astoria, Oregon, 112, 114
Astor, John Jacob, 112–13, 114, 116
Astro Materials Acquisition and Curation Office (Johnson Space Center), 441
Asturias, Spain, 22, 41, 501
Atlanta Journal-Constitution, 458
Atlantic Ocean, 93, 96, 208, 214, 257, 314, 501
Attica, 200
Auburn University, 19
Aung San Suu Kyi, 324
Auriga (constellation), 72
Auschwitz concentration camp, 114
Australia, 11, 23, 36–37, 43, 58, 60, 80, 94, 95, 96, 100, 128, 173, 246, 281, 299, 311, 314, 317, 318, 330, 351, 354, 356, 358–63, 367, 374, 376–77, 378, 379–80, 382–85, 387, 389, 390, 391–403, 404–6, 416, 419, 420, 462, 508
  Aboriginal people in, *see* Aboriginal people; *specific Aboriginal groups*
  Northern Territory of, 23, 274, 374, 375, 389, 402–3, 404–11, 419
  see also specific Australian states, cities, and regions
Awash River, 316
Ayers Rock, Australia, 408

B-24 Liberators (bomber), 18, 20
B-29 Superfortresses (bomber), 18
Bāb al-Mandab Strait, 299, 304, 316

Bache Peninsula, 138, 141, 166
Bach, Johann Sebastian, 108, 183, 192
Baffin Island, 171, 200–201, 274
Bahamas, 42–43
Bahía Darwin, 226
Bahía del Correo, 248–49, 264
Bahía Inútile (Useless Bay), Tierra del Fuego, 509
Bahía Isabel, 230, 236, 264
Baker Lake, 40
Balboa, Vasco Nuñez de, 214
Balham Valley, Antarctica, 460
Bali, 357
Ballarat, Australia, 100, 416, 417
Baltra, 232, 263
Bamyan, Afghanistan, 23
Banda Aceh, Sumatra, 172, 214, 311
Bangkok, 232
Bangladesh, 47, 374
Banks, Joseph, 58, 63, 97, 379, 421, 422
Banyjima people, 360
Barnett, Bruce, 230, 232
Barouk Forest, 65–66
Bartolomé Island, 262–63
Bartow, Rick, 252–53, 254
Bartram, William, 457
Barwick Valley, Antarctica, 460
Basket, Fuegia (Yokcushlu), 507
Bass, George, 420
Bass Strait, 420
Batavia, 381, 420
Batista, Fulgencio, 216
Baudin, Nicolas, 420
Bay of Panama, 213, 214
Bay of Sails, Antarctica, 447, 448–49
BBC, 436, 480
Beacon Supergroup, 462
Beagle, HMS, 209, 256–57, 498, 507
Beaglehole, J. C., 57, 58, 97, 220
Beagle III (ship), 232, 236–37, 238–39, 241, 248, 249, 250–51, 254, 262, 263
Bear (rescue ship), 188
Bear River, Idaho, 114, 246
Beasley, Tony, 43, 44
Beautyway (Navajo ceremony), 143, 144
Bechtel-McCone-Parsons, 18, 19
Become Ocean (Adams), 108
Bedouin people, 330
Beebe, William, 220, 247
Beethoven, Ludwig van, 183–84, 191–92, 193

Before the Horror: The Population of Hawai'i on the Eve of Western Contact (Stannard), 57, 58
behaviorally modern *Homo sapiens*, 197
Bei Dao, 245
Beistad Fjord, 165
Belgian Congo, 281
Belgium, 216
Bell 206 Jet Ranger, 487
Bellavista settlement, 221
Bellinghausen, Thaddeus von, 124
Bellot Strait, 33
Berganza (soldier dog), 214, 215
Berg Field Center, 454, 456
Bering Sea, 61, 318
Bering Strait, 61, 96
Bering, Vitus, 213
Berlanga, Fray Tomás de, bishop of Panama, 209
Berlin, Germany, 195
Bern, Switzerland, 186
Beta Crucis (Mimosa), 292
Betelgeuse (star), 73
Bettis, Stan, 122, 123
BHP Billiton Iron Ore, 358, 365, 368
Bible, 181
Bierstadt, Albert, 75
Big Dipper, 72
Big Razorback Island, 450–51
Big Skraeling Island, 155, 157, 167, 192, 194
Bindijareb people, 354, 356
Birds of East Africa (Williams and Arlott), 275
Birkenau concentration camp, 114
Birth of Venus, The (Botticelli), 262
Black Beach, Isla Santa María, 248, 264
Black Elk (Sioux Indian), 120
Black Legend of the Spanish conquistadores, 42, 118
Black Sea, 314
Blackwater Draw site (New Mexico), 146
Blake, William, 28, 103, 125
Blanding, Utah, 143
Bligh, William, 380
Blue Mountains, 390
Blue Water Vagabond (Puleston), 217
Boeing 314 Clipper (seaplane), 85
Boeke, Kees, 159, 160
Bohr, Niels, 479
Boko Haram, 215
Bolivia, 251
Bondi Beach, Sydney, 419

Boötes (constellation), 72
Borges, Jorge Luis, 364
Borneo, 109
Bornstein, Eli, 139, 140, 142, 147–48, 150, 158
Boro River, 313
Bosnia, 115
Bosnian Muslims, 115
Bosnian Serbs, 115
Bosphorus, 314
Boston University, 322
Boswell, James, 382
Botany Bay, Sydney, 60, 61, 378, 419, 420–24
Botero, Fernando, 75
Botha, P. W., 338
Botswana, 277, 278, 313, 327
Botticelli, Sandro, 262
Bougainville, Louis-Antoine de, 57
Bounty, HMS, 380–81
Bouziga, Russell, 253–54, 497, 504
Bowers, Henry "Birdie," 466, 467, 470, 471–72, 482, 483
Boxing Day tsunami (2004), 172, 311, 387
Boyle, Robert, 261
Brahmaputra River, 374
Brahms, Johannes, 107
Braun-Menéndez family, 501
Brazil, 372
Brenner Pass, 22
Bridges, E. Lucas, 496
Bridges, Thomas, 496
Brisbane, Australia, 400
British Antarctic Expedition (1910–1913), 434, 457, 466–67, 470, 471–72, 480, 482–84
British Columbia, Canada, 100–101, 417
British East India Company, 95
British navy, 96, 97
British Overseas Territories, 38
Broad Arrow Café (Port Arthur), 353, 384
mass shooting at, 384–85, 386–87
Brooks Range, Alaska, 30, 330
Browne, Janet, 259
Brown, Richard, 393, 394, 395, 402
Brunswick Peninsula, 495, 506, 510
Bryant, Charlotte, 380, 381, 382
Bryant, Emmanuel, 380, 381
Bryant, Martin, 384–85, 386–87, 388
Bryant, Mary Broad, 379–80, 381–82
Bryant, William, 379–80, 381
Buccaneer Cove, San Salvador Island, 238

Buchanan Bay, Nunavut, 134, 139, 166
Buddhas, 23
Buddhist culture, 43
Buenos Aires, Argentina, 42, 416
Bullock, Wynn, 20
Bunger Hills, East Antarctica, 459
Burke, Robert, 396
Burrup Peninsula, 357, 364, 365–66, 369, 370–71
Button, Jemmy (Orundellico), 507

Cabo Cowan, San Salvador Island, 238
Cabo de Hornos, Chile, 506
Cabo Isabela, Hispaniola, 40
Cabrilho, João Rodrigues, 79
Cage, John, 4
Calico Enterprises, 52
California, 9–10, 13, 14, 15, 17, 18, 21, 24, 38, 86, 114, 153, 341, 386
California Academy of Sciences museum (San Francisco), 230
Callahan, Harry, 20
Cambodia, 42
Cambrian period, 395
Camden, Maine, 102
Campbell, Joseph, 4, 323
Camp Clay, 186, 187–89, 200
Camp St. Regis (New York), 14–15
Camus, Albert, 125, 502
Canada, 72, 98, 138, 139, 372, 383
Canadian High Arctic, 133–43, 150–51, 155, 455
    see also High Arctic
Canal de Isabela, 264
Candy (Davies), 422
Canoga Park (Los Angeles), 13
Cantabria, 281, 370
Canterbury Museum (Christchurch), 252–53
Cantus in Memory of Benjamin Britten (Pärt), 195
Capac, Huayna, 251
Cape Armitage, Antarctica, 466, 468
Cape Crozier, Antarctica, 434, 466–72, 475
Cape Evans, Antarctica, 434, 457, 466, 467, 482, 483, 484
Cape Faraday, Canada, 172
Cape Foulweather, Oregon, 25–26, 27–28, 31, 51–53, 54–55, 56, 60, 64–65, 66–68, 72, 74–75, 78, 79, 81–82, 83–84, 85, 88, 89, 104–5, 108–9, 110, 111, 112, 125–26, 127–29, 195

Cape Froward, Chile, 497, 506
Cape Horn, Tierra del Fuego, 37, 213, 416, 486, 497
Cape MacKay, Antarctica, 466, 468, 470
Cape of Good Hope, South Africa, 381
Cape Perpetua, Oregon, 55
Cape Royds, Antarctica, 457
Cape Sabine, Pim Island, 185–86, 187–89, 214
Cape Solander, Sydney, 61
Cape Town, South Africa, 302, 381, 382, 416
Cape York Peninsula, 380, 420
Caprivi Strip (Namibia), 279
Captain Cook Bridge, 419
Caribbean, 300
Cariboo region, British Columbia, 100–101, 417
Carment, Tom, 358
Carnac, France, 160
Carnarvon Bay, Tasmania, 37, 351
Carnegie family, 113
Caroline Islands, 218
Carpentaria, Gulf of, 420
Cascade Mountains, 23, 474
Cassidy, Bill, 441
Catholics, Catholicism, 118, 348
Cavafy, Constantine, 202
Cavendish, Thomas, 507
Ceaușescu, Nicolae, 115
Celestial Empire, 335
"Cemetery Ridge," 187–89
Cenozoic era, 395
Central Intelligence Agency, 216
CERN, 60
Cerro Azul volcano, 242, 265
Cerro Rico Mountain, 251–52
Ceylon, 282
Chad, 115
Challenger (British research vessel), 213
Chandra, Vikram, 423
Charles Darwin Research Station, 221, 222
Charles V, Holy Roman Emperor, 41
Charon, 15
Charrière, Henri, 176
Châtelperronian, 304
Chatsworth, California, 10
Chatwin, Bruce, 402
Chauvet cave, 146, 366
Chechnya, 115
Cherenkov photon, 162
Cherry-Garrard, Apsley, 466, 467, 470

Chichester, Charles, 213
Chile, 38, 42, 416, 484, 496, 499, 500,
    506–11
    Punta Arenas in, *see* Punta Arenas, Chile
China, Chinese, 4, 11, 15, 35–36, 43, 93,
    109, 245, 281, 311, 324, 342, 349, 357,
    368, 374
China Clipper (Martin M-130 plane), 85,
    86, 102
Chinook Indians, 79, 91, 92, 93, 98, 104,
    107, 112, 116, 117
Choeung Ek killing field, Phnom Penh,
    349
Chongqing, China, 23, 31, 281
Christchurch, New Zealand, 252, 429, 432,
    433, 473
Christian, Fletcher, 381
Christianity, Christians, 117, 118, 215, 257,
    258, 260, 332, 333, 348, 362, 376
Christiansen, Thorlip, 186, 188
Christiansted, Saint Croix, 51, 254
Christopher (Turkana man), 270, 271–72,
    273, 275, 285, 286, 288–89, 290, 313,
    319–21, 322, 323, 326, 332, 336, 337
Chukchi Sea, 61
Chukotski Peninsula, 502
Church of England, 348
Church of Our Saviour (New York), 14
Ciudad del Rey Don Felipe, 506–7
    see also Puerto del Hombre (Port
    Famine), Chile
Civil War, U.S., 115
Clarke, Marcus, 384
Clausius, Rudolf, 68, 144
Clerke, Charles, 62
Clinch Ridge archeological site, 145
Clovis hunters, 146
Coalsack (dark nebula), 292
Coast Guard, U.S., 121
Coast Range (Oregon), 26, 53
Coats Land, Antarctica, 490
Cockburn Canal, 497
Cockburn Town, San Salvador Island, 44
Cocos Plate, 209
Coetzee, John, 423
Columbia River, 26, 79, 90, 91, 92, 94, 112,
    117, 128, 129
Columbus, Christopher, 42–43, 44, 57
Colville Indian Reservation, 94, 99
Concepción (ship), 499
Concomly (Chinook chief), 91, 98,
    116–17

Congo, 41, 60, 109, 115, 215–16
Conrad, Joseph, 57
Conservation Commission of the
    Northern Territory (CCNT), 402–3,
    404, 405, 406
Constitution, U.S., 181–82
Cook Islands, 58
Cook, James, 25–26, 27, 38, 39, 55–64, 65,
    66, 70–71, 73, 79, 81, 82, 86, 87, 88,
    89, 91, 95–98, 99, 104–6, 118, 124,
    127–28, 201, 212, 213, 220, 263–64,
    265, 282, 284, 334, 380, 396, 416, 417,
    419, 420–22, 424, 511–12
Copernicus, Nicolaus, 83, 260, 291
Cornwallis Island, 137
Cortés, Hernán, 40–41, 57, 74
Cosmic View (Boeke), 159
Crane Island, 9
Crary Science and Engineering Center,
    443–44
Creation Beings, 405, 407
Cree Indians, 98
Cretaceous period, 105, 196
CRISPR (Clustered Regularly
    Interspaced Short Palindromic
    Repeats), 263
Cro-Magnons, 193, 196, 280, 370, 371
Cuba, 41
Cudillero, Spain, 41, 371
cultural exceptionalism, 119, 310
Cumberland Bay, 492
Currey, C. H., 380, 381
Curtin University, 389, 390, 393
Custer, Elizabeth, 99, 106
Custer, George Armstrong, 99
Cutty Sark (clipper), 86

da Gama, Vasco, 57
Dalai Lama, 373
Damascus, Syria, 349
Damkjar, Eric, 151, 152, 155, 160, 201
Dampier, Australia, 358, 361, 363–66
Dampier, William, 213
Danakil Desert, Ethiopia, 299, 316
D'Angelo Bluff, 428
Dara (friend of author's mother), 10–11,
    13, 15, 18
Darién Gap, 281
Dari language, 80
Darling Range, 390
Darling River, 390
Darwin, Australia, 374–75, 402

Darwin, Charles, 82, 83, 208, 209, 210, 221, 230, 231, 242, 255, 256–61, 263, 265, 282, 283, 284, 322, 414, 416, 479, 497, 498, 507–8, 512
Darwin Cordillera, 38
Darwinists, 260
Davidson, Robyn, 402
Davies, Luke, 418–19, 422–24
Davis, John, 38, 54, 190
Dead Sea, 316
Delmas, South Africa, 245–46, 278, 279
Democritus, 260
Deneb (star), 72, 73
Depoe Bay, Oregon, 66, 74
Devil's Island (Île du Diable), 176, 178–80, 181, 182, 246
Devonian period, 395
Devon Island, 171, 174, 189–91, 455
Díaz de Vivar, Rodrigo (El Cid), 41
Dictionary of Disasters at Sea During the Age of Steam, 1824–1962 (Lloyd's of London), 88–89
Discovery Harbor, Lady Frankling Bay, 186
Discovery, HMS, 55, 128
Divine, Steve, 221–22, 224–25
Dixon, Maynard, 13
Djibouti, 299, 316
Dominican Republic, 40
Dommasch, Hans, 139, 147, 148, 150, 158
Dordogne Valley, France, 196
Dorset Mysteries, 200
Dorset people, 136, 137, 145, 148, 150, 154, 162, 163, 164, 165, 196, 199, 201, 202, 203, 452
Drake, Francis, 79, 112, 215, 507
Drake Passage, 62, 102, 103, 485, 492, 493, 496, 504
Dreamtime narratives, 403, 405–6, 407–8, 410
Dreyfus, Alfred, 176, 179
Drina Valley, Bosnia, 115
Dūābī Ghōrband, Afghanistan, 389
Du Fu, 272
Dutch East India Company, 420
Duvalier, Papa Doc, 115
Dye, Eva Emery, 99

Eaglehawk Neck, 376–77
Earhart, Amelia, 19
Early Dorset people, 146, 165, 200

Earth Knower (Dixon), 13
East Africa, 271, 276, 279–80, 314
see also specific countries, cities, and settlements
East Antarctica, 459
Easter Island, 212, 218
Eastern Washington State Historical Society, 94
Eastern Washington University, 94
Ebola virus, 304
École des Beaux-Arts (Paris), 18
Ecuador, 115, 121, 122–23, 209, 224, 248
Ecuadorian nationals, 223, 224
Edinburgh University, 467
Edo, Japan, 91, 101
Edward, Jens, 186, 187, 188
Edwards, Edward, 380, 381
Egypt, 82
El Canal de Santa Cruz, 254
El Cid (Rodrigo Díaz de Vivar), 41
elders, 53, 106, 119, 146, 198–99, 307, 311–12, 336–37, 374, 388, 404–5, 410, 413–17
see also wisdom keepers
El Día de los Compadres, 251
Elephant Island, 490, 491
Eleusinian Mysteries, 200
"El Fin" (Borges), 364
Elison, Joseph, 188
Elkins, Caroline, 119
El'lelaru (York Minster), 507
Ellesmere Island, 134, 137, 139, 166, 171, 183, 186
Ellice Islands, 282
"El pan nuestro" (Vallejo), 418
El Tío (Quechuan god), 251–52
Emma (motor schooner), 491
Encantadas, The, or Enchanted Isles (Melville), 209–10
Endeavour, HMS, 61, 63, 87, 421, 422, 511
Endurance (ship), 254, 485, 490
England, 22, 38, 79, 96, 118, 260, 353, 378, 380, 381, 382, 383, 420, 507
see also Great Britain
Enlightenment, 55, 57, 64, 91, 105, 118, 180, 417
Enola Gay (aircraft), 216
Eocene epoch, 293
Eora people, 421
epistemology, 25, 79, 212, 277, 354
Eratosthenes, 63
Eritrea, 44, 316

Eskimo Point, 186–87

Eskimos, 40, 108, 150, 171, 188, 200, 240, 330, 375
   see also Inughuit people; Inuit people; Iñupiat people

Ethiopia, 299, 301, 316, 502

Etosha Pan, 278

Europe, Europeans, 22, 57, 58, 61, 79, 93, 112, 115, 150, 153, 160, 261, 297, 299, 304, 305, 314, 419, 420, 421, 447, 496, 501

Evans, Edgar, 482

Evans, Walker, 20

Eyre, Edward, 396

Fairbanks, Alaska, 240

Falco, Orlando, 238–39, 262

Falkland Islands, 38, 102, 258, 416, 491

Falkland Islands Dependencies, 38

Fallbrook, California, 12

Famine Reach, 497

Fernandez Bay, Bahamas, 43, 44

Ferrelo, Bartolomé, 79

Fertile Crescent, 272

Field Guide to the Fishes of Galápagos, The (Merlen), 233

Fifth Symphony (Beethoven), 183–84

Figes, Eva, 133, 135, 139, 148, 157, 163, 165

Finland, 183

First Fleet, 399, 422

FitzRoy, Robert, 257, 258, 264, 507

Flinders, Matthew, 282, 396, 420

Floreana (Wittmer), 248

Floreana Island, 100
   see also Isla Santa María

Florence, Italy, 261

Florence, Oregon, 90

Florida, 10, 457

Forestier Peninsula, 376

Forlani, Paolo, 54

FORM, 358, 364–65, 367, 368, 371

Fort Conger (Nunavut, Canada), 186

Fort George (Astoria, Oregon), 91, 112

For the Term of His Natural Life (Clarke), 384

Fort Vancouver (Vancouver, Washington), 91–92

Fox, Bill, 358

Fraenkel, Gottfried, 261–62

France, French, 22, 38, 146, 176, 196, 234, 246, 253, 342, 372

Frances Langford Tea Room, 348

Franklin, John, 188

Fraser River, 101

French Alps, 19

French Guiana, 176

French Impressionists, 147

French Polynesia, 39, 219, 291

French Revolution, 335, 379

French West Africa, 281

Freud, Sigmund, 83, 260

Frimley, John, 348

Fuegians, 498

Gairdner, Meredith, 116, 117

Galápagos: Key Environments (Perry), 233

Galápagos: World's End (Beebe), 220, 247

Galápagos Islands, 24, 100, 104, 121, 122–23, 208, 209–10, 212, 213, 217, 220–32, 233–39, 241–45, 247, 248–51, 258–59, 260, 262–65, 398, 416
   see also specific islands

Galápagos National Park, 222

Galeano, Eduardo, 245, 246, 372

Gallardo, Christy, 225

Gamboa, Pedro Sarmiento de, 506–7

Gamma Crucis (Gacrux), 292

Gandhi, Mahatma, 373

García Márquez, Gabriel, 250

Garner, Helen, 422

Gatun Locks, 253–54

Gauguin, Paul, 124

genocide, 42–44, 112, 115, 126, 198, 215, 246, 328, 368, 382

George III, king of England, 61

Georges River, 419, 421

George V Coast, Antarctica, 351

Gerard Manley Hopkins: Priest and Poet (Pick), 341

Gerritsz, Hessel, 54

Ghorband River, 274

Ghost archeological site, 145

Gibraltar, Cape of, 305

Gibson Desert, Australia, 330

Giles, Ernest, 396

Giverny, France, 135, 282

global climate change, 33, 146, 160, 161, 297, 400, 459, 464, 488

Gondwanaland (supercontinent), 314

Gönnersdorf, Germany, 280

Gordon Rocks, 233

Gorgon, HMS, 381–82

Granada Hills, California, 10

Grand Canyon, 10, 13, 71
    North Rim of, 149
Grave Rib site, 165, 200
Graves Nunataks, 427–34, 435–36, 437,
    439, 440–43, 444–45, 447, 456–57,
    462–63
Great Australian Bight, 420
Great Barrier Reef, 375, 380, 381, 420, 421
Great Britain, 38, 52, 61, 62, 82, 93, 95, 97,
    100, 112, 118–19, 213, 246, 271, 324,
    334–35, 348, 377, 378, 379, 391–92,
    447, 466, 489, 507, 508
Great Dividing Range, 390
Great Lakes, 113
Great Rift Valley, 271, 281, 314–17, *315*
Great Sandy Desert, Australia, 390, 392
Great Victoria Desert, Australia, 281, 390
Greely, Adolphus, 185, 186–87, 188
Greenland, 33, 134, 139, 145, 148, 150, 165,
    166, 171–72, 174, 175, 183, 374, 472
Greenlandic Eskimos, 33, 139, 150
Greenpeace, 458
Greenwich, England, 86
Grise Fjord, 139
Grytviken (whaling station), 38–39, 492
Guam, 69, 214
Guangzhou, China, 95
Guayaquil, 248
Gulf Stream, 70
Gunn, Donald, 261–62
Guzmán, Joaquín, 115

Haa Island, 165, 166, 183
Habré, Hissène, 115
Hadar site (Ethiopia), 316
Hadza people, 374
Hamersley Range, 360, 361
Hamilton, Thomas, 386
Hammond's Illustrated Library World Atlas,
    281–82
Hanseatic (ship), 104, 490, 491–92, 494
Harare, Zimbabwe, 277, 332
Harper's Weekly, 99
Harrison, K. David, 80
Harvey, Ralph, 433
Havana, Cuba, 40
Havasupai Indians, 13
Hawai'i, 16, 94, 106, 218, 219, 251, 416
Hawaiian Islands, 57, 58, 212, 217, 219, 251
Hawaiians, 60
Hawaii Clipper (Martin M-130 plane), 86
Hawking, Stephen, 207–8

Hay, Peter, 348, 350–53, 375–76, 384, 385
Heart of Darkness (Conrad), 214
Hebei Province, China, 281
Heemskerck (ship), 420
Heidegger, Martin, 75
Heisenberg, Werner, 260
Helena, Montana, 22
Helen of Troy, 272
Heraclitus, 82
Herbert, Wally, 189–91
Hero with a Thousand Faces, The
    (Campbell), 323
Herschel, William, 76
Heyerdahl, Thor, 213
High Arctic, 24, 32–33, 133, 145, 150–51,
    152, 155, 172, 174, 274, 375, 452, 455
    see also Alexandra Fjord lowland;
    Skraeling Island
Hillary, Edmond, 68
Himachal Pradesh, India, 29
Himalayas, 29, 30, 291
Hindu Kush, 193
Hiroshima, Japan, bombing of, 3, 43–44,
    160
Hobart, Australia, 89, 347, 348, 375, 385,
    386, 387
Hōjunmaru (ship), 91, 93
Hokkaido, Japan, 80, 94, 100, 336–37
Hōkūle'a (canoe), 213, 219–20, 237, 250,
    264, 265
Hong Kong, 100, 369
Honolulu, Hawai'i, 12, 117
Honolulu Star-Advertiser, 4
Hopkins, Gerard Manley, 341
Horn of Africa, 126, 299, 474
Horsehead Nebula, 72
Horssen (ship), 381
Hotel Cabo de Hornos (Punta Arenas),
    501–2
Hotel Delfín, 217
Hotel Galápagos, 212, 215, 217, 225, 230,
    232, 263
House of Lords, British, 334
Houtman, Frederik de, 420
Huahine island, 219
Hubble: Imaging Space and Time
    (Devorkin and Smith), 75, 76
Hubble telescope, 75, 76, 77
Hudson River School, 75–76
Hudson's Bay Company (HBC), 91, 92,
    98, 112, 116
Hughes Aircraft, 18

Hughes, Howard, 18
"Human Evolution as Narrative"
    (Landau), 322–23, 324
Humboldt Current, 70
Huntford, Roland, 471
Hussein, Saddam, 216
Hut Point Peninsula, 453, 466, 468
    Pram Point on, 468
Hutu Interahamwe, 115
Huxley, Thomas Henry, 322–23

Iberian Peninsula, 41, 82
Ice Station Weddell (ISW), 485, 487,
    488–89, 499
IDP (Internally Displaced People)
    camps, 44
Île du Diable (Devil's Island), 176, 178–80,
    181, 182, 246
Île Royale, 181
Imperial Reckoning (Elkins), 119
Inca people, 167
"In Defence of the Word" (Galeano), 245
Independence I, 152
Independence II, 165
India, 29, 52, 82, 95, 314, 324, 423, 462
Indiana, 20
Indian Ocean, 96, 357, 358, 364, 365, 381
Indian Pacific (passenger train), 390
indigenous peoples, 118, 119, 167, 168,
    251–52, 374, 404, 412–16, 497–98
    see also specific indigenous groups
Indonesia, 43, 53, 109, 246, 299, 386
Industrial Revolution, 125, 256, 298
Inglefield, Greenland, 148
Inscription Point, Kurnell Peninsula, 419
Inughuit people, 150, 171, 186, 187, 188, 374
Inuit people, 40, 72–73, 139, 150, 171, 172,
    175, 185, 200–201, 274, 471
Inukitut, 141
Inuksuq Nunatak, 435, 457
Iñupiaq Eskimos, 108, 240
Iñupiat people, 240
Iowa, 21
Iran, 42, 314
Irbenskiy Proliv (ship), 502–4
Ireland, 22
Irish Ribbonists, 349
Isla Baltra, 221
Isla Darwin, 264–65
Isla Dawson, 509
Isla Española, 251
Isla Fernandina, 230–31, 242

Isla Genovesa, 211, 226–27, 238, 242
Isla Isabela, 227, 230, 234, 236, 241–42, 512
    Puerto Villamil on, *see* Puerto Villamil
Isla Marchena, 213, 248
Island Development Company, 121, 122
Isla Rábida, 227, 237–38
Isla San Cristóbal, 122, 223, 227
Isla San Salvador, 43, 44, 238, 262
Isla Santa Cruz, 213, 221–25, 226, 242, 243,
    245
    Puerto Ayora on, *see* Puerto Ayora
Isla Santa María, 209, 221, 243, 247–48,
    416
Isla Sin Nombre, 213
Isla Wolf, 264–65
isumataq, 162
Italy, 22

Jaburarra people, 365–66, 368
Jack (author's South Pole escort), 476–77
Jackal Camp (Nakirai, Kenya), 271–77,
    279–80, 283–84, 285, 286, 288–90,
    312–14, 319–21, 322, 323, 325–26, 329,
    332, 335–36, 337, 412
Jack Hills, Western Australia, 36–37, 359,
    389–90, 394–401, 402
Jade Sea, *see* Lake Turkana
Jakarta, 381
*JAMA: The Journal of the American
    Medical Association*, 160–61
James Caird (boat), 490
Jamesway shelters, 458
Janszoon, Willem, 419–20
Japan Current, 90
Japan, Japanese, 4, 5, 15, 23, 43–44, 90, 91,
    93–94, 98, 100, 101, 107, 200, 335,
    336–37, 417
Jaques, Francis L., 504
Jeffers, Robinson, 110, 111, 245, 246, 341
Jewel Box (collection of stars), 292, 312
Jimbocho district, Tokyo, 24
Johannesburg, South Africa, 245–46
Johan Peninsula, *134*, 135–36, 138–43, 148,
    166, 171, 184–89
    Eskimo Point on, 186–87
    Lakeview site on, 151–52, 155, 165
    see also Alexandra Fjord lowland;
        Skraeling Island
Johanson, Donald, 316, 322, 325
Johnson Space Center, Houston, 74, 438
    Astro Materials Acquisition and
        Curation Office at, 441

Jokel Fjord, 165
Jomo Kenyatta International Airport, 335
Jordan, 281, 314
Juárez femicides, 474
Jung, Carl, 83, 260, 324
Jupiter (planet), 439

Kabul, Afghanistan, 33–34, 172
Kalahari Desert, Botswana, 278, 306
Kalahari Gemsbok National Park (South
    Africa/Botswana), 161, 327
Kalgoorlie, Australia, 390
Kalodirr, Kenya, 318, 319
Kamba people, 23, 80, 269, 270, 271, 274,
    338
Kane Basin, 145
Kap Eiler Rasmussen, Greenland, 174
Karaca Dağ people, 181
Karijini National Park (Australia), 358,
    359–60
Karnovsky, Carolyn, 358, 361
Kata Tjuta (the Olgas) rock formation,
    Australia, 408
Kaweskar (Alacaluf) people, 496, 497,
    498, 507
Kealakekua Bay, Hawai'i, 60–61, 62, 416
Keats, John, 53
Kent, Rockwell, 29, 496
Kentucky, 22
Kenya, 23, 80, 119, 269–77, 279, 285–87,
    302, 321, 324, 326, 330, 333, 335, 411
    see also specific cities and settlements
Kenya Highlands, 324
Kenyatta, Jomo, 119, 335
Kerio River, 271, 280
Kgalagadi Transfrontier Park, see Kalahari
    Gemsbok National Park
Kikuyu people, 119
Kikamba, 80
Kimberley, Australia, 354
Kimeu, Kamoya, 269, 270–71, 272,
    273–75, 276, 277, 279–80, 285, 286,
    287, 288, 289–91, 298, 301, 312, 313,
    317–18, 319, 325–28, 329, 330, 331, 332,
    334, 336, 337–38, 339
"Kindness" (Nye), 46
King Haakon Bay, South Georgia, 490
King, Martin Luther, Jr., 373
Kitale, Kenya, 329, 339
Klein Glacier, 432, 434, 435, 439, 445, 456,
    470
Klyne, Jane, 98

Knoll, The (Antarctica), 466, 468, 469
Knuth, Eigil, 174
Koale'xoa (Chinook woman), 91
Koettlitz Glacier, 456
Kølnæs site, 174, 175, 176
Kony, Joseph, 115
Koobi Fora, 279, 280, 321
Kullu Valley, India, 29
Kupang, Indonesia, 380, 381
Kurnell Peninsula, 419
Kuroshio Current, 90
Kurrama people, 360
KWK Exemplar (ship), 369

Labrador Current, 70
La Cumbre volcano, 242, 265
Lady Franklin Bay Expedition (1881–1884),
    185, 186–87
La Gorce Mountains, 439, 445–46
Lahaina, Maui, Hawaii, 94, 100
Lahiri, Jhumpa, 372
Laing, Henry, 348
Lake Albert (Lake Mobutu), 314
Lake Bonney, 464
Lake Clifton, 354, 355–57, 358, 359, 375, 395
Lake Fryxell, 464, 465
Lake Hoare, 464
Lake Natron, 326–27
Lake Nyasa (Lake Malawi), 314
Lake of Dreams, 73
Lake Rudolf, see Lake Turkana
Lake Turkana, 271, 279, 280, 316, 329, 333
Lake Vanda, 460, 462, 463–64
Lakeview site, 151–52, 155, 165
Lake Xochimilco, 40
Lakota Indian reservation, 44
Lakota Indians, 252
Lancaster Sound, 171, 200
Landau, Misia, 322–23, 324
Lander River, 403, 404
Lange, Dorothea, 20
Lapérouse, Jean-François de Galaup,
    comte de, 213
Large Magellanic Cloud, 291
las Casas, Bartolomé de, 112
Lascaux cave, 366
Las Encantadas, 211
Last Place on Earth, The (Huntford), 471
Last Speakers, The: The Quest to Save the
    World's Most Endangered Languages
    (Harrison), 80
Las Vegas shooting (2017), 388

Late Dorset people, 137, 146, 162, 163, 201, 202, 203
Late Paleozoic, 314
Launceston, Australia, 385
Laura (childhood friend), 153
Laurasia (supercontinent), 314
LC-130 Hercules aircraft, 432, 441, 473
Leakey, Louis, 279, 316, 323, 324–25, 326, 411
Leakey, Mary, 279, 316, 324–25, 326, 411
Leakey, Meave, 316, 318–19
Leakey, Richard, 270, 276, 313, 316, 318–19, 321–22, 325, 333, 340, 341, 342
Lebanon, 65–66, 387
Léger, Alexis, 46
Lenk, Timur, 215
Leopold II of Belgium, 41, 60, 215–16
Les Îles du Salut, 176, 181
Lewis and Clark, 128
Lewis, Meriwether, 62
Liberia, 42
Life (magazine), 71
Life of Captain James Cook, The (Beaglehole), 58
light, 66–68, 69, 75–76, 115, 125–27, 135, 147–48, 167, 193, 199–200, 228, 235, 240, 291, 351, 456, 460, 465, 476, 481
Light (Figes), 133, 135, 139, 163, 165
Linnaeus, Carolus, 59
Little Skraeling Island, 155, 200
Lloyd's of London, 88–89
Local Group, 292, 487
Lodwar, Kenya, 269–71, 276, 285, 289, 290–91, 320–21, 329, 334, 338
Loima Hills, Kenya, 272, 332
Lokichar, Kenya, 271
Lokwakangole, Kenya, 269, 270, 276, 329
London, England, 117, 507
Long Island, New York, 14, 92, 100, 217
Long Island Sound, 9
Long, Richard, 160
Long Valley, The (Steinbeck), 14
López family, 41
López, Marín (Martín), 41
López Tréllez y Albierne de Asturias y Vivar, Don Eugénio, 41
Lord's Resistance Army, 115
Lorenz, Rudolf, 248
Los Angeles, California, 11, 12, 13
Louisiana, 484, 485, 499, 500
Lovell, Mary S., 19
Lower Columbia tribes, 79, 98, 116

Lucayan people, 43
Lucy skeleton, 316
Luitpold Coast, 490
Lumumba, Patrice, 216
Luritja people, 375, 405
Lyell, Charles, 82–83, 257–58
Lyra (constellation), 72

MacDonald, Ranald, 91–95, 98–99, 100–102, 104, 106–7, 116, 129, 209, 282, 335, 416–17, 512
Madagascar, 314
Madeira, 246
Madrid, Spain, 22
Magdalenian Cro-Magnons, 193, 370, 371
Magee, Alan, 102
Magellan, Ferdinand, 209, 213, 497, 499
Magellanic Clouds, 292
Mahler, Gustav, 108, 192
Maine, 37
Maingon Bay, Tasmania, 351
Makinson Inlet, 172, 173, 175
Maktaq (Inuit man), 172
Mala (Creation Being), 405, 407–8
Malacca, Strait of, 54
Malan, Rian, 245
Malibu, California, 18
Malouf, David, 422, 423
Malta, 354
Mamaroneck Harbor, New York, 9, 14, 16, 17, 27
Mammalia class, 293
Mandela, Nelson, 338, 373
Mandurah, Australia, 354
Manzanita, Oregon, 90
Maori people, 252, 254, 512
Maralinga, Australia, 391, 392
Marandoo mine (Australia), 361
Marcos, Ferdinand, 115
Margaret River, 354, 357, 359
Mariana Trench, 68, 214
Mariloula (ship), 369
Markham, Beryl, 19, 21
Marquesas Islands, 212, 217
Mars (planet), 74, 439
Marsh, Othniel Charles, 252
Martin M-130s (seaplane), 20, 85–86, 102, 104
Martin PBM Mariners (aircraft), 20
Masks of God, The (Campbell), 323
Mason Cove, Tasmania, 350, 351
Massachusetts Institute of Technology, 18

Mass in B Minor (Bach), 108
Matagorda Island, 510
Matavai Bay, Tahiti, 61
Mateens, Omar, 388
Maui, Hawai'i, 219
Mau Mau terrorism, 119
Mayewski, Paul, 472–75, 480
McCullough, Karen, 151, 152, 155, 157, 160,
    201
McDonald, Archibald, 91–92, 93, 98
McKelvey Valley, Antarctica, 460, 461, 462
McKnight, Diane, 431, 464–65
McMurdo Ice Shelf, 453–54, 466, 468
McMurdo Sound, 447, 448–51, 453, 457,
    459, 465
McMurdo Station, 430, 431, 432, 433, 434,
    436, 441, 443, 445, 447, 452, 453, 454,
    456, 457, 458, 459, 463, 464, 466,
    467–68, 470, 472, 473–74, 475, 478,
    480
    Observation Hill at, 471, 472
McMurdo Station Weather Operations
    (Mac weather ops), 433–44
Medical Museum of the Royal Naval
    Hospital (Haslar), 117
Mediterranean, 314, 324
Meekathara, Australia, 389, 392, 393
Megalithic to Subatomic (Long), 160
Melbourne, Australia, 385
Melville, Harden, 376–77
Melville, Herman, 209–10
Memory, Boat, 507
Mendaña, Álvaro de, 213
Mendel, Gregor, 208, 261
Mendelssohn, Felix, 108
Menéndez, José, 501
Mercy Corps, 388
Merlen, Godfrey, 233
Merton, Thomas, 22, 46, 324
Mesozoic era, 212, 395, 462
Métis nation, 98
Mexico, 10, 21, 96, 115, 246
Mexico City, Mexico, 40–41
Mexico, Gulf of, 21, 254
Michigan, 20, 21
Micronesia, 217, 218, 250, 299, 306
Middle East, 23, 115, 304, 305
Middle Kingdom, 335
Middle Paleolithic, 299, 304
Middle Passage, 94
Milky Way, 73, 237, 292, 487
Millstream, 363

Millstream Chichester National Park
    (Australia), 363
Mindanao, 27
Mineral Shikoku (ship), 369
Minik (Inuit man), 172
Minster, York (El'Ielaru), 507
Miocene epoch, 275, 285, 293, 296, 298,
    318–19, 331
Missouri River, 113
Mitchell, Larry, 358
Mladić, Ratko, 115
Mobutu, Joseph-Désiré (Mobutu Sese
    Seko), 216
Mohave Indians, 13
Mojave Desert, 10, 13, 27, 474
Monet, Claude, 133, 135, 147, 163, 282
Mongolia, 29
Montes Jura, 73
Montevallo College, 19
Montevideo, 258
Montezuma II, 74
Moors, 41
Moreno, Eugénio, 249–50, 254
Moriyama, Einosuke, 100, 101
Morrison, Mike, 472, 473
Moscow Art Theatre, 29
Mothers of the Disappeared in the Plaza
    de Mayo, 42
Mount Everest, 68–69, 214
Mount Howe, 428
Mount Jason, 461
Mount Whitney, 21
Mozambique Channel, 314
Mozart, Wolfgang Amadeus, 107
Mugabe, Robert, 338
Murchison, Australia, 359
Murphy, Robert Cushman, 504
Murray Hill, Manhattan, 14
Murray River, 390
Museum of the North (Fairbanks, Alaska),
    240
Muslims, 115, 332, 362
Muthaiga Country Club (Nairobi), 324
Mutitjulu (Australian indigenous
    community), 406, 407, 408, 409
Mutiwa, Nzube, 269, 270, 271, 273, 275,
    279, 287–88, 289, 313, 326, 329–30,
    332, 338, 339
Myrtle Point, Oregon, 69

Naalagiagvik, 240
Nachorugwai Desert, Kenya, 271

Nagasaki, Japan, 90, 100, 101, 246
   bombing of, 43–44, 160
Naguogugalik, Jimmy, 40
Nairobi, Kenya, 276, 279, 289, 324, 329,
   332, 333, 340–42
Nakaisieken Desert, Kenya, 302
Nakirai, Kenya, 271–77, 279–80, 283–84,
   285–87, 288–90, 298, 301, 312–14,
   319–21, 322, 323, 325–26, 329–32, 334,
   335–36, 337, 339
Nakuru, Kenya, 340
Namib Desert, Namibia, 27
Namibia, 161, 277, 278, 279, 325
Nancy (member of Antarctic excursion),
   431, 446
Nansen, Fridtjof, 189
Napadet Hills, Kenya, 272
narcofosas, 474
"Nariokotome boy" skeleton, 270
Nariokotome, Kenya, 270, 276, 279, 280,
   285, 313, 321, 325, 341
*Narrative of Arthur Gordon Pym of
   Nantucket, The* (Poe), 496
NASA, 438
Nash, June, 252
Nathaniel B. Palmer (research vessel),
   253–54, 484, 485, 486, 487, 488, 494,
   496, 497, 499, 503, 505, 509
National Archives (Washington, D.C.),
   482
National Geographic, 71, 190
National Geographic Society, 190, 326,
   327, 340
National Museums of Kenya, 274, 279,
   340, 341, 342, 343
National Science Foundation (NSF), 428,
   433, 438, 439, 447, 458, 464, 467–68,
   475, 476
   Antarctic Artists and Writers Program
      of, 459
   Office of Polar Programs at, 457, 499
*Native American in the Land of the Shogun:
   Ranald MacDonald and the Opening
   of Japan* (Schodt), 94
Native Americans, 22, 252, 375, 382, 383
   see also American Indians; see also specific
      Native American groups
Natufian settlement, 281
Nature, 36, 160, 280, 389
*Navajo Blessingway Singer: The
   Autobiography of Frank Mitchell,
   1881–1967* (Mitchell), 144

Navajo Indians, 51, 143–44, 342
navigation, 32–33, 56–57, 72, 102–6, 109,
   124–25, 133, 175, 196, 210, 217–20,
   249–50, 263–65, 280, 284, 287, 368,
   380–82, 393, 408, 420, 481, 486–87,
   507, 511
   see also specific navigators and explorers
Navy, U.S., 69, 71, 86, 253, 447
Nazca Plate, 209, 213, 242, 265
Neanderthals, 272, 295, 305–6
Needles, California, 13, 16
Nefertiti, 74
Nelson, Jack, 225, 232
Neolithic period, 160
Newall Glacier, 472, 480–81
New Caledonia, 58, 212, 246
Newfoundland, 56, 96, 420
New Guinea, 299
New Hebrides, 212
New Holland, 420
New Mexico State Penitentiary, 53–54
New South Wales, 420
New Stanley Hotel (Nairobi, Kenya), 276,
   340–41
Newton, Isaac, 260
New World, 41, 118, 372
New York, New York, 14, 15, 17, 24, 29, 41,
   113, 280–81
New York Times, 458, 499
New York Zoological Society, 247
New Zealand, 57, 217–18, 231, 251, 252,
   453, 457, 463, 468
New Zealand Daylight Time (NZDT),
   429–30
Ngeneo, Bernard, 270, 271, 275, 286, 289,
   312, 313, 332, 337
Nicholas Roerich Museum (New York), 29
Nigeria, 43, 115, 215
Night Flight (Saint-Exupéry), 19
Ninth Symphony (Beethoven), 191–92,
   193–94
Nookawarra, Australia, 393, 394
Noongar people, 354
Noose of Laurels, The (Herbert), 189
Norfolk Island, 212
Norgay, Tenzing, 68
Noriega, Manuel, 246
Norse people, 150, 155, 156, 174
North Africa, 257
North Carolina, 19
Northern Cross, 72
Northern Equatorial Current, 213

Northern Territory, Australia, 23, 274, 374, 375, 389, 402–3, 404–11, 419
  see also specific towns, settlements, and geograhic features
North Pacific Current, 90
North Pole, 162, 189–91, 253
Northridge, California, 10
North Sea, 96
North Vietnam, 246
North West Coastal Highway, 367
North West Company, 112
Northwest Passage, 56, 58, 97, 98
Northwest Territories, 138
Norway, 369
Nuestra Señora de la Pura y Limpia Concepción (ship), 40, 41, 78, 215
Nullarbor Plain, 390, 392
Nunamiut Eskimos, 30–31, 330
Nunavut, Canada, 40, 185
Nuyts, Pieter, 420
Nye, Naomi Shihab, 46
Nymphéas (Monet), 135

Oates, Titus, 457, 482
Obeyesekere, Gananath, 57, 58
Oceanic Birds of South America (Murphy), 504
Ocean of Storms, 73
"Ode to Joy" (Schiller), 192
Oglala Sioux, 120
Old Coast Road, 354–55
Oldsquaw site, 145, 199–200, 201–2
Olduvai Gorge (Tanzania), 279, 316, 323, 326, 327
Oligocene epoch, 293
Olympic Peninsula, 52, 91
Olympus Range, 459
  Bull Pass in, 461, 462
Omdurman, Sudan, 60
Omo River, 316
One Ton Depot, 482
Onga (ship), 369
Ontario, Canada, 92
On the Origin of Species (Darwin), 259
Onyx River, 461, 462–63
Oort Cloud, 253
Operation Deep Freeze, 458
Opium Wars, 335
  first, 93
Oqe (Inuit man), 171

Oregon, 22, 23, 25, 26, 54, 56, 79, 90, 105, 114, 116, 173, 195, 320, 350, 384, 386, 402, 474
Oregon Coast Highway, 52
Oregon Historical Society, 94
Oregon Historical Society Museum (Portland), 87
Oregon, University of, 22
Orienta Apartments, 17, 25
Orienta Point (New York), 9
Orientation of Animals, The: Kineses, Taxes, and Compass Reactions (Fraenkel and Gunn), 261–62
Orinoco River, 23
Orion (constellation), 72, 73, 487
Orion's Belt, 73
Orundellico (Jemmy Button), 507
Osborn, Henry Fairfield, 322
Otter Rock, 105
Our Lady of Grace school (Encino, California), 10
Ozette, Washington, 93

P-38 Lightnings (fighter plane), 18–19, 20
Pacific Coast Highway, 18
Pacific Fur Company, 112
Pacific Grove, California, 4
Pacific Northwest, 98
Pacific Ocean, 54, 55, 58, 66, 68, 69–72, 74–75, 78, 90, 93, 95, 96, 99, 100, 104, 105, 124, 208, 212, 213–14, 217, 236, 250, 253, 254, 263, 299, 334, 341, 372, 380, 507
Paddocks, Stephen, 388
Pakenham, Thomas, 289
pale chanting goshawk, 161, 164, 327
Paleoeskimo people, 134, 146, 148, 150, 151, 452–53
  see also Dorset people; Thule people
Paleolithic era, 196, 299, 304, 370
Paleozoic era, 212, 314
Palomino Ponies (Dixon), 13
Panama, 209, 246, 251, 281, 369
Panama Canal, 253–54, 484
Pan American Airways, 85
Pandora, HMS, 380, 381
Pangea (supercontinent), 314
Pangnirtung village, Baffin Island, 274
Panthalassic (superocean), 314
Panthalassic Ocean, 212
Panthalassic of Permian, 54
Papillon (Charrière), 176

Papua New Guinea, 420
Paraburdoo, Australia, 358
Paraguay, 115
Parin, Paul, 358, 360, 366
Parliament, British, 353, 376, 379
Parry Channel, 32
Pärt, Arvo, 192, 195
Pärt, Nora, 195
Partita for Violin no. 2 (Bach), 183
Pashto language, 80
Paso del Mar, 496
Paso Largo, 496
Paso Tortuoso, 496, 497
Passion According to St. John (Bach), 192
Paul (member of Antarctic excursion), 431
PBY Catalinas (seaplane), 20
Pearl Harbor, Hawai'i, 3, 5
    Japanese bombing of, 3
Peary Land, 174, 175
Peary, Robert, 189, 190–91, 253
Peel Sound, 32–33
Pelican Bay, Puerto Ayora, 212, 214, 216
Pentonville prison (London), 377
Permian period, 395
Perry, Matthew, 91, 101, 417
Perry, Roger, 233
Perseus (constellation), 72
Perth Airport, 358
Perth, Australia, 357, 359, 389, 390, 393,
    406, 410
Peru, 209, 509
Philip IV, king of Spain, 40
Philippine Clipper (Martin M-130
    plane), 86
Philippines, 115
Philippine Sea, 212, 213
Philippson, Robert, 248
Phnom Penh, Cambodia, 349
Piailug, Mau, 218, 219
Piccard, Jacques, 68, 69, 70, 71
Pidgeon, Bob, 393, 394
Pierce Junior College, 10
Pigafetta, Antonio, 496–97
Pilbara, Australia, 357, 358, 361, 364–65,
    368–69, 370, 371
Pim Island, Cape Sabine on, 185–86,
    187–89, 214
Pinar del Rio, Cuba, 41
Pine Island, 166
Pine Ridge, South Dakota, 44
Pinochet, Augusto, 42
Pintupi people, 330, 375, 405

Pisarro, Camille, 282
Pitjantjatjara people, 80, 375, 405, 406,
    407–8, 409, 412–13
Pizarro, Francisco, 43
"Place Where You Go to Listen, The"
    installation (Adams), 240
Plague, The (Camus), 502
Plaza Norte islet, 222
Plaza Sur islet, 222, 252
Pleistocene epoch, 197, 297, 316
Pliocene epoch, 275, 285, 292, 293, 296,
    298, 316, 334
Plumwood, Val, 263
Plymouth (American whaler), 100
Poe, Edgar Allan, 496
Point Dume Beach, California, 15
Point Puer, Tasmania, 37, 351–52
Point Venus, Tahiti, 39, 61, 96
Polar Continental Shelf Program (PCSP),
    138, 139, 151, 172
Polar Eskimos, *see* Inughuit people
Polaris (star), 72, 292
Police Creek, 148
Polo, Marco, 74
Polynesia, 299
Polynesian explorers, 209, 212, 213, 217–20,
    250
Pond's Bay (Pond's Inlet), Baffin Island,
    171, 172
Ponte Vecchio, Florence, 261
Port Arthur, Australia, 37, 114, 347–53, 375,
    376, 377–79, 382
    mass shooting in, 384–85, 387, 388
    Model Prison at, 353, 377, 378
Port Arthur Historic Site Management
    Authority, 349
Port Famine (Puerto del Hambre), Chile,
    496, 507–11
Port Fourchon, Louisiana, 500
Port Hedland, Australia, 358, 361,
    367–69
Port Jackson, Australia, 378, 379–80, 382,
    422
Portland, Oregon, 87, 94
Port of Los Angeles (San Pedro,
    California), 121
Port of Spain, Trinidad, 86
Portsmouth, England, 378
Port Stanley, Falkland Islands, 102, 258,
    491
Portugal, Portuguese, 22, 215, 246, 372
Post Office Bay, Isla Santa María, 100

Post Office Hill, Antarctica, 469
Potomac River, 115
Potosí, Bolivia, 251–52
Prado Museum (Madrid), 22
Prague, 104
Precambrian era, 37, 196, 356
Pre-Dorset people, 146, 165
Pretoria, South Africa, 245
Prince Henry the Navigator, 57
Prince Philip's Steps, 226
Principles of Geology (Lyell), 257–58
Proterozoic era, 395
Proulx, Annie, 423
Ptolemy, 63
Puerto Ayora, 207–9, 211–12, 213, 214,
    216–17, 220–21, 222, 223, 225, 227,
    229, 254, 263
Puerto Baquerizo Moreno, 123, 227
Puerto del Hambre (Port Famine), Chile,
    496, 507–11
Puerto Villamil, 222, 239, 242–43, 244–45,
    246–47
Puget Sound, 121
Puleston, Dennis, 217
Pulse nightclub shooting (2016), 388
Punta Arenas, Chile, 491, 494, 497, 498,
    499–502, 503, 504, 506, 508, 509, 510
Punta Cristóbal, 236, 237

Qillaq, *see* Qirdlarssuaq
Qin Dynasty, 35–36
Qin Shi Huang, emperor of China, 35
Qirdlarssuaq (Inuit man), 171–72
Quakers, 378
Quechua miners, 251–52
Queen Elizabeth Islands, 134, 139
Queen Maud Mountains, 23, 432, 439
Queensland, Australia, 400
Quito, Ecuador, 224, 263
Qur'an, 181

Rabari nomads, 330
Rajasthan, India, 330
Rapa Nui, *see* Easter Island
Raspail, Jean, 496, 497, 498, 502
Red Crescent, 33
Red Pony, The (Steinbeck), 15
Red Sea, 299
Rembang (Dutch vessel), 381
Remember (Roerich), 29–30, 31
Requiem in D Minor (Mozart), 107
Reseda, California, 10, 11

Resolute, Cornwallis Island, 137, 139, 172
Resolution, HMS, 25–26, 55, 86–88, 89,
    104
Rhee, Syngman, 383
Rhine Valley, 280
Richardson, John, 116
Río Grande, 501
Rio Tinto, 43, 368
Ritter, Friedrich, 247–48
River Avon, 253
River Dart, 189–91
River Shannon, 22
River Styx, 15
River Thames, 348
Robben Island, 338
Robinson Jeffers: Selected Poems,
    341
Robison Glacier, 439
Roca Redonda, 234–35, 264, 375
Rockefeller family, 113
Roebourne, Australia, 367, 371
Roebourne Regional Prison, 371
Roerich, Helena, 29
Roerich, Nicholas, 29–30, 31
Roggeveen, Jacob, 213
Roman Catholics, 348
Romania, 115
Romanticism, 77
Rome, Italy, 22, 41
Romero, Óscar, 373
Ross Ice Shelf, 439, *440*, 452, 453–55, 467,
    469, 471
    Windless Bight on, 468, 470
Ross Island, 434, 453, *454*, 457, 458, 466,
    468, 490
Ross Sea, 74, 439, 450
Route 91 Harvest Music festival shooting,
    388
Rowell, Barbara, 493, 494
Rowell, Galen, 493, 494
Royal Canadian Mounted Police (RCMP),
    138, 139, 140, 141, 147, 150, 151, 158,
    166
Royal College of Surgeons, 507
Royal Drambuie Society, 463, 464
Royal Navy, 96, 483, 507
Royal Society, 58
Royal Society Range, 455, 456
Royal Vanda Swim Club, 463–64
Rudd, Kevin, 384
Rushdie, Salman, 402
Russia, 29, 112, 373, 488, 503

Sag Harbor, New York, 15, 100, 209
Sagittarius (constellation), 487
Saint-Exupéry, Antoine de, 18–19, 21
Salamanca festival, 385
Salazar, António de Oliveira, 246
Salgado, Sebastião, 75
Salish Indians, 252
Samana Cay, 42–43
Samburu people, 330
Samson, Loreen, 371–72
San Cristóbal Island, *see* Isla San Cristóbal
Sandoval, Gonzalo de, 43
San Fernando Valley, California, 9–10,
    14, 21
San Francisco Bay, 86
San Gabriel Mountains, 10
San Juan, Puerto Rico, 40
San Pedro, California, 121, 122
San Salvador Island, 43, 44, 238, 262
Santa Barbara Botanic Garden, 11
Santa Cruz Island, *see* Isla Santa Cruz
Santa Monica Bay, 15
Santa Monica Mountains, 10, 11, 18
Santa Rosa, Isla Santa Cruz, 221
Santillana del Mar, Spain, 370
Santo Cristo de Burgos (Spanish
    galleon), 90
Santo Tómas, 227
São Paulo, Brazil, 43
Saqqaq culture, 165
Sarajevo, Bosnia and Herzegovina, 44
Satan Came to Eden (Strauch), 208
Satawal Island, 218
Saudi Arabia, 299
Savimbi, Jonas, 338
Schiller, Friedrich, 191–92
Schledermann, Peter, 138, 139, 140, 146,
    150, 151, 152, 155, 156–57, 160, 163,
    164–65, 166, 167, 170, 180, 183, 186,
    187, 196, 199, 200, 201, 202
Schodt, Frederik L., 94
Schuster, Carl, 25
Schutt, John, 74, 430–32, 433–34, 435,
    441–42, 444, 445, 446, 454–56
Science, 280
Scientific Revolution, 67
Scorpion Rock, 348
Scotia Sea, 496
Scotland, 386
Scott (member of Antarctic excursion),
    433, 442
Scott Base, 453

Scott Glacier, 439
Scott, Kathleen, 483, 484
Scott, Robert Falcon, 189, 434, 457, 466,
    471–72, 480, 482–84, 489
Scramble for Africa, The (Pakenham), 289
Sea of Japan, 100, 209
Seattle, Washington, 121
Seclusion Era, 93
Second International Polar Conference
    (1880), 186
Second Symphony (Mahler), 192
Selk'nam (Ona) people, 498, 500, 501
September 11, 2001, terrorist attacks, 3
Shaanxi Province, China, 35–36
Shackleton, Ernest, 38, 254, 457, 485,
    489–91, 492, 493, 494
Shakespeare, William, 176, 213
Shanghai, China, 281
Shanidar Cave, 272
Shell Oil, 43
Shoshone Indians, 114, 246
Siam, 282
    see also Thailand
Sibelius, Jean, 183
Siberia, 42, 299, 317
Sibiloi National Park (Kenya), 321
Sierra Negra volcano, 242–43, 245
Sierra Nevada, 21
Sikorsky S-42 (seaplane), 85
Siletz people, 79–80
Silk Road, 57
Silva, Noenoe, 251
Silver Bell (ship), 369
Simpson Desert, Australia, 390
Singapore, 54
Sinoto, Yosihiko, 219
Siuslaw River, 90
Sixth Extinction, 46, 85, 105, 160, 298
Skraeling Island, 134–37, 138, 140, 141,
    145–46, 150, 151, *154,* 155–58, 163, 164,
    165–66, 167, 170–71, 172, 173–76,
    180, 184, 191, 192–94, 196, 198–200,
    201–3, 457
Skylab, 476, 480
Small Magellanic Cloud, 292
Smith, Grafton Elliot, 322
Smith, Isaac, 421
Smithsonian Institution (Washington,
    D.C.), 117, 441
Smith Sound, 135, 139, 166, 171, 172
Society Islands, 217
Somaliland, Republic of, 316

Songlines, The (Chatwin), 402
South Africa, 57, 104, 161, 182, 245–46,
    277, 278, 327, 423, 462
South America, 38, 102, 221, 261, 291, 299,
    314, 462, 496, 497–98, 507, 508, 510
Southampton, Hampshire, 41
South Australia, 391–92
South China Sea, 335
South Equatorial Current, 70
Southern Cross (constellation), 72, 291,
    292, 501
Southern Mail (Saint-Exupéry), 19
Southern Ocean, 38, 61–62, 124, 420, 439,
    484, 485, 486, 504
South Georgia, 38–39, 102, 212, 490, 491,
    492, 494
South Korea, 369
South Orkney Islands, 484
South Pacific, 23, 35, 71, 96, 124, 217, 419
South Pole, 161, 162, 428, 447, 466, 471,
    472, 473–80, 481–82, 509
    British Antarctic Expedition at, 482–84
South Sandwich Islands, 38, 212
South Shetland Islands, 490
South Sudan, 44, 126, 215
South West Africa People's Organisation,
    278
Soviet Union, 53, 195, 387, 485, 488, 489
SPAD S.VII aircraft, 19
Spain, Spanish, 22, 38, 40, 41, 79, 118, 119,
    196, 209, 213, 215, 246, 251, 281, 325,
    370, 371, 507
Spanish-Cuban War (Spanish-American
    War), 41
Spar Leo (ship), 369
Specially Protected Area Number 6, *see*
    Cape Crozier, Antarctica
*Special Publication/Geological Society of
    Australia,* 36
Spitsbergen Island, 191
Spruce Goose aircraft, 18
Srebrenica, Bosnia, 115
Sri Lanka, 43, 282
Stannard, David, 57, 215
Stefansson, Vilhjalmur, 189
Steger, Will, 103, 491, 493, 494
Steinbeck, Carol, 4
Steinbeck, Elaine, 5
Steinbeck, John, 4, 5, 14, 15
Steinbeck, John (son), 5, 14, 15
Steinbeck, Thom, 5
Steinberg, Saul, 24

Stephanie (author's stepdaughter), 114, 115
Stevenson, Robert Louis, 124
Stiles Island ("the Sphinx"), 141
St. John River, 457
Stokes, Pringle, 507
Strabo, 63
Straight on Till Morning (Lovell), 19
Strait of Magellan, 484, 491, 494, *495*,
    496–97, 504, 507
Strauch, Dore, 247–48
Stromness Harbour, South Georgia, 490,
    491, 492
Structurist, 147–48
Stuart, John McDouall, 396
Sudan, 332
Suharto, 246, 383
Sullivan, Walter, 499
Sumatra, 311, 387
Sutherland, Forby, 422
Sutherland Point, Sydney, 422
Svalbard Archipelago, 191
Sverdrup Pass, 166, 183
Sverdrup site (Skraeling Island), 157–58,
    191, 193–94, 195, 202
Swan of Tuonela, The (Sibelius), 183
Swan, Susan, 423
Sydney, Australia, 60, 61, 390, 391, 417–19,
    422
    Botany Bay in, 60, 61, 378, 419, 420–24
Sydney Airport, 424
Sydney Harbor, 418
Syria, 126, 181, 316

Tahiti, 39, 61, 208, 219, 248, 381, 416
Tajikistan, 3, 387
Taliban, 389
Tanami Desert, Australia, 104, 318, 390,
    403, 404, 406, 410
Tanami Desert Wildlife Sanctuary, 403
Tanganyika, *see* Tanzania
Tanzania, 279, 281, 316, 325, 326, 374, 411
Tapia, Andrés de, 43
Tapiola (Sibelius), 183
Tasman, Abel Janszoon, 213, 282, 420
Tasmania, Australia, 37, 89–90, 350, 351,
    354, 375, 385, 386, 387, 420
    see also specific cities and towns
Tasman Peninsula, 376, 420
Tasman Sea, 351
Taylor, Charles, 42
Taylor Valley, Antarctica, 464–65
Tench, Watkin, 381–82

Tenochtitlán, Mexico, 40–41, 44, 272
    see also Mexico City, Mexico
Teresa, Mother, 373
Tertiary period, 275
Tethys Ocean, 212, 314
Tethys Seaway, 314
Texas, Gulf Coast of, 510
Thailand, 232, 282
Thair (childhood friend), 11
Theory of Mind psychology, 308, 413
Thetis (rescue ship), 188
Thijssen, François, 420
Things Fall Apart (Achebe), 289
Thiong'o, Ngũgĩ wa, 251
Thorvald Peninsula, 141, 166
300 Club, 477–78
Thule people, 134, 135, 136, 137, 145, 146,
    148, 150, 151, 152, 154, 155, 157–58, 159,
    160–61, 162, 163, 164, 165, 166, 167,
    170–71, 172, 173–74, 175–76, 179, 183,
    184–85, 192, 194, 195, 198–99, 200,
    201, 202, 284, 452, 457
Tibetan Plateau, 43
Tierra del Fuego, 246, 281, 300, 501, 502,
    509, 510, 511
Tillamook Indians, 79, 104
Timika, Indonesia, 386
Timor Sea, 299
Tinian Island, 216
Tjukurrpa, 403, 404, 408
Toelken, Barre, 143–44
Tokugawa Shogunate, 90
Tokyo, Japan, 12, 24
Tolkien, J. R. R., 367
Tom Price, Australia, 358
Tom Price railway road, 361–62
Tonga, 212
Tongan Islands, 380
Tontons Macoutes, 115
Topanga Beach, California, 10–11
Torres Strait, 420
Tortuga Bay, Santa Cruz Island, 227–29
Transantarctic Mountains, 74, 428,
    437–38, 439, 462, 471
Transitional, 165, 170
Tredinnick, Mark, 354, 357–58, 359–60,
    365, 367
Triangulum Australe (constellation), 291
Trieste (bathyscaphe), 69, 71
Truelove Lowland, 174
Truth and Reconciliation Commission,
    182

Turkana people, 269, 273, 274, 275, 277,
    285, 289–91, 317, 320, 327, 328, 330,
    332, 334
Turkey, 57, 181, 314
Turks and Caicos Islands, 40
Turkwel Lodge, 329, 338–40
Turkwel River, 270, 272, 329, 338
Tutu, Desmond, 182, 373
Tuvalu, 282
Twin Glacier, 141, 142, 143
Twin Otters, 138, 139, 151, 156, 430
Twin River, 141, 196

Ubud, Bali, 33
Uffizi Gallery (Florence), 261
Uluru-Kata Tjuta National Park,
    Australia, 406, 407–8
"Ulysses" (Tennyson), 471
Umatilla, Oregon, 252
Upper Paleolithic, 299
Ur, 146
Ursa Major (constellation), 72
U.S. Defense Mapping Agency
    Hydrographic Center, 264
U.S. Geological Survey (USGS), 428, 477
Ushuaia, Argentina, 246
Ussher, Archbishop, 258
Uttermost Part of the Earth (Bridges), 496

Vaarzon-Morel, Petronella, 402
Vallejo, César, 418
Valparaíso, Chile, 416
Vancouver Island, 91
Vanda station, 459–61, 463–64
Van Diemen's Land (Tasmania), 420
van Sheck, Grace, 18
van Sheck, Sidney, 18–19
Velázquez, Diego, 43
Venezuela, 23
Veracruz, Mexico, 40
Verdun, France, 114
Vestfold Hills, East Antarctica, 459
Victoria, Australia, 95, 385, 416
Victoria Falls (Zambia), 277, 375
Victoria Land, Antarctica, 39, 74, 447,
    458, 459, 472
Vietnam, 252, 484
Vikings, 150
Viking Ship Museum (Roskilde), 175
Virgin Islands, U.S., 41
Vitiaz Deep, 68, 69, 213
Vizcaíno, Sebastián, 79

von Richthofen, Baron (Red Baron), 19
von Wagner Bozquet, Baroness Eloise, 248
Voortrekkers, 338
Voyage of the Beagle, The (Darwin), 256, 282
Voyaging: Southward from the Strait of Magellan (Kent), 496

Wafer, Jim, 402
Wagner, Richard, 193–94
"Waiting for the Barbarians" (Cavafy), 202
Wakamba people, 326
Wake, Cameron, 472, 473, 480, 481
Walker, Alan, 276, 340, 341, 342, 343
Wallace, Alfred Russel, 82, 83, 208, 210
Wallis, Samuel, 213
Walmajarri people, 392
Walsh, Don, 68, 69, 71–72, 213
Wambua (Mangao), 270, 271, 272, 275, 285, 286–87, 289, 313, 332, 337
Wandel Sea, 174
Warlpiri people, 274, 375, 403, 404–5, 406, 410–11
Washington Post, 458
Washington State, 52, 91
Watarrka National Park, Australia, 419
Watjarri people, 359
Watling Island, *see* San Salvador Island
Webster, Mags, 358, 361
Weddell Sea, 253, 254, 485, 488–89, 490, 491, 496, 497, 499, 509
Weldon, Annamaria, 354–55, 356–57, 358, 359
West Africa, 215, 232–33, 281
West, American, 67
West Antarctica, 74
Westchester County, New York, 9
West Country (England), 215
Western Australia, 36–37, 43, 311, 357, 368
Jack Hills in, 36–37, 359, 389–90, 394–401, 402
see also specific towns and cities
Western Australia, University of, 359
Western Front, 3
Western Greenland, 374
Western Sahara desert, 21
Western Trader (freighter), 120–22, 223
West Germany, 22
West Greenland, 134, 150
Weston, Edward, 20
West with the Night (Markham), 19
Whillans, Ian, 74, 442

White Lady pictograph, 325
White, Minor, 20
Whiteside Canal, 497
White, Tim, 316
Who Will Remember the People . . . (Raspail), 496, 497, 498, 502
Wild, Frank, 491
Wilkniss, Peter, 499
Willowra, Australia, 403, 406, 410, 411
Wills, William, 396
Wilson, Edward, 466–67, 470, 471–72, 479–80, 482
Wilson Piedmont Glacier, 447
Winnipeg, Manitoba, Canada, 92
Winter Quarters Bay, Antarctica, 447–48
Winter's Tale, The (Shakespeare), 213
Winton, Tim, 422
wisdom keepers, 84, 120, 284
see also elders
Wisdom of the Desert, The (Merton), 46
Witherspoon, Gary, 144
Wittmer, Heinz, 247–48
Wittmer, Margret, 247–48, 249, 264
Wolseley, John, 419
Wommera, Australia, 391–92
Woodlawn High School (Birmingham), 19
Works Progress Administration (WPA), 19
World Trade Center (New York), 216
World War I, 19
World War II, 5, 17, 86, 90, 115, 117
Worsley, Frank, 490–91
Worst Journey in the World, The (Cherry-Garrard), 466
Wreck Bay, San Cristóbal Island, 121, 122–23
Wright Lower Glacier, 462
Wright Upper Glacier, 462
Wright Valley, Antarctica, 39, 460–61, 462, 464
Wuhan, China, 23, 31
Wyoming, 22

Xi'an, Shaanxi Province, China, 35–36

Yale University, 322
Yalgorup National Park, Western Australia, 354, 357
Yallingup, Australia, 357
Yámana (Yaghan) people, 496, 497, 498
Yamana-English: A Dictionary of the Speech of Tierra del Fuego (Bridges), 496
Yangtze River, 23, 31, 32, 281

Yaquina Bay, Newport, Oregon, 79
Yaquina Head (Oregon), 66, 74
Yaquina people, 79–80
Yelcho (steamer), 491
Yellowknife, Canada, 137–38
Yinhawangka people, 360
Yokcushlu (Fuegia Basket), 507
Young Ireland movement, 349
Yueyang, China, 31–32, 474
Yugoslavia, 281–82

Zagajewski, Adam, 192, 502
Zaire, 216
Zambezi River, 277
Zambia, 277
Zealandia, 57
Zeehaen (ship), 420
Zimbabwe, 277, 389
"Zinjanthropus," 316
Zuma Beach, California, 12,
    15, 16

A NOTE ABOUT THE AUTHOR

BARRY LOPEZ is the author of two collections of essays; several story collections; *Arctic Dreams,* for which he received the National Book Award; *Of Wolves and Men,* a National Book Award finalist; and *Crow and Weasel,* a novella-length fable. He contributes regularly to both American and foreign journals and has traveled to more than seventy countries to conduct research. He is the recipient of fellowships from the Guggenheim, Lannan, and National Science Foundations and has been honored by a number of institutions for his literary, humanitarian, and environmental work. Additional information at barrylopez.com.

A NOTE ON THE TYPE

This book was set in Adobe Garamond. Designed for the Adobe Corporation by Robert Slimbach, the fonts are based on types first cut by Claude Garamond (ca. 1480–1561). Garamond was a pupil of Geoffroy Tory and is believed to have followed the Venetian models. He gave to his letters a certain elegance and feeling of movement that won their creator an immediate reputation.

Composed by North Market Street Graphics,
Lancaster, Pennsylvania

Printed and bound by Berryville Graphics,
Berryville, Virginia

Designed by Cassandra J. Pappas